SECURITY ANALYSIS

*"Many shall be restored that now are fallen
and many shall fall that now are in honor."*

HORACE—*Ars Poetica.*

SECURITY ANALYSIS

BY

BENJAMIN GRAHAM

Investment Fund Manager; Lecturer in
Finance, Columbia University

AND

DAVID L. DODD

Assistant Professor of Finance,
Columbia University

New York WHITTLESEY HOUSE *London*

McGRAW-HILL BOOK COMPANY, INC.

20 DOC/DOC 0 9

ISBN 0-07-024496-0

In 1934, McGraw-Hill published
the first edition of *Security Analysis*, by Benjamin
Graham and David Dodd. The book was destined to become a Wall
Street classic and the cornerstone of Graham's reputation. It, along
with his teaching, other writings, and a superb record as a money
manager, created a legend that continues to be enhanced rather
than diminished with the passage of time. Graham has become the
most influential investment philosopher of our time.

Security Analysis, now in its fifth edition, is regarded around the
world as the fundamental text for the analysis of stocks and bonds,
as well as the bible of value investing. To commemorate the book's
great achievement, and to reintroduce to readers the original
ideas and language of the first edition, McGraw-Hill is
proud to publish this special reproduction
of the 1934 book.

This special reprint edition was photographed by hand from the
original pages of the first edition by Jay's Publishers Services,
Inc., Rockland, Massachusetts. A customized vertical camera,
designed to minimize distortion and to prevent damage to
rare books, was used. This edition was printed and bound
by R.R. Donnelley, Crawfordsville, Indiana.

To
ROSWELL C. McCREA

PREFACE

This book is intended for all those who have a serious interest in security values. It is not addressed to the complete novice, however, for it presupposes some acquaintance with the terminology and the simpler concepts of finance. The scope of the work is wider than its title may suggest. It deals not only with methods of analyzing individual issues, but also with the establishment of general principles of selection and protection of security holdings. Hence much emphasis has been laid upon distinguishing the investment from the speculative approach, upon setting up sound and workable tests of safety, and upon an understanding of the rights and true interests of investors in senior securities and owners of common stocks.

In dividing our space between various topics the primary but not the exclusive criterion has been that of relative importance. Some matters of vital significance, e.g., the determination of the future prospects of an enterprise, have received little space, because little of definite value can be said on the subject. Others are glossed over because they are so well understood. Conversely we have stressed the technique of discovering *bargain issues* beyond its relative importance in the entire field of investment, because in this activity the talents peculiar to the securities analyst find perhaps their most fruitful expression. In similar fashion we have accorded quite detailed treatment to the characteristics of privileged senior issues (convertibles, etc.), because the attention given to these instruments in standard textbooks is now quite inadequate in view of their extensive development in recent years.

Our governing aim, however, has been to make this a critical rather than a descriptive work. We are concerned chiefly with concepts, methods, standards, principles, and, above all, with logical reasoning. We have stressed theory not for itself alone but for its value in practice. We have tried to avoid prescribing standards which are too stringent to follow, or technical methods which are more trouble than they are worth.

The chief problem of this work has been one of perspective— to blend the divergent experiences of the recent and the remoter past into a synthesis which will stand the test of the ever enigmatic future.   While we were writing, we had to combat a widespread conviction that financial debacle was to be the permanent order; as we publish, we already see resurgent the age-old frailty of the investor—that his money burns a hole in his pocket.   But it is the conservative investor who will need most of all to be reminded constantly of the lessons of 1931–1933 and of previous collapses.   For what we shall call *fixed-value investments* can be soundly chosen only if they are approached—in the Spinozan phrase—"from the viewpoint of calamity."   In dealing with other types of security commitments, we have striven throughout to guard the student against overemphasis upon the superficial and the temporary.   Twenty years of varied experience in Wall Street have taught the senior author that this overemphasis is at once the delusion and the nemesis of the world of finance.

Our sincere thanks are due to the many friends who have encouraged and aided us in the preparation of this work.

BENJAMIN GRAHAM.
DAVID L. DODD.

NEW YORK, NEW YORK,
   May, 1934.

CONTENTS

PART VI

BALANCE-SHEET ANALYSIS. IMPLICATIONS OF ASSET VALUES

PART VII

ADDITIONAL ASPECTS OF SECURITY ANALYSIS,
DISCREPANCIES BETWEEN PRICE AND VALUE

SECURITY ANALYSIS

INTRODUCTION

THE SIGNIFICANCE OF RECENT FINANCIAL HISTORY TO THE INVESTOR AND THE SPECULATOR

Distinctive Character of the 1927–1933 Period.—Economic events between 1927 and 1933 involved something more than a mere repetition of the familiar phenomena of business and stock-market cycles. A glance at the appended chart covering the movements of the Dow-Jones averages of industrial common stocks since 1897 will show how entirely unprecedented was the extent of both the recent advance and the ensuing collapse. They seem to differ from the series of preceding fluctuations as a tidal wave differs from ordinary billows and, as such, would undoubtedly be governed by special causes and produce unparalleled effects.

Any present examination into financial principles or methods must start with a recognition of the distinctive nature of our recent experiences, and it must face and answer the numerous new questions which these experiences inspire. For clarity of treatment we may classify these questions into four groups, as they relate respectively to:

I. Speculation.
II. Investment in:
    A. Bonds and preferred stocks.
    B. Common stocks.
III. The relations between investment banking houses and the public.
IV. The human nature factor in finance.

SPECULATION

It can hardly be said that the past six years have taught us anything about speculation that was not known before. Even though the last bull and bear markets have been unexampled in recent history as regards both magnitude and duration, at bottom the experience of speculators was no different from that

1

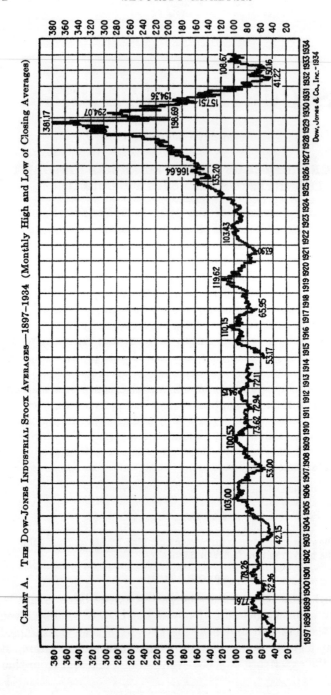

CHART A. THE DOW-JONES INDUSTRIAL STOCK AVERAGES—1897–1934 (Monthly High and Low of Closing Averages)

in all previous market cycles. However distinctive was this period in other respects, from the speculator's standpoint it would justify applying to Wall Street the old French maxim that "the more it changes, the more it's the same thing." That enormous profits should have turned into still more colossal losses, that new theories should have been developed and later discredited, that unlimited optimism should have been succeeded by the deepest despair are all in strict accord with age-old tradition. That out of the very intensity of the debacle there will arise new opportunities for large speculative gains appears almost axiomatic; and we seem to be on firm ground in repeating the old aphorisms that in speculation *when* to buy—and sell— is more important than *what* to buy, and also that almost by mathematical law more speculators must lose than can profit.

NEW AND DISTURBING PROBLEMS OF INVESTMENT

But, in the field of investment, experience since 1927 inspires questions both new and disturbing. Of these the least troublesome arise from the misuse of the term "investment" to cover the crassest and most unrestrained speculation. If that were the only cause of our investment difficulties, it could readily be cured by readopting the old-time, reasonably clean-cut distinctions between speculation and investment. But the real problem goes deeper than that of definition. It is bound up not with the grotesque failure of speculation masquerading as investment but with the scarcely less calamitous failure of investment itself, conducted in accordance with time-honored rules. It is not the wild gyrations of the common-stock averages but the precipitate decline in the bond averages (see Chart B below) which constitutes the really novel and arresting feature of recent financial history—at least from the standpoint of investment logic and practice. The heavy losses taken by conservative investors since 1928 warrant the serious question, Is there such a thing as sound and satisfactory investment? And also the secondary question, Can the investor rely upon the care and good faith of investment banking houses?

Unhappy Experience of the Bondholder since 1914.—It should be pointed out that the experience of bond investors as a class had been relatively unsatisfactory throughout the period since our entry into the World War, so that for years prior to 1928 there had been developing a realization that bonds did not afford

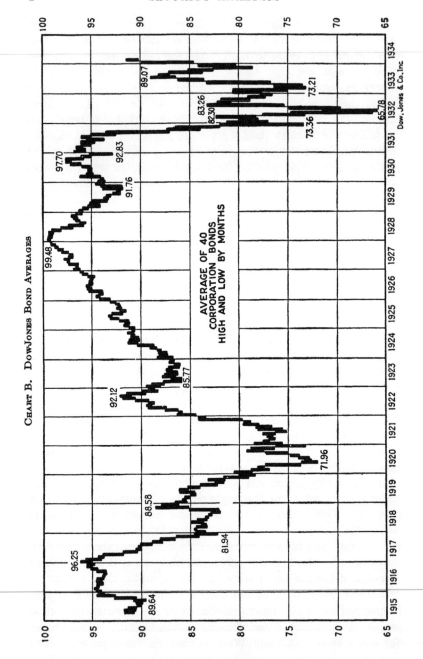

CHART B. DOW-JONES BOND AVERAGES

AVERAGE OF 40
CORPORATION BONDS
HIGH AND LOW BY MONTHS

Dow, Jones & Co., Inc.

sufficient protection against loss to compensate for the surrender of the profit element. The years 1917–1920 were marked by a tremendous decline in all bond prices as the result of war financing followed by the postwar inflation and collapse. The later recovery was not without its many individual disappointments, chiefly because the bulk of bond investment had previously been made in the railroad field and the credit of the carriers had on the whole been declining during most of the period in which numerous industrial companies had been showing extraordinary improvement. Even public-utility bonds had been adversely affected from 1919 to 1922 by the postwar increase in operating costs, as against relatively inelastic rates. On the other hand, the resultant rapid broadening of the industrial bond list had not been without its disadvantages, because many issuing companies not only failed to participate in the general improvement but even fell into difficulties. Hence the purchaser of industrial bonds found himself in a precarious field, wherein he benefited but slightly, if at all, from the brilliant successes but suffered more or less severely from the frequent failures.

It was the natural disaffection with their experience as bond owners which predisposed investors to embrace the new doctrine of common stocks as the superior form of investment—a doctrine which had a real validity within a limited range of application, but which was inevitably misapplied, with consequences too harrowing to dilate upon. Today the doctrine of common stocks as long-term investments seems discredited, but this fact does not in itself restore the bond to its old estate as the sound investment *par excellence,* nor does it explain away the unhappy history of the bondholder in recent years.

If we were to regard the record of the bond list since 1927 as indicative of what the future has in store, the considered conclusion would be warranted that sound investment as formerly conceived—meaning generally the purchase of bonds at prices close to par—no longer exists. For while it is true that a good many bond issues have come through this period without alarming depreciation in either price or quality, their number is relatively too small—nor was their superiority sufficiently manifest in advance—to warrant the belief that careful selection would have restricted commitments to this group and protected the shrewd investor against the losses suffered by others. The decline in the general bond averages was caused in part by an unwarranted lack of confidence and in part also by the depressing influence of large sales by banks intent on maintaining liquidity at all costs. But besides these temporary and psychological factors, it has reflected an undeniable and disconcerting impairment of safety in many individual cases.

The theory that a sound bond will be unaffected by a period of depression has suffered a rude shock. Margins of safety considered ample to withstand any probable shrinkage in earnings have proved inadequate; and enterprises once regarded as depression-proof are having difficulty in meeting their fixed charges. Hence if our judgment were based primarily on recent experience, we should have to advise against all investment in securities of limited value (excepting possibly short-term government bonds) and voice the dictum that both bonds and stocks should be bought only as speculations, by people who know they are speculating and who can afford to take speculative risks.

1927–1933 Period an Extreme Laboratory Test.—However, we do not accept the premise that 1927–1933 experience affords a proper norm by which to judge the future of investment. The swing of the speculative pendulum during this period was of such unprecedented amplitude as to warrant the belief that it will not recur in similar intensity for a long time to come. In other words, we should regard it more as an economic phenomenon akin to the South Sea Bubble and other isolated instances of abnormal gambling frenzy than as an indication of what the typical speculative cycle of the future will be. As a *speculative* experience, the recent cycle differed from previous ones in kind rather than in degree; but in its effects upon the *investment fabric* it had unique characteristics, seemingly of a nonrecurrent type.

We think, therefore, that from the standpoint of bond investment the last six years may be regarded as a sort of extreme laboratory test, involving degrees of stress not to be expected in the ordinary experience of the future. From this viewpoint, bond investment does not become a hopeless practice. By applying the lessons taught by this "laboratory test" to the selection of bonds, reasonably satisfactory results should be experienced in the years to come. These lessons would enjoin a more rigid insistence than heretofore upon the twofold assurances of safety—those arising from the inherent soundness and stability of the enterprise (as evidenced by the nature of the business, its relative size, its management and reputation, etc.) and equally those arising from generous margins of coverage shown by actual earnings over a sufficient period, as well as by the presence of an adequate junior equity. The strict application of these tests would no doubt result in a sharp diminution of the number of new issues which can qualify as sound investments, but it should still leave a restricted field in which bond investment can be satisfactorily practiced.

Education in Bond Selection Still Useful.—To some this conclusion may appear unduly optimistic, yet whether it is so or not will probably not make so much practical difference as would at first appear. For even if bond investment, judged over a long period, were to prove inherently unprofitable, nevertheless investment houses will continue to sell bonds and the public will proceed to buy them—just as people have always persisted in speculating, although we know that most of them must lose money. Furthermore these bond buyers would very likely be wise to make their bond investments, even though they proved moderately unremunerative, rather than to run the risks of much larger loss implicit in stock speculation. Hence education in sound and careful bond selection should be useful to the public on any hypothesis, even the pessimistic one that some final net loss is inevitable.

As this introduction is being written, the investor in high-grade bonds is faced with a new and alarming hazard, *viz.*, that inflation and consequent depreciation of the currency may impair the value of his interest and principal. However serious this danger may be, the problem it presents is essentially *temporary* in character. It relates not to the value of bonds as a channel for investing money but to the value of money itself; and it terminates as soon as the value of the monetary unit is again definitely established, whatever the new level may be. German inflation, for example, wiped out the prewar German bonds; but it had no permanent effect upon the theory of bond investment in Germany. A similar result followed the 80% depreciation of the franc. Bond investment was resumed as soon as the value of the currency was stabilized, and the principles and technique of selecting sound bonds remained exactly the same after the devaluation as before.

It is clearly preferable to own tangible property rather than money during a period when the value of money is depreciating, but this objection to money ceases the moment its value is stabilized.   Exactly the same statement applies to high-grade bonds.

Common-stock Theory a Sound Principle Misapplied.—
Investment in common stocks would seem to be in a still more questionable state than investment in bonds. Not only is the doctrine of common stocks as the best long-term investments in eclipse, but no less an authority than Lawrence Chamberlain has not hesitated to express the view that all stocks are by their nature essentially and unavoidably speculative.[1] Hence, to him the lesson of recent years is that the only sound investment is a bond. The logic of this pronouncement will be examined in some detail in an early chapter on Investment and Speculation. Dealing here with the narrower question of what is signified by recent experience, we repeat our former statement that the "common-stock insanity" was a monumental example of a *sound* principle grievously misapplied. Its history teaches us more about the nature of human beings than the nature of common stocks. Long before the "new-era" gospel was being preached, there were principles guiding the selection of common stocks for investment as distinguished from speculation. Broadly speaking, an investment stock was required to meet the same tests of safety and stability as were exacted of a bond. Common stocks passing these tests generally gave a good account of themselves as investments and in addition held possibilities of appreciation in value which were not shared by bonds.

Investment in Common Stocks Not Wholly Discredited.—In our opinion, the pyrotechnics of 1928–1933 are less destructive of the logic of this position than in the case of investment bonds. They show, of course, that stocks apparently sound could suffer an unexpected disappearance of earning power, and that as between "sound bonds" and "sound stocks" the latter group was more severely affected by the depression.

But a rigid observance of old-time canons of common-stock investment would have dictated the sale of one's holdings at a substantial profit very early in the upswing and a heroic abstinence from further participation in the market until at some point after the 1929 collapse when prices were again attractive

---

[1] See Chamberlain, Lawrence, and William W. Hay, *Investment and Speculation*, pp. xii, 8–11, 55–56, New York, 1931; especially the following from p. 55: "Common stocks, as such, are not superior to bonds as long term investments, because primarily they are not investments at all. They are speculations."

in relation to earnings and other analytical factors. No doubt this would have resulted in making repurchases too soon—as matters turned out—with consequent paper or actual losses. But whatever the net result, the fact remains that the common-stock investor, proceeding along old-time conservative lines, had opportunities of profit commensurate with any risks he ran— an advantage not possessed by the typical bond buyer. The chief weakness of these investment principles was the difficulty of adhering firmly to them in the speculative contagion of 1928 and 1929. Given a recurrence from present levels of the narrower market swings of former years, conservative diversified investment in common stocks based on careful analysis should again be productive of satisfactory results.

LOWERED STANDARDS OF INVESTMENT BANKING HOUSES

Our third question related to the status of investment banking houses and the attitude to be taken by the public toward them. Until recent years, the leading houses of issue were able successfully to combine the somewhat discordant functions of protecting their clients' interests and making money for themselves. The public was safeguarded as much for business as for ethical reasons, since a firm's reputation and continued existence depended on the soundness of the merchandise which it sold. Investment banking houses, therefore, were considered, and considered themselves, as occupying a semifiduciary relationship toward their customers. But in 1928 and 1929 there occurred a wholesale and disastrous relaxation of the standards of safety previously observed by the reputable houses of issue. This was shown in the sale of many new offerings of inferior grade, aided in part by questionable methods of presenting the facts to the public. The general collapse in values affected these unsound and unseasoned issues with particular severity, so that the losses suffered by investors in many of these flotations have been little short of appalling.

Causes of the Lowered Standards.—This general lowering of standards by investment banking firms was due to two causes, the first being the ease with which all issues could be sold, and the second being the scarcity of sound investments to sell. The latter fact arose from the new vogue of financing through common-stock offerings, as the result of which the stronger corporations not only avoided additional bond issues but even retired

large amounts of their funded debt.[1]   Hence the supply of new
bond issues complying with former strict standards of investment
quality was undergoing a sharp reduction at the very time that
the volume of funds seeking investment reached record proportions.   In previous years, when houses of issue had their choice
between selling good bonds or poor ones, they habitually chose
the good securities, even at some sacrifice of underwriting profit.
But now they had to choose between selling poor investments or
none at all—between making large profits or shutting up shop—
and it was too much to expect from human nature that under
such circumstances they would adequately protect their clients'
interests.

Problem of Regaining Public Confidence.—Investment banking houses seemingly face a difficult task in regaining the confidence of a public properly mistrustful of their motives and their
methods.   At the present time such firms are proceeding with the
utmost caution in bringing out new investment issues—a policy
dictated among other reasons by the impossibility of selling any
but the highest grade of bonds in so poor an investment market,
and by the new difficulties imposed by the Securities Act of 1933.
But if past experience is any guide, the current critical attitude
of the investor is not likely to persist; and in the next period of
prosperity and plethora of funds for security purchases, the public
will once again exhibit its ingrained tendency to forgive, and
particularly to forget, the sins committed against it in the past.
Its future protection is more likely to come not from its own
discrimination but from the chastened attitude of the houses of
issue, anxious to retain their slowly recovered prestige by avoiding
a repetition of their recent errors.

**Increased Need for Thorough Knowledge of Investment
Principles.**—But a serious obstacle to sound investment-house
policies is likely to result from the relative scarcity of new bond
issues which can meet the much more exacting requirements
enjoined, as we suggested above, by the experience of the past
15 years.   If a large popular demand for bonds should return, the
difficulty of supplying really good issues in sufficient quantity
will almost inevitably result once more in numerous offerings of
inferior caliber and unsatisfactory performance.   From this
survey, we conclude that there is greater need than before of
either a thoroughgoing knowledge of investment principles on

[1] See Appendix, Note 1, for supporting data.

the part of the individual bond buyer or else of recourse by him to advice which is both expert and disinterested.

The need for such knowledge or such expert advice is only slightly lessened by the passage of the Securities Act of 1933. This measure requires the submission of elaborate data in connection with new security issues, and it greatly extends the liability of bankers, directors, etc., for losses following misstatements or material omissions. These provisions do not assure the soundness of the security but only that the facts will be adequately disclosed.[1] The questionable character of many speculative stock offerings made after the Securities Act was passed is a striking confirmation of this fact.

THE FACTOR OF HUMAN NATURE

One of the striking features of the past five years has been the domination of the financial scene by purely psychological elements. In previous bull markets the rise in stock prices remained in fairly close relationship with the improvement in business during the greater part of the cycle; it was only in its invariably short-lived culminating phase that quotations were forced to disproportionate heights by the unbridled optimism of the speculative contingent. But in the 1921–1933 cycle this "culminating phase" lasted for years instead of months, and it drew its support not from a group of speculators but from the entire financial community. The "new-era" doctrine—that "good" stocks (or "blue chips") were sound investments regardless of how high the price paid for them—was at bottom only a means of rationalizing under the title of "investment" the well-nigh universal capitulation to the gambling fever. We suggest that this psychological phenomenon is closely related to the dominant importance assumed in recent years by intangible factors of value, *viz.*, good-will, management, expected earning power, etc. Such value factors, while undoubtedly real, are not susceptible to mathematical calculation; hence the standards by which they are measured are to a great extent arbitrary and can suffer the widest variations in accordance with the prevalent psychology. The investing class was the more easily led to ascribe reality

---

[1] "All the Act pretends to do is to require the 'truth about securities' at the time of issue, and to impose a penalty for failure to tell the truth. Once it is told, the matter is left to the investor." William O. Douglas and G. E. Bates, *The Federal Securities Act of* 1933, 43 Yale Law Journal 171, December 1933.

to purely speculative valuations of these intangibles because
it was dealing in good part with surplus wealth, to which it was
not impelled by force of necessity to apply the old-established
acid test that the principal value be justified by the income.

No Automatic Relationship between Value and Price.—There
are a number of other factors involving human nature in Wall
Street to which recent experience should lead us to pay more
serious attention than was previously accorded them. Investment theory should recognize that the merits of an issue reflect
themselves in the market price not by any automatic response
or mathematical relationship but through the minds and decisions
of buyers and sellers. Furthermore, the investors' mental
attitude not only affects the market price but is strongly affected
by it, so that the success of a commitment—properly considered—
must depend in some part on the subsequent maintenance of a
satisfactory market price. Hence in selecting an investment,
even one presumably purchased for income only, reasonable
allowance must be made for such purely market-price elements
as can be ascertained, in addition to the more primary consideration which is paid to factors of intrinsic value. (Institutions such as life insurance companies and savings banks are
much less concerned, under ordinary conditions, with the question
of the market price of their investments than are individuals.
But a cataclysm of the amplitude of that of 1931–1932 made them
rudely conscious of market valuations.)

Speculation Not a Satisfactory Substitute for Investment.—If
the field of sound investment has suffered a severe contraction,
as we suggested above, it would seem natural to turn our attention
to intelligent speculation, on the theory that a good speculation
is undoubtedly superior to a poor investment. But here again
we must recognize that the psychology of the speculator militates
strongly against his success. For, by relation of cause and effect,
he is most optimistic when prices are highest and most despondent
when they are at bottom. Hence, in the nature of things, only
the exceptional speculator can prove consistently successful, and
no one has a logical right to believe that he will succeed where
most of his companions must fail. For this reason, training in
speculation, however intelligent and thorough, is likely to prove
a misfortune to the individual, since it may lead him into market
activities which, starting in most cases with small successes,
almost invariably end in major disaster.

If investment is likely to prove unsatisfactory and speculation is certain to be dangerous, to what may the intelligent student turn? Perhaps he would be well advised to devote his attention to the field of undervalued securities—issues, whether bonds or stocks, which are selling well below the levels apparently justified by a careful analysis of the relevant facts. The opportunities in this direction have always been numerous and varied, and they are discussed at length in later chapters of this book. It is true that bargain hunting in securities is not without its pitfalls, and in recent years especially it has been subject to many disadvantages and disappointments. Yet under more normal conditions it should yield satisfactory average results, and, most important of all, it promotes a fundamentally conservative point of view, which should constitute a valuable safeguard against speculative temptations.

PART I

SURVEY AND APPROACH

CHAPTER I

THE SCOPE AND LIMITATIONS OF SECURITY ANALYSIS.
THE CONCEPT OF INTRINSIC VALUE

Analysis connotes the careful study of available facts with the attempt to draw conclusions therefrom based on established principles and sound logic. It is part of the scientific method. But in applying analysis to the field of securities we encounter the serious obstacle that investment is by nature not an exact science. The same is true, however, of law and medicine, for here also both individual skill (art) and chance are important factors in determining success or failure. Nevertheless, in these professions analysis is not only useful but indispensable, so that the same should probably be true in the field of investment and possibly in that of speculation.

In the last three decades the prestige of security analysis in Wall Street has experienced both a brilliant rise and an ignominious fall—a history related but by no means parallel to the course of stock prices. The advance of security analysis proceeded uninterruptedly until about 1927, covering a long period in which increasing attention was paid on all sides to financial reports and statistical data. But the "new era" commencing in 1927 involved at bottom the abandonment of the analytical approach; and while emphasis was still seemingly placed on facts and figures, these were manipulated by a sort of pseudo-analysis to support the delusions of the period. The market collapse in October 1929 was no surprise to such analysts as had kept their heads, but the extent of the business collapse which later developed, with its devastating effects on established earning power, again threw their calculations out of gear. Hence the ultimate result was that serious analysis suffered a double discrediting: the first—prior to the crash—due to the persistence

14

of imaginary values, and the second—after the crash—due to the disappearance of real values.

In the Introduction we expressed the view that the experiences of 1927–1933 should not be taken as a norm by which to judge the future of bond investment. The same holds true for analysis as well, and for the same reason, *viz.*, that the extreme fluctuations and vicissitudes of that period are not likely to be duplicated soon again. Successful analysis, like successful investment, requires a fairly rational atmosphere to work *in* and at least some stability of values to work *with*.

THREE FUNCTIONS OF ANALYSIS: 1. DESCRIPTIVE FUNCTION

The functions of security analysis may be described under three headings: descriptive, selective, and critical. In its more obvious form, descriptive analysis consists of marshalling the important facts relating to an issue and presenting them in a coherent, readily intelligible manner. This function is adequately performed for the entire range of marketable corporate securities by the various manuals, the Standard Statistics and Fitch services, and others. A more penetrating type of description seeks to reveal the strong and weak points in the position of an issue, compare its exhibit with that of others of similar character, and appraise the factors which are likely to influence its future performance. Analysis of this kind is applicable to almost every corporate issue, and it may be regarded as an adjunct not only to investment but also to intelligent speculation in that it provides an organized factual basis for the application of judgment for purposes of investment or speculation.

2. THE SELECTIVE FUNCTION OF SECURITY ANALYSIS

In its selective function, security analysis goes further and expresses specific judgments of its own. It seeks to determine whether a given issue should be bought, sold, retained, or exchanged for some other. What types of securities or situations lend themselves best to this more positive activity of the analyst, and to what handicaps or limitations is it subject? It may be well to start with a group of examples of analytical judgments, which could later serve as a basis for a more general inquiry.

Examples of Analytical Judgments.—In 1928 the public was offered a large issue of 6% noncumulative preferred stock of St. Louis-San Francisco Railway Company priced at 100. The record showed that in no year in the company's history had earnings been equivalent to as much as 1½ times the fixed charges and preferred dividends combined. The application of well-established standards of selection to the facts in this case would have led to the rejection of the issue as insufficiently protected.

A contrasting example: In June 1932 it was possible to purchase 5% bonds of Owens-Illinois Glass Company, due 1939, at 70, yielding 11% to maturity. The company's earnings were many times the interest requirements—not only on the average but even at that time of severe depression. The bond issue was amply covered by current assets alone, and it was followed by common and preferred stock with a very large aggregate market value, taking their lowest quotations. Here, analysis would have led to the recommendation of this issue as a strongly entrenched and attractively priced investment.

Let us take an example from the field of common stocks. In 1922, prior to the boom in aviation securities, Wright Aeronautical Corporation stock was selling on the New York Stock Exchange at only $8, although it was paying a $1 dividend, had for some time been earning over $2 a share, and showed more than $8 per share in cash assets in the treasury. In this case analysis would readily have established that the intrinsic value of the issue was substantially above the market price.

Again, consider the same issue in 1928 when it had advanced to $280 per share. It was then earning at the rate of $8 per share, as against $3.77 in 1927. The dividend rate was $2; the net-asset value was less than $50 per share. A study of this picture must have shown conclusively that the market price represented for the most part the capitalization of entirely conjectural future prospects—in other words, that the intrinsic value was far less than the market quotation.

A third kind of analytical conclusion may be illustrated by a comparison of Interborough Rapid Transit Company First and Refunding 5s with the same company's Collateral 7% Notes, when both issues were selling at the same price (say 62) in 1933. The 7% notes were clearly worth considerably more than the 5s. Each $1,000 note was secured by deposit of $1,736 face amount of 5s; the principal of the notes had matured; they were entitled either to be paid off in full or to a sale of the collateral for their benefit. The annual interest received on the collateral was equal to about $87 on each 7% note (which amount was actually being distributed to the noteholders), so that the current income on the 7s was considerably greater than that on the 5s. Whatever technicalities might be invoked to prevent the noteholders from asserting their contractual rights promptly and completely, it was difficult to imagine conditions under which the 7s would not be intrinsically worth considerably more than the 5s.

Intrinsic Value vs. Price.—From the foregoing examples it will be seen that the work of the securities analyst is not without concrete results of considerable practical value, and that it is applicable to a wide variety of situations. In all of these instances he appears to be concerned with the intrinsic value of the security and more particularly with the discovery of discrepancies between the intrinsic value and the market price. We must recognize, however, that intrinsic value is an elusive concept. In general terms it is understood to be that value which is justified by the facts, *e.g.*, the assets, earnings, dividends, definite prospects, as distinct, let us say, from market quotations established by artificial manipulation or distorted by psychological excesses. But it is a great mistake to imagine that intrinsic value is as definite and as determinable as is the market price. Some time ago intrinsic value (in the case of a common stock) was thought to be about the same thing as "book value," *i.e.*, it was equal to the net assets of the business, fairly priced. This view of intrinsic value was quite definite, but it proved almost worthless as a practical matter because neither the average earnings nor the average market price evinced any tendency to be governed by the book value.

Intrinsic Value and "Earning Power."—Hence this idea was superseded by a newer view, *viz.*, that the intrinsic value of a business was determined by its earning power. But the phrase "earning power" must imply a fairly confident expectation of certain future results. It is not sufficient to know what the past earnings have averaged, or even that they disclose a definite line of growth or decline. There must be plausible grounds for believing that this average or this trend is a dependable guide to the future. Experience has shown only too forcibly that in many instances this is far from true. This means that the concept of "earning power," expressed as a definite figure, and the derived concept of intrinsic value, as something equally definite and ascertainable, cannot be safely accepted as a *general premise* of security analysis.

Example: To make this reasoning clearer, let us consider a concrete and typical example. What would we mean by the intrinsic value of J. I. Case Company common, as analyzed, say, early in 1933? The market price was $30; the asset value per share was $176; no dividend was being paid; the average earnings for ten years had been $9.50 per share; the results for 1932 had shown a deficit of $17 per share. If we followed a customary method of appraisal, we might take the average earnings per share of common for ten years, multiply this average by ten, and arrive at an intrinsic value of $95. But let us examine the individual figures which make up this ten-year average. They are as follows:

1932	$*17.40(d)*
1931	*2.90(d)*
1930	11.00
1929	20.40
1928	26.90
1927	26.00
1926	23.30
1925	15.30
1924	*5.90(d)*
1923	*2.10(d)*
Average.......	$ 9.50

(d) Deficit.

This average of $9.50 is obviously nothing more than an arithmetical resultant from ten unrelated figures. It can hardly be urged that this average is in any way representative of *typical* conditions in the past or representative of what may be expected in the future. Hence any figure of "real" or intrinsic value derived from this average must be characterized as equally accidental or artificial.

The Role of Intrinsic Value in the Work of the Analyst.—Let us try to formulate a statement of the role of intrinsic value in the work of the analyst which will reconcile the rather conflicting implications of our various examples. The essential point is that security analysis does not seek to determine exactly what is the intrinsic value of a given security. It needs only to establish either that the value is *adequate*—e.g., to protect a bond or to justify a stock purchase—or else that the value is considerably higher or considerably lower than the market price. For such purposes an indefinite and approximate measure of the intrinsic value may be sufficient. To use a homely simile, it is quite possible to decide by inspection that a woman is old enough to vote without knowing her age, or that a man is heavier than he should be without knowing his exact weight.

This statement of the case may be made clearer by a brief return to our examples. The rejection of St. Louis-San Francisco Preferred did not require an exact calculation of the intrinsic value of this railroad system. It was enough to show, very simply from the earnings record, that the margin of value above the bondholders' and preferred stockholders' claims was too small to assure safety. Exactly the opposite was true for the Owens-Illinois Glass 5s. In this instance, also, it would undoubtedly have been difficult to arrive at a fair valuation of the business; but it was quite easy to decide that this value in any event was far in excess of the company's debt.

In the Wright Aeronautical example, the earlier situation presented a set of facts which demonstrated that the business was worth substantially more than $8 per share, or $1,800,000. In the later year, the facts were equally conclusive that the business did not have a reasonable value of $280 per share, or $70,000,-000 in all. It would have been difficult for the analyst to determine whether Wright Aeronautical was actually worth $20 or $40 a share in 1922—or actually worth $50 or $80 in 1929. But fortunately it was not necessary to decide these points in order to conclude that the shares were attractive at $8 and unattractive, intrinsically, at $280.

The J. I. Case example illustrates the far more typical common-stock situation, in which the analyst cannot reach a dependable conclusion as to the relation of intrinsic value to market price. But even here, *if the price had been low or high enough*, a conclusion might have been warranted. To express the uncertainty of the picture, we might say that it was difficult to determine in early 1933 whether the intrinsic value of Case common was nearer $30 or $130. Yet if the stock had been selling at as low as $10, the analyst would undoubtedly have been justified in declaring that it was worth more than the market price.

Flexibility of the Concept of Intrinsic Value.—This should indicate how flexible is the concept of intrinsic value as applied to security analysis. Our notion of the intrinsic value may be more or less distinct, depending on the particular case. The degree of indistinctness may be expressed by a very hypothetical "range of approximate value," which would grow wider as the uncertainty of the picture increased, *e.g.*, $20 to $40 for Wright Aeronautical in 1922 as against $30 to $130 for Case in 1933. It would follow that even a very indistinct idea of the intrinsic value may still justify a conclusion if the current price falls far outside either the maximum or minimum appraisal.

More Definite Concept in Special Cases.—The Interborough Rapid Transit example permits a more precise line of reasoning than any of the others. Here a given market price for the 5% bonds results in a very definite valuation for the 7% notes. If it were certain that the collateral securing the notes would be acquired for and distributed to the note holders, then the mathematical relationship—*viz.*, $1,736 of value for the 7s against $1,000 of value for the 5s—would eventually be established at this ratio in the market. But because of quasi-political complications in the picture, this normal procedure could not be expected with certainty. As a practical matter, therefore, it is not possible to say that the 7s are actually worth 74% more than the 5s, but it may be said with assurance that the 7s are worth *substantially more*—which is a very useful conclusion to arrive at when both issues are selling at the same price.

The Interborough issues are an example of a rather special group of situations in which analysis may reach more definite conclusions respecting intrinsic value than in the ordinary case. These situations may involve a liquidation or give rise to technical operations known as "arbitrage" or "hedging." While, viewed in the abstract, they are probably the most satisfactory field for the analyst's work, the fact that they are specialized in character and of infrequent occurrence makes them relatively unimportant from the broader standpoint of investment theory and practice.

Principal Obstacles to Success of the Analyst. *a. Inadequate or Incorrect Data.*—Needless to say, the analyst cannot be right all the time. Furthermore a conclusion may be logically right but work out badly in practice. The main obstacles to the success of the analyst's work are threefold, *viz.*, (a) the inadequacy or incorrectness of the data, (b) the uncertainties of the future, and (c) the irrational behavior of the market. The first of these drawbacks, while serious, is the least important of the three. Deliberate falsification of the data is rare; most of the misrepresentation flows from the use of accounting artifices which it is the function of the capable analyst to detect. Concealment is more common than misstatement. In the majority of cases the analyst's experience and skill should lead him to note the absence of information on an important point; but in some instances the concealment will elude detection and give rise to an incorrect conclusion.

b. *Uncertainties of the Future.*—Of much greater moment is the element of future change. A conclusion warranted by the facts and by the apparent prospects may be vitiated by new developments. This raises the question of how far it is the function of security analysis to anticipate changed conditions. We shall defer consideration of this point until our discussion of various factors entering into the processes of analysis. It is manifest, however, that future changes are largely unpredictable, and that security analysis must ordinarily proceed on the assumption that the past record affords at least a rough guide to the future. The more questionable this assumption, the less valuable is the analysis. Hence this technique is more useful when applied to senior securities (which are protected against change) than to common stocks; more useful when applied to a business of inherently stable character than to one subject to wide variations; and, finally, more useful when carried on under fairly normal general conditions than in times of great uncertainty and radical change.

c. *The Irrational Behavior of the Market.*—The third handicap to security analysis is found in the market itself. In a sense the market and the future present the same kind of difficulties. Neither can be predicted or controlled by the analyst, yet his success is largely dependent upon them both. The major activities of the investment analyst may be thought to have little or no concern with market prices. His typical function is the selection of high-grade, fixed-income-bearing bonds, which upon investigation he judges to be secure as to interest and principal. The purchaser is supposed to pay no attention to their subsequent market fluctuations, but to be interested solely in the question whether the bonds will continue to be sound investments. In our opinion this traditional view of the investor's attitude is inaccurate and somewhat hypocritical. Owners of securities, whatever their character, are interested in their market quotations. This fact is recognized by the emphasis always laid in investment practice upon *marketability*. If it is important that an issue be readily salable, it is still more important that it

command a satisfactory price.   While for obvious reasons the
investor in high-grade bonds has a lesser concern with market
fluctuations than has the speculator, they still have a strong
psychological, if not financial, effect upon him.   Even in this
field, therefore, the analyst must take into account whatever
influences may adversely govern the market price, as well as
those which bear upon the basic safety of the issue.

In that portion of the analyst's activities which relates to the
discovery of undervalued, and possibly of overvalued securities,
he is more directly concerned with market prices.   For here the
vindication of his judgment must be found largely in the ultimate
market action of the issue.   This field of analytical work may
be said to rest upon a twofold assumption: first, that the market
price is frequently out of line with the true value; and, second,
that there is an inherent tendency for these disparities to correct
themselves.   As to the truth of the former statement, there can
be very little doubt—even though Wall Street often speaks
glibly of the "infallible judgment of the market" and asserts
that "a stock is worth what you can sell it for—neither more nor
less."

The Hazard of Tardy Adjustment of Price to Value.—The
second assumption is equally true in theory, but its working out
in practice is often most unsatisfactory.   Undervaluations
caused by neglect or prejudice may persist for an inconveniently
long time, and the same applies to inflated prices caused by
overenthusiasm or artificial stimulants.   The particular danger
to the analyst is that, because of such delay, new determining
factors may supervene before the market price adjusts itself to
the value as he found it.   In other words, by the time the price
finally does reflect the value, this value may have changed considerably and the facts and reasoning on which his decision was
based may no longer be applicable.

The analyst must seek to guard himself against this danger as
best he can: in part, by dealing with those situations preferably
which are not subject to sudden change; in part, by favoring
securities in which the popular interest is keen enough to promise
a fairly swift response to value elements which he is the first
to recognize; in part, by tempering his activities to the general
financial situation—laying more emphasis on the discovery of
undervalued securities when business and market conditions are
on a fairly even keel, and proceeding with greater caution in times

of abnormal stress and uncertainty.

The Relationship of Intrinsic Value to Market Price.—The general question of the relation of intrinsic value to the market quotation may be made clearer by the appended chart, which traces the various steps culminating in the market price. It will be evident from the chart that the influence of what we call analytical factors over the market price is both *partial* and *indirect*—partial, because it frequently competes with purely speculative factors which influence the price in the opposite direction; and indirect, because it acts through the intermediary of people's sentiments and decisions. In other words, the market is not a *weighing machine*, on which the value of each issue is recorded by an exact and impersonal mechanism, in accordance with its specific qualities. Rather should we say that the market is a *voting machine*, whereon countless individuals register choices which are the product partly of reason and partly of emotion.

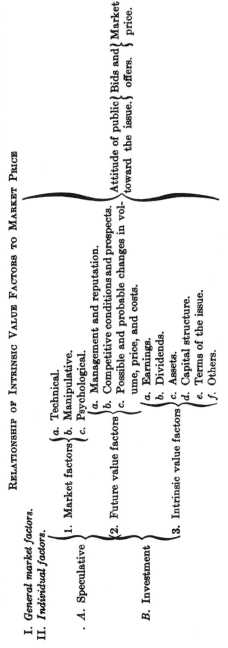

RELATIONSHIP OF INTRINSIC VALUE FACTORS TO MARKET PRICE

I. *General market factors.*
II. *Individual factors.*

A. Speculative
  1. Market factors
    a. Technical.
    b. Manipulative.
    c. Psychological.
  2. Future value factors
    a. Management and reputation.
    b. Competitive conditions and prospects.
    c. Possible and probable changes in volume, price, and costs.

B. Investment
  3. Intrinsic value factors
    a. Earnings.
    b. Dividends.
    c. Assets.
    d. Capital structure.
    e. Terms of the issue.
    f. Others.

Attitude of public toward the issue. } Bids and offers. } Market price.

ANALYSIS AND SPECULATION

It may be thought that sound analysis should produce successful results in any type of situation, including the confessedly speculative, *i.e.*, those subject to substantial uncertainty and risk. If the selection of speculative issues is based on expert study of the companies' position, should not this approach give the purchaser a considerable advantage? Admitting future events to be uncertain, could not the favorable and unfavorable developments be counted on to cancel out against each other, more or less, so that the initial advantage afforded by sound analysis will carry through into an eventual average profit? This is a plausible argument but a deceptive one; and its over-ready acceptance has done much to lead analysts astray. It is worth while, therefore, to detail several valid arguments against placing chief reliance upon analysis in speculative situations.

In the first place, what may be called the mechanics of speculation involves serious handicaps to the speculator, which may outweigh the benefits conferred by analytical study. These disadvantages include the payment of commissions and interest charges, the so-called "turn of the market" (meaning the spread between the bid and asked price), and, most important of all, an inherent tendency for the average loss to exceed the average profit, unless a certain technique of trading is followed, which is opposed to the analytical approach.

The second objection is that the underlying analytical factors in speculative situations are subject to swift and sudden revision. The danger, already referred to, that the intrinsic value may change before the market price reflects that value, is therefore much more serious in speculative than in investment situations. A third difficulty arises from circumstances surrounding the unknown factors, which are necessarily left out of security analysis. Theoretically these unknown factors should have an equal chance of being favorable or unfavorable, and thus they should neutralize each other in the long run. For example, it might be thought that a simple way to make money could be found by purchasing a number of common stocks currently earning the largest percentage on the market price and simultaneously selling those earning the smallest percentage, the idea being that helpful or harmful future changes should be about equally distributed over both groups, so that the group purchased should maintain its better aggregate showing and therefore do better in the market. But it may well be that the low price for the apparently attractive issues is due to certain important unfavorable factors which, though not disclosed, are known to those identified with the company—and *vice versa* for the issues seemingly selling above their relative value. In speculative situations, those "on the inside" often have an advantage of this kind which nullifies the premise that good and bad changes in the picture should offset each other, and which loads the dice against the analyst working with some of the facts concealed from him.

The Value of Analysis Diminishes as the Element of Chance Increases.—The final objection is based on more abstract grounds, but, nevertheless, its practical importance is very great. Even if we grant that analysis can give the speculator a mathematical advantage, it does not assure him a profit. His ventures remain hazardous; in any individual case a loss may be taken; and after the operation is concluded, it is difficult to determine whether the analyst's contribution has been a benefit or a detriment. Hence the latter's position in the speculative field is at best uncertain and somewhat lacking in professional dignity. It is as though the analyst and Dame Fortune were playing a duet on the speculative piano, with the fickle goddess calling all the tunes.

By another and less imaginative simile, we might more convincingly show why analysis is inherently better suited to investment than to speculative situations. (In anticipation of a more detailed inquiry in a later chapter, we have assumed throughout this chapter that investment implies expected safety and speculation connotes acknowledged risk.) In Monte Carlo the odds are weighted 19 to 18 in favor of the proprietor of the roulette wheel, so that on the average he wins one dollar out of each 37 wagered by the public. This may suggest the odds against the untrained investor or speculator. Let us assume that, through some equivalent of analysis, a roulette player is able to reverse the odds for a limited number of wagers, so that they are now 18 to 19 in his favor. If he distributes his wagers evenly over all the numbers, then whichever one turns up he is certain to win a moderate amount. This operation may be likened to an investment program based upon sound analysis and carried on under propitious general conditions.

But if the player wagers all his money on a single number, the small odds in his favor are of slight importance compared with the crucial question whether chance will elect the number he has chosen. His "analysis" will enable him to win a little more if he is lucky; it will be of no value when luck is against him. This, in slightly exaggerated form perhaps, describes the position of the analyst dealing with essentially speculative operations. Exactly the same mathematical advantage which practically assures good results in the investment field may prove entirely ineffective where luck is the overshadowing influence.

It would seem prudent, therefore, to consider analysis as an *adjunct* or *auxiliary* rather than as a *guide* in speculation. It is only where chance plays a subordinate role that the analyst can properly speak in an authoritative voice and accept responsibility for the results of his judgments.

3. THE CRITICAL FUNCTION OF SECURITY ANALYSIS

The principles of investment finance and the methods of corporation finance fall necessarily within the province of security analysis. Analytical judgments are reached by applying standards to facts. The analyst is concerned, therefore, with the soundness and practicability of the standards of selection. He is also interested to see that securities, especially bonds and preferred stocks, be issued with adequate protective provisions, and—more important still—that proper methods of enforcement of these covenants be part of accepted financial practice.

It is a matter of great moment to the analyst that the facts be fairly presented, and this means that he must be highly critical of accounting methods. Finally, he must concern himself with all corporate policies affecting the security owner, for the value of the issue which he analyzes may be largely dependent upon the acts of the management. In this category are included questions of capitalization set-up, of dividend and expansion policies, of managerial compensation, and even of continuing or liquidating an unprofitable business.

On these matters of varied import, security analysis may be competent to express critical judgments, looking to the avoidance of mistakes, to the correction of abuses, and to the better protection of those owning bonds or stocks.

CHAPTER II

FUNDAMENTAL ELEMENTS IN THE PROBLEM OF ANALYSIS. QUANTITATIVE AND QUALITATIVE FACTORS

In the previous chapter we referred to some of the concepts and materials of analysis from the standpoint of their bearing on what the analyst may hope to accomplish. Let us now imagine the analyst at work and ask what are the broad considerations which govern his approach to a particular problem, and also what should be his general attitude toward the various kinds of information with which he has to deal.

FOUR FUNDAMENTAL ELEMENTS

The object of security analysis is to answer, or assist in answering, certain questions of a very practical nature. Of these, perhaps the most customary are the following: What securities should be bought for a given purpose? Should issue S be bought, or sold, or retained?

In all such questions, four major factors may be said to enter, either expressly or by implication. These are:

1. The security.
2. The price.
3. The time.
4. The person.

More completely stated, the second typical question would run, Should security S be bought (or sold, or retained) at price P, at this time T, by individual I? Some discussion of the relative significance of these four factors is therefore pertinent, and we shall find it convenient to consider them in inverse order.

The Personal Element.—The personal element enters to a greater or lesser extent into every security purchase. The aspect of chief importance is usually the financial position of the intending buyer. What might be an attractive speculation for a business man should under no circumstances be attempted by a trustee or a widow with limited income. Again, United States

27

Liberty 3½s should not have been purchased by those to whom their complete tax-exemption feature was of no benefit, when a considerably higher yield could be obtained from partially taxable governmental issues.

Other personal characteristics that on occasion might properly influence the individual's choice of securities are his financial training and competence, his temperament, and his preferences. But however vital these considerations may prove at times, they are not ordinarily determining factors in analysis. Most of the conclusions derived from analysis can be stated in impersonal terms, as applicable to investors or speculators as a class.

The Time.—The time at which an issue is analyzed may affect the conclusion in various ways. The company's showing may be better, or its outlook may seem better, at one time than another, and these changing circumstances are bound to exert a varying influence on the analyst's viewpoint toward the issue. Furthermore, securities are selected by the application of standards of quality and yield, and both of these—particularly the latter—will vary with financial conditions in general. A railroad bond of highest grade yielding 5% seemed attractive in June 1931 because the average return on this type of bond was 4.32%. But the same offering made six months later would have been quite unattractive, for in the meantime bond prices had fallen severely and the yield on this group had increased to 5.86%. Finally, nearly all security commitments are influenced to some extent by the current view of the financial and business outlook. In speculative operations these considerations are of controlling importance; and while conservative investment is ordinarily supposed to disregard these elements, in times of stress and uncertainty they may not be ignored.

Security analysis, as a study, must necessarily concern itself as much as possible with principles and methods which are valid at all times—or, at least, under all ordinary conditions. It should be kept in mind, however, that the practical applications of analysis are made against a background largely colored by the changing times.

The Price.—The price is an integral part of every complete judgment relating to securities. In the selection of prime investment bonds, the price is usually a subordinate factor, not because it is a matter of indifference but because in actual practice the price is rarely unreasonably high. Hence almost entire emphasis is placed on the question whether the issue is adequately secured. But in a special case, such as the purchase of high-grade *convertible* bonds, the price may be a factor fully as important as the degree of security. This point is illustrated by the American Telephone and Telegraph Company Convertible 4½s, due 1939, which sold above 200 in 1929. The fact that principal (at par) and interest were safe beyond question did not prevent the issue from being an extremely risky purchase *at that price*— one which in fact was followed by the loss of over half its market value.[1]

In the field of common stocks, the necessity of taking price into account is more compelling, because the danger of paying the wrong price is almost as great as that of buying the wrong issue. We shall point out later that the new-era theory of investment left price out of the reckoning, and that this omission was productive of most disastrous consequences.

The Security: Character of the Enterprise and the Terms of the Commitment.—The roles played by the security and its price in an investment decision may be set forth more clearly if we restate the problem in somewhat different form. Instead of asking, (*a*) In what security? and (*b*) At what price? let us ask, (*a*) In what enterprise? and (*b*) On what terms is the commitment proposed? This gives us a more comprehensive and evenly balanced contrast between two basic elements in analysis. By the *terms* of the investment or speculation, we mean not only the price but also the provisions of the issue and its status or showing at the time.

Example of Commitment on Unattractive Terms.—An investment in the soundest type of enterprise may be made on unsound and unfavorable terms. The value of urban real estate has tended to grow steadily over a long period of years; hence it came to be regarded by many as the "safest" medium of investment. But the purchase of a preferred stock in a New York City real

---

[1] The annual price ranges for American Telephone and Telegraph Company Convertible 4½s, due in 1939, were as follows:

Year	High	Low
1929	227	118
1930	193⅜	116
1931	135	95

estate development in 1929 might have involved *terms* of investment so thoroughly disadvantageous as to banish all elements of soundness from the proposition. One such stock offering could be summarized as follows:[1]

1. *Provisions of the Issue.*—A preferred stock, ranking junior to a large first mortgage and without unqualified rights to dividend or principal payments. It ranked ahead of a common stock which represented no cash investment, so that the common stockholders had nothing to lose and a great deal to gain, while the preferred stockholders had everything to lose and only a small share in the possible gain.

2. *Status of the Issue.*—A commitment in a new building, constructed at an exceedingly high level of costs, with no reserves or junior capital to fall back upon in case of trouble.

3. *Price of the Issue.*—At par the dividend return was 6%, which was much less than the yield obtainable on real-estate second mortgages having many other advantages over this preferred stock.[2]

Example of a Commitment on Attractive Terms.—We have only to examine electric power and light financing in recent years to find countless examples of unsound securities in a fundamentally attractive industry. By way of contrast let us cite the case of Brooklyn Union Elevated Railroad First 5s, due 1950, which sold in 1932 at 60 to yield 9.85% to maturity. They are an obligation of the Brooklyn-Manhattan Transit System. The traction, or electric railway, industry has long been unfavorably regarded, chiefly because of automobile competition but also on account of regulation and fare-contract difficulties. Hence

---

[1] The financing method described is that used by the separate owning corporations organized and sponsored by the Fred F. French Company and affiliated enterprises, with the exception of some of the later Tudor City units in the financing of which interest-bearing notes, convertible par for par into preferred stock at the option of the company, were substituted for the preferred stock in the financial plan. See *The French Plan* (10th ed., December 1928) published and distributed by the Fred F. French Investing Company, Inc. See also *Moody's Manual;* "Banks and Finance," 1933, pp. 1703–1707.

[2] The real-estate enterprise from which this example is taken gave a bonus of common stock with the preferred shares. The common stock had no immediate value, but it did have a potential value which, *under favorable conditions*, might have made the purchase profitable. From the investment standpoint, however, the preferred stock of this enterprise was subject to all of the objections which we have detailed.

this security represents a comparatively unattractive *type* of enterprise. Yet the *terms* of the investment here might well make it a satisfactory commitment, as shown by the following:

1. *Provisions of the Issue.*—By contract between the operating company and the City of New York, this is a first charge on the earnings of the combined subway and elevated lines of the system, both company and city owned, representing an investment enormously greater than the size of this issue.

2. *Status of the Issue.*—Apart from the very exceptional specific protection just described, the bonds were obligations of a company with stable and apparently fully adequate earning power.

3. *Price of Issue.*—It could be purchased to yield somewhat more than the Brooklyn-Manhattan Transit Corporation 6s, due 1968, which occupy a subordinate position. (At the low price of 68 for the latter issue in 1932 its yield was 9% against 9.85% for the Brooklyn Union Elevated 5s.)

Relative Importance of the Terms of the Commitment and the Character of the Enterprise.—Our distinction between the character of the enterprise and the terms of the commitment suggests a question as to which element is the more important. Is it better to invest in an attractive enterprise on unattractive terms or in an unattractive enterprise on attractive terms? The popular view unhesitatingly prefers the former alternative, and in so doing it is instinctively, rather than logically, right. Over a long period, experience will undoubtedly show that less money has been lost by the great body of investors through paying too high a price for securities of the best regarded enterprises than by trying to secure a larger income or profit from commitments in enterprises of lower grade.

From the standpoint of analysis, however, this empirical result does not dispose of the matter. It merely exemplifies a rule that is applicable to all kinds of merchandise, *viz.*, that the *untrained buyer* fares best by purchasing goods of the highest reputation, even though he may pay a comparatively high price. But, needless to say, this is not a rule to guide the expert merchandise buyer, for he is expected to judge quality by examination and not solely by reputation, and at times he may even sacrifice certain definite degrees of quality if that which he obtains is adequate for his purpose and attractive in price. This distinction applies as well to the purchase of securities as to buying paints or watches.   It results in two principles of quite opposite character, the one suitable for the untrained investor, the other useful only to the analyst.

1. Principle for the untrained security buyer: *Do not put money in a low-grade enterprise on any terms.*
2. Principle for the securities analyst: *Nearly every issue might conceivably be cheap in one price range and dear in another.*

We have criticized the placing of exclusive emphasis on the choice of the enterprise on the ground that it often leads to paying too high a price for a good security.   A second objection is that the enterprise itself may prove to be unwisely chosen.   It is natural and proper to prefer a business which is large and well managed, has a good record, and is expected to show increasing earnings in the future.   But these expectations, though seemingly well-founded, often fail to be realized.   Many of the leading enterprises of yesterday are today far back in the ranks.   Tomorrow is likely to tell a similar story.   The most impressive illustration is afforded by the persistent decline in the relative investment position of the railroads as a class during the past two decades.   The standing of an enterprise is in part a matter of fact and in part a matter of opinion.   During recent years investment opinion has proved extraordinarily volatile and undependable.   In 1929 Westinghouse Electric and Manufacturing Company was quite universally considered as enjoying an unusually favorable industrial position.   Two years later the stock sold for much less than the net current assets alone, presumably indicating widespread doubt as to its ability to earn *any* profit in the future.

These considerations do not gainsay the principle that untrained investors should confine themselves to the best regarded enterprises.   It should be realized, however, that this preference is enjoined upon them because of the greater risk for them in other directions, and not because the most popular issues are necessarily the safest.   The analyst must pay respectful attention to the judgment of the market place and to the enterprises which it strongly favors, but he must retain an independent and critical viewpoint.   Nor should he hesitate to condemn the popular and espouse the unpopular when reasons sufficiently weighty and convincing are at hand.

QUALITATIVE AND QUANTITATIVE FACTORS IN ANALYSIS

Analyzing a security involves an analysis of the business. Such a study could be carried to an unlimited degree of detail; hence practical judgment must be exercised to determine how far the process should go. The circumstances will naturally have a bearing on this point. A buyer of a $1,000 bond would not deem it worth his while to make as thorough an analysis of an issue as would a large insurance company considering the purchase of a $500,000 block. The latter's study would still be less detailed than that made by the originating bankers. Or, from another angle, a less intensive analysis should be needed in selecting a high-grade bond yielding $4\frac{1}{2}\%$ than in trying to find a well-secured issue yielding 8% or an *unquestioned bargain* in the field of common stocks.

Technique and Extent of Analysis Should Be Limited by Character and Purposes of the Commitment.—The equipment of the analyst must include a sense of proportion in the use of his technique. In choosing and dealing with the materials of analysis he must consider not only inherent importance and dependability but also the question of accessibility and convenience. He must not be misled by the availability of a mass of data—as, for example, in the reports of the railroads to the Interstate Commerce Commission—into making elaborate studies of nonessentials. On the other hand, he must frequently resign himself to the lack of significant information because it can be secured only by expenditure of more effort than he can spare or the problem will justify. This would be true frequently of some of the elements involved in a complete "business analysis"—as, for example, the extent to which an enterprise is dependent upon patent protection or geographical advantages or favorable labor conditions which may not endure.

Value of Data Varies with Type of Enterprise.—Most important of all, the analyst must recognize that the value of a particular kind of data varies greatly with the type of enterprise which is being studied. The five-year record of gross or net earnings of a railroad or a large chain-store enterprise may afford, if not a conclusive, at least a reasonably sound basis for measuring the safety of the senior issues and the attractiveness of the common shares. But the same statistics supplied by one of the smaller oil-producing companies may well prove more deceptive than useful, since they are chiefly the resultant of two factors, *viz.*, price received and production, both of which are likely to be radically different in the future than in the past.

Quantitative vs. Qualitative Elements in Analysis.—It is convenient at times to classify the elements entering into an analysis under two headings: the quantitative and the qualitative. The former might be called the company's statistical exhibit. Included in it would be all the useful items in the income account and balance sheet, together with such additional specific data as may be provided with respect to production and unit prices, costs, capacity, unfilled orders, etc. These various items may be subclassified under the headings: (1) capitalization, (2) earnings and dividends, (3) assets and liabilities, and (4) operating statistics.

The qualitative factors, on the other hand, deal with such matters as the nature of the business; the relative position of the individual company in the industry; its physical, geographical, and operating characteristics; the character of the management; and, finally, the outlook for the unit, for the industry, and for business in general. Questions of this sort are not dealt with ordinarily in the company's reports. The analyst must look for their answers to miscellaneous sources of information of greatly varying dependability—including a large admixture of mere opinion.

Broadly speaking, the quantitative factors lend themselves far better to thoroughgoing analysis than do the qualitative factors. The former are fewer in number, more easily obtainable, and much better suited to the forming of definite and dependable conclusions. Furthermore the financial results will themselves epitomize many of the qualitative elements, so that a detailed study of the latter may not add much of importance to the picture. The typical analysis of a security—as made, say, in a brokerage-house circular or in a report issued by a statistical service—will treat the qualitative factors in a superficial or summary fashion and devote most of its space to the figures.

Qualitative Factors: Nature of the Business and Its Future Prospects.—The qualitative factors upon which most stress is laid are the nature of the business and the character of the management. These elements are exceedingly important, but they are also exceedingly difficult to deal with intelligently. Let us consider, first, the nature of the business, in which concept is included the general idea of its future prospects. Most people have fairly definite notions as to what is "a good business" and what is not. These views are based partly on the financial results, partly on knowledge of specific conditions in the industry, and partly also on surmise or bias.

During most of period of general prosperity between 1923 and 1929, quite a number of major industries were backward. These included cigars, coal, cotton goods, fertilizers, leather, lumber, meat packing, paper, shipping, street railways, sugar, woolen goods. The underlying cause was usually either the development of competitive products or services (*e.g.*, coal, cotton goods, tractions) or excessive production and demoralizing trade practices (*e.g.*, paper, lumber, sugar). During the same period other industries were far more prosperous than the average. Among these were can manufacturers, chain stores, cigarette producers, motion pictures, public utilities. The chief cause of these superior showings might be found in unusual growth of demand (cigarettes, motion pictures) or in absence or control of competition (public utilities, can makers) or in the ability to win business from other agencies (chain stores).

It is natural to assume that industries which have fared worse than the average are "unfavorably situated" and therefore to be avoided. The converse would be assumed, of course, for those with superior records. But this conclusion may often prove quite erroneous. Abnormally good or abnormally bad conditions do not last forever. This is true not only of general business but of particular industries as well. Corrective forces are usually set in motion which tend to restore profits where they have disappeared, or to reduce them where they are excessive in relation to capital.

Industries especially favored by a developing demand may become demoralized through a still more rapid growth of supply. This was true recently of radio, aviation, electric refrigeration, bus transportation, and silk hosiery. In 1922 department stores were very favorably regarded because of their excellent showing in the 1920–1921 depression; but they did not maintain this advantage in subsequent years. The public utilities were unpopular in the 1919 boom, because of high costs; they became speculative and investment favorites in 1927–1929; in 1933 fear of both inflation and rate regulation again undermined the public's confidence in them. In 1933, on the other hand, the cotton-goods industry—long depressed—forged ahead faster than most others.

The Factor of Management.—Our appreciation of the importance of selecting a "good industry" must be tempered by a realization that this is by no means so easy as it sounds. Somewhat the same difficulty is met with in endeavoring to select an unusually capable management. Objective tests of managerial ability are few and far from scientific. In most cases the investor must rely upon a reputation which may or may not be deserved. The most convincing proof of capable management lies in a superior comparative record over a period of time. But this brings us back to the quantitative data.

There is a strong tendency in the stock market to value the management factor twice in its calculations. Stock prices reflect the large earnings which the good management has produced, *plus* a substantial increment for "good management" considered separately. This amounts to "counting the same trick twice" and it proves a frequent cause of overvaluation.

The Trend of Future Earnings.—In recent years increasing importance has been laid upon the *trend of earnings*. Needless to say, a record of increasing profits is a favorable sign. Financial theory has gone further, however, and has sought to estimate future earnings by projecting the past trend into the future and then used this projection as a basis for valuing the business. Because figures are used in this process, people mistakenly believe that it is "mathematically sound." But while a trend shown in the past is a fact, a "future trend" is only an assumption. The factors that we mentioned previously as militating against the maintenance of abnormal prosperity or depression are equally opposed to the indefinite continuance of an upward or downward trend. By the time the trend has become clearly noticeable, conditions may well be ripe for a change.

It may be objected that as far as the future is concerned it is just as logical to expect a past trend to be maintained as to expect a past average to be repeated. This is probably true, but it does not follow that the trend is more useful to analysis than the individual or average figures of the past. For security analysis does not assume that a past average will be repeated, but only that it supplies a rough index to what may be expected of the future. A trend, however, cannot be used as a "rough index"; it represents a definite prediction of either better or poorer results, and it must be either right or wrong.

This distinction, important in its bearing on the attitude of the analyst, may be made clearer by the use of examples. Let us assume that in 1929 a railroad showed its interest charges earned three times on the average during the preceding seven years. The analyst would have ascribed great weight to this point as an indication that its bonds were sound. This is a judgment based on quantitative data and standards. But it does not imply a prediction that the earnings in the next seven years will average three times interest charges; it suggests only that earnings are not likely to fall so much under three times interest charges as to endanger the bonds. In nearly every actual case such a conclusion would have proved correct, despite the economic collapse that ensued.

Now let us consider a similar judgment based primarily upon the trend. In 1929 nearly all public-utility systems showed a continued growth of earnings, but the fixed charges of many were so heavy—by reason of pyramided capital structures—that they consumed nearly all the net income. Investors bought bonds of these systems freely on the theory that the small margin of safety was no drawback, since earnings were certain to continue to increase. They were thus making a clear-cut prediction as to the future, upon the correctness of which depended the justification of their investment. If their prediction were wrong—as proved to be the case—they were bound to suffer serious loss.

Trend Essentially a Qualitative Factor.—In our discussion of the valuation of common stocks, later in this book, we shall point out that the placing of preponderant emphasis on the trend is likely to result in errors of overvaluation or undervaluation. This is true because no limit may be fixed on how far ahead the trend should be projected; and therefore the process of valuation, while seemingly mathematical, is in reality psychological and quite arbitrary. For this reason we consider the trend as a *qualitative* factor in its practical implications, even though it may be stated in quantitative terms.

Qualitative Factors Resist Even Reasonably Accurate Appraisal.—The trend is, in fact, a statement of future prospects in the form of an exact prediction. In similar fashion, conclusions as to the nature of the business and the abilities of the management have their chief significance in their bearing on the outlook. These qualitative factors are therefore all of the same general character. They all involve the same basic difficulty for the analyst, *viz.*, that it is impossible to judge how far they may properly reflect themselves in the price of a given security. In most cases, if they are recognized at all, they tend to be overemphasized. We see the same influence constantly at work in the general market. The recurrent excesses of its advances and declines are due at bottom to the fact that, when values are determined chiefly by the outlook, the resultant judgments are not subject to any mathematical controls and are almost inevitably carried to extremes.

Analysis is concerned primarily with values which are supported by the facts and not with those which depend largely upon expectations. In this respect the analyst's approach is diametrically opposed to that of the speculator, meaning thereby one whose success turns upon his ability to forecast or to guess future developments. Needless to say, the analyst must take possible future changes into account, but his primary aim is not so much to *profit* from them as to *guard against* them. Broadly speaking, he views the business future as a hazard which his conclusions must encounter rather than as the source of his vindication.

Inherent Stability a Major Qualitative Factor.—It follows that the qualitative factor in which the analyst should properly be most interested is that of *inherent stability*. For stability means resistance to change and hence greater dependability for the results shown in the past. Stability, like the trend, may be expressed in quantitative terms—as, for example, by stating that the earnings of General Baking Company during 1923–1932 were never less than ten times 1932 interest charges or that the operating profits of Woolworth between 1924 and 1933 varied only between $2.12 and $3.66 per share of common. But in our opinion stability is really a qualitative trait, because it derives in the first instance from the character of the business and not from its statistical record. A stable record suggests that the business is inherently stable, but this suggestion may be rebutted by other considerations.

Examples: This point may be brought out by a comparison of two preferred-stock issues as of early 1932, *viz.*, those of Studebaker (motors) and of First National (grocery) Stores, both of which were selling above par. The two exhibits were similar, in that both disclosed a continuously satisfactory margin above preferred-dividend requirements. The Studebaker figures were more impressive, however, as the following table will indicate:

NUMBER OF TIMES PREFERRED DIVIDEND WAS COVERED

First National Stores		Studebaker	
Period	Times covered	Calendar year	Times covered
Year ended March 31, 1930.....	13.4	1929	23.3
Year ended March 31, 1929.....	8.4	1928	27.3
Year ended March 31, 1928.....	4.4	1927	23.0
15 mos. ended March 31, 1927..	4.6	1926	24.8
Calendar year, 1925...........	5.7	1925	29.7
Calendar year, 1924...........	4.9	1924	23.4
Calendar year, 1923...........	5.1	1923	30.5
Calendar year, 1922...........	4.0	1922	27.3
Annual average...........	6.3		26.2

But the analyst must penetrate beyond the mere figures and consider the inherent character of the two businesses. The chain-store grocery trade contained within itself many elements of relative stability, such as stable demand, diversified locations, and rapid inventory turnover. A typical large unit in this field, provided only it abstained from reckless expansion policies, was not likely to suffer tremendous fluctuations in its earnings. But the situation of the typical automobile manufacturer was quite different. Despite fair stability in the industry as a whole, the individual units were subject to extraordinary variations, due chiefly to the vagaries of popular preference. The stability of Studebaker's earnings could not be held by any convincing logic to demonstrate that this company enjoyed a special and permanent immunity from the vicissitudes to which most of its competitors had shown themselves subject. The soundness of Studebaker Preferred rested, therefore, largely upon a stable statistical showing which was at variance with the general character of the industry, so far as its individual units were concerned. On the other hand, the satisfactory exhibit of First National Stores Preferred was in thorough accord with what was generally thought to be the inherent character of the business. The latter consideration should have carried great weight with the analyst and should have made First National Stores Preferred appear intrinsically sounder as a fixed-value investment than Studebaker Preferred, despite the more impressive *statistical* showing of the automobile company.

Summary.—To sum up this discussion of qualitative and quantitative factors, we may express the dictum that the analyst's conclusions must always rest upon the figures and upon established tests and standards. These figures alone are not *sufficient;* they may be completely vitiated by qualitative considerations of an opposite import. A security may make a satisfactory statistical showing, but doubt as to the future or distrust of the management may properly impel its rejection. Again, the analyst is likely to attach prime importance to the qualitative element of *stability*, because its presence means that conclusions based on past results are not so likely to be upset by unexpected developments. It is also true that he will be far more confident in his selection of an issue if he can buttress an adequate quantitative exhibit with unusually favorable qualitative factors.

But whenever the commitment depends to a substantial degree upon these qualitative factors—whenever, that is, the price is considerably higher than the figures alone would justify—then the analytical basis of approval is lacking. In the mathematical phrase, a satisfactory statistical exhibit is a *necessary* though by no means a *sufficient condition* for a favorable decision by the analyst.

CHAPTER III

SOURCES OF INFORMATION

It is impossible to discuss or even to list all the sources of information which the analyst may find it profitable to consult at one time or another in his work. In this chapter we shall present a concise outline of the more important sources, together with some critical observations thereon; and we shall also endeavor to convey, by means of examples, an idea of the character and utility of the large variety of special avenues of information.

DATA ON THE TERMS OF THE ISSUE

Let us assume that in the typical case the analyst seeks data regarding: (a) the terms of the specific issue, (b) the company, and (c) the industry. The provisions of the issue itself are summarized in the security manuals or statistical services. For more detailed information regarding a bond contract the analyst should consult the indenture (or deed of trust), a copy of which may be obtained or inspected at the office of the trustee. The terms of the respective stock issues of a company are set forth fully in the charter (or articles of association). This document is not usually available, but if the stock is listed nearly all the pertinent provisions will be found given in full in the listing applications, which are readily obtainable.

DATA ON THE COMPANY

Reports to Stockholders (Including Interim News Releases).— Coming now to the *company*, the chief source of statistical data is, of course, the reports issued to the stockholders. These reports vary widely with respect to both frequency and completeness, as the following summary will show:

All important railroads supply *monthly* figures down to net after rentals (net railway operating income). Many carry the results down to the balance for dividends (net income). Many publish carloading figures *weekly*, and a few have published gross earnings

41

weekly. The pamphlet annual reports publish financial and operating figures in considerable detail.

The ruling policy of public-utility companies varies between *quarterly* and *monthly* statements. Figures regularly include gross, net after taxes, and balance for dividends. Some companies publish only a moving twelve-month total—*e.g.*, American Water Works and Electric Company (monthly), North American Company (quarterly). Many supply *weekly* or *monthly* figures of kilowatt-hours sold.

The practices followed by industrial companies are usually a matter of individual policy. In some industrial groups there is a tendency for most of the companies therein to follow the same course.

A. Monthly Statements.—Most chain stores announce their monthly sales in dollars. Prior to 1931, copper producers regularly published their monthly output. General Motors publishes monthly sales in units.

Between 1902 and 1933, United States Steel Corporation published its unfilled orders each month, but in 1933 it replaced this figure by monthly deliveries in tons. Baldwin Locomotive Works has published monthly figures of shipments, new orders, and unfilled orders in dollars. The "Standard Oil Group" of pipe-line companies publish monthly statistics of operations in barrels.

Monthly figures of net earnings are published by individual companies from time to time, but such practices have always proved sporadic or temporary (*e.g.*, Otis Steel, Mullins Manufacturing, Alaska Juneau). There is a tendency to inaugurate monthly statements during periods of improvement and to discontinue them when earnings decline. Sometimes figures by months are included in the quarterly statements—*e.g.*, United States Steel Corporation prior to 1932.

B. Quarterly Statements.—Publication of results quarterly is considered as the standard procedure in nearly all lines of industry. The New York Stock Exchange has been urging quarterly reports with increasing vigor, and has usually been able to make its demands effective in connection with the listing of new or additional securities. Certain types of businesses are considered—or consider themselves—exempt from this requirement, because of the seasonal nature of their results. These lines include sugar production, fertilizers, and agricultural implements. Seasonal fluctuations may be concealed by publishing quarterly a moving twelve-month's figure of earnings. This is done by Continental Can Company.

It is not easy to understand why all the large cigarette manufacturers and the majority of department stores should withhold their results for a full year. It is inconsistent also for a company such as Woolworth to publish sales monthly, but no interim statements of net profits. Many individual companies, belonging to practically every division of industry, still fail to publish quarterly reports. In nearly every case such interim figures are available to the management but are denied to the stockholders without adequate reason.

The data given in the quarterly statements vary from a single figure of net earnings (sometimes without allowance for depreciation or federal taxes) to a fully detailed presentation of the income account and the balance sheet, with president's remarks appended. General Motors Corporation is an outstanding example of the latter practice.

C. Semiannual Reports.—These do not appear to be standard practice for any industrial group, except possibly the rubber companies. A number of individual enterprises report semi-annually—*e.g.*, American Locomotive and American Woolen.

D. Annual Reports.—Every listed company publishes an annual report of some kind. The annual statement is generally more detailed than those covering interim periods. It frequently contains remarks—not always illuminating—by the president or the chairman of the board, relating to the past year's results and to the future outlook. The distinguishing feature of the annual report, however, is that it invariably presents the balance-sheet position.

The information given in the income account varies considerably in extent. Some reports give no more than the earnings available for dividends and the amount of dividends paid, *e.g.*, U. S. Leather Company.[1]

The Income Account.—In our opinion an income account is not reasonably complete unless it contains the following items: (1) sales, (2) net earnings (before the items following), (3) depreciation (and depletion), (4) interest charges, (5) non-operating

---

[1] Pocahontas Fuel Company appears to be the only listed enterprise which publishes an annual balance sheet only and provides no income statement of any kind.

income (in detail), (6) income taxes, (7) dividends paid, (8) surplus adjustments (in detail).

It is unfortunately true that less than half of our important enterprises supply this very moderate quota of information. Withholding of data—particularly of annual sales—is usually justified on the ground that it might be used by competitors or customers to the detriment of the company and therefore of its stockholders. Such assertions are rarely convincing, especially since they are contravened by progressive managements in every line of industry. Concealment of the sales total or the depreciation charge severely handicaps the analyst and the intelligent stockholder because it renders impossible any thoroughgoing study of the results. Nor can it be denied that the restriction of this important information to a small group identified with the management may at times be of great benefit to them and of disadvantage to the general public. The same is true of the failure to issue reports oftener than once a year.

If the stockholders of companies pursuing such archaic policies of concealment would bring sufficiently vigorous pressure upon their managements, many changes for the better could speedily be brought about.

The standard of *reasonable completeness* for annual reports, suggested above, by no means includes all the information which might be vouchsafed to shareholders. The reports of United States Steel Corporation may be taken as a model of comprehensiveness. The data there supplied embrace, in addition to our standard requirements, the following items:

1. Production and sales in units. Rate of capacity operated.
2. Division of sales as between:
     Domestic and foreign.
     Intercompany and outsiders.
3. Details of operating expenses:
     Wages, wage rates, and number of employees.
     State and local taxes paid.
     Selling and general expense.
     Maintenance expenditures, amount and details.
4. Details of capital expenditures during the year.
5. Details of inventories.
6. Details of properties owned.
7. Number of stockholders.

The Balance Sheet.—The form of the balance sheet is better standardized than the income account and it does not offer such frequent grounds for criticism as does the income account. Formerly a widespread defect of balance sheets was the failure to separate intangible from tangible fixed assets but this is now comparatively rare. An example is Allis-Chalmers Manufacturing Company.

Criticism may properly be voiced against the practice of a great many companies in stating only the *net* figure for their property account without showing the deduction for depreciation. Other shortcomings sometimes met are the failure to state the market value of securities owned—*e.g.*, Oppenheim Collins and Company in 1932; to identify "investments" as marketable or nonliquid—*e.g.*, Pittsburgh Plate Glass Company; to value the inventory at lower of cost or market—*e.g.*, Celanese Corporation of America in 1931; to state the nature of miscellaneous reserves—*e.g.*, Hazel-Atlas Glass Company; and to state the amount of the company's own securities held in the treasury—*e.g.*, American Arch Company.[1]

Periodic Reports to Public Agencies.—Railroads and most public utilities are required to supply information to various federal and state commissions. Since these data are generally more detailed than the statements to shareholders, they afford a useful supplementary source of material. A few practical illustrations of the value of these reports to commissions may be of interest.

For many years prior to 1927 Consolidated Gas Company of New York was a "mystery stock" in Wall Street because it supplied very little information to its stockholders. Great emphasis was laid by speculators upon the undisclosed value of its interest in its numerous subsidiary companies. However, complete operating and financial data relating to both the company and its subsidiaries were at all times available in the annual reports of the Public Service Commission of New York. The same situation obtained over a long period with respect to the Mackay Companies, controlling Postal Telegraph and Cable Corporation, which reported no details to its stockholders but considerable information to the Interstate Commerce Commission. A similar contrast exists between the unilluminating reports of Fifth Avenue Bus Securities Company to its share-

---

[1] Several of these points were involved in a protracted dispute between the New York Stock Exchange and Allied Chemical and Dye Corporation, which was terminated to the satisfaction of the Stock Exchange in 1933.

holders and the complete information filed by its operating subsidiary with the New York Transit Commission.

Finally, we may mention the "Standard Oil Group" of pipeline companies, which have been extremely chary of information to their stockholders. But these companies come under the jurisdiction of the Interstate Commerce Commission, and are required to file circumstantial annual reports at Washington. Examination of these reports several years ago would have disclosed striking facts about these companies' holdings of cash and marketable securities.

The voluminous data contained in the *Survey of Current Business*, published monthly by the United States Department of Commerce, have included sales figures for individual chain-store companies which were not given general publicity—*e.g.*, Waldorf System, J. R. Thompson, United Cigar Stores, Hartman Corporation, etc. Current statistical information regarding particular companies is often available in trade publications or services.

Examples: Cram's Auto Service gives weekly figures of production for each motor-car company. Willett and Gray publish several estimates of sugar production by companies during the crop year. The *Oil and Gas Journal* often carries data regarding the production of important fields by companies. The *Railway Age* supplies detailed information regarding equipment orders placed. Dow, Jones and Company estimate weekly the rate of production of United States Steel.

Listing Applications.—These have been the most important nonperiodic sources of information. The reports required by the New York Stock Exchange, as a condition to admitting securities to its list, are much more detailed than those usually submitted to the stockholders. The additional data may include sales in dollars, output in units, amount of federal taxes, details of subsidiaries' operations, basis and amount of depreciation and depletion charges. Valuable information may also be supplied regarding the properties owned, the terms of contracts, and the accounting methods followed.

The analyst will find these listing applications exceedingly helpful. It is unfortunate that they appear at irregular intervals, and therefore cannot be counted upon as a steady source of information.

Registration Statements and Prospectuses.—The Securities Act of 1933 requires that a voluminous registration statement be filed with the Federal Trade Commission in connection with the sale of new securities. These statements may be inspected in Washington or they may be obtained upon payment of a fee. The more important information in the registration statement must be included in the prospectus supplied by the underwriters to intending purchasers. There seems every likelihood that these statements and prospectuses will prove of the greatest value to the analyst and the investor.

Miscellaneous Official Reports.—Information on individual companies may be unearthed in various kinds of official documents. A few examples will give an idea of their miscellaneous character. The report of the United States Coal Commission in 1923 (finally printed as a Senate Document in 1925) gave financial and operating data on the anthracite companies which had not previously been available. Reports of the Federal Trade Commission have recently given information relative to public-utility holding companies. Some of the opinions of the Interstate Commerce Commission have contained material of great value to the analyst. Trustees under mortgages may have information required to be supplied by the terms of the indenture. These figures may be significant. For example, unpublished reports with the trustee of Mason City and Fort Dodge Railroad Company 4s, revealed that the interest on the bonds was not being earned, that payment thereof was being continued by Chicago Great Western Railroad Company as a matter of policy only, and hence that the bonds were in a far more vulnerable position than was generally suspected.

Statistical and Financial Publications.—Most of the information required by the securities analyst in his daily work may be found conveniently and adequately presented by the various statistical services. These include comprehensive manuals published annually with periodic supplements, (Poor's, Moody's); descriptive stock and bond cards, and manuals frequently revised (Standard Statistics, Fitch, Poor's); daily digests of news relating to individual companies (Standard Corporation Records, Fitch). These services have made great progress during the past 20 years in the completeness and accuracy with which they present the facts. Nevertheless they cannot be relied upon to give all the data available in the various original sources above described. Some of these sources escape them completely, and in other cases they may neglect to reproduce items of importance. It follows therefore that in any thoroughgoing study of an individual company, the analyst should consult the original reports and other documents wherever possible, and not rely upon summaries or transcriptions.

In the field of financial periodicals, special mention must be made of *The Commercial and Financial Chronicle*, a weekly publication with numerous statistical supplements. Its treatment of the financial and industrial field is unusually comprehensive; and its most noteworthy feature is perhaps its detailed reproduction of corporate reports and other documents.

Requests for Direct Information from the Company.—Published information may often be supplemented to an important extent by private inquiry of or by interview with the management. There is no reason why stockholders should not ask for information on specific points, and in many cases part at least of the data asked for will be furnished. It must never be forgotten that a stockholder is an *owner* of the business and an *employer* of its officers. He is entitled not only to ask legitimate questions but also to have them answered, unless there is some persuasive reason to the contrary.

Insufficient attention has been paid to this all-important point. The courts have generally held that a bona fide stockholder has the same right to full information as a partner in a private business. This right may not be exercised to the detriment of the corporation, but the burden of proof rests upon the management to show an improper motive behind the request or that disclosure of the information would work an injury to the business.

Compelling a company to supply information involves expensive legal proceedings and hence few shareholders are in a position to assert their rights to the limit. Experience shows, however, that vigorous demands for legitimate information are frequently acceded to even by the most recalcitrant managements. This is particularly true when the information asked for is no more than that which is regularly published by other companies in the same field.

INFORMATION REGARDING THE INDUSTRY

Statistical data respecting industries as a whole are available in abundance. The *Survey of Current Business* gives monthly figures on output, consumption, stocks, unfilled orders, etc., for many different lines. Annual data are contained in the "Statistical Abstract," the "World Almanac" and other compendiums. More detailed figures are available in the "Biennial Census of Manufacturers."

Many important summary figures are published at frequent intervals in the various trade journals. In these publications will be found also a continuous and detailed picture of the current and prospective state of the industry. Thus it is usually possible for the analyst to acquire without undue difficulty a background of fairly complete knowledge of the history and problems of the industry with which he is dealing.

CHAPTER IV

DISTINCTIONS BETWEEN INVESTMENT AND SPECULATION

The difference between investment and speculation is understood in a general way by nearly every one, but when we try to formulate it precisely, we run into perplexing difficulties. In fact something can be said for the cynic's definition that an investment is a successful speculation and a speculation is an unsuccessful investment. It might be taken for granted that United States government securities are an investment medium, while the common stock, say, of Radio Corporation of America—which in 1933 had neither dividends, earnings, nor tangible assets behind it—must certainly be a speculation. Yet operations of a definitely speculative nature may be carried on in United States government bonds (*e.g.*, by specialists who buy large blocks in anticipation of a quick rise); and on the other hand, in 1929 Radio Corporation of America common was widely regarded as an investment, to the extent in fact of being included in the portfolios of leading "Investment Trusts."

It is certainly desirable that some exact and acceptable definition of the two terms be arrived at, if only because we ought as far as possible to know what we are talking about. A more forceful reason, perhaps, might be the statement that the failure properly to distinguish between investment and speculation was in large measure responsible for the market excesses of 1928–1929 and the calamities that ensued. On this account we shall give the question a more thoroughgoing study than it usually receives. The best procedure might be first to examine critically the various meanings commonly intended in using the two expressions, and then to endeavor to crystallize therefrom a single sound and definite conception of investment.

Distinctions Commonly Drawn between the Two Terms.—The chief distinctions in common use may be listed in the following table:

	Investment	Speculation
1.	In bonds.	In stocks.
2.	Outright purchases.	Purchases on margin.
3.	For permanent holding.	For a "quick turn."
4.	For income.	For profit.
5.	In safe securities.	In risky issues.

The first four distinctions have the advantage of being entirely definite, and each of them also sets forth a characteristic which is applicable to the *general run* of investment or speculation. They are all open to the objection that in numerous individual cases the criterion suggested would not properly apply.

1. Bonds vs. Stocks.—Taking up the first distinction, we find it corresponds to a common idea of investing as opposed to speculating, and that it also has the weight of at least one authority on investment who insists that only bonds belong in that category.[1] The latter contention, however, runs counter to the well-nigh universal acceptance of high-grade preferred stocks as media of investment. Furthermore, it is most dangerous to regard the bond form as possessing inherently the credentials of an investment, for a poorly secured bond may not only be thoroughly speculative but the most unattractive form of speculation as well. It is logically unsound, furthermore, to deny investment rating to a strongly entrenched common stock merely because it possesses profit possibilities. Even the popular view recognizes this fact, since at all times certain especially sound common stocks have been rated as investment issues and their purchasers regarded as investors and not as speculators.

2 and 3. Outright vs. Marginal Purchases; Permanent vs. Temporary Holding.—The second and third distinctions relate to the customary *method* and *intention*, rather than to the innate character of investment and speculative operations. It should be obvious that buying a stock outright does not *ipso facto* make the transaction an investment. In truth the most speculative issues, *e.g.*, "penny mining stocks," *must* be purchased outright, since no one will lend money against them. Conversely, when the American public was urged during the war to buy Liberty Bonds with borrowed money, such purchases were nonetheless universally classed as investments. If strict logic were followed in financial operations—a very improbable hypothesis!—the common practice would be reversed: the safer (investment) issues

[1] Chamberlain, Lawrence, *Investment and Speculation, supra*, p. 8.

would be considered more suitable for marginal purchase, and the riskier (speculative) commitments would be paid for in full.

Similarly the contrast between permanent and temporary holding is applicable only in a broad and inexact fashion. An authority on common stocks has recently defined an investment as any purchase made with the intention of holding it for a year or longer; but this definition is admittedly suggested by its convenience rather than its penetration.[1] The inexactness of this suggested rule is shown by the circumstance that *short-term investment* is a well-established practice. *Long-term speculation* is equally well established as a rueful fact (when the purchaser holds on hoping to make up a loss), and it is also carried on to some extent as an intentional undertaking.

4 and 5. Income vs. Profit; Safety vs. Risk.—The fourth and fifth distinctions also belong together, and so joined they undoubtedly come closer than the others to both a rational and a popular understanding of the subject. Certainly, through many years prior to 1928, the typical investor had been interested above all in safety of principal and continuance of an adequate income. However, the doctrine that common stocks were the best long-term investments resulted in a transfer of emphasis from current income to future income and hence inevitably to future enhancement of principal value. In its complete subordination of the income element to the desire for profit, and also in the prime reliance it placed upon favorable developments expected in the future, the new-era style of investment—as exemplified in the general policy of the investment trusts—was practically indistinguishable from speculation. In fact this so-called investment could be accurately defined as speculation in the common stocks of strongly situated companies.

It would undoubtedly be a wholesome step to go back to the accepted idea of income as the central motive in investment, leaving the aim toward profit, or capital appreciation, as the typical characteristic of speculation. But it is doubtful whether the true inwardness of investment rests even in this distinction. Examining standard practices of the past, we find some instances in which current income was not the leading interest of a bona fide investment operation. This was regularly true, for example, of bank stocks, which until recent years were regarded as the

---

[1] Sloan, Laurence H., *Everyman and His Common Stocks*, pp. 8–9, 279 *ff.*, New York, 1931.

exclusive province of the wealthy investor. These issues returned a smaller dividend yield than did high-grade bonds, but they were purchased on the expectation that the steady growth in earnings and surplus would result in special distributions and increased principal value. In other words, it was the earnings accruing to the stockholder's credit, rather than those distributed in dividends, which motivated his purchase. Yet it would not appear to be sound to call this attitude speculative, for we should then have to contend that only the bank stocks which paid out most of their earnings in dividends (and thus gave an adequate current return) could be regarded as investments, while those following the conservative policy of building up their surplus would therefore have to be considered speculative. Such a conclusion is obviously paradoxical, and because of this fact, it must be admitted that an investment in a common stock might conceivably be founded on its earning power, without reference to current dividend payments.

Does this bring us back to the new-era theory of investment? Must we say that the purchase of low-yielding industrial shares in 1929 had the same right to be called investment as the purchase of low-yielding bank stocks in prewar days? The answer to this question should bring us to the end of our quest, but to deal with it properly we must turn our attention to the fifth and last distinction in our list—that between safety and risk.

This distinction expresses the broadest concept of all those underlying the term investment, but its practical utility is handicapped by various shortcomings. If safety is to be judged by the result, we are virtually begging the question, and come perilously close to the cynic's definition of an investment as a successful speculation. Naturally the safety must be posited in advance, but here again there is room for much that is indefinite and purely subjective. The race-track gambler, betting on a "sure thing," is convinced that his commitment is safe. The 1929 "investor" in high-priced common stocks also considered himself safe in his reliance upon future growth to justify the figure he paid and more.

Standards of Safety.—The concept of safety can be really useful only if it is based on something more tangible than the psychology of the purchaser. The safety must be assured, or at least strongly indicated, by the application of definite and well-established standards. It was this point which distinguished the bank-stock buyer of 1912 from the common-stock investor of 1929. The former purchased at price levels which he considered conservative in the light of experience; he was satisfied, from his knowledge of the institution's resources and earning power, that he was getting his money's worth in full. If a strong speculative market resulted in advancing the price to a level out of line with these standards of value, he sold his shares and waited for a reasonable price to return before reacquiring them.

Had the same attitude been taken by the purchaser of common stocks in 1928–1929, the term investment would not have been the tragic misnomer that it was. But in proudly applying the designation "blue chips" to the high-priced issues chiefly favored, the public unconsciously revealed the gambling motive at the heart of its supposed investment selections. These differed from the old-time bank-stock purchases in the one vital respect that the buyer did not determine that they were worth the price paid by the application of firmly established standards of value. The market made up new standards as it went along, by accepting the current price—however high—as the sole measure of value. Any idea of safety based on this uncritical approach was clearly illusory and replete with danger. Carried to its logical extreme, it meant that no price could possibly be too high for a good stock, and that such an issue was equally "safe" after it had advanced to 200 as it had been at 25.

A Proposed Definition of Investment.—This comparison suggests that it is not enough to identify investment with expected safety; the expectation must be based on study and standards. At the same time, the investor need not necessarily be interested in current income; he may at times legitimately base his purchase on a return which is accumulating to his credit and realized by him after a longer or shorter wait. With these observations in mind, we suggest the following definition of investment as one in harmony with both the popular understanding of the term and the requirements of reasonable precision:

An investment operation is one which, upon thorough analysis, promises safety of principal and a satisfactory return. Operations not meeting these requirements are speculative.

Certain implications of this definition are worthy of further discussion. We speak of an *investment operation* rather than an issue or a purchase, for several reasons. It is unsound to think always of investment character as inhering in an issue *per se*. The price is frequently an essential element, so that a stock (and even a bond) may have investment merit at one price level but not at another. Furthermore, an investment might be justified in a group of issues, which would not be sufficiently safe if made in any one of them singly. In other words, diversification might be necessary to reduce the risk involved in the separate issues to the minimum consonant with the requirements of investment. (This would be true, for example, of a special type of investment policy centering upon the purchase of common stocks at well below liquidating values.)

In our view it is also proper to consider as investment operations certain types of arbitrage and hedging commitments which involve the sale of one security against the purchase of another. In these operations the element of safety is provided by the combination of purchase and sale. This is an extension of the ordinary concept of investment, but one which appears to the writers to be entirely logical.

The phrases *thorough analysis, promises safety*, and *satisfactory return* are all chargeable with indefiniteness, but the important point is that their meaning is clear enough to prevent serious misunderstanding. By *thorough analysis* we mean, of course, the study of the facts in the light of established standards of safety and value. An "analysis," which recommended investment in General Electric common at a price forty times its highest recorded earnings, would be clearly ruled out, as devoid of all quality of thoroughness.

The *safety* sought in investment is not absolute or complete; the word means, rather, protection against loss under all normal or reasonably likely conditions or variations. A safe bond, for example, is one which could suffer default only under exceptional and highly improbable circumstances. Similarly, a safe stock is one which holds every prospect of being worth the price paid except under quite unlikely contingencies. Where study and experience indicate that an appreciable chance of loss must be recognized and allowed for, we have a speculative situation.

A *satisfactory return* is a wider expression than *adequate income*, since it allows for capital appreciation or profit as well as current interest or dividend yield. "Satisfactory" is a subjective term; it covers any rate or amount of return, however low, which the investor is willing to accept, provided he acts with reasonable intelligence.[1]

The conception of investment advanced above is broader than most of those in common use. Under it investment may conceivably—though not usually—be made in stocks, carried on margin, and purchased with the chief interest in a quick profit. In these respects it would run counter to the first four distinctions which we listed at the outset. But to offset this seeming laxity, we insist on a satisfactory assurance of safety based on adequate analysis. We are thus led to the conclusion that the viewpoint of analysis and the viewpoint of investment are largely identical in their scope.

[1] The purchase of United States Treasury Certificates in 1932 on a yield basis ¼ of 1% was undoubtedly an investment operation, since the buyers were consciously willing to accept so small a return. However, the quotation of 103¾ for United States Liberty Fourth 4¼s in December 1932 was a *speculative* and not an investment price, because if the government had exercised its right to call the issue early in 1933 the return to the investor would have been *less than nothing*. If the bonds were called, he would not only gain no income but would lose part of his principal.

CHAPTER V

CLASSIFICATION OF SECURITIES

Securities are customarily divided into the two main groups of bonds and stocks, with the latter subdivided into preferred stocks and common stocks. The first and basic division recognizes and conforms to the fundamental legal distinction between the creditors' position and the partners' position. The bondholder has a fixed and prior claim for principal and interest; the stockholder assumes the major risks and shares in the profits of ownership. It follows that a higher degree of safety should inhere in bonds as a class, while greater opportunity of speculative gain—to offset the greater hazard—is to be found in the field of stocks. It is this contrast, of both legal status and investment character, as between the two kinds of issues, which provides the point of departure for the usual textbook treatment of securities.

Objections to the Conventional Grouping: 1. Preferred Stock Grouped with Common.—While this approach is hallowed by tradition, it is open to several serious objections. Of these the most obvious is that it places preferred stocks with common stocks, whereas, so far as investment practice is concerned, the former undoubtedly belong with bonds. The typical or standard preferred stock is bought for fixed income and safety of principal. Its owner considers himself not as a partner in the business but as the holder of a claim ranking ahead of the interest of the partners, *i.e.*, the common stockholders. Preferred stockholders are partners or owners of the business only in a technical, legalistic sense; but they resemble bondholders in the purpose and expected results of their investment.

2. Bond Form Identified with Safety.—A weightier though less patent objection to the radical separation of bonds from stocks is that it tends to identify the *bond form* with the idea of safety. Hence investors are led to believe that the very name "bond" must carry some especial assurance against loss. This attitude is basically unsound, and on frequent occasions is responsible for serious mistakes and loss. The investor has been spared even greater penalties for this error by the rather accidental fact that fraudulent security promoters have rarely taken advantage of the investment prestige attaching to the bond form. It is true beyond dispute that bonds as a whole enjoy a degree of safety distinctly superior to that of the average stock. But this advantage is not the result of any essential virtue of the bond *form;* it follows from the circumstance that the typical American enterprise is financed with some honesty and intelligence, and does not assume fixed obligations without a reasonable expectation of being able to meet them. But it is not the obligation that creates the safety, nor is it the legal remedies of the bondholder in the event of default. *Safety depends upon and is measured entirely by the ability of the debtor corporation to meet its obligations.*

The bond of a business without assets or earning power would be every whit as valueless as the stock of such an enterprise. Bonds representing all the capital placed in a new venture are no safer than common stock would be, and are considerably less attractive. For the bondholder could not possibly get more out of the company by virtue of his fixed claim than he could realize if he owned the business in full, free and clear.[1] This simple principle seems too obvious to merit statement; yet because of the traditional association of the bond form with superior safety, the investor has often been persuaded that by the mere act of limiting his return he obtained an assurance against loss.

3. Failure of Titles to Describe Issues with Accuracy.—The basic classification of securities into bonds and stocks—or even into three main classes of bonds, preferred stocks, and common stocks—is open to the third objection that in many cases these titles fail to supply an accurate description of the issue. This is the consequence of the steadily mounting percentage of securities which do not conform to the standard patterns, but instead modify or mingle the customary provisions.

Briefly stated, these standard patterns are as follows:

I. The bond pattern comprises:
  A. The unqualified right to a fixed interest payment on fixed dates.
  B. The unqualified right to repayment of a fixed principal amount on a fixed date.

[1] See Appendix, Note 2, for a phase of the liquidation of the United States Express Company illustrating this point.

    C. No further interest in assets or profits, and no voice in the management.

II. The preferred-stock pattern comprises:

    A. A stated rate of dividend in priority to any payment on the common (hence full preferred dividends are mandatory if the common receives any dividend; but if nothing is paid on the common, the preferred dividend is subject to the discretion of the directors).

    B. The right to a stated principal amount in the event of dissolution, in priority to any payments to the common stock.

    C. Either no voting rights, or voting power shared with the common.

III. The common-stock pattern comprises:

    A. A pro rata ownership of the company's assets in excess of its debts and preferred stock issues.

    B. A pro rata interest in all profits in excess of prior deductions.

    C. A pro rata vote for the election of directors and for other purposes.

Bonds and preferred stocks conforming to the above standard patterns will sometimes be referred to as *straight bonds* or *straight preferred stocks.*

Numerous Deviations from the Standard Patterns.—However, almost every conceivable departure from the standard pattern can be found in greater or less profusion in the security markets of today. Of these the most frequent and important are identified by the following designations: *income* bonds; *convertible* bonds and preferred stocks; bonds and preferred stocks with *stock-purchase warrants* attached; *participating* preferred stocks; common stocks with *preferential features; nonvoting* common stock. Of recent origin is the device of making bond interest or preferred dividends payable either in cash or in common stock at the holder's *option.* The *callable feature* now found in most bonds may also be termed a lesser departure from the standard provision of fixed maturity of principal.

Of less frequent and perhaps unique deviations from the standard patterns, the variety is almost endless. In Note 3 of the Appendix we present a fairly comprehensive list of these, with examples of each. We shall mention here only the glaring instance of Great Northern Railway Preferred Stock which for many years has been in all respects a plain common issue; and also the resort by Associated Gas and Electric Company to the insidious and highly objectionable device of bonds convertible into preferred stock *at the option of the company*—which are, therefore, not true bonds at all.

More striking still is the emergence of completely distinctive types of securities so unrelated to the standard bond or stock pattern as to require an entirely different set of names. Of these, the most significant is the option warrant—a device which during the years prior to 1929 developed into a financial instrument of major importance and tremendous mischief-making powers. The option warrants issued by a single company— American and Foreign Power Company—attained in 1929 an aggregate market value of more than a *billion dollars*, a figure exceeding our national debt in 1914. Note 3 of the Appendix lists a number of other newfangled security forms, bearing titles such as allotment certificates, dividend participations, etc.

The peculiarities and complexities to be found in the present-day security list are added arguments against the traditional practice of pigeonholing and generalizing about securities in accordance with their *titles*. While this procedure has the merit of convenience and a certain rough validity, we think it should be replaced by a more flexible and accurate basis of classification. In our opinion, the criterion most useful for purposes of study would be the *normal behavior* of the issue after purchase—in other words its risk-and-profit characteristics as the buyer or owner would reasonably view them.

New Classification Suggested.—With this standpoint in mind, we suggest that securities be classified under the following three headings:

Class	Representative Issue
I. Securities of the fixed-value type.	A high-grade bond or preferred stock.
II. Senior securities of the variable-value type.	
A. Well-protected issues with profit possibilities.	A high-grade convertible bond.
B. Inadequately protected issues.	A lower-grade bond or preferred stock.
III. Common-stock type.	A common stock.

An approximation to the above grouping could be reached by the use of more familiar terms, as follows:

I. Investment bonds and preferred stocks.
II. Speculative bonds and preferred stocks.
    A. Convertibles, etc.
    B. Low-grade senior issues.
III. Common stocks.

The somewhat novel designations that we employ are needed to make our classification more comprehensive. This necessity will be clearer, perhaps, from the following description and discussion of each group.

Leading Characteristics of the Three Types.—The first class includes issues, of whatever title, in which prospective change of value may fairly be said to hold minor importance. The owner's dominant interest lies in the safety of his principal and his sole purpose in making the commitment is to obtain a steady income. In the second class, prospective changes in the value of the principal assume real significance. In Type A, the investor hopes to obtain the safety of a straight investment, with an added possibility of profit by reason of a conversion right or some similar privilege. In Type B, a definite risk of loss is recognized, which is presumably offset by a corresponding chance of profit. Securities included in Group II B will differ from the common-stock type (Group III) in two respects: (*a*) They enjoy an effective priority over some junior issue, thus giving them a certain degree of protection. (*b*) Their profit possibilities, however substantial, have a fairly definite limit, in contrast with the unlimited percentage of possible gain theoretically or optimistically associated with a fortunate common-stock commitment.

Issues of the fixed-value type include all *straight* bonds and preferred stocks of high quality selling at a normal price. Besides these, there belong in this class:

A. Sound convertible issues where the conversion level is too remote to enter as a factor in the purchase. (Similarly for participating or warrant-bearing senior issues.)

B. Guaranteed common stocks of investment grade.

C. Class A or prior-common stocks occupying the status of a high-grade, straight preferred stock.

On the other hand, a bond of investment grade which happens to sell at an unduly low price would belong in the second group, since the purchaser might have reason to expect and be interested in an appreciation of its market value.

Exactly at what point the question of price fluctuation becomes material rather than minor is naturally impossible to prescribe. The price level itself is not the sole determining factor. A long-term 3% bond at 60 would ordinarily belong in the fixed-value class, whereas a one-year maturity of any coupon rate selling at 80 would *not*—because in a comparatively short time it must either be paid off at a 20-point advance, or else default and probably suffer a severe decline in market value. We must be prepared, therefore, to find marginal cases where the classification (as between Group I and Group II) will depend on the personal viewpoint of the analyst or investor.

Any issue which displays the main characteristics of a common stock belongs in Group III, whether it is entitled "common stock," "preferred stock" or even "bond." The case, already cited, of American Telephone and Telegraph Company Convertible 4½s, when selling about 200, provides an apposite example. The buyer or holder of the bond at so high a level was to all practical purposes making a commitment in the common stock, for the bond and stock would not only advance together, but also decline together over an exceedingly wide price range. Still more definite illustration of this point was supplied by the Kreuger and Toll Participating Debentures at the time of their sale to the public. The offering price was so far above the amount of their prior claim that their title had no significance at all, and could only have been misleading. *These "bonds" were definitely of the common-stock type.*[1]

The opposite situation is met when issues, senior in name, sell at such low prices that the junior securities can obviously have no real equity, *i.e.*, ownership interest, in the company. In such cases, the low-priced bond or preferred stock stands virtually in the position of a common stock and should be regarded as such for purposes of analysis. A preferred stock selling at 10 cents on the dollar, for example, should be viewed not as a preferred stock at all, but as a common stock. On the one hand it lacks the prime requisite of a senior security, *viz.*, that it should be followed by a junior investment of substantial value. On the other hand, it carries all the profit features of a common stock, since the amount of possible gain from the current level is for all practical purposes unlimited.

The dividing line between Groups II and III is as indefinite as that between Groups I and II. Borderline cases can be handled without undue difficulty however, by considering them from the standpoint of either category or of both. For example, should a 7% preferred stock selling at 30 be considered a low-priced senior issue or as the equivalent of a common stock? The answer to this question will depend partly on the exhibit

[1] See Appendix, Note 4, for the terms of this issue.

of the company and partly on the attitude of the prospective buyer. If real value may conceivably exist in excess of the par amount of the preferred stock, the issue may be granted some of the favored status of a senior security. On the other hand, whether or not the buyer should consider it in the same light as a common stock may also depend on whether he would be amply satisfied with a possible 250% appreciation, or is looking for even greater speculative gain.

From the foregoing discussion the real character and purpose of our classification should now be more evident. Its basis is not the title of the issue, but the practical significance of its specific terms and status to the owner. Nor is the primary emphasis placed upon what the owner is legally entitled to demand, but upon what he is likely to get, or is justified in expecting, under conditions which appear to be probable at the time of purchase or analysis.

PART II

FIXED-VALUE INVESTMENTS

CHAPTER VI

THE SELECTION OF FIXED-VALUE INVESTMENTS

Having suggested a classification of securities by character rather than by title, we now take up in order the principles and methods of selection applicable to each group. We have already stated that the fixed-value group includes:

1. High-grade straight bonds and preferred stocks.
2. High-grade privileged issues, where the value of the privilege is too remote to count as a factor in selection.
3. Common stocks which through guaranty or preferred status occupy the position of a high-grade senior issue.

Basic Attitude toward High-grade Preferred Stocks.—By placing gilt-edged preferred stocks and high-grade bonds in a single group, we indicate that the same investment attitude and the same general method of analysis are applicable to both types. The very definite inferiority of the preferred stockholders' legal claim is here left out of account, for the logical reason that the soundness of the best investments must rest not upon legal rights or remedies but upon ample financial capacity of the enterprise. Confirmation of this viewpoint is found in the investor's attitude toward such an issue as National Biscuit Company Preferred, which for 30 years has been considered as possessing the same *essential investment character* as a good bond.[1]

Preferred Stocks Not Generally Equivalent to Bonds in Investment Merit.—But it should be pointed out immediately that issues with the history and standing of National Biscuit Preferred constitute a very small percentage of all preferred stocks. Hence, we are by no means asserting the investment equivalence of bonds and preferred stocks *in general*. On the contrary, we

[1] See Appendix, Note 5, for supporting data.

shall in a later chapter be at some pains to show that the *average* preferred issue deserves a lower rank than the average bond, and furthermore that preferred stocks have been much too readily accepted by the investing public. The majority of these issues have not been sufficiently well protected to assure continuance of dividends *beyond any reasonable doubt*. They belong properly, therefore, in the class of variable or speculative senior issues (Group II), and in this field the contractual differences between bonds and preferred shares are likely to assume great importance. A sharp distinction must, therefore, be made between the typical and the exceptional preferred stock. It is only the latter which deserves to rank as a fixed-value investment and to be viewed in the same light as a good bond. To avoid awkwardness of expression in this discussion we shall frequently use the terms "investment bonds" or merely "bonds" to represent all securities belonging to the fixed-value class.

Recent Experience Should Not Be Ignored.—As we stated in our Introduction, the extreme financial and industrial fluctuations of recent years may well call into question the fundamental logic of bond investment. Is it worth while for the investor to limit his income return and forego all prospect of speculative gain, if despite these sacrifices he must still subject himself to serious risks of loss? We have suggested in reply that the phenomena of 1927–1933 were so completely abnormal as to afford no fair basis for investment theory or practice. But we may not ignore the additional fact that ten years previously bondholders passed through another distressing period, in which bond values suffered severely from the combined effects of high interest rates and a major business depression. The next ten years may conceivably have quite a different story to tell, and one more propitious to fixed-value investment. It may be, for example, that the United States will settle down to a long period of relative economic stability, in which bonds as a whole will keep an even keel, uninfluenced by the moderate fluctuations of business and the stock market. But an analysis of investment principles cannot in sincerity advance such an hypothesis in order to ignore the implications of actual and repeated experience.

Bond Form Inherently Unattractive: Quantitative Assurance of Safety Essentials.—This experience clearly calls for a more critical and exacting attitude towards bond selection than has heretofore been considered necessary by investors, issuing houses, or authors of textbooks on investment. Allusion has already been made to the dangers inherent in the acceptance of the bond *form* as an assurance of safety, or even of smaller risk than is found in stocks. Instead of associating bonds primarily with the presumption of *safety*—as has long been the practice— it would be sounder to start with what is not presumption but fact, *viz.*, that a (straight) bond is an investment with *limited return*. In exchange for limiting his participation in future profits, the bondholder obtains a prior claim and a definite promise of payment, while the preferred stockholder obtains only the priority, without the promise. But neither priority nor promise is itself an *assurance* of payment. This assurance rests in the ability of the enterprise to fulfill its promise, and must be looked for in its financial position, record, and prospects. The essence of proper bond selection consists, therefore, in obtaining specific and convincing factors of safety in compensation for the surrender of participation in profits.

Major Emphasis on Avoidance of Loss.—Our primary conception of the bond as a commitment with limited return leads us to another important viewpoint toward bond investment. Since the chief emphasis must be placed on avoidance of loss, bond selection is primarily a negative art. It is a process of exclusion and rejection, rather than of search and acceptance. In this respect the contrast with common-stock selection is fundamental in character. The prospective buyer of a given common stock is influenced more or less equally by the desire to avoid loss and the desire to make a profit. The penalty for mistakenly rejecting the issue may conceivably be as great as that for mistakenly accepting it. But an investor may reject any number of good bonds with virtually no penalty at all, provided he does not eventually accept an unsound issue. Hence, broadly speaking, there is no such thing as being unduly captious or exacting in the purchase of fixed-value investments. The observation that Walter Bagehot addressed to commercial bankers is equally applicable to the selection of investment bonds. "If there is a difficulty or a doubt the security should be declined."[1]

Four Principles for the Selection of Issues of the Fixed-value Type.—Having established this general approach to our problem, we may now state four additional principles of more specific

[1] *Lombard Street*, p. 245, New York, 1892.

character which are applicable to the selection of individual issues:

I. *Safety is measured not by specific lien or other contractual rights, but by the ability of the issuer to meet all of its obligations.*

II. *This ability should be measured under conditions of depression rather than prosperity.*

III. *Deficient safety cannot be compensated for by an abnormally high coupon rate.*

IV. *The selection of all bonds for investment should be subject to rules of exclusion and to specific quantitative tests corresponding to those prescribed by statute to govern investments of savings banks.*

A technique of bond selection based on the above principles will differ in significant respects from the traditional attitude and methods. In departing from old concepts, however, this treatment represents not an innovation but the recognition and advocacy of viewpoints which have been steadily gaining ground among intelligent and experienced investors. The ensuing discussion is designed to make clear both the nature and the justification of the newer ideas.

I. SAFETY NOT MEASURED BY LIEN BUT BY ABILITY TO PAY

The basic difference confronts us at the very beginning. In the past the primary emphasis was laid upon the specific security, *i.e.*, the character and supposed value of the property on which the bonds hold a lien. From our standpoint this consideration is quite secondary; the dominant element must be the strength and soundness of the obligor enterprise. There is here a clear-cut distinction between two points of view. On the one hand the bond is regarded as a claim against property, on the other hand, as a claim against a business.

The older view was logical enough in its origin and purpose. It desired to make the bondholder independent of the risks of the business by giving him ample security on which to levy in the event that the enterprise proved a failure. If the business became unable to pay his claim, he could take over the mortgaged property and pay himself out of that. This arrangement would be excellent if it worked, but in practice it rarely proves to be feasible. For this there are three reasons:

1. The shrinkage of property values when the business fails.
2. The difficulty of asserting the bondholders' supposed legal rights.
3. The delays and other disadvantages incident to a receivership.

Lien Is No Guarantee against Shrinkage of Values.—The conception of a mortgage lien as a guaranty of protection independent of the success of the business itself is in most cases a complete fallacy. In the typical situation, the value of the pledged property is vitally dependent on the earning power of the enterprise. The bondholder usually has a lien on a railroad line, or on factory buildings and equipment, or on power plants and other utility properties, or perhaps on a bridge or hotel structure. These properties are rarely adaptable to uses other than those for which they were constructed. Hence if the enterprise proves a failure its fixed assets ordinarily suffer an appalling shrinkage in realizable value. For this reason the established practice of stating the original cost or appraised value of the pledged property as an inducement to purchase bonds is entirely misleading. The value of pledged assets assumes practical importance only in the event of default, and in any such event the book figures are almost invariably found to be unreliable and irrelevant. This may be illustrated by Seaboard-All Florida Railway First Mortgage 6s, selling in 1931 at 1 cent on the dollar shortly after completion of the road.[1]

Impracticable to Enforce Basic Legal Rights of Lien Holder.— In cases where the mortgaged property is actually worth as much as the debt, the bondholder is rarely allowed to take possession and realize upon it. It must be recognized that the procedure following default on a corporation bond has come to differ materially from that customary in the case of a mortgage on privately owned property. The basic legal rights of the lien holder are supposedly the same in both situations. But in practice we find a very definite disinclination on the part of the courts to permit corporate bondholders to take over properties by foreclosing on their liens, if there is any possibility that these assets may have a fair value in excess of their claim. Apparently it is considered unfair to wipe out stockholders or junior bondholders who have a potential interest in the property but are not in a position to protect it. As a result of this practice, bondholders rarely, if ever, come into actual possession of the pledged property unless its value at the time is substantially less than their claim. In most cases they are required to take new securities in a reorganized company. Sometimes the default in interest

[1] See Appendix, Note 6, for supporting data.

is cured and the issue reinstated.[1] On exceedingly rare occasions a defaulted issue may be paid off in full, but only after a long and vexing delay.[2]

Delays Are Expensive.—This delay constitutes the third objection to relying upon the mortgaged property as protection for a bond investment. The more valuable the pledged assets in relation to the amount of the lien, the more difficult it is to take them over under foreclosure, and the longer the time required to work out an "equitable" division of interest among the various bond and stock issues. Let us consider the most favorable kind of situation for a bondholder in the event of receivership. He would hold a comparatively small first mortgage followed by a substantial junior lien, the requirements of which have made the company insolvent. It may well be that the strength of the first-mortgage bondholder's position is such that at no time is there any real chance of eventual loss to him. Yet the financial difficulties of the company usually have a depressing effect on the market price of all its securities, even those presumably unimpaired in real value. As the receivership drags on, the market decline becomes accentuated, since investors are constitutionally averse to buying into a troubled situation. Eventually the first-mortgage bonds may come through the reorganization undisturbed, but during a wearisome and protracted period the owners have faced a severe impairment in the quoted value of their holdings and at least some degree of doubt and worry as to the outcome. Typical examples of such an experience can be found in the case of Missouri, Kansas and Texas Railway Company First 4s and Brooklyn Union Elevated Railroad First 5s.[3] The subject of receivership and reorganization practice, particularly as they affect the bondholder, will receive more detailed consideration in a later chapter.

Basic Principle Is to Avoid Trouble.—The foregoing discussion should support our emphatic stand that the primary aim of the bond buyer must be to avoid trouble and not to protect himself in the event of trouble. Even in the cases where the specific lien proves of real advantage, this benefit is realized under conditions which contravene the very meaning of *fixed-value* investment. In view of the severe decline in market price

[1] See Appendix, Note 7, for supporting data.
[2] See Appendix, Note 8, for supporting data.
[3] See Appendix, Note 9, for supporting data.

almost invariably associated with receivership, the mere fact that the investor must have recourse to his indenture indicates that his investment has been unwise or unfortunate. The protection that the mortgaged property offers him can constitute at best a mitigation of his mistake.

Corollaries from This First Principle. 1. *Absence of Lien of Minor Consequence.*—From Principle I there follow a number of corollaries with important practical applications. Since specific lien is of subordinate importance in the choice of high-grade bonds, the absence of lien is also of minor consequence. The debenture,[1] *i.e.*, unsecured, obligations of a strong corporation, amply capable of meeting its interest charges, may qualify for acceptance almost as readily as a bond secured by mortgage. Furthermore the debentures of a strong enterprise are undoubtedly sounder investments than the mortgage issues of a weak company. No first-lien bond, for example, enjoys a better investment rating than General Electric Company Debenture 3½s, due 1942. An examination of the bond list will probably show that the debenture issues of companies having no secured debt ahead of them will rank in investment character at least on a par with the average mortgage bond, because an enterprise must enjoy a high credit rating to obtain funds on its unsecured long-term bond.

[1] The term "debenture" in American financial practice has the accepted meaning of "unsecured bond or note." For no good reason, the name is sometimes given to other kinds of securities without apparently signifying anything in particular. There are a number of "secured debentures," *e.g.*, Chicago Herald and Examiner Secured Debenture 6½s, due 1950, and Kreuger and Toll Company Secured Debenture 5s, due 1959. Also, a number of preferred issues are called debenture preferred stock or merely debenture stock, *e.g.*, Du Pont Debenture Stock; Bush Terminal Company Debenture Stock; General Motors Corporation Debenture Stock, retired in 1930; General Cigar Company Debenture Preferred, called in 1927.

Sometimes debenture issues, properly so entitled because originally unsecured, later acquire specific security through the operation of a protective covenant, *e.g.*, New York, New Haven and Hartford Railroad Company debentures discussed in Chap. XIX. Another example was the Debenture 6½s of Fox New England Theaters, Inc., reorganized in 1933. These debentures acquired as security a block of first-mortgage bonds of the same company, which were surrendered by the vendor of the theaters because it failed to meet a guarantee of future earnings.

Observe that there is no clear-cut distinction between a "bond" and a "note" other than the fact that the latter generally means a relatively short-term obligation, *i.e.*, one maturing not more than, say, ten years after issuance.

2. *The Theory of Buying the Highest Yielding Obligation of a Sound Company.*—It follows also that if any obligation of an enterprise deserves to qualify as a fixed-value investment, then all its obligations must do so. Stated conversely, if a company's junior bonds are not safe, its first-mortgage bonds are not a desirable fixed-value investment. For if the second mortgage is unsafe the company itself is weak, and generally speaking there can be no high-grade obligations of a weak enterprise. The theoretically correct procedure for bond investment, therefore, is first to select a company meeting every test of strength and soundness, and then to purchase its highest yielding obligation, which would usually mean its junior rather than its first-lien bonds. Assuming no error were ever made in our choice of enterprises, this procedure would work out perfectly well in practice. The greater the chance of mistake, however, the more reason to sacrifice yield in order to reduce the potential loss in capital value. But we must recognize that in favoring the lower yielding first-mortgage issue, the bond buyer is in fact expressing a lack of confidence in his own judgment as to the soundness of the business—which, if carried far enough, would call into question the advisability of his making an investment in *any* of the bonds of the particular enterprise.

Example: As an example of this point, let us consider the Cudahy Packing Company First Mortgage 5s, due 1946, and the Debenture 5½s of the same company, due 1937. In June 1932 the First 5s sold at 95 to yield about 5½%, while the junior 5½s sold at 59 to yield over 20% to maturity. The purchase of the 5% bonds at close to par could only be justified by a confident belief that the company would remain solvent and reasonably prosperous, for otherwise the bonds would undoubtedly suffer a severe drop in market price. But if the investor has confidence in the future of Cudahy, why should he not buy the debenture issue and obtain an enormously greater return on his money? The only answer can be that the investor wants the superior protection of the first mortgage in the event his judgment proves incorrect and the company falls into difficulties. In that case he would probably lose less as the owner of the first-mortgage bonds than through holding the junior issue. Even on this score it should be pointed out that if by any chance Cudahy Packing Company were to suffer the reverses that befell Fisk Rubber Company, the loss in market value of the first-mortgage bonds would be fully as great as those suffered by the debentures; for in April 1932 Fisk Rubber Company First 8s were selling as low as 17 against a price of 12 for the unsecured $5\frac{1}{2}\%$ Notes. It is clear, at any rate, that the investor who favors the Cudahy first-lien 5s is paying a premium of about 15% per annum (the difference in yield) for only a *partial* insurance against loss. On this basis he is undoubtedly giving up too much for what he gets in return. The conclusion appears inescapable either that he should make no investment in Cudahy bonds or that he should buy the junior issue at its enormously higher yield. This rule may be laid down as applying to all cases where a first-mortgage bond sells at a fixed-value price (*e.g.*, close to par) and junior issues of the same company can be bought to yield a much higher return.

Senior Liens Are to Be Favored, Unless Junior Obligations Offer a Substantial Advantage.—Obviously a junior lien should be preferred only if the advantage in income return is substantial. Where the first-mortgage bond yields only slightly less, it is undoubtedly wise to pay the small insurance premium for protection against unexpected trouble.

Example: This point is illustrated by the relative market prices of Atchison Topeka and Santa Fe Railway Company General (first) 4s and Adjustment (second mortgage) 4s, both of which mature in 1995.

PRICE OF ATCHISON GENERAL 4s AND ADJUSTMENT 4s AT VARIOUS DATES

Date	Price of General 4s	Price of Adjustment 4s	Spread
Jan. 2, 1913	$97\frac{1}{2}$	88	$9\frac{1}{2}$
Jan. 22, 1915	$95\frac{1}{4}$	$86\frac{7}{8}$	$8\frac{3}{8}$
Jan. 5, 1917	$95\frac{1}{2}$	$86\frac{3}{4}$	$8\frac{3}{4}$
Feb. 28, 1919	$82\frac{3}{8}$	$75\frac{1}{8}$	$7\frac{1}{4}$
May 21, 1920	$70\frac{1}{4}$	62	$8\frac{1}{4}$
May 6, 1921	$77\frac{3}{4}$	$69\frac{3}{4}$	8
Aug. 4, 1922	$93\frac{1}{2}$	$84\frac{1}{2}$	9
Aug. 2, 1923	$89\frac{1}{4}$	$79\frac{3}{4}$	$9\frac{1}{2}$
Aug. 1, 1924	$90\frac{3}{8}$	$84\frac{1}{4}$	$6\frac{1}{8}$
Dec. 4, 1925	$89\frac{1}{4}$	$85\frac{1}{4}$	4
Jan. 3, 1930	$93\frac{1}{4}$	93	$\frac{1}{4}$
Jan. 7, 1931	$98\frac{1}{2}$	97	$1\frac{1}{2}$
June 2, 1932	81	$66\frac{1}{2}$	$14\frac{1}{2}$
June 19, 1933	93	88	5

Prior to 1924 the Atchison General 4s sold usually at about 7 to 10 points above the Adjustment 4s, and yielded about ½% less. Since both issues were considered safe without question, it would have been more logical to purchase the junior issue at its 10% lower cost. After 1923 this point of view asserted itself and the price difference steadily narrowed. During 1930 and part of 1931 the junior issue sold on numerous occasions at practically the same price as the General 4s. This relationship was even more illogical than the unduly wide spread in 1922–1923, since the advantage of the Adjustment 4s in price and yield was too negligible to warrant accepting a junior position, even assuming unquestioned safety for both liens.

Within a very short time this rather obvious truth was brought home strikingly by the widening of the spread to over 14 points during the demoralized bond-market conditions of June 1932. From the whole record it may be inferred that a reasonable differential between the two Atchison issues would be about 5 points and that either a *substantial* widening or a virtual disappearance of the spread would present an opportunity for a desirable exchange of one issue for the other.

A junior lien of Company X may be selected in preference to a first-mortgage bond of Company Y, on one of two bases:

1. The protection for the total debt of Company X is adequate and the yield of the junior lien is substantially higher than that of the Company Y issue; or
2. If there is no substantial advantage in yield, then the indicated protection for the total debt of Company X must be considerably better than that of Company Y.

Example of 2:

Issue	Price in 1929	Fixed charges earned, 1929*
Pacific Power and Light Co. First 5s, due 1955.	101	1.53 times
American Gas and Electric Co. Debenture 5s, due 2028	101	2.52 times

* Average results approximately the same.

The appreciably higher coverage of total charges by American Gas Electric would have justified preferring its junior bonds to the first-mortgage issue of Pacific Power and Light, when both were selling at about the same price.

Special Status of "Underlying Bonds."—In the railroad field an especial investment character is generally supposed to attach to what are known as "underlying bonds." These represent issues of relatively small size secured by a lien on especially important parts of the obligor system, and often followed by a series of "blanket mortgages." The underlying bond usually enjoys a first lien, but it may be a second- or even a third-mortgage issue, provided the senior issues are also of comparatively small magnitude.

Example: New York and Erie Railroad Third Mortgage Extended 4½s, due 1938, are junior to two small prior liens covering an important part of the Erie Railroad's main line. They are followed by four successive blanket mortgages on the system, and they have regularly enjoyed the favored status of an underlying bond.

Bonds of this description are thought to be entirely safe, regardless of what happens to the system as a whole. They almost always come through reorganization unscathed; and even during a receivership interest payments are usually continued as a matter of course, largely because the sum involved is proportionately so small. They are not exempt, however, from fairly sharp declines in market value if insolvency overtakes the system.

Example: Pacific Railroad of Missouri Second 5s, due 1938, which are an underlying bond of the Missouri Pacific Railroad system, continued to receive interest during the receiverships of 1915 and 1933, and their market price has been fairly well maintained under all conditions. (The Missouri Pacific *Railway* Third 4s, due 1938, also an underlying issue, but junior to the foregoing, have continued interest payments, but their price fell as low as 50 in 1932 and 1933.)

To some extent, therefore, such underlying bonds may be viewed as exceptions to our rule that a bond is not sound unless the company is sound. For the most part such bonds are owned by institutions or large investors. (The same observations may apply to certain first-mortgage bonds of operating subsidiaries of public-utility holding-company systems.)

In railroad bonds of this type, the location and strategic value of the mileage covered are of prime importance. First-mortgage bonds on nonessential and unprofitable parts of the system, referred to sometimes as "divisional liens," are not true underlying bonds in the sense that we have just used the term. Divisional first liens on worthless mileage may receive much less favorable treatment in a reorganization than blanket mortgage bonds ostensibly junior to them.

CHAPTER VII

THE SELECTION OF FIXED-VALUE INVESTMENTS: SECOND AND THIRD PRINCIPLES

II. BONDS SHOULD BE BOUGHT ON A DEPRESSION BASIS

The rule that a sound investment must be able to withstand adversity seems self-evident enough to be termed a truism. Any bond can do well when conditions are favorable; it is only under the acid test of depression that the advantages of strong over weak issues become manifest and vitally important. For this reason prudent investors have always favored the obligations of old-established enterprises which have demonstrated their ability to come through bad times as well as good.

Presumption of Safety Based upon Either the Character of the Industry or the Amount of Protection.—Confidence in the ability of a bond issue to weather depression may be based on either of two different reasons. The investor may believe that the particular business will be immune from a drastic shrinkage in earning power, or else that the margin of safety is so large that it can undergo such a shrinkage without resultant danger. The bonds of light and power companies have been favored principally for the first reason, the bonds of United States Steel Corporation subsidiaries for the second. In the former case it is the *character* of the industry, in the latter it is the *amount* of protection, which justifies the purchase. Of the two viewpoints, the one which tries to avoid the perils of depression appeals most to the average bond buyer. It seems much simpler to invest in a depression-proof enterprise than to have to rely on the company's financial strength to pull its bonds through a period of poor results.

No Industry Entirely Depression-proof.—The objection to this theory of investment is, of course, that there is no such thing as a depression-proof industry, meaning thereby one that is immune from the danger of *any* decline in earning power. It is true that the Edison companies have shown themselves subject to only minor shrinkage in profits, as compared, say, with the steel producers. But even a small decline may prove fatal if the business is bonded to the limit of prosperity earnings. Once it is admitted—as it always must be—that the industry can suffer *some* reduction in profits, then the investor is compelled to estimate the possible extent of the shrinkage and compare it with the surplus above the interest requirements. He thus finds himself in the same position as the holder of any other kind of bond, vitally concerned with the ability of the company to meet the vicissitudes of the future.

The distinction to be made, therefore, is not between industries which are *exempt* from and those which are *affected* by depression, but rather between those which are more and those which are less subject to fluctuation. The more stable the type of enterprise, the better suited it is to bond financing and the larger the portion of the supposed normal earning power which may be consumed by interest charges. As the degree of instability increases, it must be offset by a greater margin of safety to make sure that interest charges will be met; in other words, a smaller portion of total capital may be represented by bonds.

Investment Practice Recognizes Importance of Character of the Industry.—This conception has been solidly grounded in investment practice for many years. The threefold classification of enterprises—as railroads, public utilities, or industrials—was intended to reflect inherent differences in relative stability and consequently in the coverage to be required above bond interest requirements. Investors thought well, for example, of any railroad which earned its bond interest twice over, but the same margin in the case of an industrial bond was ordinarily regarded as inadequate. In the decade between 1920 and 1930, the status of the public-utility division underwent some radical changes. A sharp separation was introduced between light, heat, and power services on the one hand, and street-railway lines on the other, although previously the two had been closely allied. The trolley companies, because of their poor showing, were tacitly excluded from the purview of the term "public utility," as used in financial circles, and in the popular mind the name was restricted to electric, gas, water, and telephone companies. (Later on, promoters endeavored to exploit the popularity of the public utilities by applying this title to companies engaged in all sorts of businesses, including natural gas, ice, coal, and even storage.) The steady progress of the utility group, even in the face of the minor industrial setbacks of 1924 and 1927, led to an impressive advance in its standing among investors, so that by 1929 it enjoyed a credit rating fully on a par with the railroads. In the ensuing depression, it registered a much smaller shrinkage in gross and net earnings than did the transportation industry, and its seems logical to expect that bonds of soundly capitalized light and power companies will replace high-grade railroad bonds as the premier type of corporate investment. (This seems true to the authors despite the distinct recession in the popularity of utility bonds and stocks in 1933, due to a combination of rate reductions, governmental competition, and threatened dangers from inflation.)

COMPARISON OF RAILROAD AND PUBLIC-UTILITY GROSS AND NET WITH THE AVERAGE YIELD ON HIGH-GRADE RAILROAD AND UTILITY BONDS 1922–1932 (UNIT $1,000,000)

Year	Railroads			Public utilities				
	Gross[1]	Net[2]	Yield on railroad bonds,%[3]	Gross[4]	Gross[5]	Net[4]	Net[6]	Yield on public-utility bonds,%[3]
1922	$5,618	$ 777	4.85	$1,435	......	$ 447	....	5.46
1923	6,356	978	4.98	1,593	......	510	....	5.41
1924	5,985	987	4.78	1,691	......	546	....	5.22
1925	6,189	1,138	4.67	1,827	......	632	....	5.06
1926	6,451	1,231	4.51	1,995	......	715	....	4.90
1927	6,206	1,085	4.31	2,113	......	776	....	4.78
1928	6,185	1,195	4.34	2,230	......	869	....	4.68
1929	6,355	1,274	4.60	2,309	$5,004	1,007	$673	4.86
1930	5,342	885	4.39	2,382	4,966	1,025	632	4.65
1931	4,236	537	4.61	......	4,811	......	606	4.60
1932	3,162	332	5.99	......	4,339	......	496	5.36

Details as to sources of above data are given in the Appendix, Note 10.

[1] Railway operating revenues for all Class I railroads in the United States, including switching and terminal companies.

[2] Net railway operating income for all Class I railroads included in the above gross figures.

[3] Average yield on 15 high-grade railroad and 15 high-grade utility-company bonds, respectively, as compiled by Standard Statistics Company, Inc.

[4] Gross and net earnings of gas, electric, heat, power, traction, and water companies compiled by United States Department of Commerce. Series discontinued Dec. 31, 1930.

[5] New series (overlapping the United States Department of Commerce figures on public utility gross which were discontinued in 1930) to show recent trend. Compiled by the authors from several series published currently in the "Survey of Current Business," covering electric, gas, street-railway, and telephone companies.

[6] Net operating income of telephone companies and net income of other public utilities compiled by Federal Reserve Bank of New York.

Supposed Stability of Certain Groups Invalidated in 1931–1933.—A survey of the bond-quotation list in 1932, however, might give rise to considerable scepticism as regards the superior stability of one investment group over another. The long register of issues which had suffered severe declines in market price appeared to include impartially representatives of all three divisions. Every industrial issue selling at receivership prices could be matched by a railroad and even by a public-utility bond in the same position. This would indicate that the investor's power to protect himself against the results of depression is limited to what may be termed "normal depressions," and proves ineffectual in dealing with the catastrophic conditions of 1931–1933. However, following our suggestion that the experience of these years be viewed as a laboratory test of investment principles, some conclusions may be drawn therefrom which are not without value for the future.

**Various Causes of Bond Collapses. 1. *Excessive Funded Debt of Utilities.*—If we study the bond issues which suffered collapse in the post-bubble period, we shall observe that different causes underlay the troubles of each group. The public-utility defaults were caused not by a disappearance of earnings but by the inability of overextended debt structures to withstand a relatively moderate setback. Enterprises capitalized on a reasonably sound basis, as judged by former standards, had little difficulty in meeting bond interest. This did not hold true in the case of many holding companies with pyramided capital structures which had absorbed nearly every dollar of peak-year earnings for fixed charges and so had scarcely any margin available to meet a shrinkage in profits. The widespread difficulties of the utilities were due not to any weakness in the light and power *business*, but to the reckless extravagance of its financing methods. The losses of investors in public-utility bonds could for the most part have been avoided by the exercise of ordinary prudence in bond selection. Conversely, the unsound financing methods employed must eventually have resulted in individual collapses, even in the ordinary course of the business cycle. In consequence, the theory of investment in sound public-utility bonds appears in no sense to have been undermined by 1931–1933 experience.

2. *Stability of Railroad Earnings Overrated.*—Turning to the railroads, we find a somewhat different situation. Here the

fault appears to be that the stability of the transportation industry was overrated, so that investors were satisfied with a margin of protection which proved insufficient. It was not a matter of imprudently disregarding old established standards of safety, as in the case of the weaker utilities, but rather of being content with old standards when conditions called for more stringent requirements. Looking back, we can see that the failure of the carriers generally to increase their earnings with the great growth of the country since prewar days was a sign of a weakened relative position, which called for a more cautious and exacting attitude by the investor. If he had required his railroad bonds to meet the same tests that he applied to industrial issues, he would have been compelled to confine his selection to a relatively few of the strongly situated lines.[1] As it turned out, nearly all of these have been able to withstand the tremendous loss of traffic since 1929 without danger to their fixed charges. Whether or not this is a case of wisdom after the event is irrelevant to our discussion. Viewing past experience as a lesson for the future, we can see that *selecting railroad bonds on a depression basis* would mean requiring a larger margin of safety in normal times than was heretofore considered necessary.

3. *Perennial Instability of Industrial Earnings.*—But in the industrial list we meet a different kind of problem. Price collapses were not due primarily to unsound financial structures, as in the case of utility bonds, nor to a miscalculation by investors as to the margin of safety needed, as in the case of railroad bonds. We are confronted in many cases by a sudden disappearance of earning power, and a disconcerting question as to whether the business can survive. A company such as Gulf States Steel, for example, earned its 1929 interest charges at least $3\frac{1}{2}$ times in every year from 1922 to 1929. Yet in 1930 and 1931 operating losses were so large as to threaten its solvency.[2] Many basic industries, such as the Cuban sugar producers and our own coal mines, were depressed prior to the 1929 debacle. In the

[1] If, for example, the investor had restricted his attention to bonds of roads which in the prosperity year 1928 covered their fixed charges $2\frac{1}{2}$ times or better, he would have confined his selections to bonds of: Atchison; Canadian Pacific; Chesapeake and Ohio; Chicago, Burlington and Quincy; Norfolk and Western; Pere Marquette; Reading; and Union Pacific. (With the exception of Pere Marquette, the bonds of these roads fared comparatively well in the depression. Note, however that the above test is more stringent than the one we propose later on—average earnings = twice fixed charges.)

[2] See Appendix, Note 11, for supporting data and other examples.

past, such eclipses had always proven to be temporary, and investors felt justified in holding the bonds of these companies in the expectation of a speedy recovery. But in this instance the continuance of adverse conditions beyond all previous experience defeated their calculations and destroyed the values behind their investment.

From these cases we must conclude that even a high margin of safety in good times may prove ineffective against a succession of operating losses caused by prolonged adversity. The difficulties that befell industrial bonds, therefore, cannot be avoided in the future merely by more stringent requirements as to bond-interest coverage in normal years.

Few Industrial Bonds Maintained Investment Status in 1932–1933.—It is illuminating to examine the list of industrial companies, with bonds dealt in on the New York Stock Exchange, which maintained a strong credit standing throughout 1932 and 1933. Limiting the selection to those companies all of whose bonds maintained a price reflecting a reasonable confidence in their safety (*i.e.*, which did not sell at a price below 90 *or* else at a price to yield more than a very small fraction above 7% to maturity), the group comprises only 18 out of more than 200 companies represented in the New York Stock Exchange industrial bond section. The list is as follows:

	1932 1933 Low Price
Name of Corporation and Issue	
1. American Machine and Foundry Company:	
15-year Sinking Fund Secured 6s, due 1939	102¼
2. American Sugar Refining Company:	
15-year 6s, due 1937	98
3. Associated Oil Company:	
12-year 6% Notes, due 1935	94⅛
4. Corn Products Refining Company:	
First Sinking Fund 5s, due 1934	100⅝
5. General Baking Company:	
Debenture 5½s, due 1940	89½
6. General Electric Company:	
40-year Debenture 3½s, due 1942	93
7. General Motors Acceptance Corporation:	
Debenture 6s, due 1937	97¾
8. Humble Oil and Refining Company:	
10-year Debenture 5½s, due 1932	99
Debenture 5s, due 1937	94
9. International Business Machine Corporation:	
Computing-Tabulating-Recording Company First 6s, due 1941	104

10. Liggett and Myers Tobacco Company:
   Debenture 5s, due 1951 . . . . 96½
   Debenture 7s, due 1944 . . . . 115⅛
11. P. Lorillard Company:
   Debenture 5s, due 1951 . . . . 81¼
   Debenture 7s, due 1944 . . . . 101⅝
12. National Sugar Refining Company:
   Warner Sugar Refining Company First 7s, due 1941 . . . . . . 97½
13. Pillsbury Flour Mills Company:
   First 6s, due 1943 90
14. Smith (A. O.) Corporation:
   10-year 6½s, due 1933 . . . . 95½
15. Socony-Vacuum Corporation:
   General Petroleum Corporation First 5s, due 1940 . . . . 95¾
   Standard Oil Company (N.Y.) Debenture 4½s, due 1951 . . . 82
   White Eagle Oil and Refining Company Debenture 5½s,
      due 1937 . . . . 96½
16. Standard Oil Company of Indiana:
   Pan American Petroleum and Transport Company Con-
      vertible 6s, due 1934 100
   Sinclair Crude Oil Purchasing Company 5½s, due 1938 . . . 91¾
   Sinclair Pipe Line Company 20-year 5s, due 1942 . . . . 89⅛
17. Standard Oil Company of New Jersey:
   20-year 5% Debentures, due 1946 . . . . 98¾
18. United States Steel Corporation:
   Illinois Steel Company Debenture 4½s, due 1940 . . . . 90¾
   Tennessee Coal, Iron and Railroad Company General 5s,
      due 1951 93

The majority of these companies are of outstanding importance in their respective industries. This point suggests that *dominant size* is a trait of considerable advantage in dealing with exceptionally unfavorable developments in the industrial world, which may mean in turn that industrial investments should be restricted to large companies. The evidence, however, may be objected to on the ground of having been founded on an admittedly abnormal experience. But even in the supposedly prosperous period between 1922 and 1929, the bonds of smaller industrial enterprises did not prove a dependable medium of investment. There were many instances wherein an apparently well-established earning power suffered a sudden disappearance.[1] In fact these unpredictable variations were sufficiently numerous to suggest the conclusion that there is an inherent lack of stability in the small or medium-sized industrial enterprise, which makes

[1] See Appendix, Note 12, for examples.

them ill-suited to bond financing. A tacit recognition of this weakness has been responsible in part for the growing adoption of conversion and subscription-warrant privileges in connection with industrial-bond financing. To what extent such embellishments can compensate for insufficient safety will be discussed in our chapters on Senior Securities with Speculative Features. But in any event the widespread resort to these profit-sharing artifices seems to confirm our view that bonds of smaller industrial companies are not well qualified for consideration as fixed-value investments.

Unavailability of Sound Bonds No Excuse for Buying Poor Ones.—However, if we recommend that straight bond investment in the industrial field be confined to companies of dominant size, we face the difficulty that such companies are few in number and many of them have no bonds outstanding. It may be objected further that such an attitude would severely handicap the financing of legitimate businesses of secondary size and would have a blighting effect on investment-banking activities. The answer to these remonstrances must be that no consideration can justify the purchase of unsound bonds at an investment price. The fact that no good bonds are available is hardly an excuse for either issuing or accepting poor ones. Needless to say, the investor is never forced to buy a security of inferior grade. At some sacrifice in yield he can always find issues that meet his requirements, however stringent; and, as we shall point out later, attempts to increase yield at the expense of safety are likely to prove unprofitable. From the standpoint of the corporations and their investment bankers, the conclusion must follow that if their securities cannot properly qualify as straight investments, they must be given profit-making possibilities sufficient to compensate the purchaser for the risk he runs.

Conflicting Views on Bond Financing.—In this connection, observations are in order regarding two generally accepted ideas on the subject of bond financing. The first is that bond issues are an element of weakness in a company's financial position, so that the elimination of funded debt is always a desirable object. The second is that when companies are unable to finance through the sale of stock it is proper to raise money by means of bond issues. In the writers' view both of these widespread notions are quite incorrect. Otherwise there would be no really sound basis for any bond financing. For they imply that only weak companies should be willing to sell bonds—which, if true, would mean that investors should not be willing to buy them.

Proper Theory of Bond Financing.—The proper theory of bond financing, however, is of quite different import. A reasonable amount of funded debt is of advantage to a prosperous business, because the stockholders can earn a profit above interest charges through the use of the bondholders' capital. It is desirable for both the corporation and the investor that the borrowing be limited to an amount which can safely be taken care of under all conditions. Hence, from the standpoint of sound finance, there is no basic conflict of interest between the strong corporation which floats bonds and the public which buys them. On the other hand, whenever an element of unwillingness or compulsion enters into the creation of a bond issue by an enterprise, these bonds are *ipso facto* of secondary quality and it is unwise to purchase them on a straight investment basis.

Unsound Policies Followed in Practice.—Financial policies followed by corporations and accepted by the public have for many years run counter to these logical principles. The railroads, for example, have financed the bulk of their needs through bond sales, resulting in an overbalancing of funded debt as against stock capital. This tendency has been repeatedly deplored by all authorities, but accepted as inevitable because poor earnings made stock sales impracticable. But if the latter were true, they also made bond purchases inadvisable. It is now quite clear that investors were imprudent in lending money to carriers which themselves complained of the necessity of having to borrow it.

While investors were thus illogically lending money to weak borrowers, many strong enterprises were paying off their debts through the sale of additional stock. But if there is any thoroughly sound basis for corporate borrowing, then this procedure must also be regarded as unwise. If a reasonable amount of borrowed capital, obtained at low interest rates, is advantageous to the stockholder, then the replacement of this debt by added stock capital means the surrender of such advantage. The elimination of debt will naturally simplify the problems of the management, but surely there must be some point at which the return to the stockholders must also be considered. Were this not so, corporations would be constantly raising money from their owners and they would never pay any part of it back in dividends.

It should be pointed out that the mania for debt retirement in 1927–1929 has had a disturbing effect upon our banking situation, since it eliminated most of the good commercial borrowers and replaced them by second-grade business risks and by loans on stock collateral, which were replete with possibilities of harm.

Significance of the Foregoing to the Investor.—The above analysis of the course of industrial bond borrowing in the last decade is not irrelevant to the theme of this chapter, *viz.*, the application of depression standards to the selection of fixed-value investments. Recognizing the necessity of ultra-stringent criteria of choice in the industrial field, the bond buyer is faced by a further narrowing of eligible issues due to the elimination of funded debt by many of the strongest companies. Clearly his reaction must not be to accept the issues of less desirable enterprises, in the absence of better ones, but rather to refrain from any purchases on an investment basis if the suitable ones are not available. It appears to be a financial axiom that whenever there is money to invest, it is invested; and if the owner cannot find a good security yielding a fair return, he will invariably buy a poor one. But a prudent and intelligent investor should be able to avoid this temptation, and reconcile himself to accepting an unattractive yield from the best bonds, in preference to risking his principal in second-grade issues for the sake of a larger coupon return.

Summary.—The rule that bonds should be bought on the basis of their ability to withstand depression has been part of an old investment tradition. It was nearly lost sight of in the prosperous period culminating in 1929, but its importance was made painfully manifest during the following collapse. The bonds of reasonably capitalized electric and gas companies have given a satisfactory account of themselves during this period, and the same is true—to a lesser degree—of the relatively few railroads which showed a large margin above interest charges prior to 1930. In the industrial list, however, even an excellent past record has in many cases proved undependable, especially where the company is of small or moderate size. For this reason, the investor would seem to gain better protection against adverse developments by confining his industrial selections to companies which meet the two requirements of (1) dominant size, and (2) substantial margin of earnings over bond interest.

III. THIRD PRINCIPLE: UNSOUND TO SACRIFICE SAFETY FOR YIELD

In the traditional theory of bond investment a mathematical relationship is supposed to exist between the interest rate and the degree of risk incurred. The interest return is divided into two components, the first constituting "pure interest"—*i.e.*, the rate obtainable with *no* risk of loss—and the second representing the premium obtained to compensate for the risk assumed. If, for example, the "pure interest rate" is assumed to be 3%, then a 4% investment is supposed to involve one chance in a hundred of loss, while the risk incurred in an 8% investment would be five times as great, or 1 in 20. (Presumably the risk should be somewhat less than that indicated, to allow for an "insurance profit.")

This theory implies that bond-interest rates are closely similar to insurance rates, and that they measure the degree of risk on some reasonably precise actuarial basis. It would follow that, by and large, the return from high- and low-yielding investments should tend to equalize, since what the former gain in income would be offset by their greater percentage of principal losses, and *vice versa*.

No Mathematical Relationship between Yield and Risk.— This view, however, seems to us to bear little relation to the realities of bond investment. Security prices and yields are not determined by any exact mathematical calculation of the expected risk, but they depend rather upon the *popularity* of the issue. This popularity reflects in a general way the investors' view as to the risk involved, but it is also influenced largely by other factors, such as the degree of familiarity of the public with the company and the issue (seasoning) and the ease with which the bond can be sold (marketability).

It may be pointed out further that the supposed actuarial computation of investment risks is out of the question theoretically as well as in practice. There are no experience tables available by which the expected "mortality" of various types of issues can be determined. Even if such tables were prepared, based on long and exhaustive studies of past records, it is doubtful whether they would have any real utility for the future. In life insurance the relation between age and mortality rate is well defined and changes only gradually. The same is true, to a much lesser extent, of the relation between the various types of structures and the fire hazard attaching to them. But the relation between different kinds of investments and the risk of loss is entirely too indefinite and too variable with changing conditions, to permit of sound mathematical formulation. This is particularly true because investment losses are not distributed fairly evenly in point of time, but tend to be concentrated at intervals, *i.e.*, during periods of general depression. Hence the typical investment hazard is roughly similar to the conflagration or epidemic hazard, which is the exceptional and incalculable factor in fire or life insurance.

Self-insurance Generally Not Possible in Investment.—If we were to assume that a precise mathematical relationship does exist between yield and risk, then the result of this premise should be inevitably to recommend the lowest yielding—and therefore the safest—bonds to all investors. For the individual is not qualified to be an insurance underwriter. It is not his function to be paid for incurring risks; on the contrary it is to his interest to pay others for insurance against loss. Let us assume a bond buyer has his choice of investing $1,000 for $30 per annum without risk, or for $80 per annum with 1 chance out of 20 each year that his principal would be lost. The $50 additional income on the second investment is mathematically equivalent to the risk involved. But in terms of *personal requirements*, an investor cannot afford to take even a small chance of losing $1,000 of principal in return for an extra $50 of income. Such a procedure would be the direct opposite of the standard procedure of *paying* small annual sums to protect property values against loss by fire and theft.

The Factor of Cyclical Risks.—The investor cannot prudently turn himself into an insurance company and incur risks of losing his principal in exchange for annual premiums in the form of extra-large interest coupons. One objection to such a policy is that sound insurance practice requires a very wide distribution of risk, in order to minimize the influence of luck and to allow maximum play to the law of probability. The investor may endeavor to attain this end by diversifying his holdings, but as a practical matter he cannot approach the division of risk attained by an insurance company. More important still is the danger that many risky investments may collapse together in a depression period, so that the investor in high-yielding issues will find a period of large income (which he will probably spend) followed suddenly by a deluge of losses of principal.

It may be contended that the higher yielding securities on the whole return a larger premium above "pure interest" than the degree of risk requires; in other words, that in return for taking the risk, investors will in the long run obtain a *profit* over and above the losses in principal suffered. It is difficult to say definitely whether or not this is true. But even assuming that the high coupon rates will, in the great aggregate, more than compensate on an *actuarial* basis for the risks accepted, such bonds are still undesirable investments from the *personal* standpoint of the average investor. Our arguments against the investor turning himself into an insurance company remain valid even if the insurance operations all told may prove profitable. The bond buyer is neither financially nor psychologically equipped to carry on extensive transactions involving the setting up of reserves out of regular income to absorb losses in substantial amounts suffered at irregular intervals.

Risk and Yield Are Incommensurable.—The foregoing discussion leads us to suggest the principle that income return and risk of principal should be regarded as *incommensurable*. Practically speaking, this means that acknowledged risks of losing principal should not be offset merely by a high coupon rate, but can be accepted only in return for a corresponding opportunity for enhancement of principal, *e.g.*, through the purchase of bonds at a substantial discount from par, or possibly by obtaining an unusually attractive conversion privilege. While there may be no real *mathematical* difference between offsetting risks of loss by a higher income or by a chance for profit, the *psychological* difference is very important. The purchaser of low-priced bonds is fully aware of the risk he is running; he is more likely to make a thorough investigation of the issue and to appraise carefully the chances of loss and of profit; finally—most important of all—he is prepared for whatever losses he may sustain, and his profits are in a form available to meet his losses. Actual investment experience, therefore, will not favor the purchase of the typical high-coupon bond offered at about par, wherein, for example, a 7% interest return is imagined to compensate for a distinctly inferior grade of security.[1]

[1] In an exceptional year such as 1921 strongly entrenched bonds were offered bearing a 7% coupon, due to the prevailing high money rates.

Fallacy of the "Business Man's Investment."—An issue of this type is commonly referred to in the financial world as a "business man's investment" and is supposedly suited to those who can afford to take some degree of risk. Most of the foreign bonds floated between 1923 and 1929 belonged in that category. The same is true of the great bulk of straight preferred stock issues. According to our view, such "business man's investments" are an illogical type of commitment. The security buyer who can afford to take some risk should seek a commensurate opportunity of enhancement in price and pay only secondary attention to the income obtained.

Reversal of Customary Procedure Recommended.—Viewing the matter more broadly, it would be well if investors reversed their customary attitude toward income return. In selecting the grade of bonds suitable to their situation, they are prone to start at the top of the list, where maximum safety is combined with lowest yield, and then to calculate how great a concession from ideal security they are willing to make for the sake of a more attractive income rate. From this point of view, the ordinary investor becomes accustomed to the idea that the type of issue suited to his needs must rank somewhere below the very best, a frame of mind which is likely to lead to the acceptance of definitely unsound bonds, either because of their high income return or by surrender to the blandishments of the bond salesman.

It would be sounder procedure to start with minimum standards of safety, which all bonds must be required to meet in order to be eligible for further consideration. Issues failing to meet these minimum requirements should be automatically disqualified as straight investments, regardless of high yield, attractive prospects, or other grounds for partiality. Having thus delimited the field of eligible investments, the buyer may then apply such further selective processes as he deems appropriate. He may desire elements of safety far beyond the accepted minima, in which case he must ordinarily make some sacrifice of yield. He may also indulge his preferences as to the nature of the business and the character of the management. But, essentially, bond selection should consist of working upward from definite minimum standards rather than working downward in haphazard fashion from some ideal but unacceptable level of maximum security.

CHAPTER VIII

SPECIFIC STANDARDS FOR BOND INVESTMENT

IV. FOURTH PRINCIPLE: DEFINITE STANDARDS OF SAFETY MUST BE APPLIED

Since the selection of high-grade bonds has been shown to be in good part a process of exclusion, it lends itself reasonably well to the application of definite rules and standards designed to disqualify unsuitable issues. Such regulations have in fact been set up in many states by legislative enactment to govern the investments made by savings banks and by trust funds. In most such states, the banking department prepares each year a list of securities which appear to conform to these regulations and are therefore considered "legal," *i.e.*, eligible for purchase under the statute.

It is our view that the underlying idea of fixed standards and minima should be extended to the entire field of straight investment, *i.e.*, investment for income only. These legislative restrictions are intended to promote a high average level of investment quality and to protect depositors and beneficiaries against losses from unsafe securities. If such regulations are desirable in the case of institutions, it should be logical for individuals to follow them also. We have previously challenged the prevalent idea that the ordinary investor can afford to take greater investment risks than a savings bank, and need not therefore be as exacting with respect to the soundness of his fixed-value securities. The experience of 1927 to 1933 undoubtedly emphasizes the need for a general tightening of investment standards, and a simple method of attaining this end might be to confine all straight-bond selections to those which meet the legal tests of eligibility for savings banks or trust funds. Such a procedure would appear directly consonant with our fundamental principle that straight investments should be made only in issues of unimpeachable soundness, and that securities of inferior grade must be bought only on an admittedly speculative basis.

90

New York Savings-bank Law as a Point of Departure.—As a matter of *practical policy*, an individual bond buyer is likely to obtain fairly satisfactory results by subjecting himself to the restrictions which govern the investment of savings banks' funds. But this procedure cannot be seriously suggested as a *general principle of investment*, because the legislative provisions are themselves far too imperfect to warrant their acceptance as the best available theoretical standards. The acts of the various states are widely divergent; most of them are antiquated in important respects; none is entirely logical or scientific. The legislators did not approach their task from the viewpoint of establishing criteria of sound investments for universal use; consequently they felt free to impose arbitrary restrictions on savings-bank and trust funds, which they would have hesitated to prescribe for investors generally. The New York statute, generally regarded as the best of its class, is nevertheless marred by a number of evident defects. In the formulation of comprehensive investment standards, the New York legislation may best be used, therefore, as a guide or point of departure, rather than as a final authority. The ensuing discussion will follow fairly closely the pattern set forth in the statutory provisions (as they existed in 1934); but these will be criticized, rejected, or amplified, wherever such emendation appears desirable.

GENERAL CRITERIA PRESCRIBED BY THE NEW YORK STATUTE

The specific requirements imposed by the statute upon bond investments may be classified under seven heads, which we shall proceed to enumerate and discuss:

1. The *nature* and *location* of the business or government.
2. The *size* of the enterprise, or the issue.
3. The *terms* of the issue.
4. The *record* of solvency and dividend payments.
5. The relation of *earnings* to interest requirements.
6. The relation of the *value* of the property to the funded debt.
7. The relation of *stock* capitalization to the funded debt.

NATURE AND LOCATION

The most striking features of the laws governing savings-bank investments is the complete exclusion of bonds in certain broad categories. The New York provisions relative to investment in bonds may be summarized as follows:

Admitted	Excluded
United States government, state, and municipal bonds.	Foreign government and foreign corporation bonds.
Railroad, gas, electric, and telephone bonds.	Street railway and water bonds.
Bonds secured by first mortgages on real estate.	All industrial bonds.
	Bonds of financial companies (investment trusts, credit concerns, *etc.*).

The Fallacy of Blanket Prohibitions.—The legislature was evidently of the view that bonds belonging to the excluded categories are essentially too unstable to be suited to savings-bank investment. If this view is entirely sound, it would follow from our previous reasoning that all issues in these groups are unsuited to conservative investment generally. Such a conclusion would involve revolutionary changes in the field of finance, since a large part of the capital now regularly raised in the investment market would have to be sought on an admittedly speculative basis.

In our opinion, a considerable narrowing of the investment category is in fact demanded by the unsatisfactory experience of bond investors over a fairly long period. Nevertheless, there are strong objections to the application of blanket prohibitions of the kind now under discussion. Investment theory should be chary of easy generalizations. Even if full recognition is given, for example, to the unstable tendencies of industrial bonds, as discussed in Chap. VII, the elimination of this entire major group from investment consideration would seem neither practicable nor desirable. The existence of a fair number of industrial issues (even though a small percentage of the total) which have maintained an undoubted investment status through the severest tests, would preclude investors generally from adopting so drastic a policy. Moreover, the confining of investment demand to a few eligible types of enterprise is likely to make for scarcity, and hence for the acceptance of inferior issues merely because they fall within these groups. This has in fact been one of the unfortunate results of the present legislative restrictions.

Individual Strength May Compensate for Inherent Weakness of a Class.—It would seem a sounder principle, therefore, to require a stronger exhibit by the *individual* bond to compensate for any weakness supposedly inherent in its *class*, rather than to seek to admit all bonds of certain favored groups and to exclude all bonds of others. An industrial bond may properly be required to show a larger margin of earnings over interest charges and a smaller proportion of debt to going-concern value than would be required of an obligation of a gas or electric enterprise. The same would apply in the case of traction bonds. In connection with the exclusion of *water-company bonds* by the New York statute, it should be noted that this group is considered by most other states to be on a par with gas, electric, and telephone obligations. There seems to be no good reason for subjecting them to more stringent requirements than in the case of other types of public-service issues.

Obligations of Foreign Governments.—In dealing with foreign-government debts we are confronted with a different situation. Such issues respond in but small degree to financial analysis, and investment therein is ordinarily based on general considerations, such as confidence in the country's economic and political stability and the belief that it will faithfully endeavor to discharge its obligations. To a much greater extent, therefore, than in the case of other bonds, an opinion may be justified or even necessitated as to the *general desirability* of foreign-government bonds for fixed-value investment.

The Factor of Political Expediency.—Viewing objectively the history of foreign-bond investment in this country since it first assumed importance during the World War, it is difficult to escape an unfavorable conclusion on this point. In the final analysis, a foreign-government debt is an unenforceable contract. If payment is withheld, the bondholder has no direct remedy. Even if specific revenues or assets are pledged as security, he is practically helpless in the event that these pledges are broken. It follows that while a foreign-government obligation is in theory a claim against the entire resources of the nation, the extent to which these resources are actually drawn upon to meet the external debt burden is found to depend in good part on political expediency. The grave international dislocations of the post-war period made some defaults inevitable, and supplied the pretext for others. In any event, because nonpayment has become a familiar phenomenon, its very frequency has removed much of the resultant obloquy. Hence the investor has, seemingly, less reason than of old to rely upon herculean efforts being made by a foreign government to live up to its obligations during difficult times.

The Foreign-trade Argument.—It is generally argued that a renewal of large-scale international lending is necessary to restore world equilibrium. More concretely, such lending appears to be an indispensable adjunct to the restoration and development of our export trade. But the investor should not be expected to make unsound commitments for idealistic reasons or to benefit American exporters. As a *speculative operation*, the purchase of foreign obligations at low prices, such as prevailed in 1932, might prove well justified by the attendant possibilities of profit; but these tremendously depreciated quotations are in themselves a potent argument against later purchases of new foreign issues at a price close to 100% of face value, no matter how high the coupon rate may be set.

The Individual-record Argument.—It may be contended, however, that investment in foreign obligations is essentially similar to any other form of investment in that it requires discrimination and judgment. Some nations deserve a high credit rating based on their past performance, and these are entitled to investment preference to the same degree as are domestic corporations with satisfactory records. The legislatures of several states have recognized the superior standing of Canada by authorizing savings banks to purchase its obligations, and Vermont has accepted also the dollar bonds of Belgium, Denmark, Great Britain, Holland, and Switzerland.

A strong argument in the contrary direction is supplied by the appended list of the various countries having debts payable in dollars, classified according to the credit rating indicated by the market action of their bonds during the severe test of 1932.

1. Countries whose bonds sold on an investment basis: Canada, France, Great Britain, Netherlands, Switzerland.

2. Countries whose bonds sold on a speculative basis: Argentina, Australia, Austria, Bolivia, Brazil, Bulgaria, Chile, China, Colombia, Costa Rica, Cuba, Czechoslovakia, Denmark, Dominican Republic, Esthonia, Finland, Germany, Guatemala, Greece, Haiti, Hungary, Japan, Jugoslavia, Mexico, Nicaragua, Panama, Peru, Poland, Rumania, Russia, Salvador, Uruguay.

3. Borderline countries: Belgium, Ireland, Italy, Norway, Sweden.

Of the five countries in the first or investment group, the credit of two, *viz.*, France and Great Britain, was considered speculative in the preceding depression of 1921–1922. Out of 42 countries represented, therefore, only three (Canada, Holland,

and Switzerland) have enjoyed an unquestioned investment rating during the twelve years ending in 1932.

Twofold Objection to Purchase of Foreign-government Bonds.— This evidence suggests that the purchase of foreign-government bonds is subject to a twofold objection of *generic* character: theoretically, in that the basis for credit is fundamentally intangible; and practically, in that experience with the foreign group has been preponderantly unsatisfactory. Apparently it will require a considerable betterment of world conditions, demonstrated by a fairly long period of punctual discharge of international obligations, to warrant a revision of this unfavorable attitude toward foreign bonds as a class.

Canadian issues may undoubtedly be exempted from this blanket condemnation, both on their record and because of the closeness of the relationship between Canada and the United States. Individual investors, for either personal or statistical reasons, may be equally convinced of the high credit standing of various other countries, and will therefore be ready to purchase their obligations as high-grade investments. Such commitments may prove to be fully justified by the facts; but for some years, at least, it would be well if the investor approached them in the light of *exceptions* to a general rule of avoiding foreign bonds, and required them accordingly to present exceptionally strong evidence of stability and safety.

Bonds of Foreign Corporations.—In *theory*, bonds of a corporation, however prosperous, cannot enjoy better security than the obligations of the country in which the corporation is located. The government, through its taxing power, has an unlimited prior claim upon the assets and earnings of the business; in other words, it can take the property away from the private bondholder and utilize it to discharge the national debt. But in actuality, distinct limits are imposed by political expediency upon the exercise of the taxing power. Accordingly we find instances of corporations meeting their dollar obligations even when their government is in default.[1]

Foreign-corporation bonds have an advantage over governmental bonds in that the holder enjoys specific legal remedies in the event of nonpayment, such as the right of foreclosure. Consequently it is probably true that a foreign company is under greater *compulsion* to meet its debt than is a sovereign

---

[1] See Appendix, Note 13, for examples.

nation. But it must be recognized that the conditions resulting in the default of government obligations are certain to affect adversely the position of the corporate bondholder. Restrictions on the transfer of funds may prevent the payment of interest in dollars even though the company may remain amply solvent.[1] Furthermore, the distance separating the creditor from the property, and the obstacles interposed by governmental decree, are likely to destroy the practical value of his mortgage security. For these reasons the unfavorable conclusions reached with respect to foreign-government obligations as fixed-value investments must be considered as applicable also to foreign-corporation bonds.

SIZE

The bonds of very small enterprises are subject to objections which disqualify them as media for conservative investment. A company of relatively minor size is more vulnerable than others to unexpected happenings, and it is likely to be handicapped by the lack of strong banking connections or of technical resources. Very small businesses, therefore, have never been able to obtain public financing and have depended on private capital, those supplying the funds being given the double inducement of a share in the profits and a direct voice in the management. The objections to bonds of undersized corporations apply also to tiny villages or microscopic townships, and the careful investor in municipal obligations will ordinarily avoid those below a certain population level.

The establishment of such minimum requirements as to size necessarily involves the drawing of arbitrary lines of demarcation. There is no mathematical means of determining exactly at what point a company or a municipality becomes large enough to warrant the investor's attention. The same difficulty will attach to setting up any other quantitative standards, as for example the margin of earnings above interest charges, or the relation of stock or property values to bonded debt. It must be borne in mind, therefore, that all these "critical points" are necessarily rule-of-thumb decisions, and the investor is free to use other amounts if they appeal to him more. But however arbitrary the standards selected may be, they are undoubtedly of great practical utility in safeguarding the bond buyer from inadequately protected issues.

[1] See Appendix, Note 14, for examples.

Provisions of New York Statute.—The New York statute has prescribed various standards as to minimum size in defining investments eligible for savings banks. As regards municipal bonds, a population of not less than 10,000 is required for states adjacent to New York, and of 30,000 for other states. Railroads must either own 500 miles of standard-gauge line or else have operating revenues of not less than $10,000,000 per annum. Unsecured and income bonds of railroad companies are admitted only if (among other special requirements) the *net income* available for dividends amounts to $10,000,000. For gas and electric companies, gross revenues must have averaged $1,000,000 per year during the preceding five years; but in the case of telephone bonds, this figure must be $5,000,000. There are further provisions to the effect that the size of the bond issue itself must be not less than $1,000,000 for gas and electric companies, and not less than $5,000,000 in the case of telephone obligations.

Some Criticisms of These Requirements.—The figures of minimum gross receipts do not appear well chosen from the standpoint of bond investment in general. The alternative tests for railroads, based on either mileage or revenues, is confusing and unnecessary. The $10,000,000-gross requirement by itself is too high; it would have eliminated, for example, the Bangor and Aroostock Railroad, one of the few lines to make a satisfactory exhibit during the 1930–1933 depression as well as before. Equally unwarranted is the requirement of $5,000,000 gross for telephone concerns, as against only $1,000,000 for gas and electric utilities. This provision would have ruled out the bonds of Tri-State Telephone and Telegraph Company prior to 1927, although they were then (and since) obligations of unquestioned merit. We believe that the following proposed requirements for minimum size, although by necessity arbitrarily taken, are in reasonable accord with the realities of sound investment:

	Minimum Requirement of Size
Municipalities	25,000 population
Public-utility enterprises	$2,000,000 gross
Railroad systems	$3,000,000 gross
Industrial companies	$5,000,000 gross

Industrial Bonds and the Factor of Size.—Since industrial bonds are not eligible for savings banks under the New York law, no minimum size is therein prescribed. We have expressed the view that industrial obligations may be included among high-grade investments provided they meet stringent tests of safety. The experience of the past decade indicates that dominant or at least substantial size affords an element of protection against the hazards of instability to which industrial enterprises are more subject than are railroads or public utilities. A cautious investor, seeking to profit from recent lessons, would apparently be justified in deciding to confine his purchases of fixed-value bonds to perhaps the half dozen leading units in each industrial group, and also perhaps in adding the suggested minimum requirement of a $5,000,000 annual turnover.

Such minimum standards may be criticized as unduly stringent, in that if they were universally applied (which in any event is unlikely) they would make it impossible for sound and prosperous businesses of moderate size to finance themselves through straight bond issues. It is conceivable that a general stabilization of industrial conditions in the United States may invalidate the conclusions derived from the recent period of extreme variations. But until such a tendency in the direction of stability has actually demonstrated itself, we should favor a highly exacting attitude toward the purchase of industrial bonds *at investment levels.*

Large Size Alone No Guarantee of Safety.—These recommendations on the subject of minimum size do not imply that enormous dimensions are in themselves a guarantee of prosperity and financial strength. The biggest company may be the weakest if its bonded debt is disproportionately large. Moreover, in the railroad, public-utility, and municipal groups, no practical advantage attaches to the very largest units as compared with those of medium magnitude. Whether the gross receipts of an electric company are twenty millions or a hundred millions has, in all probability, no material effect on the safety of its bonds; and similarly a town of 75,000 inhabitants may deserve better credit than would a city of several millions. It is only in the industrial field that we have suggested that the bonds of a very large enterprise may be inherently more desirable than those of middle-sized companies; but even here a thoroughly satisfactory statistical showing on the part of the large company is necessary to make this advantage a dependable one.

Other Provisions Rejected.—The New York statute includes an additional requirement in respect to unsecured railroad bonds, *viz.*, that the *net* earnings after interest charge must equal ten million dollars. This does not appear to us to be justified, since we have previously argued against attaching particular significance to the possession or lack of mortgage security. There is a certain logical fallacy also in the further prescription of a minimum size for the bond issue itself in the case of public utilities. If the enterprise is large enough as measured by its gross business, then the smaller the bond issue the easier it would be to meet interest and principal requirements. The legislature probably desired to avoid the inferior marketability associated with very small issues. In our view, the element of marketability is generally given too much stress by investors; and in this case we do not favor following the statutory requirement with respect to the size of the issue as a general rule for bond investment.

CHAPTER IX

SPECIFIC STANDARDS FOR BOND INVESTMENT
(*Continued*)

THE PROVISIONS OF THE ISSUE

Under this heading come such features as the security of the bonds, the conditions affecting interest payments, and the date of maturity. Conversion and similar privileges, specified in the indenture, are, of course, important in themselves, but they do not enter into the determination of standards for the selection of *fixed-value* investments.

Under the New York statute, only bonds secured by mortgage are eligible in the public-utility group. However, debenture (unsecured) railroad bonds are admitted, provided the earnings and dividend record meet stiffer requirements than are set forth for mortgage issues. The statute also permits the purchase of income bonds (*i.e.*, those on which the obligation to pay interest is dependent upon earnings) on the same basis as debentures.

Obsolete and Illogical Restrictions.—In our opinion this set of restrictions is quite out of date and illogical. In view of our emphatic argument in Chap. VI against attaching predominant weight to specific security, it must be clear that we do not favor the exclusion of any group of unsecured bond issues *per se*, or even the establishment of any *sharply defined* standards or requirements which favor secured bonds over debentures.

Needless to say, if other things are equal it is better to own a first-mortgage bond than a second-mortgage bond, and it is better to have a second lien than no mortgage at all. Consequently the intelligent investor will always be somewhat more exacting in his requirements (especially as to earnings coverage) when dealing with a debenture obligation than with a first-mortgage issue. But the amount of preference to be accorded the latter type is a matter for individual judgment or inclination, and not for a quantitative rule.

Income Bonds in Weaker Position Than Debentures.—While the New York statute is unduly severe in its categorical exclusion of all unsecured public-utility issues, its acceptance of railroad *income* issues on the same basis as railroad *debentures* is fully as objectionable for the opposite reason. The provisions of income bonds vary greatly among the different issues, the basic distinction being between those on which interest *must* be paid if earned and those over which the directors have a greater or lesser measure of discretion. Generally speaking, income bonds are allied more closely to preferred stocks than to ordinary fixed obligations. We shall consider them, accordingly, in our chapter on preferred stocks, in which we shall set forth the need for especial caution and strictness in the selection of this type of security for *straight investment*.

Standards of Safety Should Not Be Relaxed Because of Early Maturity.—Investors are inclined to attach considerable importance to the *maturity date* of an issue, because of its bearing on whether it is a short- or long-term security. A short maturity, carrying with it the right to repayment soon after purchase, is considered an advantageous feature from the standpoint of safety. Consequently, investors are prone to be less exacting in their standards when purchasing notes on bonds due in a short time (say, up to three years) than in their other bond selections.

In our opinion this distinction is unsound. A near maturity means a problem of refinancing for the company as well as a privilege of repayment for the investor. The noteholder cannot count on the mere fact of maturity to assure this repayment. The company must either have the cash available (which happens relatively seldom) or else an earning power and financial position which will permit it to raise new funds. Corporations frequently sell short-term issues because their credit is too poor at the time to permit of a long-term flotation at a reasonable rate. Such a practice frequently results in trouble for the company, and therefore for the investor, at maturity.

For these reasons we advise against the drawing of distinctions between long- and short-term issues, resulting in any relaxation of standards of safety in the selection of issues of the latter type.

RECORD OF INTEREST AND DIVIDEND PAYMENTS

Bonds purchased on an investment basis should have behind them a sufficiently long record of successful operation and of financial stability on the part of the issuer. New enterprises and those recently emerged from financial difficulties are not entitled to the high credit rating essential to justify a fixed-value investment. A similar disqualification would logically apply to states or municipalities which have failed to meet their obligations punctually at any time over a preceding period of years.

Provisions of New York Statute.—The New York statute recognizes this criterion and gives it concrete expression as follows: Bonds of states other than New York are eligible if the state has not defaulted on interest or principal payments during the previous *ten years.* For municipalities outside of New York State, the period is *twenty-five years;* for railroads, *six years;* for gas, electric, and telephone companies, *eight years.*

With respect to bonds of corporations, however, the requirements as to *earnings coverage*—to be discussed under the next heading—should adequately take care of the question of past record. The time covered by the earnings requirement is only a little shorter than the periods above suggested, and hence it would seem an unnecessary complication to exact a past-solvency test in addition to an earnings test.

Civil obligations, on the other hand, are not sold on the basis of an earnings record. Consequently the investor is compelled to attach primary importance to a satisfactory history of punctual payment. The requirement on this point set forth in the New York statute would no doubt appear reasonable to the average investor.

We cannot recommend such a rule of investment, however, without considering the results that would follow from its general adoption. If *all* purchases of municipal bonds required a clean record for 25 years, how could any township float a bond issue during the first quarter-century of its existence? And similarly, if a state or city has been driven into default, how will it finance itself during the 10 or 25 years, respectively, needed to restore its obligations to the eligible list? In the case of corporations, such financing might be accomplished on a speculative basis, through the sale of stock, or convertible bonds, or even bonds at a large discount. But such methods are not open to municipalities. The difficulty is met in actual practice by raising the coupon rate on the obligations of states or municipalities with inferior credit. Certain cities in Florida, for example, emerging from financial embarrassment, will perhaps be able to attract new funds by offering a 6% coupon rate in contrast with 3% paid by New York State. But this solution of the problem runs counter to the principle, previously developed, that a high coupon rate is not adequate compensation for the assumption of substantial risk of principal. In other words, it would be a mistake to buy a municipal obligation for its high yield, if it is recognized as inferior in grade and subject to more than a nominal possibility of default.

A Dilemma and a Suggested Solution.—We are faced therefore by a dilemma, since the theoretically correct attitude of the bond buyer would render impossible the necessary financing of many municipalities. Viewing the matter realistically, it may be dismissed with the observation that there will always be enough undiscriminating investors on hand to absorb the bonds of any town or village which offers a seemingly attractive rate. Consequently the logical and careful bond buyer can avoid such issues without fatal results to borrowers having second-rate credit.

This disposition of the dilemma is too cynical to be entirely satisfactory. The ideal solution would probably lie in setting up some especially stringent quantitative tests to compensate for the failure by a municipality to meet the twenty-five year requirement of punctual payment. If a city has fallen into financial difficulties, it must rehabilitate itself by reducing its expenditures, or by raising its tax rate and other revenue, or possibly by a compulsory scaling down of its debt, corresponding to a corporate reorganization. By such means the town may place its finances on an entirely new and sound basis entitling it to a satisfactory credit rating in spite of its previous default. But the prudent investor will accord such a credit rating only after a careful study of the financial exhibit, including such items as the relation of expenditures and total debt, on the one hand, to population, property values and revenues, on the other. The bond buyer should expect to obtain a higher than standard yield on municipal obligations of this character, in repayment *not for the assumption of special risk, but for the effort required to satisfy himself of the soundness of the issue.*

A similar attitude should be taken towards newly organized civil bodies, where only a short record of debt service is available.[1]

---

[1] The technique of analysis of state or municipal finances is elaborate and it does not lend itself to dependable short cuts. An adequate treatment of

The Dividend Record.—The statutes governing legal investments have traditionally laid great stress upon a satisfactory record of dividend payments by the issuing enterprise. In most states a bond is eligible only if the company has paid regular dividends in certain minimum amounts for at least five years. This requirement is evidently based on the theory that since corporations exist in order to pay dividends, only those which do in fact pay dividends may be said to be really successful and therefore suitable for bond investment.

Dividend Record Not Conclusive Evidence of Financial Strength.—It may not be denied that dividend-paying concerns as a class are more prosperous than non-dividend payers. But this fact would not in itself justify the summary condemnation of all the bonds of non-dividend-paying enterprises. An exceedingly strong argument *against* such a rule lies in the fact that the payment of dividends is only an *indication* of financial strength; and not only does it fail to afford any *direct advantage* to the bondholder, but it may often be injurious to his interests by reducing the corporation's resources. In actual practice the dividend provisions of the statutes governing legal investments have at times had consequences directly opposite to those intended. Railroad companies in a weak financial position have improvidently continued dividend payments for the particular purpose of maintaining their bonds on the eligible list, so that the very practice supposed to indicate strength behind the bond has in reality undermined its safety.

The Role of the Dividend Record in Bond Investment.—The evidence given by the balance sheet and income account must be regarded as a more dependable clue to the soundness of an enterprise than is the record of dividend payments. It seems best therefore to dispense with all hard and fast rules on the latter point in determining the suitability of bond issues for straight investment. But the failure of a company to pay dividends when the earnings appear satisfactory should properly cause an intending bond buyer to scrutinize the situation with

---

the subject would lie outside of the purview of this book or the competence of the authors. We refer the reader to treatments of the subject in standard works on investment such as Hastings Lyon, *Investment,* pp. 56–179, New York, 1926; Lawrence Chamberlain, and George W. Edwards, *The Principles of Bond Investment,* Part II, pp. 123–273, New York, 1927; Ralph E. Badger, *Investment: Principles and Practices,* pp. 594–684, New York, 1928.

more than usual care, in order to discover whether the policy of the directors is due to weak elements in the picture not yet reflected in the income account. We might also point out incidentally that the bonds of dividend-paying companies possess a certain mechanical advantage in that their owners may receive a definite and perhaps timely warning of impending trouble by the later passing of the dividend; and being thus placed on their guard, they may be able to protect themselves against serious loss. Bonds of non-dividend-paying concerns are at a certain disadvantage in this respect, but in our opinion this may be adequately offset by the exercise of somewhat greater caution on the part of the investor.

The New York statute is somewhat more progressive than those of other states in its treatment of the dividend question. Railroads are required alternatively either to have paid dividends of a certain amount in five out of the last six years, or failing this, to meet more stringent requirements as to coverage of fixed charges. Public-utility companies are required either to have paid certain dividends in each of the five preceding years, or else to have earned an amount equal thereto. This provision falls into the error of the other statutes by possibly impelling payment of unearned dividends. The progressive idea appears in the converse side of the provision, which permits nonpayment of dividends so long as they are earned.

RELATION OF EARNINGS TO INTEREST REQUIREMENTS

The present-day investor is accustomed to regard the ratio of earnings to interest charges as the most important specific test of safety. It is to be expected therefore that any detailed legislation governing the selection of bond investments would be sure to include minimum requirements in respect to this cardinal factor. Nevertheless the majority of the statutes cover this point in only a fragmentary and inadequate manner. The legislatures have relied to a considerable extent on their requirements as to the company's dividend record to assure a satisfactory earning power.[1] As we have just pointed out, this

---

[1] Vermont, for example, permits investment in bonds of New England railroads without any earnings test; in the case of other roads the fixed charges must not exceed 20% of the *gross* business. A record of continuous dividend payments is required in both cases.

criterion is open to serious objection. The superiority of the New York statute is manifest chiefly in two provisions: first, its recognition of the prime importance of an adequate earnings record; and secondly, its consistent treatment of a company's *total* fixed charges as an indivisible unit.

Requirements of the New York Law.—The requirements of the New York law with respect to *earnings coverage* may be summarized as follows:

In the case of railroad-mortgage bonds (or collateral-trust bonds equivalent thereto) and railroad-equipment obligations, the company must have earned its fixed charges $1\frac{1}{2}$ times in five out of the six years immediately preceding, and also in the latest year. If dividends have not been paid as stipulated, then the period is set at nine out of the ten preceding years.

In the case of other kinds of railroad bonds, *e.g.*, debentures, income obligations, etc., the fixed charges (including interest on income bonds, if any) must be earned *twice* in both the latest year and in five out of the six preceding years. In this category, the requirement as to dividend payments is apparently absolute, and no substitute therefor is admitted.

In the case of gas, electric, and telephone bonds, the *average* earnings for the past five years must have equalled *twice* the *average* total-interest charges, and the same coverage must have been shown in the latest year.

Three Phases of the Earnings Coverage : 1. Method of Computation.—In analyzing these statutory provisions, three elements deserve consideration. The first is the *method* of computing the earnings coverage; the second is the *amount* of coverage required; and the third is the *period* required for the test.

The Prior-deductions Method.—Various methods are in common use for computing and stating the relation of earnings to interest charges. One of these (which may be called the Prior-deductions Method) is thoroughly objectionable. Nevertheless it has been followed by the majority of issuing houses in their circulars offering junior bonds for sale, because it makes for a deceptively strong exhibit. The procedure consists of first deducting the prior charges from the earnings and then calculating the number of times the junior requirements are covered by the balance. The following illustration will show both the method itself and its inherent absurdity:

Company *A* has $10,000,000 of first-mortgage, 5% bonds and $5,000,000 of debenture 6% bonds.

Its average earnings are $1,400,000.
Deduct interest on first 5s   500,000 earned 2.8 times

Balance for debenture 6s $ 900,000
Interest on debenture 6s $ 300,000 earned 3 times

A circular offering the 6% debenture issue would be likely to state that "as shown above" the interest charges are covered three times. It should be noted, however, that the interest on the first 5s is covered only 2.8 times. The implication of these figures would be that the junior issue is better protected than the senior issue, which is clearly absurd. The fact is that the results shown for junior bonds by this prior-deductions method are completely valueless and misleading, and intelligent investors should make strenuous objections to its use in offering circulars.

The Cumulative-deductions Method.—The second procedure may be called the Cumulative-deductions Method. Under this method, interest on a junior bond is always considered in conjunction with prior and equivalent charges. In the example given, the interest on the debenture 6s would be computed as earned 1¾ times, found by dividing the combined charges of *both* issues, namely $800,000, into the available earnings of $1,400,000. The first-mortgage interest, however, would be said to be earned 2.8 times, since bond interest *junior* to the issue analyzed is left out of consideration in this method. The majority of investors would regard this point of view as entirely sound, and the procedure has been specifically prescribed by a number of states in their enactments governing the eligibility of bonds for savings-bank investment.[1]

The Total-deductions or "Over-all" Method.—In a previous chapter, however, we have emphasized the primary importance of a company's ability to meet *all* its fixed obligations, because insolvency resulting from default on a junior lien invariably reacts to the disadvantage of the prior-mortgage bondholders.

[1] See, for example, *Maine*, Sec. 27, Chap. 57 of Revised Statutes, as amended by Chap. 222 of Public Laws 1931, subsections VI, VII, and VIII, dealing with obligations of steam railroads, public utilities, and telephone companies. Similar provisions are to be found in the Vermont statute relative to public-utility bonds. New Hampshire permits the cumulative-deductions method for all corporate bonds except that, rather strangely, it requires the total-deductions method in the case of the bonds of telephone and telegraph companies.

An investor can be sure of his position only if the total-interest charges are well covered. Consequently, the conservative and therefore advisable way of calculating interest coverage should always be by the "total-deductions method," *i.e.*, the controlling figure should be the number of times that *all* interest charges are covered. This would mean that the *same earnings ratio would be used in analyzing all the bonds of any company*, whether they are senior or junior liens. In the example above given, the ratio would be 1¾, as applied to either the first 5s or the debenture 6s. In bond circulars and annual reports this method is now commonly referred to as the "over-all basis" for computing interest coverage.[1]

There is no reason, of course, why the coverage for a senior bond should not be computed by the cumulative-deductions method also, and if this coverage is very large it may properly be regarded as an added argument in favor of the issue. But our recommendation is that in applying any *minimum* requirement designed to test the company's strength, the total fixed charges should always be taken into account. The New York statute holds consistently to this very stand, and in our opinion it deserves to be approved and followed.

2. Minimum Requirements for Earnings Coverage.—The preference accorded by the New York statute to railroad bonds over public-utility issues is no longer justified, and the more recent record of both groups suggests that their relative positions should be reversed. It is necessary, also, to add a minimum figure for industrial bonds, which should clearly be set higher than for either utilities or rails. Taking these factors into account, we should recommend the following minimum requirements for the coverage of total fixed charges:

Public utilities	1¾ times
Railroads	2 times
Industrials	3 times

[1] The phrases: "earnings ratio," "times interest earned," and "earnings coverage," all have the same significance. The statement that "interest is covered 1¾ times" is more readily understood than the equivalent expression, sometimes used, that "the factor of safety is 75%," and we should advise the consistent use of the former type of expression. Some authorities (*e.g.*, Moody's "*Manual of Investments*" prior to 1930) have used the expression "margin of safety" to mean the ratio of the balance after interest to the earnings available for interest.

Example: If interest is covered 1¾ times the margin of safety becomes ¾ ÷ 1¾ = 42⁶⁄₇%.

3. The Period Comprised by the Earnings Test.—Our summary of the New York provisions regarding earnings coverage pointed out that the five-year average is used in the case of utility issues. For railroad bonds, however, the stipulated minimum margin must be shown in five separate years out of the latest six. In all instances, the minimum must be met in the year immediately preceding the date of investment.

Requirements such as the last two are easy to promulgate, but they are poorly suited to the realities of bond investment in an economic world subject to recurring years of serious depression. If it should be characteristic of business in general to experience eight prosperous or average years followed by two unprofitable ones, the effect of these rules would be to encourage investment in bonds (at high prices) during good times, and to impel their sale (at low prices) during depressions.

In our view, the only practical *rigid* application of a minimum-earnings standard must be to the *average* results over a period of time. A five-year average, as prescribed by the statute in the case of public-utility bonds, would seem too short under many circumstances, and we should suggest a *seven-year period* as a more suitable standard.

Other Phases of the Earnings Record.—There are, of course, a number of other aspects of the earnings picture to which the investor would do well to pay attention. Among these are the *trend*, the *minimum* figure, and the *current* figure. The importance of each of these cannot be gainsaid, but they do not lend themselves effectively to the application of hard and fast rules. In this case, as in the matter of mortgage security previously discussed, a distinction must be drawn between the few factors which can successfully be embraced by definite and universally applicable rules, and the many other factors which resist such exact formulation but must nevertheless be taken into account by the *judgment* of the investor.

Unfavorable Factors May Be Offset.—The practical method of dealing with elements of the latter type may be illustrated in this matter of the earning exhibit. The investor must *demand* an average at least equal to the minimum standard. In addition, he will be *attracted* by: (*a*) a rising trend of profits; (*b*) an especially good current showing; and (*c*) a satisfactory margin over interest charges in *every* year during the period studied. If a bond is deficient in any one of these three aspects, the result should not necessarily be to condemn the issue, but rather to exact an average earnings coverage well in excess of the minimum, and to require closer attention to the general or qualitative elements in the situation.   If the trend has been unfavorable, or the latest figure alone has been decidedly poor, the investor should certainly not accept the bond unless the average earnings have been substantially above the minimum requirement—and unless also he has reasonable grounds for believing that the downward trend or the current slump is not likely to continue indefinitely.   Needless to say, the *amount* by which the average must be advanced in order to offset an unfavorable trend or current exhibit is a matter within the discretion of the investor to determine, and cannot be developed into any set of mathematical formulas.

CHAPTER X

SPECIFIC STANDARDS FOR BOND INVESTMENT
(Continued)

THE RELATION OF THE VALUE OF THE PROPERTY TO THE FUNDED DEBT

In our earlier discussion (Chap. VI) we pointed out that the soundness of the typical bond investment depends upon the ability of the obligor corporation to take care of its debts, rather than upon the value of the property on which the bonds have a lien. This broad principle naturally leads directly away from the establishment of any *general* tests of bond safety based upon the value of the mortgaged assets, where this value is considered apart from the success or failure of the enterprise itself.

Stating the matter differently, we do not believe that in the case of the ordinary corporation bond—whether railroad, utility, or industrial—it would be advantageous to stipulate any minimum relationship between the value of the physical property pledged (taken at either original or reproduction cost) and the amount of the debt. In this respect we are in disagreement with statutory provisions in many states (including New York) which reflect the traditional emphasis upon property values. The New York law, for example, will not admit as eligible a gas, electric, or telephone bond, unless it is secured by property having a value 66⅔% in excess of the bond issue. This value is presumably book value, which either may be the original dollar cost less depreciation, or it may be some more or less artificial value set up as a result of transfer or reappraisal.

Special Types of Obligations: 1. Equipment Obligations.—It is our view that the book value of public-utility properties—and of railroads and the typical industrial plant as well—is no guidance in determining the safety of the bond issues secured thereon. There are, however, various *special* types of obligations, the safety of which is in great measure dependent upon the assets securing them, as distinguished from the going-

111

concern value of the enterprise as a whole. The most characteristic of these, perhaps, is the railroad-equipment trust certificate, secured by title to locomotives, freight cars, or passenger cars, and by the pledge of the lease under which the railroad is using the equipment. The investment record of these equipment obligations is very satisfactory, particularly because even the most serious financial difficulties of the issuing road have very rarely prevented the prompt payment of interest and principal.[1] The primary reason for these good results is that the specific property pledged is removable and usable by other carriers. Consequently it enjoys an independent salable value, similar to automobiles, jewelry, and other chattels on which personal loans are made. Even where there might be great difficulty in actually selling the rolling stock to some other railroad at a reasonable price, this mobility still gives the equipment obligation a great advantage over the mortgages on the railroad itself. Both kinds of property are essential to the operation of the line, but the railroad bondholder has no alternative save to permit the receiver to operate his property, while the holder of the equipment lien can at least threaten to take the rolling stock away. It is the possession of this *alternative* which in practice has proved of prime value to the owner of equipment trusts because it has virtually compelled the holders even of the first mortgages on the road itself to subordinate their claim to his.

It follows that the holder of equipment-trust certificates has two separate sources of protection, the one being the credit and success of the borrowing railway, the other being the value of the pledged rolling stock. If the latter value is sufficiently in excess of the money loaned against it, he may be able to ignore the first or credit factor entirely, in the same way as a pawnbroker ignores the financial status of the individual to whom he lends money and is content to rely exclusively on the pledged property.

The conditions under which equipment trusts are usually created supply a substantial degree of protection to the purchaser. The legal forms are designed to facilitate the enforcement of the lienholder's rights in the event of nonpayment. In practically all cases at least 20% of the cost of the equipment is

---

[1] See Appendix, Note 15, for information on the investment record of such issues.

provided by the railway, and consequently the amount of the equipment obligations is initially not more than 80% of the value of the property pledged behind them. The principal is usually repayable in 15 equal annual installments, beginning one year from issuance, so that the amount of the debt is reduced more rapidly than ordinary depreciation would require.

The protection accorded the equipment-trust holder by these arrangements has been somewhat diminished in recent years, due partly to the drop in commodity prices which has brought reproduction (and therefore, salable) values far below original cost, and also to the reduced demand for equipment, whether new or used, because of the smaller traffic handled. During the 1930–1933 depression certain railroads in receivership (*e.g.*, Seaboard Air Line and Wabash) required holders of maturing equipment obligations to extend their maturities for a short period. These maneuvers suggest that the claim of "almost absolute safety" frequently made on behalf of equipment issues will have to be moderated somewhat; but it cannot be denied that this form of investment enjoys a positive and substantial advantage through the realizability of the pledged assets.[1]

2. **Collateral-trust Bonds.**—Collateral-trust bonds are obligations secured by the pledge of stocks or other bonds. In the typical case, the collateral consists of bonds of the obligor company itself, or of the bonds or stocks of subsidiary corporations. Consequently the realizable value of the collateral is usually dependent in great measure on the success of the enterprise as a whole. But in the case of the collateral-trust issue of investment companies, a development of recent years, the holder may be said to have a primary interest in the market value of the pledged securities, so that it is quite possible that by virtue of the protective conditions in the indenture, he may be completely taken care of under conditions which mean virtual extinction for the stockholders. This type of collateral-trust bond may therefore be ranked with equipment-trust obligations as exceptions to our general rule that the bond buyer must place his chief reliance on the success of the enterprise and not on the property specifically pledged.

Going behind the form to the substance, we may point out that this characteristic is essentially true also of investment-

[1] See Appendix, Note 16, for comment and supporting data.

trust *debenture* obligations. For it makes little practical difference whether the portfolio is physically pledged with a trustee, as under a collateral-trust indenture, or whether it is held by the corporation subject to the claim of the debenture bondholders. In the usual case the debentures are protected by adequate provisions against increasing the debt, and frequently also by a covenant requiring the market price of the company's assets to be maintained at a stated percentage above the face amount of the bonds.

Example: The Reliance Management Corporation Debenture 5s, due 1954, are an instance of the working of these protective provisions. The enterprise as a whole was highly unsuccessful, as is shown vividly by a decline in the price of the stock from 69 in 1929 to 1 in 1933. In the case of the ordinary bond issue, such a collapse in the stock value would have meant almost certain default and large loss of principal. But here the fact that the assets could be readily turned into cash gave significance to the protective covenants behind the debentures. It made possible and compelled the repurchase by the company of more than three-quarters of the issue, and it even forced the stockholders to contribute additional capital to make good a deficiency of assets below the indenture requirements. This resulted in the bonds selling as high as 88 in 1932 when the stock sold for only $2\frac{1}{2}$.

In Chap. XVIII, devoted to protective covenants, we shall discuss the history of a collateral-trust bond issue of an investment company (Financial Investing Company), and we shall point out that the intrinsic strength of such obligations is often impaired—unnecessarily, in our opinion—by hesitation in asserting the bondholders' rights.

3. Real-estate Bonds.—Of much greater importance than either of the two types of securities just discussed is the large field of real-estate mortgages and real-estate mortgage bonds. The latter represent participations of convenient size in large individual mortgages. There is no doubt that in the case of such obligations the value of the pledged land and buildings is of paramount importance. The ordinary real-estate loan made by an experienced investor is based chiefly upon his conclusions as to the fair value of the property offered as security. It seems to us, however, that in a broad sense the values behind real-estate mortgages are *going-concern values, i.e.,* they are derived fundamentally from the earning power of the property, either actual or presumptive. In other words, the value of the pledged asset is not something *distinct* from the success of the enterprise (as is possibly the case with a railroad-equipment trust certificate), but is rather *identical* therewith.

This point may be made clearer by a reference to the most typical form of real-estate loan, a first mortgage on a single-family dwelling house. Under ordinary conditions a home costing $10,000 would have a rental value (or an equivalent value to an owner-tenant) of some $1,200 per year, and would yield a net income of about $800 after taxes and other expenses. A 6% first-mortgage loan on the savings-bank basis, *i.e.*, 60% of value, or $6,000, would therefore be protected by a normal *earning power* of over twice the interest requirements. Stated differently, the rental value could suffer a reduction of over one-third before the ability to meet interest charges would be impaired. Hence the mortgagee reasons that regardless of the ability of the then owner of the house to pay the carrying charges, he could always find a tenant or a new purchaser who would rent or buy the property on a basis at least sufficient to cover his 60% loan. (By way of contrast, it may be pointed out that a typical *industrial plant*, costing $1,000,000 and bonded for $600,000, could not be expected to sell or rent for enough to cover the 6% mortgage if the issuing company went into bankruptcy.)

Property Values and Earning Power Closely Related.—This illustration shows that under normal conditions obtaining in the field of *dwellings, offices*, and *stores*, the property values and the rental values go hand in hand. In this sense it is largely immaterial whether the lender views mortgaged property of this kind as something with salable value or as something with an earning power, the equivalent of a going concern. To some extent this is true also of vacant lots and unoccupied houses or stores, since the salable value of these is closely related to the *expected* rental when improved or let. (It is emphatically not true, however, of buildings erected for a special purpose, such as factories, etc.)

Misleading Character of Appraisals.—The foregoing discussion is important in its bearing on the correct attitude that the intending investor in real-estate bonds should take towards the property values asserted to exist behind the issues submitted to him. During the great and disastrous development of the real-

estate mortgage-bond business between 1923 and 1929, the only
datum customarily presented to support the usual bond offering—
aside from an estimate of future earnings—was a statement of
the *appraised value* of the property, which almost invariably
amounted to some 66⅔% in excess of the mortgage issue. If
these appraisals had corresponded to the salable values which
experienced buyers of or lenders on real estate would place upon
the properties, they would have been of real utility in the selection
of sound real-estate bonds. But unfortunately they were purely
artificial valuations, to which the appraisers were willing to
attach their names for a fee, and whose only function was to
deceive the investor as to the protection which he was receiving.

The method followed by these appraisals was the capitalization on a liberal basis of the rental expected to be returned
by the property. By this means, a typical building which cost
$1,000,000, including liberal financing charges, would immedi-
ately be given an "appraised value" of $1,500,000. Hence a
bond issue could be floated for almost the entire cost of the
venture so that the builders or promoters retained the equity
(*i.e.*, the ownership) of the building, without a cent's investment,
and in many cases with a goodly cash profit to boot.[1] This
whole scheme of real-estate financing was honeycombed with
the most glaring weaknesses, and it is a sad commentary on the
lack of principle, penetration, and ordinary common sense on
the part of all parties concerned that it was permitted to reach
such gigantic proportions before the inevitable collapse.[2]

Abnormal Rentals Used as Basis of Valuation.—It was indeed
true that the scale of rentals prevalent in 1928–1929 would
yield an abundantly high rate of income on the cost of a new
real-estate venture. But this condition could not properly
be interpreted as making a new building immediately worth
50% in excess of its actual cost. For this high income return
was certain to be only temporary, since it could not fail to
stimulate more and more building, until an oversupply of space

---

[1] The 419–4th Avenue Corporation (Bowker Building) floated a $1,230,000
bond issue in 1927 with a paid-in capital stock of only $75,000. (By the
familiar process, the land and building which cost about $1,300,000 were
appraised at $1,897,788.) Default and receivership in 1931–1932 were
inevitable.

[2] See Appendix, Note 17, for report of Real Estate Securities Committee
of the Investment Bankers Association of America commenting on defaults
in this field.

caused a collapse in the scale of rentals. This overbuilding was the more inevitable because it was possible to carry it on without risk on the part of the owner, who raised all the money needed from the public.

Debt Based on Excessive Construction Costs.—A collateral result of this overbuilding was an increase in the cost of construction to abnormally high levels. Hence even an apparently conservative loan made in 1928 or 1929, in an amount not exceeding two-thirds of *actual cost*, did not enjoy a proper degree of protection, because there was the evident danger (subsequently realized) that a sharp drop in construction costs would reduce fundamental values to a figure below the amount of the loan.

Weakness of Specialized Buildings.—A third general weakness of real-estate-bond investment lay in the entire lack of discrimination as between various types of building projects. The typical or standard real-estate loan was formerly made on a home, and its peculiar virtue lay in the fact that there was an indefinitely large number of prospective purchasers or tenants to draw upon, so that it could always be disposed of at some moderate concession from the current scale of values. A fairly similar situation is normally presented by the ordinary apartment house, or store, or office building. But when a structure is built for some *special* purpose, such as a hotel, garage, club, hospital, church, or factory, it loses this quality of rapid disposability, and *its value becomes bound up with the success of the particular enterprise for whose use it was originally intended.* Hence mortgage bonds on such structures are not actually real-estate bonds in the accepted sense, but rather *loans extended to a business;* and consequently their safety must be judged by all the stringent tests surrounding the purchase of an industrial obligation.

This point was completely lost sight of in the rush of real-estate financing preceding the collapse in real-estate values. Bonds were floated to build hotels, garages, and even hospitals, on very much the same basis as loans made on apartment houses. In other words, an appraisal showing a "value" of one-half to two-thirds in excess of the bond issue was considered almost enough to establish the safety of the loan. It turned out, however, that when such new ventures proved commercially unsuccessful and were unable to pay their interest charges, the "real-estate" bondholders were in little better position than the holders of a mortgage on an unprofitable railroad or mill property.[1]

Values Based on Initial Rentals Misleading.—Another weakness should be pointed out in connection with apartment-house financing. The rental income used in determining the appraised value was based on the rentals to be charged at the outset. But apartment-house tenants are accustomed to pay a substantial premium for space in a new building, and they consider a structure old, or at least no longer especially modern and desirable, after it has been standing a very few years. Consequently, under normal conditions the rentals received in the first years are substantially larger than those which can conservatively be expected throughout the life of the bond issue.

Suggested Rules of Procedure.—From this detailed analysis of the defects of real-estate-bond financing in the past decade, a number of specific rules of procedure may be developed to guide the investor in the future.

In the case of single-family dwellings, loans are generally made directly by the mortgage holder to the owner of the home, *i.e.,* without the intermediary of a real-estate mortgage *bond* sold by a house of issue. But an extensive business has also been transacted by mortgage companies (*e.g.,* Lawyers Mortgage Company, Title Guarantee and Trust Company) in guaranteed mortgages and mortgage-participation certificates, secured on such dwellings.

Where investments of this kind are made, the lender should be certain: (*a*) that the amount of the loan is not over 66⅔% of the value of the property, as shown either by actual recent cost or by the amount which an experienced real-estate man would consider a fair price *to pay* for the property; and (*b*) that this cost or fair price does not reflect recent speculative inflation and does not greatly exceed the price levels existing for a long period previously. If so, a proper reduction must be made in the maximum relation of the amount of mortgage debt to the current value.

The more usual real-estate mortgage *bond* represents a participation in a first mortgage on a new apartment house or office building. In considering such offerings the investor should ignore the conventional "appraised values" submitted and

[1] See Appendix, Note 18, for example (Hudson Towers).

demand that the actual cost, fairly presented, should exceed the amount of the bond issue by at least 50%. Secondly, he should require an estimated income account, conservatively calculated to reflect losses through vacancies and the decline in the rental scale as the building grows older. This income account should forecast a margin of at least 100% over interest charges, after deducting from earnings a depreciation allowance to be actually expended as a sinking fund for the gradual retirement of the bond issue.

Issues termed "first-*leasehold* mortgage bonds" are in actuality second mortgages. They are issued against buildings erected on leased land and the ground rent operates in effect as a first lien or prior charge against the entire property. In analyzing such issues the ground rent should be added to the bond-interest requirements to arrive at the total interest charges of the property. Furthermore, it should be recognized that in the field of real-estate obligations the advantage of a first mortgage over a junior lien is much more clean-cut than in an ordinary business enterprise.[1]

In addition to the above quantitative tests, the investor should be satisfied in his own mind that the location and type of the building are such as to attract tenants and to minimize the possibility of a large loss of value through unfavorable changes in the character of the neighborhood.[2]

Real-estate loans should not be made on buildings erected for a special or limited purpose, such as hotels, garages, etc. Commitments of this kind must be made in the venture itself, considered as an individual business. From our previous discussion of the standards applicable to a high-grade industrial-bond purchase, it is difficult to see how any bond issue on a new hotel, or the like, could logically be bought on a straight invest-

[1] See Appendix, Note 19, for examples and comment.

[2] One of the few examples of a conservatively financed real-estate-bond issue extant in 1933 is afforded by the Trinity Buildings Corporation of New York First 5½s, due 1939, secured on two well-located office buildings in the financial district of New York City. This issue was outstanding in the amount of $4,300,000, and was secured by a first lien on land and buildings assessed for taxation at $13,000,000. In 1931, gross earnings were $2,230,-000 and the net after depreciation was about six times the interest on the first-mortgage bonds. In 1932, rent income declined to $1,653,000, but the balance for first-mortgage interest was still about 3½ times the requirement. In September 1933 these bonds sold close to par.

ment basis. All such enterprises should be financed at the outset by private capital, and only after they can show a number of years of successful operation should the public be offered either bonds or stock therein.[1]

[1] The subject of guaranteed real-estate mortgage issues is treated in Chap. XVII.

CHAPTER XI

SPECIFIC STANDARDS FOR BOND INVESTMENT
(*Concluded*)

RELATION OF STOCK CAPITALIZATION TO BONDED DEBT

The amount of stock and surplus following or junior to a bond issue expresses the same fact as the excess of resources over indebtedness. This can be seen at once from the following condensed typical balance sheet:

Assets, less current liabilities (net assets)........ $1,000,000

	$1,000,000
Bonded debt	$ 600,000
Stock and surplus (stock equity)	400,000
	$1,000,000

The resultant simple formula is as follows:

$$\frac{\text{Stock equity}}{\text{Bonded debt}} = \frac{\text{net assets}}{\text{bonded debt}} - 1$$

Standards Prescribed by the New York Law.—If we are studying balance-sheet figures, therefore, we can look either at the net assets or at the stock equity to determine the indicated coverage or margin above the principal amount of the debt. The New York statute governing investments of savings banks employs both approaches in its regulations with respect to public-utility bonds. It stipulates: (1) that the mortgage debt in question, plus all underlying mortgage debt, shall not exceed 60% of the value of the mortgaged property; and (2) that the capital stock shall be equal to at least two-thirds of the mortgage debt. It will readily be observed from the typical balance sheet just given that these two requirements are broadly equivalent. Where a company has a substantial *unsecured* indebtedness, however, it might meet requirement 1 and not requirement 2, so that in such cases the second stipulation supplies an added protection. This point may be illustrated by the following example:

121

Mortgaged property..	$10,000,000	Mortgage debt.......	$ 6,000,000
Working capital......	1,000,000	Debentures..........	3,000,000
(Unmortgaged).......	....	Stock and surplus.....	2,000,000
	$11,000,000		$11,000,000

In this case the mortgage debt is only 60% of the pledged property but the stock equity is much less than two-thirds of the mortgage debt. Hence the latter bonds would not be eligible.

It should be noted that the New York statute considers only the par or stated value of the stock issues (including, of course, both preferred and common), and it does not give credit for the book surplus, which is part of the stockholders' equity. The theory behind this restriction may be that the surplus is legally distributable to the stockholders, and cannot therefore be counted on as a permanent protection for the bondholders. In actuality, however, a utility company's surplus is almost invariably invested to a large extent in fixed assets and is not distributable in cash. Hence, if tests of this kind are to be required, the *stock-and-surplus* figure would appear more logical than the stock issue alone.

Equity Test of Doubtful Merit in the Case of Utilities.—We are inclined to question whether any substantial advantage is gained in the ordinary case by applying the property or stock-issue test to public-utility bonds. It is unlikely to give any indication of safety or lack of safety not already shown by the earnings record. In some few instances, perhaps, the income exhibit may be satisfactory but the asset coverage unduly small, and the latter point may suggest that since the company is earning an exceptionally high rate on its investment, it is vulnerable to unfavorable rate regulation. The primary difficulty, however, lies in the lack of dependability of the balance-sheet figures of property values (and hence of stock equity) as an indication either of the actual cash investment or of the reproduction value which may be designated as the rate base. Under present conditions, moreover, there appears to be no sound reason for exacting a property-value or stock-equity test for public-utility bonds and none for railroad bonds.

There is, of course, no objection to the application of this stock-equity test (based on book figures) to both railroad and public-utility obligations, as an added precaution, either regularly, or in special cases where there is reason to doubt the reliability of the earnings record as a measure of the future ability to meet bond interest. If this test is applied, it should be pointed out that a maximum ratio of 60% of debt to 40% of stock and surplus is proportionately more severe than a minimum earnings ratio of 1¾ times interest charges. It would be more consistent, therefore, to admit a bonded debt as high as 75% of the property value, or three times the amount of the stock and surplus.

Importance of a Real-value Coverage behind a Bond Issue.— Our principal objection to the property-value criterion arises from the undoubted fact that the *book valuations of fixed assets* are practically worthless as indications of the safety of a bond. But on the other hand we are convinced that a substantial margin of *going-concern value* over funded debt is not only important but even vitally necessary to assure the soundness of a fixed-value investment. Before paying standard prices for bonds of any enterprise, whether it be a railroad, a telephone company, or a department store, the investor must be convinced that the business is worth a great deal more than it owes. In this respect the bond buyer must take the same attitude as the lender of money on a house or a diamond ring, with the important difference that it is *the value of the business as an entity* which the investor must usually consider, and not that of the separate assets.

Going-concern Value and Earning Power.—"The value of the business as an entity" is most often entirely determined by its earning power. This explains the overshadowing significance that has come to be attached to the income exhibit, for the latter reveals not only the ability of the company to meet its *interest* charges, but also the extent to which the going value of the business may be said to exceed the *principal* of the bond issue. It is for this reason that most investors have come to regard the earnings record as the only statistical or quantitative test necessary in the selection of bond issues. All other criteria commonly employed are either qualitative or subjective (*i.e.*, involving personal views as to the management, prospects, etc.).

While it is desirable to make the tests of safe bonds as simple and as few as possible, their reduction to the single criterion of the margin of earnings over interest charges would seem to be a dangerous oversimplification of the problem. The earnings during the period examined may be nonrepresentative, either because they resulted from definitely temporary conditions, favorable or the reverse, or because they were presented in such a way as not to reflect the true income. These conditions are particularly likely to occur in the case of industrial companies, which are subject both to greater individual vicissitudes and to a smaller degree of accounting supervision than is true of railroads and utilities.

Shareholders' Equity Measured by Market Value of Stock Issues—a Supplemental Test.—We feel, therefore, that it is essential, in the case of industrial bonds at least, to supplement the earnings test by some other quantitative index of the margin of going-concern value above the funded debt. The best criterion that we are able to offer for this purpose is the ratio of the *market value* of the capital stock to the total funded debt. Strenuous objections may, of course, be leveled against using the market price of stock issues as a proof of anything, in view of the extreme and senseless variations to which stock quotations are notoriously subject. Nevertheless, with all its imperfections, the market value of the stock issues is generally recognized as a better index of the fair going value of a business than is afforded by the balance-sheet figures or even the ordinary appraisal.[1]

Note carefully that we are proposing the use of stock prices for the restricted purpose only of ascertaining whether or not a substantial equity exists behind the bond issue. This is by no means tantamount to stating that the price is always an exact measure of the fair or intrinsic value. The market-price test is suggested as a *rough index* or clue to the existing values, and it is to be employed only as a supplement—but an important supplement—to the more carefully scrutinized figures supplied by the earnings record.

The utility of the market-price test in extreme cases is unquestionable. The presence of a stock equity with market value many times as large as the total debt carries a strong assurance of the safety of the bond issue,[2] and conversely, an exceedingly small stock equity at market prices must call the soundness of the bond into serious question. The determination of the

---

[1] The liquidating value, arising chiefly from the net quick assets, may at times exceed the market price, but this point is seldom of significance in the selection of *high-grade investments*.

[2] See our discussion of Fox Film Corporation 6% Notes, as of December 1933 in Chap. L.

EXAMPLES OF NORMAL RELATIONSHIP BETWEEN INTEREST COVERAGE AND STOCK-VALUE RATIO

Item	North American Co.	Atchison, Topeka & Santa Fe Railway Co.	Pillsbury Flour Mills, Inc.
Year ended..............	June 30, 1932	June 30, 1932	June 30, 1932
Balance for interest charges.......	$ 46,000,000	$ 31,200,000	$1,236,000
Interest charges.......	24,600,000*	13,000,000	406,000
Times earned........	1.87	2.40	3.04
Balance for dividends........	$ 21,400,000	$ 18,200,000	$ 830,000
Preferred dividends........	1,800,000	6,200,000	
Balance for common........	19,600,000	12,000,000	830,000
Per share........	$2.73	$5.00	$1.51
Bonded debt........	$450,000,000*	$310,000,000	$6,770,000
Preferred stock at market........	600,000 sh. @ 45—$ 27,000,000	1,240,000 sh. @ 67—$ 83,000,000	
Common stock at market........	7,170,000 sh. @ 33— 237,000,000	2,430,000 sh. @ 50— 121,500,000	550,000 sh. @ 15—$8,250,000
Total stock at market........	$264,000,000	$204,500,000	$8,250,000
Ratio of stock to bonds........	0.59 to 1	0.66 to 1	1.22 to 1

* Includes preferred stock of subsidiary companies and requirements thereof (see Chap. XIII).

market value of the stock equity, and its comparison with the total amount of funded debt, is a well-established feature of bond analysis, and it is frequently included in bond-offering circulars (when the showing made is satisfactory). We recommend that this calculation be made a standard element in the procedure of bond selection, at least for industrial issues; and that minimum requirements under this heading be set up which will serve as a secondary quantitative test of safety.

Minima for the Stock-equity Test.—What should be the normal minimum relationship between stock values and funded debt? To assist in answering this question, we present a summarized exhibit of a public utility, a railroad, and an industrial company, as of September 1932. Although market conditions at that time were in general abnormal, the examples were so selected as to provide price-earning relationships approximating those obtaining prior to the inflation of 1928–1929, except that a somewhat more liberal valuation is now deservedly placed on the per-share earnings of light and power common stocks.

The examples shown on page 125 illustrate a rough arithmetical proportion which usually exists between the interest coverage on the one hand and the stock-to-bond ratio on the other. This is set forth in the following table:

Type of enterprise	Minimum number of times fixed charges earned (Interest coverage)	Minimum ratio of stock value to bonded debt (Stock-value ratio)
Public utilities.	1¾	$1 of stock to $2 of bonds
Railroads......	2	$1 of stock to $1.50 of bonds
Industrials.....	3	$1 of stock to $1 of bonds

As we have previously intimated, if the above relationships *always* obtained, there would be no reason to apply both tests, since the passing of one would assure the passing of the other. Such is not the case, however, and we must accordingly consider what is implied when the stock-value ratio gives a substantially different indication from the interest coverage.

Significance of Unusually Large Stock-value Ratio.—Let us assume first that the interest coverage is just above the minimum but the stock-value ratio is considerably higher than the minimum. The result is obviously to strengthen the fairly favorable impression gathered from the earnings exhibit alone, and therefore to instill correspondingly greater confidence in the soundness of the issue.

Example: P. Lorillard Company Debenture 5s, due 1951 and 7s, due 1944.

Interest charges earned 1931...................			3.7 times
5-year average (1927–1931)...................			2.5 times
7-year average (1925–1931)...................			3.2 times
Total bonded debt......................................			$19,800,000
Stock value, Sept. 1932.......	113,000 sh. of Pfd.	@ 105	$11,900,000
	1,900,000 sh. of Com.	@ 17	32,300,000
		Total stock equity.......	$44,200,000
Stock-value ratio......................................			2¼ to 1

Here the market value of the shareholders' equity is a distinctly reassuring sign and would undoubtedly offset any question raised by the rather substandard showing made by the five-year average of interest coverage.

Significance of a Subnormal Stock-value Ratio.—The opposite case is that in which the interest coverage may be called satisfactory but the stock-value ratio is substantially below the minimum requirement.

Examples: The problem presented here can be better understood by a study of the appended exhibit of Inland Steel Company First 4½s in September 1932 and its comparison with the exhibits of two other bond issues.

From the standpoint of the interest coverage, the showing of the Inland Steel bonds might well be considered adequate. While the 1931–1932 results are bad, no better could be expected under the general conditions then prevailing. In line with the rule suggested in Chap. IX, the high coverage in normal years and the 6½-year average well exceeding the minimum requirement may be accepted as compensating for the current losses.

The stock-value ratio in the case of Inland Steel, however, is not satisfactory, since there was considerably less than a dollar of stock at market price for each dollar of bonds. It may be objected that the price of Inland Steel shares was abnormally low, due to the general undervaluation of stocks in 1932, and that it was no more indicative of the soundness of Inland Steel bonds than were the 1932 earnings. This *may* be true, of course, but the difficulty is that the bond buyer is required to be *sure*

COMPARATIVE EXHIBIT OF THREE BOND ISSUES

Item	Inland Steel 4½s, due 1978 and 1981 Price 82, yield 5.6 %	Crucible Steel 5s, due 1940 Price 60, yield 13.4 %	General Baking 5¼s, due 1940 Price 97, yield 6 %
Annual interest charge..........	$ 1,890,000	$ 675,000	$ 313,500
Earned for interest by years:			
1932 (first half)...........	496,000(d)	1,348,000(d)	2,271,000
1931...................	3,126,000	1,539,000(d)	5,151,000
1930...................	7,793,000	4,542,000	5,433,000
1929...................	13,042,000	8,364,000	7,240,000
1928...................	10,569,000	5,849,000	7,597,000
1927...................	7,482,000	5,844,000	7,784,000
1926...................	7,851,000	6,787,000	6,321,000
6½-year average..........	$ 7,595,000	$ 4,400,000	$ 6,430,000
Interest coverage..........	4.6 times*	7.1 times*	20.1 times
Bonded debt..........	$42,000,000	$13,500,000	$ 5,700,000
Stock value:			
Preferred..........	....	250,000 sh. @ 30 = $7,500,000	91,000 sh. @ 110 = $10,000,000
Common...........	1,200,000 sh. @ 20 = $24,000,000	450,000 sh. @ 17 = $7,650,000	1,595,000 sh. @ 15 = $24,000,000
Total stock value...........	$24,000,000	$15,100,000	$34,000,000
Stock-value ratio...........	0.57 to 1	1.12 to 1	6 to 1

* Adjusted for changes in the funded debt during the period.

that the stock is undervalued in order to justify his choice of the bond. To say that you will buy a company's bond for investment because you are certain that its stock is selling too low, appears to us to be a basically unsound piece of reasoning. It is a case of "heads you lose, tails you don't win." If you are right in your judgment of the stock value, it would certainly be more profitable to buy the stock than the bonds; if you are wrong about the shares, *i.e.*, if they are not really undervalued, then you have made an unattractive bond investment. If stock buying is outside your program, as well it might be, then sound logic requires that you turn your attention to the bonds of some other company which do not require a decision on your part that the stock price is unduly low in order to justify the purchase of the bond issue.

The Crucible Steel and General Baking exhibits are presented along with that of Inland Steel in order to support our reasoning that the purchase of the latter's bonds on a 5.60% basis in September 1932 should not have appeared particularly attractive at that time. It will be noted that the Crucible Steel bonds show up on the whole about as well from the standpoint of earnings, and definitely better as far as the stock-value ratio is concerned. Yet the amortized yield on the Crucible issue was more than twice as high as that obtainable from the Inland Steel 4½s. However, since the Inland Steel Company enjoys an unusually favorable reputation, some investors are likely to insist that they would rather intrust their money to Inland Steel than to Crucible Steel regardless of the comparative statistical showing. These investors' judgment *may* turn out to be entirely right on this point; but the difficulty is again that to justify the purchase of Inland Steel bonds, they must assume that the market is *both* valuing Inland Steel shares too low, and valuing Crucible Steel shares too high. This would appear to us to be entirely too complicated and doubtful an assumption to have to make in connection with so relatively simple a matter as the investment of money at a 5½% return during 1932.

Turning to the comparison between Inland Steel 4½s and General Baking 5½s, showing about the same yield, we cannot fail to be impressed by the enormous statistical superiority of the latter issue, as shown by both the earnings coverage and stock-value-ratio tests. The General Baking example is given to establish the point that even under the subnormal conditions of September 1932 the investor could still find reasonably yielding issues which measured up to the strictest quantitative tests for bond selection.

We advert once more to our controlling principle that the bond investment is a *negative* art. This discussion was not intended to imply that the Inland Steel 4½s were a poor investment—the contrary is in all probability the case—but we wish to point out that a logical examination of the picture at the time would not have led to an affirmative verdict for that issue.

Stock-value Ratio for Railroad and Public-utility Companies.— In the case of industrial companies the stock-value ratio may be easily calculated. Railroads and public utilities, however, are likely to present various complications. In addition to the bonded debt as shown in the balance sheet, it may also be necessary to consider rental obligations equivalent to debt and preferred stocks of subsidiaries ranking ahead of parent company bonds. These difficulties militate somewhat against the use of the stock-value ratio test for railroad and utility bonds. It is true also that this test is not quite so necessary in these two fields as in the case of industrial companies, for reasons mentioned previously in this chapter. If a careful investor desires to apply the stock-value ratio test to utility and railroad issues, he may do so most satisfactorily by *capitalizing the fixed charges*. This procedure will be described in the next chapter, which deals with special factors in the analysis of railroad and utility bonds.

Stock-value Test Not to Be Modified to Reflect Changing Market Conditions.—The question arises: To what extent should the stock-value-ratio test be modified to reflect changing market conditions? It would seem proper to expect, and therefore to demand, a higher relative market value for the stock behind a bond issue when times are good than during a depression. If $1 of stock to $1 of bonds is taken as the "normal" requirement for an industrial company, would it not be sound to demand, say, a $2-to-$1 ratio when stock prices are inflated, and conversely to be satisfied with a 50-cent-to-$1 ratio when quotations are far below intrinsic values? But this suggestion is impracticable for two reasons, the first being that it implies that the bond buyer can recognize an unduly high or low level of stock prices, which is far too complimentary an assumption. The second is that it would require bond investors to act with especial caution when things are booming and with greater confidence when times are hard. This is a counsel of perfection which it is not in human nature to follow. Bond buyers are people, and they cannot be expected to escape entirely either the enthusiasm of bull markets or the apprehensions of a severe depression.

We should not propose a rule, therefore, by which investors are to require a larger than usual stock-value ratio when prices are high; for such advice will not be followed. Nor shall we propose the opposite rule for bear markets, particularly because by diligent search it will always be possible to find some investments that meet all the normal tests even under depressed conditions, as our General Baking Company example will bear out.

SUMMARY OF MINIMUM QUANTITATIVE REQUIREMENTS SUGGESTED FOR FIXED-VALUE INVESTMENT

1. Size of obligor:

Municipalities: population............................	30,000
Public utilities: gross revenues.......................	$2,000,000
Railroads: gross revenues............................	3,000,000
Industrials: gross revenues...........................	5,000,000

2. Interest coverage:

Public-utility bonds: (7-yr. average)...................	1¾	times
Railroad bonds: (7-yr. average).......................	2	times
Industrial bonds: (7-yr. average).....................	3	times
Real-estate bonds: (dependable estimate)..............	2	times

3. Value of property:

Real-estate bonds: Fair value of property (based on actual sales in a noninflated market) must be 50% more than the bond issue.
Investment trust bonds: Similar ratio, using market value of assets.

4. Market value of the stock issues:

Public utilities......................	50% of the bonded debt
Railroads...........................	66⅔% of the bonded debt
Industrials..........................	100% of the bonded debt

CHAPTER XII

SPECIAL FACTORS IN THE ANALYSIS OF RAILROAD AND PUBLIC-UTILITY BONDS

RAILROAD-BOND ANALYSIS

The selection of railroad bonds can be made a process of extreme complexity. The reports of the carriers to the Interstate Commerce Commissions contain voluminous data on the financial and physical condition of the railroads which supply material for elaborate analysis. A really thorough study of a railway report would devote attention to the following items, among others:

1. Financial:
    a. Composition and trend of operating revenue.
    b. Ratio of maintenance expenditures to gross.
    c. Relative amount and trend of transportation expenses.
    d. Character of "other income."
    e. Coverage for, and relative growth of, interest and other deductions.

2. Physical:
    a. Location.
    b. Amount of double and third track.
    c. Weight of rail.
    d. Character of ballast.
    e. Amount and capacity of equipment owned.

3. Operating:
    a. Character and density of traffic.
    b. Average haul and average rate received.
    c. Train load.
    d. Fuel costs.
    e. Train- and car-mile operating costs.
    f. Maintenance charges per unit of equipment.

In addition to the above items affecting the railroad as a whole, a special study can be made of the mileage covered by the mortgage lien under consideration.[1]

[1] Elaborate graphic portrayal of railroad mortgage liens, the specific trackage covered, etc., together with supporting data and descriptions, are provided by White and Kemble's *Atlas and Digest of Railroad Mortgages*, covering all of the railroads of major importance in the United States.

Elaborate Technique of Analysis Not Necessary for Selection of High-grade Bonds.—Comprehensive analyses of this kind are actually made by the investment departments of large financial institutions which purchase railroad bonds. They are, however, not only clearly beyond the competence of the individual investor, but in our opinion they are hardly consistent with the true nature of high-grade bond investment. The selection of a fixed-value security for limited-income return should be, relatively, at least, a simple operation. The investor must make certain by quantitative tests that the income has been amply above the interest charges and that the current value of the business is well in excess of its debts. In addition, he must be satisfied in his own judgment that the character of the enterprise is such as to promise continued success in the future, or more accurately speaking, to make failure a highly unlikely occurrence.

These tests and this expression of judgment should not require a highly elaborate technique of analysis. If the investor in railroad bonds must weigh such factors as a favorable trainload trend as against a poor diversification of traffic handled, he is called upon to exercise penetration and skill out of all proportion to the reward offered, *viz.*, a fixed income return of from 4 to 6%. He would certainly be better advised to buy United States government securities, which yield a lower return but are safe beyond question, or else to let one of the large savings banks invest his money for him with the aid of its extensive statistical staff.

Recommended Procedure.—The complexities associated with railroad-bond analysis have arisen naturally—but in our view, rather illogically—from the wealth of data available for study. The fact that a mass of figures is obtainable does not mean that it is necessary, or even advantageous, to dissect them. We recommend that the buyer of high-grade railroad bonds confine his quantitative study to the coverage of fixed charges (with due attention to the trend of earnings and the adequacy of

---

More exhaustive study of the character and volume of traffic originating on and transported over particular sections of the road securing individual mortgage issues is greatly facilitated by examination of the "Freight Traffic Density Charts" and data assembled by H. H. Copeland and Son of New York City which are distributed privately by them among a large group of investment institutions.

maintenance expenditures) and to the amount of the stock equity. If he desires to be particularly careful, he will probably be better advised to *increase* his minimum requirements on these two points, rather than to extend his statistical tests to numerous other features of the annual reports.

It may make our viewpoint clearer if we add that such elaborate analyses may at times be of real value to the purchaser of *speculative* railroad bonds or stocks, as aids to his judgment of what the future will bring. But the whole *raison d'être* of *fixed-value* investment is opposed to any *primary* reliance upon surmises as to the future, since the field for exercising such judgment must logically be among those issues which offer possibilities of gain as a reward for being right, commensurate with the penalties attached to being wrong.

Technical Aspects of Railroad-income Analysis.—The application of the interest-coverage test to railroad bonds involves a few technical questions which require attention. Railways have various kinds of fixed charges which are obligations equivalent to bond interest and which clearly should be included with such interest in calculating the margin of safety. There are also certain deductions which partake to some extent of the nature of fixed charges and to some extent also of operating expenses. Furthermore, there are credits designated as "other income," such as bond interest received, which may properly be considered as offsets to interest paid—at least for the purpose of comparison with other roads. In the following schedule we allocate the more important items of this character that are encountered in railroad statements.

1. Bond interest and equivalent charges.
   a. Interest on funded and unfunded debt.
   b. Rent for leased lines.
   c. Joint-facility rents (net debit).

2. Deductions midway between fixed charges and operating expense.
   a. Hire of equipment (net debit).
   b. Miscellaneous rents and miscellaneous deductions.

3. Credits that may be partially offset against bond interest (in order of dependability).
   a. Bond interest received; rent for leased lines and joint-facility rents (net credit).
   b. Hire of equipment (net credit); dividends received.
   c. Miscellaneous nonoperating income.

Methods of Computing Fixed-charge Coverage.—Considerable argument might be indulged in as to the most scientific way of handling all these items in order to arrive at the best formulation of the fixed charges. The matter may be simplified, however, by bearing in mind that the bond buyer is not interested in exactitude, but rather in reasonable accuracy. After all, the data he is dealing with represent past history, the sole value of which is to serve as a hint or clue to the future. For such a purpose refinement of calculation is of little benefit. We suggest that for railroad bonds the necessities of the case with respect to interest coverage may be met by setting up a double test, and requiring that the minimum margin be shown by each. The method proposed is as follows:

Test A. Number of times Fixed Charges are earned:

Fixed Charges = Gross Income − Net Income.

$$\text{Times Fixed Charges earned} = \frac{\text{Gross Income}}{\text{Gross Income} - \text{Net Income}}$$

NOTE: "Gross Income" is the "Net after Rents" plus "Other Income." "Net Income" is the balance available for dividends.[1]

Test B. Number of times Net Deductions are earned:

Net Deductions = Railway Operating Income − Net Income.

$$\text{Times Net Deductions earned} = \frac{\text{Railway Operating Income}}{\text{Railway Operating Income} - \text{Net Income}}$$

NOTE: "Railway Operating Income" is the same as a "Net after Taxes," *i.e.* the gross revenues minus operating expenses and taxes.

It is necessary to apply only one of these two tests, *viz.*, the more stringent one, which may readily be identified by inspection. The rule is as follows: If gross income exceeds net after taxes, apply the fixed-charges test (Test *A*). If net after taxes exceeds gross income, apply the net-deductions test (Test *B*). The application of these alternative tests will be clear from the examples as shown on page 136.

The Pennsylvania Railroad's reports offer an exceptional case, in that the larger part of its substantial other income is a direct offset against the fixed charges. These other-income items

---

[1] The figure for fixed charges as computed by Standard Statistics Company excludes some of the minor items, which are subtracted from gross income first, under the caption of "miscellaneous deductions." Our method is simpler, but the Standard Statistics calculation will give almost the same result, so that if their results are available they may as well be used.

CALCULATION OF MARGIN OF SAFETY FOR RAILROAD BONDS
(Unit $1,000; calendar year 1931)

Item	Chesapeake & Ohio	Chicago Great Western	Northern Pacific
1. Gross revenue.....................	$119,552	$20,108	$62,312
2. Net after taxes (railway operating income)...........................	35,417	4,988	3,403
3. Equipment and joint-facility rents....	dr. 88	dr. 2,417	cr. 3,398
4. Net after rents (net railway operating income)..........................	35,329	2,571	6,801
5. Other income.....................	2,269	196	16,853
6. Gross income.....................	$ 37,598	$ 2,767	$23,654
7. Interest and other fixed charges.......	10,902	1,866	14,752
8. Balance for dividends...............	$ 26,696	$ 901	$ 8,902

Chesapeake and Ohio, 1931

Gross income exceeds net after taxes. Therefore use fixed-charges test (Test *A*).

$$\text{Fixed charges earned} = \frac{(6)}{(6) - (8)} = \frac{37,598}{10,902} = 3.45 \text{ times}$$

Chicago Great Western, 1931

Net after taxes exceeds gross income. Therefore use net-deductions test (Test *B*).

$$\text{Net deductions earned} = \frac{(2)}{(2) - (8)} = \frac{4,988}{4,087} = 1.22 \text{ times}$$

Northern Pacific, 1931

Gross income exceeds net after taxes. Therefore use fixed-charges test (Test *A*).

$$\text{Fixed charges earned} = \frac{23,654}{14,752} = 1.60 \text{ times}$$

NOTES ON THE ABOVE TESTS

1. Chesapeake and Ohio represents the typical exhibit in which the results of both tests would have pointed to the same conclusion—in this case to the presence of a satisfactory margin of safety for the bonds.

2. In the case of Chicago Great Western, Test *A*, which is ordinarily applied, would not adequately reflect the burden of the unusually large rental deductions. Their effect is shown by Test *B*, and in accordance with our suggestion this less favorable result should be the one considered by the investor.

3. Northern Pacific presents the opposite situation. Its other income has been exceptionally large as compared with the bond interest, so that in most years the net deductions figure out as a credit. In this case the investor should follow the results of Test *A*, and consider Test *B* as a secondary indication of strength.

consist of interest and guaranteed dividends received on securities of the system itself which are owned by the parent company, so that the same items appear later as interest and rentals paid. In 1932 these offsetting amounts totalled some $28,000,000. They should properly be eliminated from the statement altogether. The effect of their inclusion was to reduce the indicated coverage under the fixed-charges test, as the following will show:

1932	Fixed-charges test		Net-deductions test
	As reported	As corrected	
Gross income.....	$95,500,000	$67,500,000	Net after taxes $61,000,000
Fixed charges.....	82,000,000	54,000,000	Net deductions 47,500,000
Time earned......	1.17	1.25	.... 1.28
Times earned, 7-year average....	1.77	2.25	.... 2.42

In this case the net-deductions test afforded a fairer criterion than the fixed-charges test uncorrected. Where an especially careful analysis is to be made, the reported figures should be adjusted as above indicated, on the basis of the available facts.

Bearing of Maintenance Expenditures upon Fixed-charge Coverage.—There are two important items in railroad accounting which are subject in some degree to arbitrary determination by the management, and which may therefore be treated in any one year in such a manner as to produce deceptively favorable or unfavorable results. The first of these is the maintenance account. If unduly small amounts are spent on upkeep of road and equipment, the net profits are thereby increased at the expense of the property, and the balance reported as available for fixed charges does not fairly represent the earning power during the period under review. Bond buyers might do well to examine the maintenance ratio (*i.e.*, the percentage of gross revenues expended on upkeep of way and rolling stock) in order to make sure that it is not suspiciously below standard. At this writing, no very definite ideas appear to be current as to what constitutes a *normal* maintenance ratio. For want of a better index, we suggest that the actual average expenditures during the five years ending in 1930 be regarded as an approximate normal. This amounts to between 32 and 34.5% of gross for the various geographical regions into which the railways of the country have been divided.[1] A substantial discrepancy between the average for an individual road and these ruling figures would call for additional investigation on the part of the analyst or investor.

Nonrecurring Dividend Receipts.—A second item which sometimes repays scrutiny is that of Dividends Received. When a railroad controls subsidiary companies, it is possible to draw out accumulated profits at irregular intervals in the form of special dividends paid to the parent company. The effect of such transactions is to overstate the actual earning power of the parent company for the year in which the subsidiary's special dividend was received.[2]

Excessive Maintenance and Undistributed Earnings of Subsidiaries.—Railroad reports will also disclose the opposite situation at times, *viz.*, excessive maintenance expenditures or the existence of large current earnings of subsidiaries not paid over to the parent company. The effect of such accounting is to understate the true earning power of the carrier examined. Matters of this kind are of considerable interest in the analysis of stock values, but the bond buyer's concern with such factors is of secondary character. In general he should not permit them to reverse an otherwise unfavorable verdict as to the safety of the bond, but he should recognize that their presence gives added attractiveness to bond issues which show adequate security without taking them into account.[3]

PUBLIC-UTILITY BOND ANALYSIS

The popularity of public-utility securities between 1926 and 1929 resulted in an enormous increase in the amount of such financing, but this increased quantity was accompanied by a definite retrogression in the standards of quality and in the methods of presentation employed by the issuing houses. Investment bankers, including some of the highest reputation, followed entirely indefensible practices in their offering circulars, in order to make the issues appear safer than they actually were. Of these objectionable devices, the most important were: (1) the

---

[1] See Appendix, Note 20, for detailed tabulation of these ratios by regions and examples of deviations therefrom.

[2] This and allied phases of accounting having to do with income of a nonrecurring character are considered in detail in Chaps. XXXI to XXXIII (see especially Chap. XXXIII, p. 382, where several examples are given).

[3] See Appendix, Note 21, for examples.

application of the term "public utility" to industrial operations; (2) the use of the prior-deductions method of stating the earnings coverage; and (3) the ignoring of depreciation in calculating the net earnings available for bond interest.

1. Abuse of the Term "Public Utility."—Just what constitutes a public-utility enterprise may be the subject of some controversy. In its strict definition it would be any enterprise supplying an essential service to the public, subject to the terms of a franchise and to continuous regulation by the state. (While steam railroads are in fact a public-utility undertaking, it is convenient and customary to place them in a separate category.) From the investment standpoint, the most important idea associated with a public utility is that of *stability*, based first upon the rendering of an exclusive and indispensable service to a large number of customers, and, secondly, upon the legal right to charge a rate of compensation sufficient to yield a fair return on the invested capital.

It must be borne in mind that this stability is relative rather than absolute, since it is not immune from basic changes or unexpected vicissitudes. Twenty years ago the leading type of utility was the street railway; but this industry is now subject to such severe competition from other forms of local transportation that in most communities it is not practicable to set the fare high enough to return reasonable earnings on the actual investment. Furthermore, during the war inflation period of 1918–1920 the light and power companies suffered keenly from rising labor and material costs together with difficulties and delays in obtaining permission to advance rates proportionately. These hardships had for a time an adverse effect upon the popularity of all utility investments, but the subsequent brilliant expansion of both gross and net earnings of gas, electric, water, and telephone companies speedily restored their securities to favor.

It is to three of these services, *viz.*, gas, electric, and telephone, that the utility investments of savings banks are restricted by the New York statute. We have remarked previously (page 93) that this category may properly be widened to include companies supplying water to communities of substantial size.

Pseudo-utilities.—But in the heyday of public-utility-bond flotations, this popular label was used by banking houses to promote the sale of many issues which partook only partially at best of the true character of public utilities and which may well be stigmatized as "pseudo-utilities." Companies selling ice, operating taxicabs, or owning cold-storage plants, became suddenly "affected with a public interest" to an extent permitting them to bond themselves for the major portion of their property investment and to sell these bonds to investors as public-utility securities. In most instances the enterprises so financed represented a combination of small gas, electric, or telephone establishments with the ice or cold-storage business, in such a way as to confuse or mislead the public as to the true nature of the investment offered. An outstanding and unfortunate precedent for this hybrid form of organization was set many years ago by the Cities Service Company, which combined a large bona fide public-utility network with an equally large venture in the production, refining, and marketing of oil.

Natural Gas.—The period preceding the 1929 crash was marked also by the sudden transmutation of natural gas from a branch of the oil industry into "one of the country's leading public utilities." Up to that time, natural gas had been used mainly as industrial fuel and as raw material for the production of gasolines and carbon black. Improvements in pipe-line construction permitted the transport of this gas over long distances to urban centers where it replaced considerable quantities of manufactured gas. Promoters and banking houses were quick to exploit the popular appeal of this new "utility"; and by the use of this designation an enormous total of natural-gas bond financing was successfully foisted on the public. As in the case of the ice plants, considerable recourse was had to the device of combining a natural-gas development with small bona fide utility properties. In many cases, the sale of these bonds under the guise of public-utility investments was a gross abuse of the public confidence, because the bulk of the natural-gas output was being taken for manufacturing use and the business was subject to all the hazards of the fuel industry.

The above exposition should make it plain that there are utilities and "utilities," and that investors must not take stability for granted because an issue is marketed under this popular title. In particular they should shun these hybrid mixtures of electric or telephone services with *industrial* activities, because at bottom every such combination represents an attempt to sail under false colors.[1]

[1] See Appendix, Note 22, for examples.

2. Use of the Prior-deductions Method of Calculating Coverage.—We have already indicated (pages 106–107) the fallacy involved in the calculation of interest coverage after the deduction of prior charges. In our opinion, the continued use of this method by houses of issue must be considered a severe reflection on their business probity, because they cannot fail to be aware that the formula is a misleading one, but they nevertheless employ it for the purpose of creating a deceptive appearance of safety.

3. Omission of Depreciation Charges in Calculating Coverage. It is equally difficult to discover any satisfactory reason for the widespread failure of the bond-offering circulars to deduct the depreciation allowance before computing the interest coverage. Depreciation is a real and vital element in the operating expense of a public utility. In the case of the typical well-established company, a good part of the annual-depreciation reserve is actually expended for the renewal of worn-out or obsolete equipment, so that it cannot be claimed that depreciation is a mere bookkeeping concept which need not be taken seriously. There is naturally room for a divergence of opinion with respect to the proper amount of depreciation to charge in any situation; but if proper attention were given to the extremely important element of obsolescence, it is hardly likely that the allowance made by the typical holding company will be found excessive, and in fact it is more likely to understate the true depreciation.[1]

In the writers' opinion, the cavalier omission of depreciation charges in the statement of earnings applicable to bond interest comes perilously close to outright misrepresentation of the facts.[2] A device fully as misleading is illustrated by the offering in 1924 of Cities Service Power and Light Company 6s, due in 1944. In this case, the indenture was so drawn as to require a *minimum*

[1] See the pungent comments on this head by William Z. Ripley, *Main Street and Wall Street*, pp. 172–175 and 333–336, especially the latter, Boston, 1927. See also Chap. XXXIV of the text for a further discussion of depreciation charges.

[2] This pernicious practice is encouraged, however, by the loosely drawn provisions governing investments by savings banks in public-utility bonds in various states, which apply the earnings test before deducting depreciation. In Vermont, for example, depreciation is deducted in determining the net income of telephone companies, but *not* in the case of gas, electric, water, and traction companies. See Appendix, Note 23, for comments by various committees of the Investment Bankers Association of America with respect to the manner of handling depreciation charges in bond circulars.

charge for depreciation and maintenance amounting to much less than the sums actually expended and reserved by the various operating subsidiaries. In the bond prospectus the earnings were stated after deductions for depreciation "assumed at rates in the Indenture securing these bonds," which in plain language meant that the true depreciation was greatly understated in calculating the margin of safety behind the bond issue. This piece of financing is commented on further below.

Recommended Procedure.—It is emphatically recommended that the intending purchaser of a public-utility bond issue make sure that a normal depreciation charge has been deducted from earnings, before he accepts the reported statement of interest coverage. Based upon the reports of many such companies, it would seem that an allowance amounting to less than 6% of gross may be viewed with suspicion as probably inadequate. In fact, the conservatively minded might be justified in applying a minimum figure of 8% of gross. Depreciation actually accrues, of course, as a percentage of the property account and not of the revenues. But since there is a fairly constant relationship between the investment and the gross receipts (about $4 of property for $1 of revenue) the adequacy of the depreciation allowance may be conveniently judged by reference to the gross revenues.

Examples Showing Need for Critical Examination of Offering Circular.—The following actual example illustrates in rather extreme fashion the necessity for a critical examination of bankers' circulars offering public-utility bonds.

Utilities Service Company Convertible Debenture 6½s, due 1938, offered in 1928 at 99½, yielding 6.55%. The presentation in the offering circular may be summarized as follows:

Amount of issue....$3,000,000
Business..........Operates 20 telephone companies, and 4 ice companies.
Value of property..$12,500,000 after depreciation, equal to $1,650 per $1,000-bond after deducting prior obligations.

Earnings	Year ended May 31, 1928
Gross	$3,361,000
Net before depreciation	969,000
Prior deductions	441,000
Balance for debentures	528,000
Interest on debentures	195,000
Balance for stock	333,000

"Balance as above is equal to 2.71 times interest on this issue."

Criticism of This Offering Circular.—1. The *business* is a combination of utility (telephone) and industrial (ice) operations, but it is bonded more heavily than a 100% utility enterprise could safely stand, the total debt being 84% of the appraised property value. The proportion of gross and net contributed by the ice business is not stated and must therefore be assumed to be substantial.[1]

2. The omission of the depreciation charge from the earnings statement is so misleading as to appear almost fraudulent. Depreciation reserves by telephone companies absorb a large percentage of gross receipts. In the case of the American Telephone and Telegraph System this percentage averages about 15%; and the same deduction was actually made by the chief subsidiary of the Utilities Service Company (Lima Telephone Company). If depreciation at the rate of 15% of gross is charged against the total revenue, the amount so to be deducted would be $500,000, and *would leave practically no earnings available for the debenture interest.* In other words, instead of covering the debenture interest 2.71 times as stated, the company would be failing to earn the interest charges by a large deficit.

The ice operations would carry a smaller depreciation charge than 15% of gross, but this advantage should be offset by the greater margin of safety required for an industrial business. Furthermore, if the net valuation of $12,500,000 placed on the property is accepted, then in any event the annual depreciation deduction should not be less than 4% or $500,000.

3. The calculation of interest coverage in the circular made by the prior-deductions method would indicate that the debentures were better protected than the prior liens. (They "earned their interest" 2.71 times, while senior interest was covered 2.20 times.)

Assuming a *low* depreciation charge of $300,000 per annum, and presenting the interest deductions properly, the exhibit of this bond offering should be restated as follows:

Gross...	$3,361,000
Net before depreciation........................	969,000
Depreciation (estimated).......................	300,000
Balance for interest............................	669,000
Total interest charges.........................	636,000
Balance for dividends..........................	33,000
Interest charges earned.......................	1.05 times

[1] Figures subsequently published show that the ice business made up more than half of the total business.

4. The statement that there was $1,650 of property value behind each $1,000 debenture is based upon a similarly misleading method. The aggregate bonded debt was $10,500,000 against $12,500,000 of appraised value, so that the appraisal showed only $1,190 of value behind each $1,000 of *total* debt.[1]

Example: It may be illuminating also to make a similar critical examination of the advertisement offering Cities Service Power and Light Company Secured 6s, due 1944, at 96 to yield 6.35%, as published in April 1926. The earnings data covering the calendar year 1925 were presented substantially as follows:

Gross, including other income................	$49,662,000
Net after operating expenses and taxes........	19,096,000

Deduct:	
Fixed charges and preferred dividends of subsidiaries...........................	10,102,000
Depreciation ("assumed at rates in the indenture securing these bonds")..............	1,574,000
Minority interest........................	209,000

Income applicable to interest of Cities Service Power and Light.........................	7,211,000
Interest on this issue......................	1,466,000

"Income applicable to interest charges, as shown above, was over 4.9 times maximum annual interest requirements on Series A bonds of $1,466,250, and over 4.1 times maximum annual interest charges of $1,736,250 on all outstanding funded debt of Cities Service Power and Light Company."

This circular was misleading in two important respects: first in employing the prior-deductions method for computing the earnings coverage on the bonds offered; and secondly, in using an artificial and quite inadequate basis of depreciation. A study of the application to list this issue on the New York Stock Exchange shows that the operating subsidiaries actually made appropriations for replacements amounting to $5,214,000 for the year ending June 30, 1925. This was almost four times the arbitrary rates set up in the indenture. A revision of the offering circular, to conform with the actual situation in respect to depreciation, and with the proper method of stating interest coverage, will show the following exhibit:

---

[1] In 1932 the Utilities Service Company went into receivership and the debenture bondholders lost their entire investment.

Gross.....................................	$49,662,000
Net, after minority interest..................	19,189,000
Depreciation for year ending June 30, 1925.....	5,214,000
Balance for fixed charges....................	13,975,000
Interest and preferred dividends of subsidiaries.	10,102,000
Interest charges of parent company...........	1,736,000
Total fixed charges.........................	11,838,000
Balance for parent company dividends.........	2,137,000
Fixed charges earned.......................	1.18 times

This showing is very different indeed from a coverage of 4.1 or 4.9 times interest as indicated in the offering circular.

Deduction of Federal Taxes in Computing Interest Coverage.— The federal income tax is imposed upon profits after subtracting interest paid. Hence earnings available for interest should properly be shown before deducting the federal tax. In corporate reports to stockholders it is customary to reverse this order, and in most cases the amount of the tax is not shown. But in analyzing the exhibit of a bond issue, it should not be necessary to revise the income statements by adding back the federal taxes, actual or estimated. The reason is that the result produced by such revision can very rarely make enough difference to affect the apparent eligibility of the bond issue for investment; furthermore, the error, such as it is, lies on the side of understatement— which is by no means objectionable in the selection of investment bonds. In general, the analyst should refrain from elaborate computations or adjustments which are not needed to arrive at the conclusion he is seeking.

In bond-offering circulars, the income available for interest is usually stated before deduction of federal tax, in order to make the best showing permissible. This cannot properly be objected to, except sometimes in the case of offerings of bond issues of public-utility holding companies. Such bonds are usually junior to the preferred stocks of subsidiary companies, and the federal tax must be computed and deducted before these dividends are paid. Hence, objection may fairly be leveled against a presentation such as was made in the offering circular of Cities Service Power and Light Debenture 5½s in November 1927, wherein the earnings applicable for interest on the holding company's bonds were stated before deducting federal taxes of the system.

CHAPTER XIII

OTHER SPECIAL FACTORS IN BOND ANALYSIS

"Parent Company Only" vs. Consolidated Return.—Both bond-offering circulars and annual reports almost invariably present the earnings statement of a public-utility holding-company system in a *consolidated* form, *i.e.*, they start with the gross revenues of the operating subsidiaries and carry the figures down through operating expenses, depreciation, fixed charges, and preferred dividends of subsidiaries, until they arrive at the balance available for the parent company's interest charges, and finally at the amount earned on its common stock. There is also published, largely as a matter of form, the income account of the parent company only, which starts with the dividends received by it from the operating subsidiaries and therefore does not show the latter's interest and preferred dividend payments to the public. The interest coverage shown by the income account of the parent company only is an example of the prior-deductions method, and consequently it will almost always make a better showing for the parent company's bonds than will be found in the consolidated report. The investor should pay no attention to the "parent company only" figures, and insist upon a completely consolidated income account.

Example: The following example will illustrate this point:

STANDARD GAS AND ELECTRIC SYSTEM, 1931

Item	"Parent company only"	Consolidated results
Gross revenues...............	$16,790,000	$159,070,000
Balance for fixed charges.......	16,514,000	57,190,000
Fixed charges................	4,739,000	42,226,000
Balance for parent-company stocks...................	11,775,000	14,964,000
Fixed charges earned.........	3.48 times	1.36 times

The parent company did not receive in dividends the full amount earned by its subsidiaries, but even with this smaller

146

income the prior-deductions method results in a much larger indicated coverage for the parent-company bond interest on the basis of its own results than on a consolidated basis.

Dividends on Preferred Stocks of Subsidiaries.—In a holding-company system the preferred stocks of the important operating subsidiaries are in effect senior to the parent company's bonds, since interest on the latter is met chiefly out of dividends paid on the subsidiaries' common stocks. For this reason subsidiary preferred dividends are always included in the fixed charges of a public-utility holding-company system. In other words, these fixed charges consist of the following items, in order of seniority:

1. Subsidiaries' bond interest.
2. Subsidiaries' preferred dividends.
3. Parent company's bond interest.

This statement assumes that all the subsidiary companies are of substantially the same relative importance to the system. An individual subsidiary which happens to be unprofitable may discontinue preferred dividends and even bond interest, while at the same time the earnings of the other subsidiaries may permit the parent company to continue its own interest and dividend payments. In such a case, which is somewhat exceptional, the unprofitable subsidiary's charges are not really senior to the parent company's securities. This point is discussed at the end of Chap. XVII.

The fixed charges should also properly include any annual rentals paid for leased property which are equivalent to bond interest or guaranteed dividends. In the majority of holding-company reports this practice is followed.

The holder of preferred shares of an important operating subsidiary has to all intents and purposes a claim which is as fixed and enforceable on the system's earnings as have the owners of the parent company's bonds. But if the parent company becomes insolvent, then the owners of the underlying preferred issues no longer occupy the strategic position of bondholder, since they cannot compel the operating subsidiary to continue paying its preferred dividends.

Example: New York Water Service Corporation Preferred may be cited as an example. The company is an operating subsidiary of Federal Water Service Corporation, which in turn was a subsidiary of Tri-Utilities Corporation. Dividends on this issue and on Federal Water Service Preferred ranked as fixed charges of the Tri-Utilities system. When the latter company was unable to meet interest on its debentures and went into receivership in August 1931 dividends on these underlying preferred issues were promptly discontinued, although both were apparently earned and the income of New York Water Service Corporation actually showed an increase over the previous year.

Minority Interest in Common Stock of Subsidiaries.—The earnings applicable to minority stock are usually deducted in the income statement *after* the parent company's bond interest, and hence the former item does not reduce the margin of safety as generally computed. We prefer to subtract the minority interest *before* calculating the interest coverage. Exact treatment would require a prorating of deductions, but this involves needlessly burdensome calculations. When the minority interest is small, as is true in most cases, the difference between the various methods is inconsequential. When the minority interest is fairly large, analysis will show that the customary procedure gives a margin of safety somewhat higher than is strictly accurate, whereas our method errs moderately in the opposite direction, and hence should be preferred by conservative investors.[1]

"Capitalization of Fixed Charges," for Railroads and Utilities. In the previous chapter we pointed out certain difficulties in the way of arriving at a fair statement of the ratio of stock to debt in the case of railroads and public utilities. Debt may be represented not only by bond issues but also by guaranteed stocks, annual rental obligations, and effectively also by nonguaranteed preferred stocks of operating subsidiaries. In computing the interest coverage these items are taken care of by using the omnibus figure of fixed charges, instead of merely the bond interest. It is suggested that a good approximation of the *principal* amount of all these obligations may quickly be obtained by multiplying the fixed charges by some appropriate figure— say 20 in the case of railroads, and 18 in the case of a public utility.

This means that we assume the "effective debt" of a railroad to be that figure, 5% on which would produce the annual fixed charges. Similarly we assume that the effective debt of a public-utility company is that figure, 5.5% on which would yield its annual fixed charges. This process is called "capitalizing the

[1] See Appendix, Note 24, for a calculation under the three methods applied to the report of United Light and Railways Company for 1932.

fixed charges." The rate of capitalization is fixed somewhat higher for utilities than for rails to reflect the actual situation with respect to their capital obligations in the aggregate.

Technique Illustrated: Railroads.—In the case of railroad issues we have suggested that the earnings coverage be applied to either the Net Deductions or the Fixed Charges (as previously defined), whichever are larger. In the same way the larger of these two items should be used as the base for computing the principal amount of the road's "effective debt." The technique to be followed is illustrated herewith:

Examples:

New York, New Haven and Hartford Railroad

A. Net deductions (1932)................... $ 18,511,000
B. Fixed charges (1932).................... 17,403,000
Net deductions capitalized at 5%............ $370,000,000
(Funded debt shown on balance sheet—$258,000,000)
Preferred stock: 490,000 sh. @ 50 (July 1933).. $ 24,500,000
Common stock: 1,570,000 sh. @ 22 (July 1933)   34,500,000
                                               _____
Total market value of stock issues.......... $ 59,000,000
Stock-to-bond ratio—1 to 6¼
Net deductions earned, 1932................. 0.93 times
Net deductions earned, 7-yr. average........ 1.57 times

Chesapeake and Ohio Railway

A. Net deductions (1932)................... $ 9,870,000
B. Fixed charges (1932).................... 10,760,000
Fixed charges, capitalized at 5%............ $215,000,000
Bonded debt shown on balance sheet........ 222,000,000
Common stock: 7,650,000 sh. @ 38 (July 1933)   291,000,000
Stock-to-bond ratio—1 to ¾ (*i.e.*, $1 of stock to 76 cents of
   bonds)
Fixed charges earned, 1932................. 3.21 times
Fixed charges earned, 7-yr. average......... 3.80 times

Conclusions Based on Foregoing.—The "effective debt" of the New Haven was computed from the net deductions (which are larger than the fixed charges, because they include a substantial debit for equipment rentals, etc.). This effective debt is considerably more than that shown in the balance sheet. With the preferred and common stocks together selling in July 1933 for less than a sixth of the true debt, it is evident that the bonds had an insufficient stock equity at the time. If the *prospects* were considered favorable there might be good reason to buy the common stock for large capital appreciation. But no such possibility attached to the 6% bonds selling at 92, and consequently the purchase of this issue could not be supported by sound analysis.

The Chesapeake and Ohio exhibit, on the other hand, supplies a stock-value ratio which fully confirms the satisfactory showing of the earnings coverage. If the investor were satisfied with the prospects of this road, he would then be justified in buying its bonds (e.g., the Refunding and Improvement 4½s selling at 92½) since these meet both quantitative tests in satisfactory fashion.

Technique Illustrated: Utilities.—The capitalization of fixed charges in the case of a public-utility holding company is illustrated as follows:

Examples:

1932	American Water Works & Electric Co.	Columbia Gas & Electric Corp.
Subsidiary bond interest..............	$ 8,700,000	$ 3,191,000
Subsidiary pfd. dividends............	5,646,000	2,513,000
Parent-company bond interest.......	1,295,000	(Net) 6,265,000
Total fixed charges...............	$15,641,000*	$ 11,969,000
Fixed charges capitalized at 5½ %.....	284,700,000	217,600,000
Market value of preferred stock (July 1933).	200,000 sh. @ 75 = $15,000,000	940,000 sh. @ 80 = $ 75,200,000
		40,000 sh. @ 70 = $ 2,800,000
		130,000 sh. @ 104 = $ 13,500,000
Market value of common stock (July 1933).	1,750,000 sh. @ 25 = $44,000,000	11,609,000 sh. @ 20 = $232,000,000
Total market value of stock issues.....	$59,000,000	$323,500,000
Stock-to-bond ratio.................	1 to 4.8	1 to 0.7
Fixed charges earned, 1932..........	1.24 times	2.44 times
Fixed charges earned, 5-year average..	1.40 times	3.66 times

* Very small minority interest left out of account.

Interpretation.—In these two examples the stock-to-bond ratios confirm the indications of the earnings ratios. The American Water Works and Electric Company is seen to carry far too heavy a weight of senior obligations in proportion to the amount of stock; while Columbia Gas and Electric is conservatively capitalized in this respect. These analyses would compel the avoidance of American Water Works bonds by investors, but they would accord a very satisfactory quantitative standing to the Columbia Gas bonds. (The purchase of the latter would nevertheless require *also* that the investor be reasonably satisfied as to the prospects and other qualitative elements in the Columbia Gas picture.)

In July 1933 American Water Works 6% bonds, due 1975, sold at 84 and Columbia Gas 5% bonds, due 1952, sold at 83. If choice were to be made between the two issues, our principles of bond investment would definitely require the sacrifice of the 1% in coupon return in order to obtain adequate instead of inadequate safety.

THE WORKING-CAPITAL FACTOR IN THE ANALYSIS OF INDUSTRIAL BONDS

For reasons already explained, a company's statement of its *fixed assets* will not ordinarily carry much weight in determining the soundness of its bonds. But the *current-asset position* has an important bearing upon the financial strength of nearly all industrial enterprises, and consequently the intending bond purchaser should give it close attention. It is true that industrial bonds which meet the stringent tests already prescribed will in nearly every instance be found to make a satisfactory working-capital exhibit as well, but a separate check is nevertheless desirable in order to guard against the exceptional case.

Current assets (termed also "liquid," "quick," or "working" assets) include cash, marketable securities, receivables, and merchandise inventory.[1] These items are either directly equivalent to cash, or are expected to be turned into cash, through sale or collection, in the ordinary course of business. To conduct its operations effectively, an industrial enterprise must possess a substantial excess of current assets over current liabilities, the latter being all debts payable within a short term. This excess is called the working capital, or the net current assets.

Three Requisites with Respect to Working Capital.—In examining the quick-asset situation, an industrial bond buyer should satisfy himself on three counts, *viz.:*

1. That the cash holdings are ample.
2. That the ratio of quick assets to current liabilities is a strong one.
3. That the working capital bears a suitable proportion to the funded debt.

[1] Some authorities exclude inventories from "quick assets," but include them in "current assets." A distinction of this kind may be useful, but it is not yet standard.

It is not feasible to fix definite minimum requirements for any one of these three factors, especially since the normal working-capital situation varies widely with different types of enterprise. It is generally held that quick assets should be at least double the current liabilities, and a smaller ratio would undoubtedly call for further investigation. We suggest an additional standard requirement, *viz.*, that the working capital be at least equal to the amount of the bonded debt. This is admittedly an arbitrary criterion, and in some cases it may prove unduly severe. But it is interesting to note that in the case of every one of the industrial issues which maintained their investment rank marketwise throughout 1932, as listed on pages 81–82, the working capital exceeded the total of bonds.[1]

In contrast with the emphasis laid upon the current-asset position of industrial concerns, relatively little attention has been paid to the working capital shown by railroads, and none at all to that of public utilities. The reason for this is twofold. Neither railways nor utilities have the problem of financing the production and carrying of merchandise stocks or of extending large credits to customers. Furthermore, these companies have been accustomed to raising new capital periodically for expansion purposes, in the course of which they readily replenish their cash account if depleted. Because new financing is easily obtainable by prosperous companies of this type, even an excess of current liabilities over quick assets has not been considered a serious matter. Recent experience indicates the desirability of substantial cash holdings by a railroad to meet emergency developments, and the bond buyer might do well to favor those public utilities also which maintain a strong working-capital position.

[1] General Baking reached this position *during* 1932. Including General Baking, 13 of the 18 companies showed *cash* assets alone exceeding their funded debt.

CHAPTER XIV

THE THEORY OF PREFERRED STOCKS

That the typical preferred stock represents an unattractive *form* of investment contract is hardly open to question. On the one hand, its principal value and income return are both limited; on the other hand, the owner has no fixed, enforceable claim to payment of either principal or income. It may be said that preferred stocks combine the limitations of creditorship (bonds) with the hazards of partnership (common stocks). Yet despite these strong theoretical objections, the preferred stock has developed into a major factor in our financial scheme, and has evidently succeeded in commending itself to the American investor. In 1932 there were about 440 different preferred issued listed on the New York Stock Exchange as against some 800 common stocks. In 1929 the value of the listed preferred shares exceeded 8½ billion dollars and was about half as great as that of listed corporation bonds.

The Verdict of the Market Place.—In the subsequent market collapse, the price of these shares suffered a drastic shrinkage, as is shown by the following comparative figures:

AVERAGE OF ALL LISTED ON THE NEW YORK STOCK EXCHANGE

Type of security	High price, 1929	Low price, 1932
United States corporate bonds.....	95.33	52.68
Preferred stocks................	84.99	25.38
Common stocks.................	89.94	10.59

These figures establish the fact that under the "laboratory test" conditions of 1929–1932, preferred stocks as a whole proved considerably more vulnerable than bonds to adverse conditions. Whether results shown in such unprecedented circumstances afford any guide to normal investment practice is not at all certain, but this performance lends added sharpness to the one contrast between the theoretical weakness of preferred stocks

153

and their wide popularity. A thoroughgoing analysis would seem to be called for, in order to determine the true merits of preferred shares as a practical medium of investment.

Basic Difference between Preferred Stocks and Bonds.—The essential difference between preferred stocks and bonds is that payment of preferred dividends is entirely discretionary with the directors, whereas payment of bond interest is compulsory. Preferred dividends must indeed be paid as long as any disbursements are being made on the common shares; but since directors have the power to suspend common dividends at any time the preferred stockholder's right to income is at bottom an entirely contingent one. However, if a company's earnings are regularly far in excess of preferred-dividend requirements, payment is usually made as a matter of course; and in such instances, the absence of an enforceable claim to dividends does not seem to be of real importance. This explains the existence of a relatively small number of high-grade preferred issues which are considered equivalent in quality to sound bonds and sell at comparable prices.

At the opposite extreme are the cases in which corporations are *unable* to pay anything, whether it be on bonds or on preferred stock. In such situations the bondholder's legal right to receive interest results not in payment but in receivership and foreclosure. As we have previously pointed out, the practical value of these remedies is doubtful, and in most instances it may fairly be said that the position of a bond in default is little better than that of a nondividend-paying preferred stock without bonds ahead of it.

At both extremes therefore, the contractual superiority of bonds over preferred stocks is not of substantial value. This fact has led many investors to believe that *as a general rule* the bond form has no real advantage over the preferred stock form. Their line of reasoning runs that "if the company is good, its preferred stock is as good as a bond; and if the company is bad, its bonds are as bad as a preferred stock."

Weakness Because of the Discretionary Right to Omit Dividends.—This point of view is highly inexact, because it fails to take into account the wide middle region occupied by companies neither unqualifiedly "good" nor unqualifiedly "bad," but subject to variations and uncertainties in either direction. If it could be assumed that directors will always pay preferred dividends when *possible* (and hence will suspend payment only under conditions which would compel default of interest if the issue were a bond), then even in the intermediate situations the preferred stockholder's status would not be greatly inferior to the bondholder's. But in actual fact this is not the case, because directors frequently exercise their discretion to withhold preferred dividends when payment is by no means impossible but merely inconvenient or inexpedient. It is considered an approved financial policy to sacrifice the preferred stockholder's present income to what he is told is his future welfare; in other words, to retain cash available for dividends in the treasury to meet future emergencies or even for future expansion.

Even if it be conceded that such a practice may ultimately be advantageous to the preferred stockholder, the fact remains that it subjects his income to a hazard not present in the case of a similarly situated bond. If such a hazard is at all substantial, it automatically disqualifies the preferred issue as a fixed-value investment, because it is the essence of such investments that the income must be considered entirely dependable. Stating the point more concretely, any preferred stock subject to a real danger of dividend reduction or suspension will fluctuate widely in market value. It is a point worth noting that in all cases where the dividend *could* be continued, but instead is withheld "for the sake of the stockholders' future advantage," the quoted price suffers a severe decline, indicating that the investment market does not agree with the directors as to what is really in the best interests of the preferred stockholders.

Conflicts of Interest.—Any investor would rather have his income continued, even at possible risk to the future of the business. There is evidently a basic disagreement, amounting almost to a logical contradiction, between what the investor considers to be his individual advantage (*viz.*, the continuance of his income at all costs) and what he seems willing to admit may be sound corporate policy (*viz.*, the suspension of dividends for the sake of the future). In this connection, the question of a possible conflict of interest between the preferred and the common stockholders is of undoubted importance. Withholding preferred dividends may be of distinct advantage to the common stock. The directors are legally required to represent the interests of all stockholders impartially, but since in fact they are most often elected by the common stockholders they tend to act primarily in the latter's behalf.   Directors have also grown accustomed to consider the interests of the enterprise itself, as an entity apart from the interests of its owners—*i.e.*, the stockholders—and they frequently pursue policies with the apparent purpose and result of strengthening the corporation at the actual expense of its proprietors.   This paradoxical viewpoint may perhaps be explained in part by the customary close connection between corporate directors and the salary-drawing officers.[1]

Form of Preferred Contract Often Entails Real Disadvantage. *Example:* Whatever the reason or justification may be, the fact remains that preferred stockholders are subject to the danger of interruption of dividend payments under conditions which would not seriously threaten the payment of bond interest. This means that the *form* of the preferred stockholder's contract will often entail a real disadvantage.   A striking illustration of this fact is afforded by the case of United States Steel Corporation Preferred, which is probably the largest senior stock issue in the world, and was for many years thoroughly representative of those preferred shares which enjoyed a high investment rating. In 1931—although the depression was well advanced—this issue sold at a price to yield only 4.67%, and it was thought to occupy an impregnable position as a result of the accumulation of enormous sums out of the earnings during the preceding 30 years and their application to the improvement of manufacturing facilities, the enlargement of working capital, and the retirement of nearly all the bonded debt.   Yet immediately thereafter, a single year of operating losses jeopardized the preferred dividend to such an extent as to destroy nearly two-thirds of its market price and undermine completely its standing as a prime investment.   In the following year the dividend was reduced to $2 annually.

Weakness of Contractual Position Illustrated.—These disastrous developments were due, of course, to the unprecedented losses of 1932–1933.   But if it had not been for the weakness of the preferred stock *form*, the holder of these shares would have had little reason to fear the discontinuance of his income. In other words, if he had possessed a *fixed* claim for interest instead of a contingent claim for dividends, he could have relied with confidence on the corporation's enormous resources to take

[1] See our further discussion of this point in Chap. XLIV on Stockholder-management Relationships.

care of its obligations. In support of this contention, a brief comparison is appended of the market action of Inland Steel 4½s (previously discussed) with that of United States Steel Preferred.

Period	U.S. Steel Pfd.		Inland Steel 4½% bonds, due 1978	
	Price	Yield, %	Price	Yield, %
High price, 1931........	150	4.67	97¾	4.62
Low price, 1932.........	51½	13.59	61	7.54
High price, Jan. 1933....	67	10.45	81	5.67

Both of these issues were subject to the same adverse business conditions, but the *contractual* weakness of United States Steel Preferred was responsible for the loss of an investment position which the Inland bonds were able to retain without serious difficulty (except for a brief period of utter demoralization in the bond market).

Yield and Risk.—Admitting once again that 1932–1933 experience constitutes a "laboratory test" of abnormal severity, should the behavior of preferred stocks under it be regarded nevertheless as so unsatisfactory as to disqualify them in general as fixed-value investments? In considering this question, it should be noted that a small number of preferred issues maintained an investment rating even at the worst moments of 1932. The proponents of preferred shares will contend, moreover, that under *normal* variations in business conditions the higher yield of this group will compensate for such inferiority as exists in their safety as compared with bonds. This is an argument which always appeals to the investor in good times, when the increased income is an actuality and the risk to principal seems a remote contingency. In bad times there is perhaps an opposite disposition to consider only the shrinkage of principal suffered and to forget about the higher income received in the years preceding.

To present a broader view of this question, we revert to our previous discussion of bonds with varying degrees of safety, in which we arrived at the principle that risk and income return are at bottom *incommensurable*. If this statement is valid for bonds, it must apply with equal force to preferred stocks. This means that it is not sound procedure to purchase a preferred stock at an investment price (*e.g.*, close to par) when the presence of a substantial risk to principal is recognized, but when this risk is expected to be offset by an attractive dividend return. It would follow from this principle that the only preferred stock which can properly be bought for investment would be one which in the purchaser's opinion carries no appreciable risk of dividend suspension.

Qualification of High-grade Preferred Stocks.—What must be the qualifications of such a preferred stock? In the first place, it must meet all the minimum requirements of a safe bond. In the second place, it must exceed these minimum requirements by a certain added margin to offset the discretionary feature in the payment of dividends, *i.e.*, the margin of safety must be so large that the directors may always be expected to declare the dividend as a matter of course. Thirdly, the stipulation of inherent stability in the business itself must be more stringent than in the case of a bond investment, because a company subject to alternations between large profits and temporary losses is likely to suspend preferred dividends during the latter periods even though its *average* earnings may far exceed the annual requirements.

The above reasoning suggests conclusions that correspond to the actual behavior of preferred shares in 1932. These conclusions are, not that preferred stocks must, *per se*, be excluded from the investment category, but rather that such severe specific requirements must be imposed upon them as to make the number of eligible issues comparatively small. The list shown on page 159 comprises all of the preferred stocks listed on the New York Stock Exchange which maintained a price equal to a 7% return or less at all times during 1932.[1] There are appended also, certain quantitative data bearing on the degree of safety enjoyed by each of these issues.

Sound Preferred Issues Exceptional.—This list of preferred stocks comprises only 5% of the total number of issues listed on the New York Stock Exchange in 1932. This small percentage bears out our thesis that a sound preferred stock,

---

[1] We exclude Standard Oil Export Corporation 5% Preferred, Pittsburgh, Fort Wayne and Chicago Railway 7% Preferred, and other guaranteed preferred issues, since they occupy substantially the position of a debenture bond of the guarantor.

LISTED PREFERRED STOCKS WHICH MAINTAINED AN INVESTMENT PRICE-LEVEL THROUGHOUT 1932–1933

Name of company	Low price in 1932–1933	Rate in dollars	Yield at low price, %	Funded debt	Number of times preferred dividends (and fixed charges) were earned		Ratio of lowest market value of common to preferred and bonds combined
					Average 1927–1931	Minimum	
General Electric (Par $10) (Cum.)	109⅝	0.60	5.6	Yes	19.19	13.75	5.09
Eastman Kodak (Cum.)	104¾	6.00	5.7	No	51.91	36.25	12.30
Duquesne Light (Cum.)	85	5.00	5.9	Yes	3.67	3.19	*
Public Service Electric & Gas (Cum.)	83	5.00	6.0	Yes	2.58	2.02	...†
United States Tobacco (Non-cum.)	115	7.00	6.1	No	9.00	6.67	8.42
Procter & Gamble (2d Preferred) (Cum.)	81	5.00	6.2	Yes	17.96‡	15.35‡	4.59
Norfolk & Western R.R. (Non-cum.)	65	4.00	6.2	No	5.96	4.57	0.69
G. W. Helme (Non-cum.)	113⅜	7.00	6.3	Yes	8.12	7.67	2.61
American Tobacco (Cum.)	95¼	6.00	6.3	No	10.48§	7.26§	3.87∥
Ingersoll-Rand (Cum.)	94	6.00	6.4	No	39.50	(d)	6.06
Standard Brands (Cum.)	110	7.00	6.4	No	18.00¶	15.86	9.90
Kansas City Power & Light (Cum.)	90¾	6.00	6.6	Yes	2.92	1.99	**
Otis Elevator (Cum.)	90	6.00	6.7	No	17.25	11.32	3.08
American Snuff (Non-cum.)	90	6.00	6.7	No	8.46	7.98	2.55
National Biscuit (Cum.)	101	7.00	6.9	No	11.31	9.37	4.76
Consolidated Gas Co. of New York (Cum.)	72½	5.00	6.9	Yes	2.92	2.70	0.70
Liggett & Myers (Cum.)	100	7.00	7.0	No	7.06	6.20	2.09
Brown Shoe (Cum.) (Sinking Fund)	100	7.00	7.0	Yes	5.63	5.04	1.65
Pacific Telephone & Telegraph (Cum.)	85¼	6.00	7.0	No	1.96	1.61	0.72
Corn Products Refining (Cum.)	99½	7.00	7.0	Yes	6.90	5.73	2.18
Island Creek Coal (Par $1) (Cum.)	85	6.00	7.06	No	13.58	9.28	2.62

* Not computed. All common stock owned by Philadelphia Company. No market.

† Not computed. All common stock owned by Public Service Company of New Jersey.

‡ Fiscal years ending June 30. Including year ended June 30, 1932, average would have been 14.75 and minimum 6.34.

§ Treating annual rental of $2,500,000 payable to Tobacco Products Corporation (N. J.) as an operating expense. Treating it as a fixed charge, average would have been 6.33 and minimum 4.52.

∥ Tobacco Products Corporation (N. J.) 6½% bonds, due in 2022, secured by lease of brands to American Tobacco Company, not included with funded debt of the latter. Including this item as funded debt of American Tobacco the stock-value ratio would be 2.31:1.

¶ Average for 2½ years since organization of company.

** Not computed. All common stock owned by Continental Gas and Electric Corporation.

while not an impossibility, is an exceptional phenomenon. It may be called exceptional not only in the numerical sense, but also from a more theoretical standpoint. In practically every instance in the above list, the preferred stock could have been replaced by a bond issue without affecting in any material degree the soundness of the corporation's capital structure. This means that the company itself derived no important advantage through having preferred stock outstanding instead of bonds, and on the other hand it suffered important disadvantages through income-tax liability and also because of the higher cost of its senior capital.[1] Stating the matter differently, in order that a preferred stock may be thoroughly sound, the burden it imposes must be so light that the company may just as readily carry that burden in the form of a bond obligation.

We are led therefore, to the final conclusion that not only are sound preferred stocks exceptional but in a certain sense they must be called anomalies or mistakes, because they are preferred issues which should really be outstanding as bonds. Hence the preferred stock *form* lacks basic justification, from an investment standpoint, in that it does not offer mutual advantages to both the issuer and the owner. Wherever the issuing business derives a real benefit from its discretionary right to suspend dividends, then the owner does not possess a fixed-value investment. And conversely, when the issue is a high-grade one, then the issuer derives no such benefit.

High-grade Preferred Stocks Usually Seasoned Issues.—In support of the above conclusion, it should be observed that high-grade industrial preferred issues have almost always reached this position as the result of many years of prosperous growth by the corporation *after* the preferred stock was first created. Exceedingly few preferred shares are so strongly entrenched at the *time of original sale* as to meet the stringent requirements needed for a full investment rating. For when a corporation is able to make as strong a showing as we require, it will nearly always prefer to do its financing through a relatively small bond issue, at a low interest rate and with substantial income-tax saving. This does not apply to the public-utility companies since, for reasons probably related to the "legal investment" status of their bond issues, they prefer to carry a portion of their

---

[1] Bond interest is deductible from earnings before arriving at the profit subject to income tax, but preferred dividends may not be so deducted.

senior financing in the form of stock. (Thus, four of the five high-grade utility preferred stocks included in the above list were floated in recent years.) But the *industrial* preferred shares in this list present an entirely different picture. Only one out of the 15 issues was actually sold to the public within the past 20 years, and even this exception (Procter and Gamble Company 5% Preferred) was floated to replace an older preferred issue at a lower dividend rate. The General Electric Company senior shares are the result of a stock-dividend plan, but the 13 other issues originated long ago and owe their investment status to the prosperous years which followed.

From the above reasoning, we state the practical rule that the purchase of new industrial preferred stock offerings on an investment basis is almost invariably a mistake of judgment, for it is a rare phenomenon to find such an offering thoroughly well secured.[1]

Origin of the Popularity of Preferred Stocks.—At the beginning of this discussion, we referred to the prominent role that preferred stocks have played in financing American corporations. But if our subsequent analysis is correct in concluding that this form of straight investment is fundamentally unsound, it may be asked why this unsoundness was not long ago convincingly demonstrated by the actual experience of investors. The answer is that the great popularity of preferred stocks developed during a 15-year period which rather accidentally favored the typical preferred stockholder against the typical bondholder. At the beginning of this period, just before the World War, the majority of preferred stocks were industrial issues and most of these were admittedly speculative in character, selling at substantial discounts from par. The tremendous prosperity and growth of our larger enterprises during the war, and during the years subsequent to 1922, effected a great improvement in the status and hence in the market price of many of the leading industrial preferred stocks. Within the same time, railroad and traction obligations, which constituted the main portion of the *bond list*, were subjected to influences of a generally adverse character. Investors, observing that the typical preferred stock was behaving better than the typical bond, drew the natural

[1] The purchase of preferred shares *at a large discount from par*, or having conversion or other profit-sharing features, represents a commitment of an entirely different sort, and will be discussed in our later chapters on Senior Securities with Speculative Features.

but erroneous inference that preferred stocks in general were intrinsically as sound as bonds.

Poor Record Shown by Extensive Study of Preferred Issues.— More detailed investigation will show that the popularity of preferred stocks rested upon the excellent performance of a comparatively *small number* of old-established, and prominent industrial issues. During the latter part of the period under review, the much more numerous *new* flotations of industrial preferred stocks, sold on the strength of this very popularity, did not fare so well. A study was made under the direction of the Harvard School of Business Administration, covering all the new preferred-stock offerings from January 1, 1915 to January 1, 1920 which ranked between $100,000 and $25,000,000 in size (607 issues in all). This showed that the average price of 537 issues for which quotations were obtainable on January 1, 1923, had declined to a figure 28.8% below the original offering price (from 99 to 70½), so that their purchasers had suffered a shrinkage in principal greater than the total income received. The conclusions drawn from this inductive study were highly unfavorable to preferred stocks as a form of straight investment.[1]

A More Recent Study.—A more recent investigation published by the Bureau of Business Research of the University of Michigan leads its author to a quite different opinion.[2] His "tests" of preferred stocks *preceded by bond issues* (both railroad and industrial) indicate clearly that senior shares of this type do not offer a satisfactory medium of investment. But with respect to industrial preferred stocks *not preceded by bonds*, the author's tests bring him to the opposite conclusion. Of these, he asserts that "they appear to meet the most exacting investment tests" and also that diversified investment in such issues would seem to "provide both a degree of safety for principal and an income return greater than that achieved by industrial or railroad bonds."

---

[1] Quotations were not obtainable, even from the issuing houses, for 70 out of 607 issues. Hence the loss to the investor was undoubtedly greater than that indicated by the 537 cases studied statistically. For further details of this study see: Arthur S. Dewing, "The Role of Economic Profits in the Return on Investments," *Harvard Business Review*, Vol. I, pp. 451, 461–462; Arthur S. Dewing, *Financial Policy of Corporations*, Book vi, Chap. 2, pp. 1198–1199, New York, 1926.

[2] Rodkey, Robert G., *Preferred Stocks as Long-term Investments*, Ann Arbor, University of Michigan Press, 1932.

The deduction that it is better to buy preferred stocks *without* rather than *with* bonds ahead of them is undoubtedly sound, since the latter group is clearly more vulnerable to adverse developments. But in our view the methods followed in this investigation are open to so many objections as to deprive its other conclusions of all practical value.[1] One feature of the study, however, deserves particular comment. The detailed figures show in striking fashion that the *stability* of nearly every preferred stock considered was directly dependent upon an *increase* in the value of the common stock. The preferred stockholder had a satisfactory investment only while the common stock was proving a profitable speculation. As soon as any common stock declined in market value below the original price, the preferred shares did likewise.

An investment subject to such conditions is clearly unwise. It is a case of: "Heads, the common stockholder wins; tails, the preferred stockholder loses." One of the basic principles of investment is that the safety of a security with limited return must never rest primarily upon the future *expansion* of profits. If the investor is positive that this expansion will take place, he should obviously buy the common stock and participate in its profits. If, as must usually be the case, he cannot be so certain of future prosperity, then he should not expose his capital to a risk of loss (by buying the preferred stock) without compensating opportunities for enhancement.

[1] For a brief statement of Mr. Rodkey's approach and of the objections thereto see Appendix, Note 25.

CHAPTER XV

TECHNIQUE OF SELECTION OF PREFERRED STOCKS FOR INVESTMENT

Our discussion of the theory of preferred stocks led to the practical conclusion that an investment preferred issue must meet all the requirements of a good bond, with an extra margin of safety to offset its contractual disadvantages. In analyzing a senior stock issue, therefore, the same tests should be applied as we have previously suggested and described with respect to bonds.

More Stringent Requirements Suggested.—In order to make the quantitative tests more stringent, it should be sufficient to increase the minimum earnings coverage above that prescribed for the various bond groups. The criteria we propose are as follows:

MINIMUM AVERAGE-EARNINGS COVERAGE

Class of enterprise	For investment bonds	For investment preferred stocks
Public Utilities........	1¾ times fixed charges	2 times fixed charges plus preferred dividends
Railroads.............	2 times fixed charges	2½ times fixed charges plus preferred dividends
Industrials...........	3 times fixed charges	4 times fixed charges plus preferred dividends

It does not appear to us to be necessary to make corresponding advances in the minimum figures for size of the enterprise or for the stock-value ratio, because of the secondary importance of these tests as compared with the earnings coverage. The margins of safety above suggested are materially higher than those hitherto accepted as adequate, and it may be objected that we are imposing requirements of unreasonable and prohibitive stringency. It is true that these requirements would have disqualified a large part of the preferred-stock financing done in the years prior to 1931, but such severity would have been of

164

benefit to the investing public. A general stabilization of business and financial conditions may later justify a more lenient attitude towards the minimum earnings coverage, but until such stabilization has actually been discernible over a considerable period of time the attitude of investors towards preferred stocks must remain extremely critical and exacting.

Referring to the list of preferred stocks given on page 159, it will be noted that in the case of all the *industrial* issues the stock-value ratio at its lowest exceeded 1.6 to 1, and also that the average earnings coverage exceeded 5.6 times.

Mere Presence of Funded Debt Does Not Disqualify Preferred Stocks for Investment.—It is proper to consider whether an investment rating should be confined to preferred stocks not preceded by bonds. That the absence of funded debt is a desirable feature for a preferred issue goes without saying; it is an advantage similar to that of having a first mortgage on a property instead of a second mortgage. It is not surprising, therefore, that preferred stocks without bonds ahead of them have *as a class* made a better showing than those of companies with funded debt. But from this rather obvious fact it does not follow that *all* preferred stocks with bonds preceding are unsound investments, any more than it can be said that *all* second-mortgage bonds are inferior in quality to *all* first-mortgage bonds. Such a principle would entail the rejection of all public-utility preferred stocks (since they invariably have bonds ahead of them) although these are better regarded as a group than are the "nonbonded" industrial preferreds. Furthermore, in the extreme test of 1932, a substantial percentage of the preferred issues which held up were preceded by funded debt.[1]

To condemn a powerfully entrenched security such as General Electric preferred because it has an infinitesimal bond issue ahead of it, would be the height of absurdity. This example should illustrate forcibly the inherent unwisdom of subjecting investment selection to hard and fast rules of a *qualitative* character. In our view, the presence of bonds senior to a preferred stock is a fact which the investor must take carefully into account, impelling him to greater caution than he might otherwise exercise; but if the company's exhibit is sufficiently

---

[1] Out of the 21 such issues listed on p. 159 11 were preceded by bonds, *viz.*, five public utilities, one railroad, and five (out of 15) industrials.

impressive the preferred stock may still be accorded an investment rating.

Total-deductions Basis of Calculation Recommended.—In calculating the earnings coverage for preferred stocks with bonds preceding, it is absolutely essential that the bond interest and preferred dividend be taken *together*. The almost universal practice of stating the earnings on the preferred stock separately (in dollars per share) is exactly similar to, and as fallacious as, the prior-deductions method of computing the margin above interest charges on a junior bond. If the preferred stock issue is much smaller than the funded debt, the earnings per share will indicate that the preferred dividend is earned more times than is the bond interest. Such a statement must either have no meaning at all, or else it will imply that the preferred dividend is safer than the bond interest of the same company, an utter absurdity.[1]

EXAMPLES OF CORRECT AND INCORRECT METHODS OF CALCULATING
EARNINGS COVERAGE FOR PREFERRED STOCKS

A. Colorado Fuel and Iron Company: 1929 figures

Earned for bond interest	$3,978,000
Interest charges	1,628,000
Preferred dividends	160,000
Balance for common	2,190,000

Customary but incorrect statement		Correct statement	
Interest charges earned	2.4 times	Interest charges earned	2.4 times
Preferred dividend earned	14.7 times	Interest and preferred dividends earned	2.2 times
Earned per share of preferred	$117.50		

NOTE: The above statement of earnings on the preferred stock alone is either worthless or dangerously misleading.

B. Warner Bros. Pictures, Inc.:   Year ended
                                                            Aug. 30, 1930

Earned for interest	$12,553,000
Interest charges	5,478,000
Preferred dividends	403,000
Balance for common	6,672,000

[1] See Appendix, Note 26, for comment upon neglect of this point by writers of textbooks on investment.

Customary but incorrect statement		Correct statement	
Interest charges		Interest charges	
earned..........	2.3 times	earned............	2.3 times
Preferred dividends		Interest and preferred	
earned..........	17.5 times	dividends earned...	2.1 times
Earned per share of			
preferred........	$68.61		

C. West Penn Electric Company:

	1931 figures
Gross...	$35,739,000
Net before charges...............................	14,405,000
Fixed charges (include preferred dividends of subsidiaries).....................................	8,288,000
Dividends on 7% and 6% preferred issues...........	2,268,000
Dividends on Class *A* stock (junior to 6% and 7% Pfd.).	415,000
Balance for common.............................	3,434,000

Customary but incorrect statement			Correct statement	
Times interest or dividends earned	Earned per share		Times earned	
Fixed charges... 1.74 times			Fixed charges.. 1.74 times	
6% and 7% preferred (combined)... 2.70 times	$17.93		Charges and preferred dividends ... 1.36 times	
Class *A*........ 9.28 times	64.96		Fixed charges, preferred dividends, and Class *A* dividends....... 1.31 times	

The West Penn Electric Company Class *A* stock is in reality a second preferred issue. In this example the customary statement makes the preferred dividend appear safer than the bond interest; and because the Class *A* issue is small, it makes this second preferred issue appear much safer than either the bonds or the first preferred. The correct statement shows that the Class *A* requirements are covered 1.31 times instead of 9.28 times, a tremendous difference. The erroneous method of stating the earnings coverage was probably responsible in good part for the high price at which the Class *A* shares sold in 1931 (105¼). In 1932 they declined to 25, but curiously enough sold repeatedly above the 7% preferred issue. Evidently some investors were still misled by the per-share earnings figures, and imagined the second preferred safer than the first preferred.

An Apparent Contradiction Explained.—Our principles of preferred-dividend coverage lead to an apparent contradiction, *viz.*, that the preferred stockholders of a company must require a larger minimum coverage than the bondholders of the same company, yet by the nature of the case the actual coverage is bound to be smaller. For in any corporation the bond interest alone is obviously earned with a larger margin than the bond interest and preferred dividends combined. This fact has created the impression among investors (and their advisers) that the tests of a sound preferred stock must necessarily be less stringent than those of a sound bond.[1] But this is not true at all. The real point is that where a company has both bonds and preferred stock the preferred stock can be safe enough *only if the bonds are much safer than necessary.* Conversely, if the bonds are only just safe enough, the preferred stock cannot be sound. This is illustrated by two examples, as follows:

Year	Liggett & Myers Tobacco Co.		Commonwealth & Southern Corp.	
	Number of times interest earned	Number of times int. and pfd. dividend earned	Number of times fixed charges earned	Number of times fixed charges and pfd. dividend earned
1930	15.2	7.87	1.84	1.48
1929	13.9	7.23	1.84	1.55
1928	12.3	6.42	1.71	1.44
1927	11.9	6.20	1.62	1.37
1926	11.2	5.85	1.52	1.31
1925	9.8	5.14	1.42	1.28

The Liggett and Myers preferred-dividend coverage (including, of course, the bond interest as well) is substantially above

[1] See, for example, the following quotations from R. E. Badger, *Investment Principles and Practices*, New York, 1928:

"Similarly, it is a general rule that, on the average, the interest on industrial bonds should be covered at least three times, in order that the bond should be considered safe" (pp. 306–307).

"To the author's mind, an industrial preferred stock should be regarded as speculative unless combined charges and dividend requirements are earned at least twice over a period of years" (p. 309).

"One is probably safe in stating that, where combined charges are twice earned, including interest charges on the bonds of the holding company, the presumption is in favor of the soundness of such holding company issue. Likewise, where combined prior charges and preferred dividend requirements are earned 1.5 times, the presumption is in favor of such preferred stock of the holding company" (p. 414).

our suggested minimum of four times. The bond-interest coverage alone is therefore far in excess of the smaller minimum required for it, *viz.*, three times. On the other hand, the Commonwealth and Southern fixed-charge coverage in 1930 was just about at the proposed minimum of 1¾ times. This meant that while the various bonds *might* qualify for investment, the 6% preferred stock could not possibly do so, and the purchase of that issue at a price above par in 1930 was an obvious mistake.

"Dollars-per-share" Formula Misleading.—When a preferred stock has no bonds ahead of it, the earnings may be presented either as so many dollars per share or as so many times dividend requirements. The second form is distinctly preferable, for two reasons. The more important one is that the use of the "dollars-per-share" formula in cases where there are no bonds is likely to encourage its use in cases where there are bonds. Security analysts and intelligent investors should make special efforts to avoid and decry this misleading method of stating preferred-dividend coverage, and this may best be accomplished by dropping the dollars-per-share form of calculation entirely. As a second point, it should be noted that the significance of the dollars earned per share is dependent upon the rate of dividend carried by the preferred stock. Earnings of $20 per share would be much more favorable for a $5 preferred issue selling at 80 than for an $8 preferred selling at 125. In the one case the dividend is earned four times, in the other, 2½ times. The dollars-per-share figure loses all comparative value when the par value is less than $100, or when there is no-par stock with a low dividend rate per share. Earnings of $18.60 per share in 1931 on S. H. Kress and Company 6% Preferred (par $10) are of course far more favorable than earnings of $20 per share on some 7% preferred stock, par $100.

Calculation of the Stock-value Ratio.—The technique of applying this test to preferred stocks is in all respects similar to that of the earnings-coverage test. The bonds, if any, and the preferred stock must be taken together and the total compared with the market price of the common stock only. When calculating the protection behind a bond, the preferred issue is part of the stock equity; but when calculating the protection behind the preferred shares, the common stock is now, of course, the only junior security. In cases where there are both a first and second preferred issue, the second preferred is added to the common stock in calculating the equity behind the first preferred.

EXAMPLE OF CALCULATION OF STOCK-VALUE RATIOS FOR PREFERRED STOCKS
Procter and Gamble Company

Capitalization	Face amount	Low price 1932	Value at low price in 1932
Bonds..........................	$10,500,000		
8% pfd. (1st pfd.)..............	2,250,000	@ 140	$ 3,150,000
5% pfd. (2d pfd.)...............	17,156,000	@ 81	13,900,000
Common (number of shares).....	6,410,000	@ 20	128,200,000

A. Stock-value ratio for bonds

$$\frac{3{,}150{,}000 + 13{,}900{,}000 + 128{,}200{,}000}{10{,}500{,}000} = 13.8{:}1$$

B. Stock-value ratio for 1st pfd.

$$\frac{13{,}900{,}000 + 128{,}200{,}000}{10{,}500{,}000 + \quad 3{,}150{,}000} = 10.4{:}1$$

C. Stock-value ratio for 2d pfd.

$$\frac{128{,}200{,}000}{10{,}500{,}000 + 3{,}150{,}000 + 13{,}900{,}000} = 4.6{:}1$$

Should the market value of the common stock be compared with the *par* value or the *market* value of the preferred? In the majority of cases it will not make any vital difference which figure is used. There are, however, an increasing number of no-par-value preferreds (and also a number like Island Creek Coal Company Preferred and American Zinc, Lead and Smelting Company Preferred in which the real par is entirely different from the stated par).[1] In these cases an equivalent would have to be constructed from the dividend rate. Because of such instances and also those where the market price tends to differ materially from the par value (*e.g.*, Norfolk and Western Railway Company 4% Preferred or Procter and Gamble Company 8% preferred in 1932), it would seem the better rule to use the *market price* of preferred stocks regularly in computing stock-value ratios. On the other hand the regular use of the *face value* of bond issues, rather than the market price, is recommended, because it is much more convenient and does not involve the objections just discussed in relation to preferred shares.

[1] Island Creek Coal Preferred has a stated par of $1 and American Zinc Preferred has a stated par of $25, but both issues carry a $6 dividend and they are entitled to $120 per share and $100 per share respectively in the event of liquidation. Their true par is evidently $100.

Noncumulative Issues.—The theoretical disadvantage of a noncumulative preferred stock as compared with a cumulative issue is very similar to the inferiority of preferred stocks in general as compared with bonds. The drawback of not being able to compel the payment of dividends on preferred stocks generally is almost matched by the handicap in the case of noncumulative issues of not being able to receive in the future the dividends withheld in the past. This latter arrangement is so patently inequitable that new security buyers (who will stand for almost anything) object to noncumulative issues, and for many years new offerings of straight preferred stocks have almost invariably had the cumulative feature.[1] Noncumulative issues have generally come into existence as the result of reorganization plans in which old security holders have been virtually forced to accept whatever type of security was offered them. But in recent years the preferred issues created through reorganization have been preponderantly cumulative, though in some cases this provision becomes operative only after a certain interval. Austin Nichols and Company $5 Preferred, for example, was issued under a Readjustment Plan in 1930 and became cumulative in 1934. Missouri-Kansas-Texas Railroad Company Preferred, created in 1922, became cumulative in 1928.

Chief Objection to Noncumulative Provision.—One of the chief objections to the noncumulative provision is that it permits the directors to withhold dividends even in good years, when they are amply earned, the money thus saved inuring to the benefit of the common stockholders. Experience shows that noncumulative dividends are seldom paid unless they are necessitated by the desire to declare dividends on the common; and if the common dividend is later discontinued, the preferred dividend is almost invariably suspended soon afterwards.[2]

[1] The only important "straight," noncumulative preferred stock sold to stockholders or the public since the war was St. Louis-San Francisco Railway Company Preferred. In the case of Illinois Central Railroad Company Noncumulative Preferred, the conversion privilege was the overshadowing inducement at the time of issue.

[2] Kansas City Southern Railway Company 4% Noncumulative Preferred, which paid dividends between 1907 and 1929 while the common received nothing, is an outstanding exception to this statement. St. Louis Southwestern Railway Company 5% Noncumulative Preferred received full dividends during 1923–1929 while no payments were made on the common; but for a still longer period preferred dividends, although earned, were wholly or partially withheld (and thus irrevocably lost).

Example: St. Louis-San Francisco Railway Company affords a typical example. No dividends were paid on the (old) preferred issue between 1916 and 1924, although the dividend was fully earned in most of these years. Payments were not commenced until immediately before dividends were initiated on the common; and they were continued (on the new preferred) less than a year after the common dividend was suspended in 1931.

The manifest injustice of such an arrangement led the New Jersey courts (in the United States Cast Iron Pipe case)[1] to decide that if dividends are earned on a noncumulative preferred stock but not paid, then the holder is entitled to receive such amounts later before anything can be paid on the common. This meant that in New Jersey a noncumulative preferred stock was given a cumulative claim on dividends to the extent that they were earned. The United States Supreme Court however, handed down a contrary decision (in the Wabash Railway case)[2] holding that while the noncumulative provision may work a great hardship on the holder, he has nevertheless agreed thereto when he accepted the issue. This is undoubtedly sound law, but the inherent objections to the noncumulative provision are so great (chiefly because of the opportunity it affords for unfair policies by the directors) that it would seem to be advisable for the legislatures of the several states to put the New Jersey decision into statutory effect by prohibiting the creation of completely noncumulative preferred stocks, requiring them to be made cumulative at least to the extent that the dividend is earned. This result has been attained in a number of individual instances through insertion of appropriate charter provisions.[3]

Features of the List of 21 Preferred Issues of Investment Grade.—Out of some 440 preferred stocks listed on the New York Stock Exchange in 1932, only 40, or 9%, were noncumulative.

---

[1] *Day v. United States Cast Iron Pipe and Foundry Company,* 94 N.J. Eq. 389, 123 Atl. 546 (1924), *aff'd.* 96 N.J. Eq. 738, 126 Atl. 302 (1925); *Moran v. United States Cast Iron Pipe and Foundry Company,* 95 N.J. Eq. 389, 123 Atl. 546 (1924), *aff'd,* 96 N.J. Eq. 698, 126 Atl. 329 (1925).

[2] *Wabash Railway Company et al. v. Barclay et al.,* 280 U.S. 197 (1930), reversing *Barclay v. Wabash Railway,* 30 Fed. (2d) 260 (1929). See discussion in A. A. Berle, Jr., and G. C. Means, *The Modern Corporation and Private Property,* pp. 190–192.

[3] See, for example, the provisions of United Stores Corporation Class *A* Convertible Preferred; Aeolian Company 6% Class *A* Preferred; United States Lines Company Convertible Second Preferred.

Of these, 29 were railroad or street-railway issues and only 11 were industrial issues. The reader will be surprised to note, however, that out of only 21 preferred stocks selling continuously on an investment basis in 1932, no less than four were noncumulative. Other peculiarities are to be found in this favored list, and they may be summarized as follows (see page 159):

1. Both the number of noncumulative issues and the number of preferred stocks preceded by bonds are proportionately higher among the 21 "good" companies than in the Stock Exchange list as a whole.
2. The industry best represented is the *snuff* business, with three companies.
3. Miscellaneous peculiarities:
   a. Only one issue has a sinking fund provision.
   b. One issue is a second preferred (Procter and Gamble).
   c. One issue has a par value of only $1 (Island Creek Coal).
   d. One issue is callable at close to the lowest market price of 1932–1933 (General Electric).

Matters of Form, Title, or Legal Right Relatively Immaterial.— We trust that no overzealous exponent of the inductive method will conclude from these figures either: (1) that noncumulative preferreds are superior to cumulative issues; or (2) that preferreds preceded by bonds are superior to those without bonds; or (3) that the snuff business presents the safest opportunity for investment. The real significance of these unexpected results is rather the striking confirmation they offer to our basic thesis that matters of form, title, or legal right are relatively immaterial, and that the showing made by the individual issue is of paramount importance. If a preferred stock could always be expected to pay its dividend without question, then whether it is cumulative or noncumulative would become an academic question solely, in the same way that the inferior contractual rights of a preferred stock as compared with a bond would cease to have practical significance. Since the dividend on United States Tobacco Company Preferred was earned more than sixteen times in the depression year 1931—and since, moreover, the company had been willing to buy in a large part of the preferred issue at prices ranging up to $125 per share—the lack of a cumulative provision caused the holders no concern at all. This example must of course, be considered as exceptional; and as a point of practical investment policy we should suggest that no matter how impressive may be the exhibit of a noncumulative preferred stock, it would be better to select a cumulative issue for purchase in order to enjoy better protection in the event of unexpected reverses.[1]

Amount Rather Than Mere Presence of Senior Obligations Important.—The relatively large number of companies in our list having bonds outstanding is also of interest, as demonstrating that it is not the mere *presence* of bonds, but rather the *amount* of the prior debt which is of serious moment. In three cases the bonds were outstanding in merely a nominal sum, as the result of the fact that nearly all of these companies had a long history, so that some of them carried small residues of old bond financing.[2]

By a coincidence all three of the noncumulative industrial preferred stocks in our list belong to companies in the *snuff* business. This fact is interesting, not because it proves the investment primacy of snuff, but because of the strong reminder it offers that the investor cannot safely judge the merits or demerits of a security by his personal reaction to the kind of business in which it is engaged. An outstanding record for a long period in the past, plus strong evidence of inherent stability, plus the absence of any concrete reason to expect a substantial change for the worse in the future, afford probably the only sound basis available for the selection of a *fixed-value* investment. The miscellaneous peculiarities in our list (mentioned under 3, above) are also useful indications that matters of form or minor drawbacks have no essential bearing on the quality of an investment.

[1] See, for example, the record of American Car and Foundry Company 7% Noncumulative Preferred. For many years prior to 1928 this issue sold higher than United States Tobacco Company 7% Noncumulative Preferred. By 1929 it had completed 30 years of uninterrupted dividend payments, during the last 20 of which its market price had never fallen below 100. Yet in 1932 the dividend was passed and the quotation declined to 16. Similarly, Atchison, Topeka and Santa Fe Railway Company Preferred, a 5% noncumulative issue, paid full dividends between 1901 and 1932 and was long regarded as a gilt-edged investment. As late as 1931 the price reached 108¼, *within a half-point of the highest level in its history*, and a yield of only 4.6%. The very next year the price fell to 35, and in the following year the dividend was reduced to a $3 basis.

[2] These companies were General Electric, American Tobacco, and Corn Products Refining. The University of Michigan study by Rodkey recognizes this point in part by certain exclusions of bond issues amounting to less than 10% of capital and surplus.

CHAPTER XVI

INCOME BONDS AND GUARANTEED SECURITIES

I. INCOME BONDS

The contractual position of an income bond (sometimes called an adjustment bond) stands midway between that of a straight bond and a preferred stock. Practically all income obligations have a definite maturity, so that the holder has an unqualified right to repayment of his principal on a fixed date. In this respect his position is entirely that of the ordinary bondholder. However, it should be pointed out that income bonds are almost always given a long maturity date, so that the right of repayment is not likely to be of practical importance in the typical case studied. In fact we have discovered only one instance of income bondholders actually having received repayment of their principal in full by reason of maturity.[1]

Interest Payment Sometimes Wholly Discretionary.—In the matter of interest payments some income bonds are almost precisely in the position of a preferred stock, because the directors are given practically complete discretion over the amounts to be paid to the bondholders. The customary provisions require that interest be paid to the extent that income is available, but many indentures permit the directors to set aside whatever portion of the income they please for capital expenditures or other purposes, before arriving at the "available" balance. In the case of the Green Bay and Western Railroad Company

[1] This was a $500,000 issue of Milwaukee Lake Shore and Western Income 6s, issued in 1881, assumed by the Chicago and Northwestern in 1891, and paid off at maturity in 1911. St. Louis-San Francisco Railway Company Income 6s and Adjustment 6s were both *called* for repayment at par in 1928, which was 32 and 27 years, respectively, prior to their maturity. This proved fortunate for the bondholders since the road went into receivership in 1932. The history of the 'Frisco between its emergence from receivership in 1916 and its subsequent relapse into receivership in 1932 is an extraordinary example of the heedlessness of both investors and speculators, who were induced by a moderate improvement, shown in a few years of general prosperity, to place a high rating on the securities of a railroad with a poor previous record and a topheavy capital structure.

Income Debentures "Series B," the amounts paid out between 1922 and 1931, inclusive, aggregated only 6% although the earnings were equal to only slightly less than 22%.   The more recent indentures (*e.g.*, Chicago, Milwaukee, St. Paul and Pacific Railroad Company Adjustment 5s) tend to place definite limits on the percentage of earnings which may be withheld in this manner from the income bondholders; but a considerable degree of latitude is usually reserved to the directors.   It may be said that individual income-bond issues may be found illustrating almost every step in the range of variation between straight preferred stocks and ordinary bonds.

Low Investment Rating of Income Bonds as a Class.—Since the contractual rights of income bonds are always more or less superior to those of preferred stocks, it might be thought that a greater proportion of income bonds than of preferred stocks would deserve an investment rating.   Such is not the case, however.   In fact we know of only one income obligation which has maintained an investment standing continuously over any length of time, *viz.*, Atchison Topeka and Santa Fe Railway Company Adjustment 4s, due 1995.   We have here a contrast between theory and actuality, the reason being, of course, that income bonds have been issued almost exclusively in connection with corporate reorganizations and have therefore been associated with companies of secondary credit standing.   The very fact that the interest payments are dependent on earnings implies the likelihood that the earnings may be insufficient.   Preferred-stock dividends are equally dependent upon earnings, but the same implication is not associated with them.   Hence the general investment status of income bonds as a class is seen to have been governed by the circumstances under which they are created rather than by the legal rights which attach to them.   To use an analogy: If it had been the general practice here, as in England, to avoid mortgage-bond issues wherever possible, using them only where doubtful credit made this protection necessary, then we might find that mortgage bonds in general would occupy an investment position distinctly inferior to that of debenture bonds.

Increased Volume of Income Bonds Probable.—Looking forward, it may be true that in the future income obligations will show a larger proportion of investment issues than will be found among preferred stocks.   The numerous reorganizations growing out of the 1930–1933 depression are creating a large new crop of income bonds, and some of these companies may later so improve their position as to place their income obligations in the investment class, as happened to the Atchison, Topeka and Santa Fe after its reorganization in 1895. There is also the point, so far overlooked, that income bonds effect a substantial saving in corporation taxes as compared with preferred stocks, without important offsetting disadvantages. Some strong companies may be led to replace their present preferred stocks—or to do their new financing—by income obligations, for the sake of this tax saving, in the same way as they are now creating artificially low par values for their shares to reduce the transfer taxes thereon. A development of this kind in the future might result in a respectable number of income-bond issues deserving to rank as fixed-value investments.[1]

Calculations of Margins of Safety for Income Bonds.—The technique of analyzing an income-bond exhibit is identical with that for a preferred stock. Computations of earnings on the issue taken separately must, of course, be rigorously avoided, although such calculations are given by the statistical agencies.

We suggest that the minimum earnings coverage recommended in the preceding chapter for preferred stocks be required also for income bonds when selected as fixed-value investments.

Example: The following analysis of the Missouri-Kansas-Texas Railroad Company income account for 1930 will illustrate the proper method of dealing with all the senior securities of a company having adjustment bonds. It also shows how the two methods of figuring the fixed charges of a railroad system (discussed in Chap. XII) are to be applied to the analysis of income bonds and preferred stock.

Missouri-Kansas-Texas Railroad Company, Calendar Year 1930
(All dollar figures in thousands)

Gross revenue..	$45,949
Railway operating income (net after taxes)........	13,353
Gross income (net after rents, plus other income)....	12,009
Fixed charges (fixed interest and other deductions)..	4,230
Balance for adjustment interest..................	7,779
Adjustment interest...........................	696

[1] The Associated Gas and Electric Company used the device of "bonds" convertible into preferred stock *at the option of the company,* and obtained this tax saving without the burden of a fixed bond obligation. The income-bond form would have been far less misleading to the ordinary investor than this extraordinary invention.

Missouri-Kansas-Texas Railroad Company, Calendar Year 1930
(Continued)

Balance for dividends (net income)................	7,083
Preferred dividends............................	4,645
Balance for common............................	2,438

Net after taxes exceeds gross income.   Hence use net-deductions test.
Net deductions = difference between net after taxes and balance for adjust-
          ment interest
         = \$13,353 − \$7,779.

Times earned

Net deductions $\qquad\qquad\qquad\qquad$ = \$ 5,574 $\quad \dfrac{\$13,353}{\$\ 5,574} = 2.40$

Net deductions and adjustment interest $\quad$ = $\quad$ 6,270 $\quad \dfrac{\$13,353}{\$\ 6,270} = 2.14$

Net deductions, adjustment interest and pre- = \$10,915 $\quad \dfrac{\$13,353}{\$10,915} = 1.22$
ferred dividends

Note that interest on income or adjustment bonds is not part of the total interest charges when calculating the coverage for the *fixed-interest* bonds.   In this respect the position of an income bond is exactly that of a preferred stock.   Note also that the statement made by the statistical services that 57.29% was earned on the M-K-T Adjustment 5s. (*i.e.,* that the "interest was covered" more than eleven times) is valueless or misleading.

Significance of These Figures for the Investor in Early 1931.— The 1930 earnings were somewhat lower than the ten-year average and could apparently be viewed as a fair indication of the normal earning power of M-K-T.   The coverage for the preferred stock was clearly inadequate from any investment standpoint.   The coverage for the adjustment-bond interest on the more conservative basis (the net-deductions method) was below our minimum requirement of 2½ times, so that this issue would not have qualified for investment.   The coverage for the fixed-bond interest was substantially above our minimum and indicated a satisfactory degree of protection.

Naturally the disastrous decline of earnings in 1931–1933 could not have been foreseen or fully guarded against.   The market price of M-K-T fixed obligations suffered severely in 1932; but since the company's debt structure was relatively conservative, it did not come so close to insolvency as the majority of other carriers.   In fact, the 1932 and 1933 interest was paid on the adjustment bonds, although such payment was not obligatory.

Senior Income Bonds.—There are a few instances of income bonds which are senior in their lien to other bonds bearing fixed interest. The Atchison Adjustment 4s are the best known example, being followed by 4% fixed-interest debenture issues which have regularly sold at a lower price. The situation holds true also with respect to St. Louis Southwestern Railway Company Second Income 4s.[1] While the theoretical status of such bonds is rather confusing, the practical procedure called for is, obviously, to treat the interest thereon as part of the company's *fixed charges.*

II. GUARANTEED ISSUES

No special investment quality attaches to guaranteed issues as such. Inexperienced investors may imagine that the word "guaranteed" carries a positive assurance of safety; but, needless to say, the value of any guaranty depends strictly upon the financial condition of the guarantor. If the guarantor has nothing, the guaranty is worthless. In contrast with the attitude of the financial novice, Wall Street displays a tendency to *underestimate* the value of a guaranty, as shown by the lower prices often current for guaranteed issues in comparison with the debentures or even the preferred stock of the guarantor. This sophisticated distrust of guarantees dates back to the Kanawha and Hocking Coal and Coke Company case in 1915, when the guarantor railroad endeavored to escape its liability by claiming that the guaranty, made in 1901, was beyond its corporate powers and hence void. This attempt at evasion, encouraged by the outcome of antitrust suits in the Ohio and federal courts, in the end proved completely unsuccessful; but it cast a shadow over the value of all guarantees, from which they have not completely emerged even after 20 years.[2] We know of no important case in which a solvent company has escaped the consequences of its guaranty through legal technicalities.[3]

[1] An unusual case is afforded by Wabash Railway Noncumulative Income Debenture 6s, due 1939, interest on which is payable "from net income." Although called debentures, they are secured by a direct lien and have priority over the Wabash Railway Refunding and General Mortgage. Although entitled by their terms only to noncumulative interest dependent on earnings, this interest was paid regularly from 1916 through 1933, despite the fact that the company entered receivership in 1931 and defaulted upon the junior-mortgage (fixed) interest in 1932.

[2] See Appendix, Note 27, for a condensed history of this famous case.

[3] See Appendix, Note 28, for further comment and examples bearing on this statement.

Status of Guaranteed Issues.—If a company guarantees interest, dividend, or principal payments, its failure to meet this obligation will expose it to receivership. The claim against the guarantor ranks equally with any unsecured debt of the company, so that guaranteed issues deserve the same rating as a debenture bond of the guarantor and a better rating than its preferred stock. A guaranteed issue may also be entitled to an investment rating because of its own position and earning power independent of the guaranty. In such cases the guaranty may add to its security, but it cannot detract therefrom even if the guarantor company itself is in bad straits.

Examples: The Brooklyn Union Elevated Railroad 5s (see page 30) were guaranteed by the Brooklyn Heights Railroad Company, which went into receivership in 1919; but the bond came through the reorganization unscathed because of its own preferred position in the Brooklyn Rapid Transit System. Similarly U. S. Industrial Alcohol Company Preferred dividends were guaranteed by Distilling Company of America; the latter enterprise became bankrupt, but the Alcohol Company was easily able to continue the dividend out of its own earnings and later to retire the preferred issue at 125.

A common or preferred stock fully guaranteed by another company has the status of a bond issue as far as the guarantor is concerned. If the guaranty proves worthless, it would naturally return to the position of a stock—usually a weak issue, but possibly a strong one, as in the case of U. S. Industrial Alcohol Company Preferred just mentioned. A similar situation obtains with respect to income bonds of one company guaranteed by another (*e.g.*, Chicago, Terre Haute, and Southeastern Railway Company Income 5s, *guaranteed* by the Chicago, Milwaukee, St. Paul and Pacific Railroad Company).

Exact Terms of Guaranty Are Important.—The exact terms of a guaranty have obviously a vital influence upon its value. A guaranty of interest only is likely to be much less significant than a guaranty of principal as well.

Example: Philippine Railway Company First 4s, due 1937, for example, are guaranteed as to interest only by the Philippine government, the bonds of which are designated as instrumentalities of the United States. There seems no doubt at all that the *interest* will be paid promptly up to maturity. But the absence of a guaranty of principal, and the uncertainty as to what will happen at maturity because of the poor earnings of the railroad itself, have served to keep the price of the issue at a very low level. (The lowest price in 1932 was 16½. At this level, however, they were clearly selling too low if it is assumed that the bonds would have *any value* at maturity.)

A similar disadvantage attaches to a guaranty of dividends running for a limited period.

Example: The actual working out of such a situation was shown in the case of American Telegraph and Cable Company common stock, which was guaranteed as to 5% dividends only for 50 years from 1882 by the Western Union Telegraph Company under a lease terminating in 1932. Because of the long record of dividend payments, investors came finally to consider the dividend as a fixture, and as late as 1922 the stock sold at 70. But in the meantime the strategic or trade value of the leased cable properties was rapidly diminishing, so that the value of the stock at the expiration of the lease was likely to be very small. A settlement was made in 1930 with Western Union under which the American Telegraph and Cable stockholders received the equivalent of about $20 for the principal of their stock.[1]

Example: A rather unusual example of the importance of the exact terms of a guaranty was supplied by Pratt and Whitney Preferred (retired in 1928). According to the security manuals, the dividend on this issue was "guaranteed" by its parent company, Niles-Bement-Pond. But in fact the Niles company agreed to make up unpaid dividends on Pratt and Whitney Preferred only to the extent that Niles had earnings available therefor after payment of its own preferred dividends. Hence no dividends were received by Pratt and Whitney Preferred stockholders from November 1924 to June 1926 without any claim being enforceable against Niles-Bement-Pond. In view of the possibility of such special provisions, particular care must be exercised to obtain complete information regarding the terms of a guaranty before purchasing any security on the strength thereof.

Joint and Several Guarantees.—Such guarantees are given by more than one company to cover the same issue, and each company accepts responsibility not only for its pro rata share but also for the share of any other guarantor who may default. In other words, each guarantor concern is potentially liable for the

[1] An alert investor might have taken warning of this possibility from statements contained in the annual reports of Western Union, starting with 1913, wherein this company's own holdings of American Telegraph and Cable stock were written down annually towards an estimated value of $10 per share in 1932.

entire amount of the issue. Since two or more sponsors are better than one, bonds bearing a joint and several guarantee are likely to have special advantages.

Example: The most familiar class of issues backed by such a guaranty are the bonds of union railroad stations. An outstanding example is supplied by Kansas City Terminal Railway Company First 4s, due 1960, which are guaranteed jointly and severally by no less than 12 railroads all of which use the company's facilities. The 12 guarantors are as follows: Atchison, Alton, Burlington, St. Paul, Great Western, Rock Island, Kansas City Southern, M-K-T, Missouri Pacific, 'Frisco, Union Pacific, and Wabash.

The value of each of these individual guarantees has varied greatly from road to road and from time to time, but at least three of the companies have consistently maintained sufficient financial strength to assure a Terminal bondholder that his obligation would be met without difficulty. Investors have not fully appreciated the superior protection accorded by the combined responsibility of the 12 carriers as compared with the liability of any one of them singly. The price record shows that the Kansas City Terminal Railway Company 4s frequently sold at no higher prices than representative issues of individual guarantor companies which later turned out to be of questionable soundness, whereas at no time was the safety of the Terminal bond ever a matter of doubt.[1]

It would seem good policy for investors, therefore, to favor bonds of this type, which carry the guaranty of a *number* of substantial enterprises, in preference to the obligations of a single company.

Federal Land Bank Bonds.—A somewhat different aspect of the joint and several guarantee appears in the important case of the Federal Land Bank bonds, which are secured by deposit of farm mortgages. The obligations of each of the 12 separate banks are guaranteed by the 11 others, so that each Federal Land Bank bond is in reality a liability of the entire system. When these banks were organized, there was created concurrently a group of Joint Stock Land Banks which also issued bonds, but the obligations of one Joint Stock Bank were not guaranteed by the others.[2] Both sets of land banks were under United

---

[1] See Appendix, Note 29, for supporting data.

[2] The word "Joint" in the title referred to the ownership of the stock by various interests, but it may have created an unfortunate impression among

States government supervision and the bonds of both were made exempt from federal taxation. Practically all of the stock of the Federal Land Banks was subscribed for originally by the United States government (which, however, did not assume liability for their bonds); the Joint Stock Land Bank shares were privately owned.

At the inception of this dual system, investors were disposed to consider the federal supervision and tax exemption as a virtual guarantee of the safety of the *Joint Stock* Land Bank bonds, and they were therefore willing to buy them at a yield only ½% higher than that returned by the *Federal* Land Bank bonds. In comparing the nonguaranteed Joint Stock bonds with the mutually guaranteed federal bonds, the following observations might well have been made:

1. Assuming the complete success of the farm-loan system, the guarantee would be superfluous, since each bond issue separately would have enjoyed ample protection.

2. Assuming complete failure of the system, the guarantee would prove worthless, since all the banks would be equally insolvent.

3. For any intermediate stage between these two extremes, the joint and several guarantee might prove extremely valuable. This would be particularly true as to bonds of a farm-loan district subjected to extremely adverse conditions of a *local* character.

In view of the fact that the farm-loan system was a new and untried undertaking, investors therein should have assured themselves of the largest possible measure of protection. Those who in their eagerness for the extra ½% of income return dispensed with the joint guarantee committed a patent mistake of judgment.[1]

---

investors that there was a joint responsibility by the group of banks for the liabilities of each.

[1] A number of the Joint Stock bond issues defaulted during 1930–1932, a large proportion sold at receivership prices, and all of them declined to a speculative price level. On the other hand, not only were there no defaults among the Federal Land Bank bonds, but their prices suffered a relatively moderate shrinkage, remaining consistently on an investment level. This much more satisfactory experience of the investor in the Federal Land Bank bonds was due in good part to the additional capital subscribed by the United States government to these Banks, and to the closer supervision to which they were subjected, but the joint and several guarantee undoubtedly proved of considerable benefit.

Note also that Joint Stock Land Bank bonds were made legal investments for trust funds in many states, and remained so during 1932–1933 despite their undoubtedly inadequate security.

CHAPTER XVII

GUARANTEED SECURITIES (*Continued*)

GUARANTEED REAL-ESTATE MORTGAGES AND MORTGAGE BONDS

The practice of guaranteeing securities reached its widest development in the field of real-estate mortgages. These guarantees are of two different types: the first being given by the corporation engaged in the sale of the mortgages or mortgage participations (or by an affiliate); the second and more recent form being the guaranty given by an independent surety company, which assumes the contingent liability in return for a fee.

The idea underlying real-estate-mortgage guarantees is evidently that of insurance. It is to the mortgage holder's advantage to protect himself, at some cost in income return, against the possibility of adverse developments affecting his particular property (such as a change in the character of the neighborhood). It is within the province of sound insurance practice to afford this protection in return for an adequate premium, provided of course, that all phases of the business are prudently handled. Such an arrangement will have the best chance of success if:

1. The mortgage loans are conservatively made in the first instance.

2. The guaranty or surety company is large, well managed, independent of the agency selling the mortgages, and has a diversification of business in fields other than real estate.

3. Economic conditions are not undergoing fluctuations of abnormal intensity.

The collapse in real-estate values after 1929 was so extreme as to contravene the third of these conditions. Accordingly the behavior of real-estate-mortgage guarantees during this period may not afford a really fair guide to their future value. Nevertheless, some of the characteristics which they revealed are worthy of comment.

This Business Once Conservatively Managed.—In the first place a striking contrast may be drawn between the way in which the business of guaranteeing mortgages had been conducted

prior to about 1924 and the lax methods which developed thereafter, during the very time that this part of the financial field
was attaining its greatest importance.

If we consider the policies of the leading New York City
institutions which guaranteed real-estate mortgages (*e.g.*, Bond
and Mortgage Guarantee Company, Lawyers Mortgage Company), it is fair to say that for many years the business was
conservatively managed. The amount of each mortgage was
limited to not more than 60% of the value, carefully determined;
large individual mortgages were avoided; and a fair diversification
of risk, from the standpoint of location, was attained. It is true
that the guarantor companies were not independent of the selling
companies, nor did they have other types of surety business.
It is true also that the general practice of guaranteeing mortgages
due only three to five years after their issuance contained the
possibility, later realized, of a flood of maturing obligations at a
most inconvenient time. Nevertheless, the prudent conduct
of their activities had enabled them successfully to weather
severe real-estate depressions such as occurred in 1908 and 1921.

New and Less Conservative Practices Developed.—The building boom which developed during the new era was marked by
an enormous growth of the real-estate-mortgage business and
of the practice of guaranteeing obligations of this kind. New
people, new capital, and new methods entered the field. Several
small local concerns which had been in the field for a long period
were transformed into highly aggressive organizations doing a
gigantic and nation-wide business. Great emphasis was laid
upon the long record of success in the past, and the public was
duly impressed—not realizing that the size, the methods, and
the personnel were so changed that they were in fact dealing
with a different institution. In a previous chapter we pointed
out how recklessly unsound were the methods of financing real-
estate ventures during this period. The weakness of the mortgages themselves applied equally to the guarantees which were
frequently attached thereto for an extra consideration. The
guarantor companies were mere subsidiaries of the sellers of
the bonds. Hence, when the crash came, the value of the
properties, the real-estate-bond company, and the affiliated
guarantor company all collapsed together.

Evil Effects of Competition and Contagion.—The rise of the
newer and more aggressive real-estate-bond organizations had

a most unfortunate effect upon the policies of the older concerns. By force of competition they were led to relax their standards of making loans. New mortgages were granted on an increasingly liberal basis, and when old mortgages matured, they were frequently renewed in a larger sum. Furthermore, the face amount of the mortgages guaranteed rose to so high a multiple of the capital of the guarantor companies that it should have been obvious that the guaranty would afford only the flimsiest of protection in the event of a general decline in values.

When the real-estate market broke in 1931, the first consequence was the utter collapse of virtually every one of the newer real-estate-bond companies and their subsidiary guarantor concerns. As the depression continued, the older institutions gave way also. The holders of guaranteed mortgages or participations therein (aggregating about $3,000,000,000 guaranteed by New York title and mortgage companies alone) found that the guaranty was a mere name and that they were entirely dependent upon the value of the underlying properties. In most cases these had been mortgaged far more heavily than reasonable prudence would have permitted. Apparently only a very small fraction of the mortgages outstanding in 1932 were created under the conservative conditions and principles that had ruled up to, say, eight years previously.

Guarantees by Independent Surety Companies.—During the 1924–1930 period several of the independent surety and fidelity companies extended their operations to include the guaranteeing of real-estate mortgages for a fee or premium. Theoretically, this should have represented the soundest method of conducting such operations. In addition to the strength and general experience of the surety company there was the important fact that such a guarantor, being entirely independent, would presumably be highly critical of the issues submitted for its guaranty. But this theoretical advantage was offset to a great extent by the fact that the surety companies began the practice of guaranteeing real-estate-mortgage bonds only a short time prior to their debacle, and they were led by the general overoptimism then current to commit serious errors in judgment. In most cases the resultant losses to the suretor were greater than it could stand; several of the companies were forced into receivership (notably National Surety Company), and holders of bonds with such guarantees failed to obtain full protection.

Outlook for the Future.—Looking forward to the return of normal conditions in the real-estate field, it seems certain that mortgage bonds will again be sold to the public, and no doubt the business of guaranteeing these mortgages will again be developed. For reasons above stated, the giving of such guarantees by independent surety companies appears theoretically the sounder practice. But the sad experience of surety companies in the field is likely to keep them out of it, and it would seem more probable that the practice will be revived in its original form, *viz.*, through guarantees assumed by companies in the real-estate-mortgage business. In such event, the buyers of these mortgage bonds or participations must attach chief importance to the conservatism of the issuing company and relatively slight weight to its guarantee.

LEASEHOLD OBLIGATIONS EQUIVALENT TO GUARANTEES

The property of one company is often leased to another for a fixed annual rental sufficient to pay interest and dividends on the former's capital issues. Frequently the lease is accompanied by a specific guaranty of such interest and dividend payments, and in fact the majority of guaranteed corporate issues originate in this fashion.[1] But even if there is no explicit guaranty, a lease or other contract providing fixed annual payments will supply the equivalent of a guaranty on the securities of the lessee company.

Examples: An excellent instance of the value of such an arrangement is afforded by the Westvaco Chlorine Products Corporation $5\frac{1}{2}$s, issued in 1927 and maturing in 1937. The Westvaco Company agreed to sell part of its output to a subsidiary of Union Carbide and Carbon Corporation, and the latter enterprise guaranteed that monthly payments would be made to the trustee sufficient to take care of the interest and retirement of the $5\frac{1}{2}\%$ bonds. In effect this arrangement was a guaranty of interest and principal of the Westvaco issue by Union Carbide and Carbon, a very strong concern. By reason of this protection

---

[1] For example Pittsburgh, Fort Wayne and Chicago Railway Company Preferred and Common receive 7% dividends under a 999-year lease to the Pennsylvania Railroad Company. These dividends are also guaranteed by the Pennsylvania.

and the continuous purchases for redemption made thereunder, the price of the issue was maintained at 99 or higher throughout 1932–1933. This contrasts with a decline in the price of West-vaco common stock from 116½ in 1929 to 3 in 1932.

Another interesting example is supplied by the Tobacco Products Corporation of New Jersey 6½s, due 2022. The properties of this company are leased to American Tobacco Company under a 99-year contract, expiring also in 2022, providing for annual payments of $2,500,000 (with the privilege to the lessee to settle by a lump-sum payment equivalent to the then present value of the rental, discounted at 7% per annum). By means of a sinking-fund arrangement these rental payments are calculated to be sufficient to retire the bond issue in full prior to maturity, in addition to taking care of the interest. Assuming the legality of the lease to be beyond question, these Tobacco Products 6½s must be considered as the equivalent of fixed obligations of American Tobacco Company. As such they rank ahead of American Tobacco Preferred, dividends on which, of course, are not a fixed charge. When the bonds were created in 1931 the investing public was either sceptical of the validity of the lease or—more probably—was not familiar with this situation, for American Tobacco Preferred sold at a much higher relative price than the Tobacco Products bonds. The following data will support this statement:

Issue	1931–1933 range			
	Low price	Yield, %	High price	Yield, %
Tobacco Products 6½s....	73	8.90	102½	6.34
American Tobacco Co. 6% Preferred..............	95¼	6.30	120	5.00

Guaranteed Issues Frequently Undervalued.—This example illustrates the fairly frequent undervaluation of guaranteed or quasi-guaranteed issues as compared with other securities of the guarantor enterprise. A well-known instance was that of San Antonio and Aransas Pass Railway Company First 4s, due 1943, guaranteed as to principal and interest by Southern Pacific Company. Although these enjoyed a mortgage security in addition to the guaranty they regularly sold at prices yielding higher returns than did the unsecured obligations of the Southern Pacific.[1]

Examples: A more striking contrast was afforded by the price of Barnhart Bros. and Spindler Company First and Second Preferred (both guaranteed as to principal and dividends by American Type Founders Company) in relation to the price of the guarantor's own preferred stock which is not a fixed obligation. Additional examples of this point are afforded by the

COMPARATIVE PRICES AND YIELDS OF GUARANTEED SECURITIES AND SECURITIES OF THE GUARANTOR

Issue	Date	Price	Yield, %
San Antonio & Aransas Pass 1st 4s/1943 (GTD)..........................	Jan. 2, 1920	56¼	8.30
Southern Pacific Co. Debenture 4s/1929.	Jan. 2, 1920	81	6.86
Barnhart Bros. & Spindler 7% 1st Pfd. (GTD)..........................	1923 low price	90	7.78
Barnhart Bros. & Spindler 7% 2d Pfd. (GTD)..........................	1923 low price	80	8.75
American Type Founders 7% Pfd......	1923 low price	95	7.37
Huyler's of Delaware 7% Pfd. (GTD)..	April 11, 1928	102½	6.83
Schulte Retail Stores 8% Pfd.........	April 11, 1928	129	6.20
Armour of Delaware 7% Pfd. (GTD)...	Feb. 13, 1925	95⅛	7.36
Armour of Illinois 7% Pfd............	Feb. 13, 1925	92⅞	7.54

[1] Dewing, in his *Financial Policy of Corporations*, pp. 152–153, New York, 1926, makes the following statement with respect to guaranteed bonds:

"True, there may be instances in which a holding or controlling corporation will maintain the interest or rental on an unprofitable subsidiary's bonds, for strategic reasons only; the rule holds good almost always that the strength of a guaranteed bond is no greater than that of the corporation issuing it and the earning capacity of the property directly covered by it."
In an appended footnote he adds " . . . Yet its (San Antonio and Aransas Pass Railway's) strategic importance to the Southern Pacific Company's lines is such that the guarantor company very wisely meets the bond interest deficit."

It seems clear to us that these statements misinterpret the essential character of the obligation under a guarantee. Southern Pacific met the San Antonio and Aransas Pass bond interest deficit, not out of "wisdom" but by compulsion. The strength of a guaranteed bond *may be very much greater than* that of the corporation issuing it, because that strength rests upon the dual claim of the holder against *both* the issuing corporation and the guarantor.

price of Huyler's of Delaware, Inc., Preferred, guaranteed by Schulte Retail Stores Corporation, as compared with the price of Schulte Preferred; and by the price of Armour and Company of Delaware guaranteed preferred, as compared with the preferred stock of the guarantor company, Armour and Company of Illinois. Some comparative quotations relating to these examples are given on page 189.

It is obvious that in cases of this sort advantageous exchanges can be made from the lower yielding into the higher yielding security with no impairment of safety; or else into a much better secured issue with little sacrifice of yield, and sometimes with an actual gain.[1]

INCLUSION OF GUARANTEES AND RENTALS IN THE CALCULATION OF FIXED CHARGES

All obligations equivalent to bond interest should be included with a company's interest charges when calculating the coverage for its bond issues. This point has already been explained in some detail in connection with railroad fixed charges, and it was touched upon briefly in our discussion of public-utility bonds. The procedure in these groups offers no special difficulties. But in the case of certain types of industrial companies, the treatment of rentals and guarantees may offer confusing variations. This question is of particular moment in connection with retail enterprises, theater companies, etc., in which rent or other obligations related to buildings occupied may be an important element in the general picture. Such a building may be owned by the corporation and paid for by a bond issue, in which case the obligation will be fully disclosed in both the balance sheet and the income account. But if another company occupies a similar building under long-term lease, no separate measure of the rental obligation appears in the income account and no indication thereof can be found in the balance sheet. The second company may *appear* sounder than the first, but that is only because its obligations are undisclosed; essentially, both companies are carrying a similar burden. Conversely, the outright ownership of premises free and clear carries an important advantage (from the standpoint of preferred stock, particularly)

---

[1] In Note 30 of the Appendix will be found a concise discussion of certain interesting phases of guarantees and rentals, as illustrated by the N.Y. and Harlem Railroad and the Mobile and Ohio Railroad situations.

over operation under long-term lease, although the capitalization set-up will not reveal this advantage.

Examples: If Interstate Department Stores Preferred had been compared with The Outlet Company Preferred in 1929 the two exhibits might have appeared closely similar; the earnings coverage averaged about the same, and neither company showed any bond or mortgage liability. But Outlet's position was in actuality by far the stronger, because it owned its land and buildings while those of Interstate (with a minor exception) were held under lease. The real effect of this situation was to place a substantial fixed obligation ahead of Interstate Department Stores Preferred which did not exist in the case of Outlet. In the chain-store field a similar observation would apply to a comparison of J. C. Penney Preferred and S. H. Kress Preferred in 1932; for the latter company owned more than half of its store properties, while nearly all the Penney locations were leased.

Lease Liabilities Generally Overlooked.—The question of liability under long-term leases received very little attention from the financial world until its significance was brought home rudely in 1931 and 1932, when the high level of rentals assumed in the preceding boom years proved intolerably burdensome to many merchandising companies.

Example: The influence of this factor upon a supposed investment security is shown with striking force in the case of United Cigar Stores Preferred. This issue, and its predecessor, had for many years shown every sign of stability and had sold accordingly at a consistently high level. For 1928 the company reported "no funded debt" and earnings equal to about seven times the preferred dividend. Yet so crushing were the liabilities under its long-term leases (and to carry properties acquired by subsidiaries), that in 1932 bankruptcy was resorted to and the preferred stock was menaced with extinction.

Such Liabilities Complicate Analysis.—It must be admitted that in the case of companies where the rental factor is important, its obtrusion has badly complicated the whole question of bond or preferred stock analysis. Unfortunately there is no method, short of an exhaustive investigation, by which the investor in an industrial company of this type can obtain an accurate idea of the liability existing under leaseholds. Nor can exact comparisons be made between companies having bond issues outstanding

against owned properties and other concerns having leasehold
obligations instead. The investor must deal with the individual
case as well as he can. If the liabilities involved are evidently
of minor character (*e.g.*, rent of office space by the typical manufacturing company) the matter may be dismissed without careful
study. But if the undisclosed obligations under leases may
possibly be of major dimensions, then the investor must either
investigate the situation thoroughly or else pass by the securities
of such a company in favor of another involving less complicated
problems.[1]

SUBSIDIARY COMPANY BONDS

The bonds of a subsidiary of a strong company are generally
regarded as well protected, on the theory that the parent company
will take care of all its constituents' obligations. This viewpoint
is encouraged by the common method of setting up consolidated
income accounts, under which all the subsidiary bond interest
appears as a charge against all the combined earnings, ranking
ahead of the parent company's preferred and common stocks.
If, however, the parent concern is not contractually responsible
for the subsidiary bonds, by guaranty or lease (or direct assumption), this form of statement may prove to be misleading. For
if a particular subsidiary proves unprofitable, its bond interest
may conceivably not be taken care of by the parent company,
which may be willing to lose its investment in this part of its
business and turn it over to the subsidiary's bondholders. Such
a development is unusual, but the possibility thereof was forcibly
demonstrated in 1932–1933 by the history of United Drug
Company 5s, due 1953.

Examples: United Drug was an important subsidiary of Drug,
Inc., which had regularly earned and paid large dividends, gained
chiefly from the manufacture of proprietary medicines and other
drugs. In the first half of 1932, the consolidated income account
showed earnings equal to ten times the interest on United Drug
5s, and the record of previous years was even better. While
this issue was not assumed or guaranteed by Drug, Inc., investors
considered the combined showing so favorable as to assure the
safety of the United Drug 5s beyond question. But United Drug

[1] For example, Loew's, Inc., Debenture 6s, due in 1941, are apparently
well covered, but the extent of the company's rental commitments is
unknown, and convincing analysis of the issue is thereby rendered
difficult.

owned, as part of its assets and business, the stock of Louis K. Liggett Company, which operated a large number of drug stores and which was burdened by a high-rental problem similar to that of United Cigar Stores. In September 1932 Liggett's notified its landlords that unless rents were reduced it would be forced into bankruptcy.

This announcement brought rudely home to investors the fact that the still prosperous Drug, Inc., was not assuming responsibility for the liabilities of its (indirect) subsidiary, Liggett's, and they immediately became nervously conscious of the fact that Drug, Inc., was not responsible for interest payments on United Drug 5s either. Sales of these bonds resulting from this discovery depressed the price from 93 earlier in the year down to 42. At the latter figure, the $40,000,000 of United Drug 5s were quoted at only $17,000,000, although the parent company's stock was still selling for more than $100,000,000 (3,500,000 shares at about 30). In the following year the "Drug, Inc., System" was voluntarily dissolved into its component parts—an unusual development—and the United Drug Co. resumed its entirely separate existence.

During the period 1914-1921 the New Amsterdam Gas Company First 5s, due 1948, had a somewhat similar history. As an obligation of a subsidiary of Consolidated Gas Company of New York the bonds were highly thought of by investors and for many years sold close to par. The public ignored the fact, however, that the bonds were not guaranteed by the parent company and that New Amsterdam Gas Company was not earning its interest charges. In the depression of 1920-1921 the unfavorable *contractual* position of this issue asserted itself and the price declined to 58.

Consolidated Traction Company of New Jersey First 5s were obligations of a large but unprofitable subsidiary of Public Service Corporation of New Jersey. The bonds were not guaranteed by the parent company. When they matured in 1933 many of the holders accepted an offer of 65 for their bonds made by the parent company.

Separate Analysis of Subsidiary Interest Coverage Essential.— These examples suggest that just as investors are prone to underestimate the value of a guaranty by a strong company, they sometimes make the opposite mistake and attach undue significance to the fact that a company is controlled by another. From the standpoint of fixed-value investment, nothing of importance may be taken for granted. Hence a subsidiary bond should not be purchased on the basis of the showing of its parent company, unless the latter has assumed direct responsibility for the bond in question. In other cases the exhibit of the subsidiary itself can afford the only basis for the acceptance of its bond issues.[1]

If the above discussion is compared with that on page 146, it will be seen that investors in bonds of a *holding company* must insist upon a consolidated income account, in which the subsidiary interest—whether guaranteed or not—is shown as a prior charge; but that purchasers of unguaranteed *subsidiary* bonds cannot accept such consolidated reports as a measure of their safety, and must require a statement covering the subsidiary alone. These statements may be obtainable only with some difficulty, as was true in the case of United Drug 5s, but they must nevertheless be insisted upon.

[1] As a practical matter, the financial interest of the parent company in its subsidiary, and other business reasons, may result in its protecting the latter's bonds even though it is not obligated to do so. This would be a valid consideration, however, only in deciding upon a purchase on a *speculative* basis (*i.e.*, carrying a chance of principal profit), but would not justify buying the bond at a full investment price. Concretely stated, it might have made United Drug 5s an excellent speculation at 45, but they were a poor investment at 93.

CHAPTER XVIII

PROTECTIVE COVENANTS AND REMEDIES OF SENIOR SECURITY HOLDERS

In this and the two succeeding chapters we shall consider the provisions usually made to protect the rights of bond owners and preferred stockholders against impairment, and the various lines of action which may be followed in the event of nonfulfillment of the company's obligations. Our object here, as throughout this book, is not to supply information of a kind readily available elsewhere, but rather to subject current practices to critical examination and to suggest feasible improvements therein for the benefit of security holders generally.

Indenture or Charter Provisions Designed to Protect Holder of Senior Securities.—The contract between a corporation and the owners of its bonds is contained in a document called the *indenture* or *deed of trust*. The corresponding agreements relating to the rights of preferred stockholders are set forth in the Articles, or Certificate, of Incorporation. These instruments usually contain provisions designed to prevent corporate acts injurious to senior security holders and to afford remedies in case of certain unfavorable developments. The more important occurrences for which such provision is almost always made may be listed under the following heads:

1. In the case of bonds:
    a. Nonpayment of interest, principal, or sinking fund.
    b. Default on other obligations, or receivership.
    c. Issuance of new secured debt.
    d. Dilution of a conversion (or subscription) privilege.
2. In the case of preferred stocks:
    a. Nonpayment of (cumulative) preferred dividends for a period of time.
    b. Creation of funded debt or a prior stock issue.
    c. Dilution of a conversion (or subscription) privilege.

A frequent, but less general, provision requires the maintenance of working capital at a certain percentage of the bonded debt of industrial companies. (In the case of investment-trust or

195

holding-company bonds it is the market value of all the assets which is subject to this provision).

The remedies provided for bondholders in cases falling under 1*a* and 1*b* above are fairly well standardized. Any one of these untoward developments is designated as an "event of default" and permits the trustee to declare the principal of the bond issue due and payable in advance of the specified maturity date. The provisions therefor in the indenture are known as "acceleration clauses." Their purpose in the main is to enable the bondholders to assert the full amount of their claim in competition with the other creditors.

Contradictory Aspects of Bondholders' Legal Rights.—In considering these provisions from a critical standpoint, we must recognize that there are contradictory aspects to the question of the bondholders' legal rights. Receivership is a dreaded term in Wall Street; its advent means ordinarily a drastic shrinkage in the price of all the company's securities, including the bonds for the "benefit" of which the receivership was instituted. As we pointed out in a former chapter, the market's appraisal of a bond in default is no higher on the whole, and perhaps lower, than that of a non-dividend-paying preferred stock of a solvent company.

The question arises, therefore, whether the bondholders might not be better off if they did not have any enforceable claim to principal or interest payments *when conditions are such as to make prompt payment impossible.* For at such times the bondholder's legal rights apparently succeed only in ruining the corporation without benefiting the bondholder. As long as the interest or principal is not going to be paid anyway, would it not be to the interest of the bondholders themselves to postpone the date of payment and keep the enterprise out of the courts?

If we reflect upon this question, we may reach the conclusion that the bondholder would be in a sounder position with either *better* or *fewer* legal rights. Under present practice he has the power to make trouble for everybody, including himself, but he is not given the right of swift and inexorable enforcement of his just claim. He can get a receiver appointed in 24 hours, but frequently he cannot get either his money or the property in five years.[1] Sometimes the delay seems unavoidable because

---

[1] Receivers were appointed for Wickwire Spencer Steel Company in 1927. At the end of 1933 the receivers were still operating the property, while the first-mortgage bondholders had received no payments in the interim.

of complicated and conflicting liabilities. In many cases, however, the receivership is prolonged in order to protect the stockholders' interest, or "equity," in the business—on the ground that a prompt sale of the property would bring an inadequate price and unfairly wipe out the junior issues.

Considerable argument might be indulged in on the question whether the stockholders are entitled to such protection against cruel and rapacious creditors, or whether this practice amounts to taking away from the investors the very safeguards in return for which they gave up all participation in the company's surplus profits. But whichever contention is the more plausible, the fact remains that the present procedure is of advantage to neither the bondholder nor the stockholder. In most cases default means only a vast and pervasive uncertainty, which threatens extinction to the stockholders but fails to promise anything specific to the bondholders.

Tendency of Securities in Receivership To Sell below Their Fair Value.—As a result there is a general tendency for the securities of companies in receivership to sell below their fair value in the aggregate; and also a tendency for illogical relationships to be established between the price of a bond issue in default and the price of the junior stock issues.

Examples: The Fisk Rubber Company case is an excellent example of the former point; the Studebaker Corporation situation in September 1933 illustrates the latter.

MARKET VALUE OF FISK RUBBER SECURITIES IN APRIL 1932

$7,600,000 First 8s @ 16	$ 1,200,000
8,200,000 Debenture 5½s @ 11	900,000
Stock issues	Nominal
Total market value of the company	$ 2,100,000

BALANCE SHEET, JUNE 30, 1932

Cash	$ 7,687,000
Receivables (less reserve of $1,425,000)	4,838,000
Inventories (at lower of cost or market)	3,216,000
	$15,741,000
Accounts Payable	363,000
Net current assets	$15,378,000
Fixed assets (less $8,400,000 depreciation)	23,350,000

The company's securities were selling together for less than one-third of the cash alone, and for only one-seventh of the net quick assets, allowing nothing for the fixed property.[1]

STUDEBAKER CORPORATION, SEPTEMBER 1933

Issue	Face amount	Market price	Market value
10-year 6% notes and other claims.....	$22,000,000	40	$ 8,800,000
Preferred stock......................	5,800,000	27	$ 1,500,000
Common stock (shares)..............	*2,464,000*	6	14,700,000
Total value of stock issues........	....	..	$16,200,000

The company's debt, selling at 40 cents on the dollar, was entitled to *prompt payment in full* before the stockholder received anything. Nevertheless, the market placed a much larger value upon the stock issues than upon the prior debt.

Prompt and Rigorous Enforcement Would Benefit the Bondholder.—It is clear that some better methods should be developed for protecting the bondholder than the usual prolonged and expensive receivership. Improvement in procedure may be worked out along either of two general lines. It may take the form of a more rigorous enforcement of the bondholders' contractual rights. The effort would be made to imitate as closely as possible the technique of an ordinary commercial receivership. Emphasis would be laid upon a speedy realization for creditors by either liquidation or reorganization. The stockholders would receive no consideration other than being allowed a rather brief period to raise sufficient cash to pay the company's debts in full. The question whether the property is "worth" more than the debts would be tested solely by the ability of the receiver or the stockholders promptly to find some one to buy it for a greater sum.

If this pattern were followed the bondholder would benefit through a quicker realization, a shorter period of uncertainty, and particularly through the unqualified assertion of his prior claim against the assets. Under such a procedure, the anomaly shown by the market price of Fisk Rubber obligations would be

[1] As pointed out in Chap. L, below, the Fisk Rubber 8s later proved to be worth close to 100 and the 5½s more than 70.

almost impossible, because there would be no question, first, that a creditor could get in cash his share of the realizable quick assets, and secondly, that such payment would be made in short order. Nor would the equal anomaly of the Studebaker quotations be possible if this procedure were standard. For again it would be clear without question that the stock would have no value at all unless the notes were paid in full, and also that such payment would have to be arranged for in a comparatively short space of time.

Voluntary Readjustment as an Alternative to Receivership.—A second line of improvement in reorganization procedure lies in the substitution of a voluntary readjustment for the tortuous and expensive machinery of the bankruptcy courts. The business makes a composition of some kind with its creditors, and thus receivership is averted. It is usually in the interest of bondholders to make such a deal when there is small likelihood of obtaining a substantial cash payment through liquidation. For if it is to their advantage that the business continue as a going concern, it would probably be better to adjust their claims voluntarily on some equitable basis, rather than subject the enterprise to the manifold injuries of receivership. An arrangement can ordinarily be arrived at which will give the creditors close to complete ownership of the business; and the small interest which must be conceded to the stockholders is worth giving up to save the company from the courts.

Examples: At the end of 1931 Radio-Keith-Orpheum Corporation, needing funds to meet pressing obligations, found ordinary financing impossible. The stockholders ratified a plan under which in effect they surrendered 75% of their stock interest, which was given in turn as a bonus to those who supplied the $11,600,000 required by purchasing debenture notes. (Continued large losses, however, forced the company into receivership a year later.)

In 1933 Fox Film Corporation effected a recapitalization of the same general type. The stockholders gave up over 80% of their holdings, and this stock was in turn *exchanged* for nearly all of approximately $40,000,000 of 5-year notes and bank debt.

The Kansas City Public Service Company readjustment plan, also consummated in 1933, was designed to meet the simpler problem of reducing interest charges during a supposedly temporary period of subnormal earnings. It provided that the coupon rate on the 6% first-mortgage bonds should be reduced to 3% during the four years 1933–1936, restored to 6% for 1937–1938, and advanced to 7% for 1939–1951, thus making up the 12% foregone in the earlier years. A substantial sinking fund, contingent upon earnings, was set up to retire the issue gradually and to improve its market position.

It was obvious that the Kansas City Public Service bondholders were better off to accept temporarily the 3% which could be paid rather than to insist on 6% which could not be paid and thereby precipitate a receivership. (The previous receivership of the enterprise, terminated in 1926, had lasted six years.) In this case the stockholders were not required to give up any part of their junior interest to the bondholders in return for the concessions made. While theoretically some such sacrifice and transfer would be equitable, it was not of much practical importance here because any stock bonus given to the bondholders would have had a very slight market value. It should be recognized as a principle, however, that the waiving of any important right by the bondholders entitles them to some *quid pro quo* from the stockholders—in the form either of a contribution of cash to the enterprise or of a transfer of some part of their claim on future earnings to the bondholders.[1]

Changes Introduced by New Bankruptcy Legislation.—The amendments to the National Bankruptcy Act, passed by Congress in 1933 and pending in 1934, establish a new procedure midway between voluntary readjustment and the ordinary reorganization through receivership. A company unable to meet its obligations may be reorganized under a plan accepted by the holders of 66⅔% of each of the issues affected, if the plan is approved by the court (and by the Interstate Commerce Commission or the state public service commission in the case of a railroad or public utility). This machinery, when operative, will avoid receivership, sale at foreclosure, the payment of nonassenting security holders, and consequently a good part of the usual delay and expense of corporate reorganizations. The Schackno Act, passed by the New York State Legislature,

---

[1] The reorganization of Industrial Office Building Company in 1932–1933 is a remarkable example of the conversion of fixed-interest bonds into income bonds without sacrifice of any kind by the stockholders. A detailed discussion of this instance is given in the Appendix, Note 31.

established a similar procedure affecting real-estate mortgage bonds.[1]

The new legislation deals effectively—perhaps somewhat too effectively[2]—with the vexing problem of the nonassenting minority bondholder which has been the chief obstacle in the way of voluntary readjustments.

Critique of Functions and Attitude of Bond Trustees.—There is a crying need for a change in the functions and attitude of bond trustees. Under present practice the trust companies which assume these duties are not trustees at all, in any true sense, but rather "agents of the bondholders." As a general rule they take no action on their own initiative, but only when directed to do so (and fully indemnified) by a certain percentage of the bondholders. Indentures say practically nothing about the duties of the trustee, but a great deal about his immunities and indemnification. Trusteeship however, connotes more than a mere following of instructions. It should imply the use of judgment and initiative in protecting the interests of the *cestuis que trustent*, or beneficiaries.

Bond trustees are generally conscientious and painstaking in the discharge of their routine duties, nor can they be accused of any lack of desire to protect the bondholders. But they are guided by a traditional attitude which is inherently opposed to positive action. According to this viewpoint a much more serious responsibility attaches to doing something than to doing nothing. The result is that where the trustee is given discretion to act, it will scarcely ever exercise this discretion. For it is much more afraid of taking some action for which it might later be open to criticism than it is of any loss which the bondholders may suffer through its refraining from action. The latter course it can always defend on the ground that if the bondholders want their rights asserted they can have this done by making written demand upon the trustee as provided in the indenture.

Example from Investment-trust Field.—An example taken from the investment-trust field will show how the inclination of the trustee to avoid positive action on its own initiative operates to

---

[1] At this writing the constitutionality of the Schackno Act is under determination by the courts.

[2] The denial of the right of appraisal, even on the basis of liquidating values, to the dissenting creditor may work out unfairly at times in the case of industrial reorganizations.

deprive the bondholder of the safeguards which he is apparently justified in counting on when he makes his commitment.

Financial Investing Company of New York sold two 5% collateral-trust issues, due respectively in 1932 and 1940. These bonds were secured by deposit with the trustee of listed securities, diversified in accordance with stringent requirements. The company covenanted to maintain such collateral at a value of at least 120% of the outstanding bonds. The trustee was empowered: (1) to give notice to the corporation in the event the required margin was impaired; (2) to declare the principal due if the deficiency was not remedied within thirty days; and (3) to sell the collateral in such event and apply the proceeds to payment of principal and interest.

These covenants appeared to give the bondholders practically the same protection as is enjoyed by a bank making a collateral loan on marketable securities. If the stipulated margin became impaired and was not made good, the collateral could be sold out to satisfy the loan. The only important difference appeared to be the allowance in the bond indenture of a thirty-day period to restore the margin to the required percentage.

But the actual history of the Financial Investing issues was strikingly at variance with that of the typical collateral loan made by banks during the same period. In October 1931, the margin fell below 20% and the trustee advised the corporation of this "event of default." The margin was not made good within the thirty days, but the collateral was not sold. In August 1932 the bid price for the bond fell as low as 20. In October 1932 the principal of one issue matured and was not paid. This event compelled action; the collateral securing both issues was sold out; and in January 1933, *15 months* after the "margin call," the bondholders finally received about 65 cents on the dollar.

Explanation of This Unsatisfactory Outcome.—We see here a wide discrepancy between the apparently effectual safeguards accorded the bondholders in their indenture and the highly unsatisfactory results which they actually experienced—*viz.*, a substantial loss, a long delay, and a particularly harrowing shrinkage in market value during the interim. What is the explanation? Was it inertia or carelessness on the part of the trustee? Superficially it might well seem so; yet in fact the trustee gave much time and thought to this situation. But its efforts were controlled—and vitiated—by the established principle of bond trusteeship, *viz.*, "Never do anything that any one might possibly criticize, unless requested to do so by bondholders in the manner specified in the indenture." In the case of Financial Investing 5s, the trustee could be compelled to act upon request in writing from the holders of 30% of the bonds, accompanied by the usual indemnities. The trustee hesitated to sell the collateral promptly on its own initiative, because if the market recovered later it might be accused by the stockholders of having unwarrantably wiped them out. It appears also that for a similar reason some of the bondholders were opposed to the sale of the collateral after its value had fallen below the par amount of the issue.

It is not difficult to show that these objections to carrying out the protective provisions of the indenture were basically unsound. In fact, if they were tenable there would be no excuse for having these provisions in the indenture. If we analyze this incident as a whole, we see that the unsatisfactory results flowed from a combination of:

1. The lack of clearly established rules of procedure to enforce the terms of an indenture.

2. A typical body of bondholders with little financial acumen and less initiative.

3. A basis of trusteeship under which the trustees look to these inert and unreasoning bondholders for guidance, instead of guiding them.

Unsound Justification Offered for Perfunctory Character of Trusteeship.—The failure of trustees to take aggressive action on behalf of the bondholders is generally explained and justified on the ground that the standard compensation is too small to warrant anything but perfunctory activities. In the writers' view this argument is without merit. It is true that trustees receive only a modest fee for the purely formal services rendered while everything goes smoothly. But there is nothing to prevent their receiving adequate compensation for any discretionary action which they may take in the event of default. As a matter of fact they do receive extra allowances from the court in all receivership cases, and these allowances are by no means ungenerous in view of the mechanical nature of most of the work they perform. Furthermore this work is largely a duplication of services rendered by the protective committee; and the same is true to an even greater extent of the services of their counsel, for which liberal fees are also paid.

The Problem of the Protective Committee.—The present practice, under which the initiative is taken by a protective committee acting independently of the trustee, is open to more serious objections than that based upon the additional expense. Since 1929 the general status of protective committees has become uncertain and most unsatisfactory. Formerly it was taken for granted that the investment bankers who floated the issue would organize a protective committee in the event of default. But in recent years there has been a growing tendency to question the propriety or desirability of such action. Bondholders may lack faith in the judgment of the issuing house, or they may question its ability to represent them impartially because of other interests in or connections with the enterprise; or they may even consider the underwriters as legally responsible for the losses incurred. The arguments in favor of *competent* representation by agencies *other* than the houses of issue are therefore quite convincing. The difficulty lies however, in securing such competent representation. With the original issuing houses out of the picture, anybody can announce himself as chairman of a protective committee and invite deposits. The whole procedure has become unstandardized and open to serious abuses. Duplicate committees often appear; an undignified scramble for deposits takes place; persons with undesirable reputations and motives can easily inject themselves into the situation.

Recommended Reform.—The whole procedure could be clarified and standardized if the trustee under the indenture were expected to assume the duty of *actively* protecting the bond issue. The large institutions which hold these positions have the facilities, the experience, and the standing required for the successful discharge of such a function. There seems no good reason, in the ordinary case, why the trustee should not itself organize the protective committee, with one of its executive officers as chairman and with the other members selected from among the larger bondholders or their nominees. The possible conflict of interest between the trustee as representative of all the bondholders and the protective committee as representative of the depositing holders only, will be found on analysis rarely to be of more than technical and minor consequence. Such a conflict, if it should arise, could be solved by submission of the question to the court. As stated before, there is no difficulty about awarding sufficient compensation to the trustee and its counsel for their labors and accomplishment on behalf of the bondholders.

This arrangement envisages effective cooperation between the trustee and a group of bondholders who in the opinion of the trustee are qualified to represent the issue as a whole. The best arrangement might be to establish this bondholders' group at the time the issue is sold, *i.e.*, without waiting for an event of default to bring it into being, in order that there may be from the very start some responsible and interested agency to follow the affairs of the corporation from the bondholders' standpoint, and to make objections, if need be, to policies which may appear to threaten the safety of the issue. Reasonable compensation for this service should be paid by the corporation. This would be equivalent in part to representation of the bondholders on the board of directors. If the time were to arrive when the group would have to act as a protective committee on behalf of the bondholders, their familiarity with the company's affairs should prove of advantage.

CHAPTER XIX

PROTECTIVE COVENANTS (*Continued*)

Prohibition of Prior Liens.—A brief discussion is desirable regarding certain protective provisions other than those dealing with the ordinary events of default. (The matter of safeguarding conversion and other participating privileges against dilution will be covered in the chapters dealing with Senior Securities with Speculative Features.) Dealing first with mortgage bonds, we find that indentures almost always prohibit the placing of any new prior lien on the property. Exceptions are sometimes made in the case of bonds issued under a reorganization plan, when it is recognized that a prior mortgage may be necessary to permit raising new capital in the future.

Example: In 1926, Chicago, Milwaukee, St. Paul and Pacific Railroad Company issued $107,000,000 of Series *A* Mortgage 5% bonds and, junior thereto, $185,000,000 of Convertible Adjustment Mortgage 5s, in exchange for securities of the bankrupt Chicago, Milwaukee and St. Paul Railway Company. The indentures permitted the later issuance of an indefinite amount of First and Refunding Mortgage Bonds, which would rank ahead of the Series *A* Mortgage 5s.[1]

Equal-and-ratable Security Clause.—When a bond issue is unsecured it is almost always provided that it will share equally in any mortgage lien later placed on the property.

Example: The New York, New Haven and Hartford Railroad Company sold a number of debenture issues between 1897 and 1908. These bonds were originally unsecured, but the indentures provided that they should be equally secured with any mortgage subsequently placed upon the property. In 1920 a first and refunding mortgage was authorized by the stockholders; consequently the earlier issues have since been equally secured with bonds issued under the new mortgage. They still carry the title of "debentures," but this is now a misnomer. (There is, however, an issue of 4% debentures, due in 1957, which did not

---

[1] In 1933 the St. Paul was granted permission to issue some of the new first and refunding bonds, to be held as collateral for short-term loans made by the United States government.

carry this provision and hence are unsecured. In 1932 these sold at 30, compared with a low of 40 for the secured "debenture" 4s, due 1956.)[1]

Purchase-money Mortgages.—It is customary to permit without restriction the assumption of *purchase-money mortgages.* These are liens attaching only to new property subsequently acquired and their assumption is not regarded as affecting the position of the other bondholders. The latter supposition is not necessarily valid, of course, since it is possible thereby to increase the ratio of total debt of the enterprise to the total shareholder's equity in a manner which might jeopardize the position of the existing bondholders.

Subordination of Bond Issues to Bank Debt in Reorganization. In the case of bonds or notes issued under a reorganization plan it is sometimes provided that their claim shall be junior to that of present or future bank loans. This is done to facilitate bank borrowings which otherwise could be effected only by the pledging of receivables or inventories as security. An example of this arrangement is afforded by Aeolian Company Five-Year Secured 6% Notes, due in 1937, which were issued under a capital readjustment plan in partial exchange for the Guaranteed 7% Preferred Stock of the company. The notes were subordinated to $400,000 of bank loans.

Safeguards against Creation of Additional Amounts of the Same Issue.—Nearly all bonds or preferred issues enjoy adequate safeguards in respect to the creation of additional amounts of the issue. The customary provisions require a substantial margin of earnings above the requirements of the issue as thus enlarged. For example, additional New York Edison Company First Lien and Refunding Gold Bonds may not be issued, except for refunding purposes, unless consolidated net earnings for a recent 12-month period have been at least twice the annual interest charges on all underlying bonds outstanding, including those to be issued. In the case of Otis Steel Company First Mortgages, 6 due 1941, the required ratio is $2\frac{1}{2}$ times.[2]

---

[1] In exceptional cases, debenture obligations are entitled to a *prior lien* on the property in the event that a subsequent mortgage is placed thereon.

Example: National Radiator Corporation Debenture $6\frac{1}{2}$s, due 1947, and the successor corporation's income debenture 5s, due 1946.

[2] For similar provisions in the case of preferred stocks see Consolidated Gas Company of New York $5 Preferred, North American Edison Company

Provisions of this kind with reference to earnings-coverage are practically nonexistent in the railroad field, however. Railroad bonds of the blanket-mortgage type more commonly restrict the issuance of additional bonds through a provision that the total funded indebtedness shall not exceed a certain ratio to the capital stock outstanding, and by a limitation upon the emission of new bonds to a certain percentage of the *cost* or *fair value* of newly acquired property. (See, for example, the Baltimore and Ohio Railroad Company Refunding and General Mortgage Bonds and the New York, New Haven and Hartford Railroad Company First and Refunding Bonds.) In the older bond issues it was customary to close the mortgage at a relatively small fixed amount, thus requiring that additional funds be raised by the sale of junior securities. This provision gave rise to the favorably situated "underlying bonds" to which reference was made in Chap. VI.

In the typical case additional issues of mortgage bonds may be made only against pledge of new property worth considerably more than the increase in debt. (See, for examples: Youngstown Sheet and Tube Company First Mortgage 5s, of which $50,000,-000 unissued bonds are reserved to finance 75% of the cost of additions or improvements to the mortgaged properties; Pacific Telephone and Telegraph Company Refunding 5s, due 1952, which may be issued in further amounts to finance additions and betterments up to 75% of the actual cost thereof; Pere Marquette Railway Company First-mortgage bonds, which may be issued up to 80% of the cost or fair value, whichever is the lower, of newly constructed or acquired property.)

These safeguards are logically conceived and almost always carefully observed. Their practical importance is less than might appear, however, because in the ordinary instance the showing stipulated would be needed anyway in order to attract buyers for the additional issue.

Working-capital Requirements.—The provisions for maintaining working capital at a certain percentage of the bonded debt are by no means standardized. They appear only in industrial bond indentures. The percentage varies and the penalties for nonobservance vary also. In some cases the result is merely the prohibition of dividend payments until the proper level of

---

$6 Preferred, Ludlum Steel Company $6.50 Preferred, Gotham Silk Hosiery Company 7% Preferred.

working capital is restored. In other cases, the principal of the bond issue may be declared due.

Examples: The indenture securing National Acme Company First 6s, due 1942, forbids the payment of any cash dividend unless the net quick assets are 150% of the bonds then outstanding. In the case of Lane Bryant, Inc., 6s, due 1940, the figure is set at 125%. The U. S. Rubber Company 6½% notes, due 1940, are protected by a more stringent provision which prohibits any dividend payments that would reduce the current assets below twice the total debt, exclusive of the mortgage bonds, and prohibits *common* dividend payments that would reduce current assets below all the indebtedness of any description. As a result of these requirements, preferred dividends were omitted in May 1928. The indenture securing General Baking Company Debenture 5½s, due 1940, stipulates that net current assets must exceed 40% of the bonded debt as a condition to common dividends. The smaller figure in this case is probably due to the fact that the baking business does not require a large working capital.

American Machine and Foundry Company 6s, due 1939, have a twofold provision, the first prohibiting dividends unless net current assets equal 150% of the outstanding bond issue, and the second requiring unconditionally that the net current assets be maintained at 100% of the face value of outstanding bonds. In the case of U. S. Radiator Corporation 5s, due 1938, the company agreed at all times to maintain net working capital equal to 150% of the outstanding funded debt. In the case of Amalgamated Laundries 6½s, due 1936, the covenant was of a somewhat different type—the company agreeing to maintain current assets equal to twice current liabilities during the life of the bonds.

It would appear to be sound theory to require regularly some protective provisions on the score of working capital in the case of industrial bonds. We have already suggested that an adequate ratio of net quick assets to funded debt be considered as one of the specific criteria in the selection of industrial bonds. This criterion should ordinarily be set up in the indenture itself, so that the bondholder will be entitled to the *maintenance* of a satisfactory ratio throughout the life of the issue and to an adequate remedy if the figure declines below the proper point.

The prohibition of dividend payments under such conditions is sound and practicable. But the more stringent penalty, which terms a deficiency of working capital "an event of default" is not likely to prove effective or beneficial to the bondholder. The objection that receivership harms rather than helps the creditors applies with particular force in this connection. Referring to the U. S. Radiator 5s, mentioned above, we may point out that the balance sheet of January 31, 1933, showed a default in the 150% working-capital requirement. (The net current assets were $2,735,000, or only 109% of the $2,518,000 bond issue.) Nevertheless, the trustee took no steps to declare the principal due, nor was it asked to do so by the required number of bondholders. In all probability a receivership invoked for this reason would have been considered as highly injurious to the bondholders' interests. But this attitude would mean that the provision in question should never have been included in the indenture.[1]

Voting Control as a Remedy.—We suggest, therefore, that if a more stringent remedy than mere prohibition of dividends is desired in the event of working-capital impairment, this remedy consist of a transfer of voting control from the stockholders to the bondholders. Voting by bondholders is a rare phenomenon in this country, though by no means unknown. (An example is the right of holders of Third Avenue Railway Adjustment 5s, due 1960, to vote for directors, giving that issue control of the company *if the right were exercised*.)[2] But there is sound reason for vesting voting rights in the bondholders if the company fails to observe some specific covenant in the indenture. This arrangement is certainly preferable to the right to have a receiver appointed, which the lawyers persist in inserting in deeds of trust, although common sense should tell them that their remedy is usually worse than the disease.

[1] Similar situations existed in 1933 with respect to G. R. Kinney (shoe) Company 7½s, due 1936, and Budd Manufacturing Company First 6s, due 1935. Early in 1934, the U. S. Radiator Corporation asked the debenture holders to modify the provisions respecting both working-capital maintenance and sinking-fund payments. No substantial *quid pro quo* was offered for these concessions. Characteristically, the reason given by the company itself for this move was not that the bondholders were entitled to some remedial action, but that the "technical default under the indenture" interfered with projected bank borrowings by the company.

[2] By a succession of agreements, voting control of Mobile and Ohio Railroad Company is exercised by the holders of that company's general 4% bonds, due 1938, each $500 bond having one vote. (Most of these bonds are owned by Southern Railway.)

Protective Provisions for Investment-trust Issues.—In the case of investment-trust bonds, however, stringent protective provisions and stringent remedies are practicable and therefore desirable. Such bonds are essentially similar to the collateral loans made by banks on marketable securities. As a protection for these bank loans, it is required that the market value of the collateral be maintained at a certain percentage in excess of the amount owed. In the same way the lenders of money to an investment trust should be entitled to demand that the value of the portfolio continuously exceed the amount of the loans by an adequate percentage, *e.g.*, 25%. If the market value should decline below this figure the investment trust should be required to take the same action as any other borrower against marketable securities. It should either put up more money (*i.e.*, raise more capital from the stockholders) or sell out securities and retire debt with the proceeds, in an amount sufficient to restore the proper margin.

The disadvantages that inhere in bond investment generally justify the bond buyer in insisting upon every possible safeguard. In the case of investment-trust bonds, a very effective measure of protection may be assured by means of the covenant to maintain the market value of the portfolio above the bonded debt. Hence investors in investment-trust issues should demand this type of protective provision, and—what is equally important—they should require its strict enforcement. While this stand will inflict hardship upon the stockholders when market prices fall, this is part of the original bargain, in which the stockholders agreed to take most of the risk in exchange for the surplus profits.[1]

A survey of bond indentures of investment trusts discloses a signal lack of uniformity in the matter of these protective provisions. Most of them do require a certain margin of asset value over debt as a condition to the sale of additional bonds. The required ratio of net assets to funded debt varies from 120% (*e.g.*, General American Investors) to 250% (*e.g.*, Niagara Shares Corporation). The more usual figures are 150, 175, or 200%. A similar restriction is placed upon the payment of cash dividends.

[1] If the market value of the assets falls below 100% of the funded debt a condition of insolvency would seem to be created which entitles the bondholders to insist upon immediate remedial action. For otherwise the stockholders would be permitted to speculate on the future with what is entirely the bondholders' capital.

The ratio required for this purpose varies from 125% (*e.g.*, Domestic and Foreign Investors) to 175% (which must be shown to permit cash dividends on Central States Electric Corporation common). The modal figure is probably 140 or 150%.

But the majority of issues do not require at all times and unconditionally the maintenance of a minimum excess of asset value above bonded indebtedness. Examples of such a covenant may indeed be given—*e.g.*, International Securities Corporation of America, Investment Company of America, and Reliance Management Corporation, all of which agree to maintain a 25% margin above the amount of their bond issues. The foregoing discussion strongly suggests, however, that this type of provision should not be exceptional, but universal.[1]

SINKING FUNDS

In its modern form a sinking fund provides for the periodic retirement of a certain portion of a senior issue through payments made by the corporation. The sinking fund acquires the security by call, by means of sealed tenders, or by open-market purchases made by the trustee or the corporation. In the latter case the corporation turns in the bonds to the sinking fund in lieu of cash. The sinking fund usually operates once or twice a year, but provisions for quarterly and even monthly payments are by no means unusual. In the case of many bond issues, the bonds acquired by the sinking fund are not actually retired but are "kept alive," *i.e.*, they draw interest and these interest sums are also used for sinking-fund purchases, thus increasing the latter at a compounded rate.

Example: An important instance of this arrangement is supplied by the two issues of United States Steel Sinking Fund 5s, originally totalling $504,000,000. Bonds of the junior issue, listed on the New York Stock Exchange, were familiarly known in the bond market as "Steel Sinkers." By adding the interest on bonds in the fund, the annual payments grew from $3,040,000 in 1902 to $11,616,000 in 1928. (The following year the entire outstanding amounts of these issues were retired or provided for.)

---

[1] The offering circular of Alleghany Corporation Collateral Trust 5s, due 1949, indicated that a coverage of 150% would be compulsory. Yet the indenture provided that failure to maintain this margin would not constitute an event of default, but would result only in the prohibition of dividends and in the impounding by the trustee of the income from the pledged collateral.

Benefits.—The benefits of a sinking fund are of a twofold nature. The continuous reduction in the size of the issue makes for increasing safety and the easier repayment of the balance at maturity. Also important is the support given to the market for the issue through the repeated appearance of a substantial buying demand. Nearly all industrial bond issues have sinking funds; the public-utility group shows about as many with as without; in the railroad list sinking funds are exceptional. But in recent years increasing emphasis has been laid upon the desirability of a sinking fund, and few long-term senior issues of any type are now offered without such a provision.[1]

Indispensable in Some Cases.—Under some circumstances a sinking fund is absolutely necessary for the protection of a bond. This is true in general when the chief backing of the issue consists of a wasting asset. Bonds on mining properties invariably have a sinking fund, usually of substantial proportions and based upon the tonnage mined. A sinking fund of smaller relative size is regularly provided for real-estate-mortgage bonds. In all these cases the theory is that the annual depletion or depreciation allowances should be applied to the reduction of the funded debt.

Examples: A special example of importance is the large Interborough Rapid Transit Company First and Refunding 5% issue, due 1966, which is secured mainly by a lease on properties which belong to the City of New York. Obviously it was essential to provide through a sinking fund for the retirement of the entire issue by the time the lease expires in 1967, since the corporation would then be deprived of most of its assets and earning power. Similarly with Tobacco Products 6½s, due in 2022, which depend for their value entirely upon the annual payments of $2,500,000 made by American Tobacco Company under a lease expiring in 2022.

The absence of a sinking fund under conditions of this kind invariably leads to trouble.

Examples: Federal Mining and Smelting Company supplied the unusual spectacle of a mining enterprise with a large preferred-stock issue ($12,000,000); and furthermore the preferred stock had no sinking fund. Declaration of a $10 dividend on the

[1] During 1933 the Interstate Commerce Commission strongly recommended that railways adopt sinking funds to amortize their existing debt. The Chicago and Northwestern Railway thereupon announced a plan of this kind, the details of which are not particularly impressive.

common in 1926 led to court action to protect the preferred stock against the threatened breakdown of its position through depletion of the mines coupled with the distribution of cash earnings to the junior shares. As a result of the litigation the company refrained from further common dividends and devoted its surplus profits to reducing the preferred issue.

Iron Steamboat Company General Mortgage 4s, due 1932, had no sinking fund, although the boats on which they were a lien were obviously subject to a constant loss in value. These bonds to the amount of $500,000 were issued in 1902 and were a second lien on the entire property of the company (consisting mainly of seven small steamboats operating between New York City and Coney Island), junior to $100,000 of first-mortgage bonds. During the years 1909 to 1925, inclusive, the company paid dividends on the common stock aggregating in excess of $700,000 and by 1922 had retired all of the first-mortgage bonds through the operation of the sinking fund for that issue. At this point the 4s, due 1932, became a first lien upon the entire property. In 1932 when the company went into bankruptcy the entire issue was still outstanding. The mortgaged property was sold at auction in February 1933 for $15,050, a figure resulting in payment of less than 1 cent on the dollar to the bondholders. An adequate sinking fund might have retired the entire issue out of the earnings which were distributed to the stockholders.

When the enterprise may be regarded as permanent, the absence of a sinking fund does not necessarily condemn the issue. This is true not only of most high-grade railroad bonds, and of many high-grade utility bonds, but also of most of the select group of old-line industrial preferred stocks which merit an investment rating, e.g., National Biscuit Preferred, which has no sinking fund. From the broader standpoint, therefore, sinking funds may be characterized as invariably desirable, and sometimes but not always indispensable.

Serial Maturities as an Alternative.—The general object sought by a sinking fund may be obtained by the use of serial maturities. The retirement of a portion of the issue each year by reason of maturity corresponds to the reduction by means of sinking-fund purchases. Serial maturities are relatively infrequent, their chief objection resting probably in the numerous separate market quotations which they entail. In the equipment-trust field, however, they are the general rule. This exception may be explained by the fact that insurance companies and other financial institutions are the chief buyers of equipment obligations, and for their special needs the variety of maturity dates proves a convenience. Serial maturities are also frequently employed in state and municipal financing.

Problems of Enforcement.—The enforcement of sinking-fund provisions of a bond issue presents the same problem as in the case of covenants for the maintenance of working capital. Failure to make a sinking-fund payment is regularly characterized in the indenture as an event of default, which will permit the trustee to declare the principal due and thus bring about receivership. The objections to this "remedy" are obvious and we can recall no instance in which the omission of sinking-fund payments, unaccompanied by default of interest, was actually followed by enforcement of the indenture provisions. When the company continues to pay interest but claims to be unable to meet the sinking fund, it is not unusual for the trustee and the bondholders to withhold action and merely to permit arrears to accumulate. More customary is the making of a formal request to the bondholders by the corporation for the postponement of the sinking-fund payments. Such a request is almost invariably acceded to by the great majority of bondholders, since the alternative is always pictured as insolvency. This was true even in the case of Interborough Rapid Transit 5s, for which—as we have pointed out—the sinking fund was an essential element of protection.[1]

The suggestion made in respect to the working-capital covenants, *viz.*, that voting control be transferred to the bondholders in the event of default, is equally applicable to the sinking-fund provision. In our view that would be distinctly preferable to the present arrangement under which the bondholder must either do nothing to protect himself or else take the drastic and calamitous step of compelling receivership.

The emphasis we have laid upon the proper kind of protective provisions for industrial bonds should not lead the reader to believe that the presence of such provisions carries an assurance of safety. This is far from the case. The success of a bond

[1] The plan of voluntary readjustment proposed in 1922 postponed sinking-fund payments on these bonds for a five-year period. About 75% of the issue accepted this modification.

investment depends primarily upon the success of the enterprise, and only to a very secondary degree upon the terms of the indenture. Hence the seeming paradox that the senior securities that have fared best in the depression have on the whole quite unsatisfactory indenture or charter provisions. The explanation is that the best issues as a class have been the oldest issues, and these date from times when less attention was paid than now to protective covenants.

In the Appendix, Note 32, we present two examples of the opposite kind (Willys-Overland Company First 6½s, due 1933, and Berkey and Gay Furniture Company First 6s, due 1941) wherein a combination of a strong statistical showing with all the standard protective provisions failed to safeguard the holders against a terrific subsequent loss. But while the protective covenants we have been discussing do not *guarantee* the safety of the issue, they nevertheless *add* to the safety, and are therefore worth insisting upon.

CHAPTER XX

PREFERRED-STOCK PROTECTIVE PROVISIONS. MAINTENANCE OF JUNIOR CAPITAL

Preferred stocks are almost always accorded certain safeguards against the placing of new issues ahead of them. The standard provision prohibits either a prior stock or a mortgage-bond issue except upon approval by vote of two-thirds or three-fourths of the preferred stock. The prohibition is not made absolute because conditions are always within contemplation under which the preferred stockholders may find it to their advantage to authorize the creation of a senior issue. This may be done because new financing through a bond issue is necessary to avoid receivership. An example is afforded by Eitingon-Schild Company in 1932. According to the provisions of the 6½% First Preferred stock the company could not create a mortgage, lien, or charge on any of its property, except purchase-money obligations, extensions of existing mortgages, and pledge of liquid assets to secure loans made in the ordinary course of business. Because of the precarious financial condition of the company in 1932 the preferred stockholders authorized certain financial rearrangements, including the creation of a $5,500,000 issue of 5% debentures containing certain provisions the effect of which was to create a special charge against fixed properties.

Protection against Creation of Unsecured Debt Desirable.— It is a common practice to give preferred stockholders no control over the creation of *unsecured debt*. This point is exemplified by the American Metal Company, which in 1930 issued $20,000,-000 of debenture notes without vote of the preferred stockholders, but in 1933 was compelled to ask for their approval of the possible pledging of collateral to refund the notes at maturity. This distinction appears to us to be unsound, since unsecured debt is just as much a threat to a preferred stock as is a mortgage obligation. It does seem illogical to provide, as is usually done, that preferred stockholders may forbid the issuance of new preferred shares ranking ahead of or equivalent to theirs, and

217

also of any secured indebtedness, but that they have nothing to say about the creation of a debenture bond issue, however large.

Presumably this exclusion arose from the desire to permit bank borrowing for ordinary business purposes; but this point may be taken care of by a specific stipulation to that effect—just as the standard provision now used permits the pledge of assets to secure "loans made in the ordinary course of business" without requiring preferred stockholders' consent.[1]

The preferred stockholders' vote is rather frequently availed of to permit the issuance of an equal-ranking or even a prior security which is to be exchanged for the preferred stock itself under a recapitalization plan, the latter usually being designed to dispose of accumulated dividends. By giving the new issue equality with or priority over the old, stockholders who might otherwise be inclined to reject the composition are almost compelled to accept it.

Examples: In 1930 Austin Nichols and Company had 7% preferred stock outstanding on which dividends of $21 per share had accumulated. The company offered to exchange each share for one share of $5 Cumulative Prior *A* stock plus 1.2 shares of common. By vote of the preferred stockholders accepting the plan, the new Prior *A* stock was made senior to the old preferred. As a result, about 99% of the latter was turned in for exchange. International Paper and Fisk Rubber made similar adjustments of back dividends on the preferred in 1917 and 1925, respectively. In these cases, additional preferred stock was issued ranking equally with the old shares.

The Aeolian Company had a 7% cumulative preferred stock, dividends on which were guaranteed by a parent company, Aeolian, Weber Piano and Pianola Company, which owned all the common. A recapitalization plan was devised in 1932 for the purpose of reducing the fixed annual payment from 7 to 3%.

---

[1] It should be noted, however, that there is a growing tendency in recent years to protect preferred stockholders against the creation of debenture bonds by requiring their approval of the issuance of any "bonds, notes, debentures or other evidence of indebtedness maturing later than one year from the date of their issue." See for examples: the Kendall Company $6 Participating Preferred; Ludlum Steel Company $6.50 Convertible Preferred; Melville Shoe Corporation 6% First Preferred; the Gamewell Company $6 Preferred; A. M. Byers Company 7% Preferred. Among the older issues Loose-Wiles Biscuit Company 7% First Preferred has this type of protection.

Under this plan, each $100 share of preferred stock was offered in exchange for $50 face value of 6% secured notes and $50 of new 6% nonguaranteed preferred stock, cumulative only to the extent that dividends were earned. Both the new secured notes and the new 6% preferred stock ranked ahead of the old 7% preferred stock, practically all of which accepted the offer of exchange.

Preferred Stock Sinking Funds.—Very few public-utility or railroad preferred-stock issues have a sinking-fund provision. But in the case of industrial preferred-stock offerings sinking funds have become the general rule. The advantages which bonds derive from a sinking fund are equally applicable to preferred stocks. Furthermore, in view of the weak contractual position of preferred stocks, which we have frequently emphasized, there is the more reason for the buyer to insist on special protective arrangements of this kind. But while a sinking fund is thus a highly desirable feature of a preferred issue, its presence is no assurance nor is its absence a negation of adequate safety. The list of 21 preferred stocks which maintained an investment status throughout 1932–1933 (given in Chap. XIV) contains only one issue with a sinking-fund provision. As previously explained, this paradox is due to the fact that nearly all the strong industrial preferreds are old established issues, and the sinking fund is a relatively recent development.

The amount of the sinking fund is usually fixed at a certain percentage of the maximum amount of preferred stock outstanding, 3% being perhaps the most frequent figure. Less often the amount is based on a percentage of profits. There are a number of variations and technicalities of a descriptive nature, which we shall not detail. In most cases the payment of the sinking fund is obligatory, provided: (1) preferred dividends have been paid in full or "provided for," and (2) there remain surplus profits equal to the sinking-fund requirement.

A small number of preferred stocks are protected by an agreement to maintain net current assets, usually at 100% of the preferred issue or 100% of the preferred stock plus bond issues. In some cases the penalty for nonobservance is merely a prohibition of common dividends (*e.g.*, Sidney Blumenthal and Company); while in other cases voting control passes to the preferred stock (*e.g.*, A. G. Spalding and Bros. where working capital must equal 125% of the preferred issue).

Voting Power in the Event of Nonpayment of Dividends.—The second general type of protective provision for preferred stocks relates to voting power accruing in the event of nonpayment of dividends. As far as we know, these stipulations apply only to cumulative issues. The arrangement varies with respect to when the voting power becomes effective and to the degree of control bestowed. In a few cases (*e.g.*, Aeolian Company old 7% Preferred and Royal Baking Powder Company 6% Preferred) the voting right accrues after one dividend is omitted. At the other extreme, the right becomes effective only after eight quarterly payments are in default (*e.g.*, Brunswick-Balke-Collender Company). The customary period allowed is one year. The right conferred upon the preferred stock may be: (1) to vote exclusively for the directors; (2) to elect separately a majority of the board; (3) to elect separately a minority of the board; or (4) to vote share for share with the common stock.

Example of (1): McKesson and Robbins, Inc., Preferred Stock received the sole right to elect the directors upon omission of the fourth quarterly dividend in December 1932.

Example of (2): In 1933 Hahn Department Stores Preferred obtained the right to elect a majority of the board, because of the omission of four quarterly dividends.

Example of (3): Universal Pictures First Preferred has the right to elect two directors in the event of default of six quarterly dividends. Brooklyn and Queens Transit Corporation Preferred may elect one-third of the board if all arrears are not paid up within a year after any quarterly dividend is omitted.

Example of (4): City Ice and Fuel Preferred votes share for share with the common in the event of nonpayment of four quarterly dividends.[1]

The value of the last arrangement would seem to depend a good deal on whether the preferred stock is larger or smaller than the common issue. If larger, the share-for-share voting right could give the issue effective control; but in most cases the preferred issue is smaller, and hence this voting right is likely to prove ineffective.

---

[1] An unusual variation of this idea is found in the case of Du Pont "Nonvoting Debenture Stock" (a preferred issue). The holders are given the right to vote equally with the common stock in the event that the earnings for any calendar year fall below 9% on the debenture stock issue. They receive exclusive voting power if dividends are in default for six months.

Noncumulative Issues Need Greater Protection.—The practices outlined above merit certain other criticisms of a more general nature. In the first place, while it is taken for granted that these special voting provisions should apply to cumulative preferred stocks only, the exclusion of noncumulative issues seem to us to be most illogical. Their holders have certainly a greater reason to demand representation in the event of nonpayment, because they have no right to recover the lost dividends in the future. In our view it should be established as a financial principle that *any* preferred stock which is not paying its full dividend currently should have some separate representation on the board of directors.

On the other hand we do not consider it proper to deprive the common stock of all representation when preferred dividends are unpaid. Complete domination of the board by the preferred stockholders may lead to some practices distinctly unfair to the common stock, *e.g.*, perpetuation of preferred-stock control by unnecessarily refraining from paying up back dividends in full. An alert minority on the board of directors, even though powerless in the actual voting, may be able to accomplish a great deal in preventing unfair or unsound practices.

A General Canon Regarding Voting Power.—From the foregoing discussion, a general canon with respect to voting power may readily be formulated. The standard arrangement should give every preferred and every common issue the separate right to *elect some directors under all circumstances.* It would be logical for the common stock to elect the majority of the board as long as preferred dividends were regularly paid; and equally logical that whenever the full dividend was not paid, on either a cumulative or noncumulative preferred issue, the right to choose the majority of the board should pass to the preferred stockholders.

Adequate protection for preferred issues should require that voting control pass to the holders in the event not only of default in dividends but also of nonpayment of the sinking fund or the failure to maintain working capital as stipulated. A few charters, *e.g.*, those of Bayuk Cigars and A. G. Spalding, afford this threefold remedial right to the preferred stockholders. In our view, this practice should be standard instead of exceptional.

Value of Voting Control by Preferred Stock May be Questioned.—Viewing the matter realistically, it must be admitted that the vesting of voting control in holders of a preferred issue does not necessarily prove of benefit to them. In some cases, perhaps, no effective use can be made of this privilege; in other cases the holders are too inert—or too poorly advised—to protect their interests even though they have power to do so. These practical limitations may be illustrated by a case in point, *viz.*, the Maytag Company.

In 1928 this enterprise (manufacturing washing machines) was recapitalized, and issued the following securities:

> 100,000 shares of $6 Cumulative First Preferred.
> 320,000 shares of $3 Cumulative Preference (Second Preferred).
> 1,600,000 shares of common.

Approximately 80% of all these shares were received by the Maytag family. Through investment bankers they sold to the public their holdings of first and second preferred. This netted them individually (*i.e.*, not the company) the sum of about $20,000,000. They retained control of the business through their ownership of common stock. The charter provided that neither preferred issue should have voting rights unless four quarterly dividends were defaulted on either. In that case both issues, voting together as a single class, would have the right to elect a majority of the directors. In 1932 dividends were omitted on both classes of preferred. Voting control consequently passed to the holders of these issues early in 1933.

Peculiarly enough, the only change made during the 1932-1933 period in the board of directors was the *resignation* of the single member who—as partner of one of the issuing houses—had presumably represented the preferred stockholders. All of the five directors remaining were operating officials and closely identified with the common-stock ownership. In the meantime the price of the two preferred issues declined to 15 and 3⅛ respectively, as compared with original offering prices of 101 and 50.

Reviewing the situation, we see private owners of a business selling a preferred claim against its profits for a very large sum, which they retained individually. To protect the public's stake in the enterprise, the preferred issues were given voting control in the event of continued nonpayment of dividends. This event occurred, and with it a catastrophic decline in the value of the shares. But the new voting control was not exercised, and the board of directors remained dominated, even more completely than before, by those owning the common stock.

Wall Street's attitude toward this incident would be that, since the management of the company was honest and capable, a change in the directorate would be unnecessary and even unwise. In our opinion this reasoning misses the basic point. No doubt the operating management should remain unchanged; possibly—though by no means certainly—directors representing the preferred stock would follow the same financial policies in matters affecting the senior issues as would be followed by a board identified with the common stock. But the crux of the matter is that these decisions should actually be made by a board of directors of which the majority has been selected by the preferred stockholders in accordance with their rights. Regardless of whether a change in the board would result in any change in policy, the directors should be chosen as provided in the articles of incorporation. For otherwise the voting provision is entirely meaningless. It becomes merely a phrase to persuade the preferred-stock buyer into believing he has safeguards which are in fact nonexistent.[1]

Recommended Procedure in Such Cases.—In the writers' view the proper procedure in cases such as the Maytag situation is perfectly clear. The preferred stockholders individually have no satisfactory means of going about the nomination and election of directors to represent them. This duty should devolve upon the issuing houses, and they should discharge it conscientiously. They should: (1) obtain a list of the preferred stockholders of record; (2) advise them of their new voting rights; and (3) recommend to them a slate of directors and request their proxies to vote for these nominees. The directors suggested should, of course, be as well qualified as possible for their posts. They must be free from any large interest in or close affiliation with the common stock, and it would be desirable if they were themselves substantial owners of preferred shares.

It is quite possible, none the less, that the directors chosen by the preferred stockholders will be incompetent, or for other reasons fail to represent their interests properly. But this is not a valid argument against the possession and the exercise of voting power by preferred stockholders. The same objection

[1] It should be added that the dividends on Maytag $6 Preferred were resumed in October 1933 and accumulations discharged in 1934.

applies to voting rights of common stockholders—and of citizens. The remedy is not disenfranchisement but education.

Maintenance of Adequate Junior Capital.—We wish to call attention finally to a protective requirement for both bondholders and preferred stockholders which is technically of great importance, but which frequently is not taken care of in indentures or charter provisions. The point referred to is the maintenance of an adequate amount of junior capital. We have previously emphasized the principle that such junior capital is an indispensable condition for any sound fixed-value investment. No loan could prudently be made to a business at 5 or 6% interest unless the business were worth a considerable amount over and above the amount borrowed. This is elementary and well understood. But it is not generally realized that the corporation laws permit the *withdrawal* of substantially all the capital and surplus *after* the loan has been made. This can be done by the legal process of reducing the capital to a nominal sum and distributing the amount of the reduction to the stockholders. Such a maneuver the creditors are powerless to prevent unless they have specifically guarded against it in their loan contract.

Danger in the Right to Reduce Stated Capital.—Let us attempt to bring this point home by a hypothetical example. A company is engaged in the business of lending money on installment accounts. It has $2,100,000 of capital and surplus. Ostensibly for the purpose of expanding its operations, it borrows $2,000,000 by sale of a 20-year 5% debenture bond issue. The earnings and stock equity appear to provide sufficient protection for the bonds. Business subsequently falls off and the company has a substantial amount of unused cash. The stockholders vote to reduce the capital to $100,000 (in theory it might be reduced to $1); and they receive back $2,000,000 in cash, as a return of capital.

In effect the stockholders have recovered their capital with the cash supplied by the bondholders, but they retain ownership and control of the business together with the right to receive all profits above 5%. The bondholders find themselves in the absurd position of having provided all the capital and having thereby assumed all the risk of loss, without any share in the profits above ordinary interest. Such a development would be most unfair; but apparently it can be carried out legally unless the indenture of the bond issue specifically prevented it by stipulating that no distributions could be made to the stockholders which would reduce the capital and surplus below a certain figure.

The removal of the bondholders' "cushion" by its direct withdrawal in cash—as in our hypothetical example—is a rare, perhaps unexampled, occurrence. But a corresponding situation does actually arise in practice through a combination of large operating losses followed by a reduction in capital to wipe out the consequent balance sheet deficit.

Examples: In Chap. XXXVIII we refer to an extraordinary example of this kind, namely the Interborough-Metropolitan case. Here the stated capital was reduced by stockholders' action to eliminate a huge profit-and-loss deficit. Following this action, earnings of a distinctly temporary character were disbursed in dividends, instead of being conserved for the benefit of the bondholders, who later suffered a tremendous loss. To effect the capital reduction under the laws then existing, a "merger" with a dummy corporation was resorted to. The same artifice has been used several times since in connection with recapitalization schemes, *e.g.*, Central Leather Company in 1926 and Kelly-Springfield Tire Company in 1932.

As the result of losses sustained during the depression of the 1930s numerous reductions of capitalization have been voted by the stockholders. These actions have been taken without consulting the bondholders. Most of such reductions have been effected by changes from no-par shares to shares of a low par value. Frequently this has been accompanied by write-offs of intangible assets or mark-downs of fixed assets. Such write-downs of asset values on one side of the balance sheet and capital on the other are of no special significance from the bondholders' standpoint, except possibly in the fact that they may permit unduly low depreciation charges and therefore unduly liberal dividend payments. But in most of these cases a substantial sum also has effectively been transferred from capital to surplus and thus made available to absorb future operating losses and to facilitate the resumption of dividends before past losses have been made up.

For example, Remington Rand, Inc., changed its common stock from no par to $1 par and thereby, together with cancellation of shares held by the company itself, reduced the stated value of the common from $17,133,000 to $1,291,000. It applied $7,800,000 of this reduction to write down its intangible assets,

$2,300,000 additional to mark down its plant account, and $400,000 for miscellaneous write-downs and reserves. This left about $5,350,000 actually transferred from capital to surplus.

Similar reductions were made by New York Shipbuilding Corporation, Servel, Inc., Warner Bros. Pictures, Inc., H. F. Wilcox Oil and Gas Company, Thermoid Company. National Acme Company reduced the par value of its capital stock twice, from $50 to $10 in 1924 and from $10 to $1 in 1933. The result was a telescoping of its stated capital from $25,000,000 into $500,000. In the case of Capital Administration Company not only was the stated value of the common stock reduced, but the $3 cumulative preferred stock was also given a fictitiously low par value of $10.

Some Issues Protected against This Danger.—Fortunately for the bondholders in some of these cases, the indentures contain provisions prohibiting dividends or other distributions to the stockholders unless there is an adequate margin of resources above the indebtedness. In the case of Remington Rand Debenture 5½s, a threefold protection is supplied by the terms of the trust indenture, *viz.*:

1. Cash dividends may be paid only out of earned surplus.
2. Cash dividends may be paid only if net tangible assets after deducting the dividend in question shall equal at least 175% of the funded debt.
3. No stock may be retired, in excess of $3,500,000, except out of additional paid-in capital or earned surplus.

The last provision is directed against the reduction of junior capital by buying in preferred or common stock. It would be more satisfactory if it prohibited the *acquisition* (rather than the *retirement*) of the company's own stock.

Protective provisions of these various kinds appear in many but by no means all indentures. (They are absent, for example, in the case of New York Shipbuilding and Servel bonds, to name two of the companies which reduced their stated capital by stockholders' vote.) From the foregoing discussion, it should be clear that these covenants are essential to the proper safeguarding of a bond issue. Conscientious issuing houses and intelligent investors should insist on their inclusion in all indentures.

Anomalous Position of Preferred Stocks in This Connection.— The position of preferred stocks in this matter is a somewhat peculiar one. Their holders have the same interest as have bondholders in the maintenance of an adequate amount of junior capital. But losses which result in a balance-sheet deficit will legally prevent the payment not only of common dividends but of preferred dividends as well. Hence the preferred stockholders are likely to be very anxious for a reduction in the stated value of the common stock, which will eliminate the profit-and-loss deficit and permit the resumption of dividends on their own shares. In such cases their interest in maintaining an adequate amount of junior capital is offset by their greater desire to make dividends possible. (At the close of 1921, for example, losses taken by Montgomery Ward had created a profit-and-loss deficit of $7,700,000, which had compelled suspension of the preferred dividend. Accordingly holders of this issue welcomed a reduction in the stated value of the common stock from $28,300,000 to $11,400,000, which eliminated the balance-sheet deficit and thus permitted the resumption of the preferred dividends and discharge of the accumulations.)

This situation has even been exploited by the common stockholders to compel large concessions from the preferred holders in connection with a profit-and-loss deficit. A notorious example is the Central Leather reorganization plan, resulting in the formation of a successor company, U.S. Leather. As the price of their vote in favor of reducing the stated capital, the common stockholders forced the preferred holders to waive their back dividends and to reduce their cumulative right to future dividends.

Preferred Stocks Need Both Specific Protective Provisions and Voting Power for Their Protection.—These considerations confirm our previously expressed criticisms of the preferred stock *form* as an investment medium. It is not particularly difficult to safeguard these issues against the *withdrawal* of junior capital; this is frequently done and should always be done.[1] But to deal satisfactorily from the preferred stockholders' standpoint with conditions resulting in a profit-and-loss deficit is a difficult matter. It requires, above all, complete control of the corporation's policies by directors representing the preferred issue. This serves to emphasize the importance of adequate voting power for preferred stockholders in the event of nonpayment of dividends.

[1] For example, the charter of General American Investors Company, Inc., prohibits any dividend or other distribution on the common which will reduce net assets below $150 per share of preferred stock. The charter of Interstate Department Stores, Inc., requires the consent of holders of two-thirds of the preferred stock to any distribution to the holders of common stock of capital or surplus resulting from any statutory reduction of capital.

CHAPTER XXI

SUPERVISION OF INVESTMENT HOLDINGS

Traditional Concept of "Permanent Investment."—Not many years ago "permanent investment" was one of the stock phrases of finance. It was applied to the typical purchase by a conservative investor, and may be said to have embraced three constituent ideas: (1) intention to hold for an indefinite period; (2) interest solely in annual income, without reference to fluctuations in the value of principal; and (3) freedom from concern over future developments affecting the company. A sound investment was by definition one which could be bought, put away, and forgotten except on coupon or dividend dates.

This traditional view of high-grade investments was first seriously called into question by the unsatisfactory experiences of the 1920–1922 depression. Large losses were taken on securities which their owners had considered safe beyond the need of examination. The ensuing seven years, while generally prosperous, affected different groups of investment issues in such divergent ways that the old sense of complete security—with which the term "gilt-edged securities" was identified—suffered an ever-increasing impairment. Hence even before the market collapse of 1929, the danger ensuing from neglect of investments previously made, and the need for periodic scrutiny or supervision of all holdings, had been recognized as a new canon in Wall Street. This principle, directly opposed to the former practice, is frequently summed up in the dictum, "There are no permanent investments."

Periodic Inspection of Holdings Necessary—but Troublesome.—That the newer view is justified by the realities of fixed-value investment can scarcely be questioned. But it must be frankly recognized also that this same necessity for supervision of all security holdings implies a rather serious indictment of the whole concept of fixed-value investment. If risk of loss can be minimized only by the exercise of constant supervisory care, in addition to the painstaking process of initial choice, has not such investment become more trouble than it is worth?

Let it be assumed that the typical investor, following the conservative standards of selection herein recommended, will average a yield of 5% on a diversified list of corporate securities. This 5% return appears substantially higher than the 3% obtainable from United States government bonds, and also more attractive than the 3½ or 4% offered by savings banks. Nevertheless, if we take into account not only the effort required to make a proper selection but also the greater efforts entailed by the subsequent repeated check-ups, and if we then add thereto the still inescapable risk of depreciation or definite loss, it must be confessed that a rather plausible argument can be constructed against the advisability of fixed-value investments in general. The old idea of permanent, trouble-free holdings was grounded on the not illogical feeling that if a limited-return investment could not be regarded as trouble-free it was not worth making at all.

Alternative Programs.—Objectively considered, investment experience of the last decade undoubtedly points away from the fixed-value security field and into the direction of either (1) United States government bonds or savings-bank deposits, or (2) admittedly speculative operations, with endeavors to reduce risk and increase profits by means of skillful effort. But having stated this conclusion, as demanded by truth and logic, we hasten to point out that its practical value is highly questionable, because it is in basic conflict with both the psychology and the financial abilities of the investing public. People will not reconcile themselves to a 3% return on their capital; consequently they will prefer to expend considerable energy and to incur slight risks in order to raise the yield to 5%. This applies also to the difference between a savings bank's 3½ or 4% return as opposed to an investment list's 5% return, reinforced by the feeling that the savings banks themselves must necessarily run some risks in investing their depositors' money, and also by a rather unsound confidence on the part of the investor in his ability to handle his own money as effectively as a savings bank can do it for him.[1]

The second alternative, *viz.*, to speculate instead of investing, is entirely too dangerous for the typical person who is building

[1] If we were to assume that all the 5% investors decided to entrust their funds to savings banks, the problems of fixed-value investment would exist as before, except that they would have to be met by financial institutions exclusively instead of by both institutions and individuals.

up his capital out of savings or business profits. The disadvantages of ignorance, of human greed, of mob psychology, of trading costs, of weighting of the dice by insiders and manipulators, will in the aggregate far overbalance the purely theoretical superiority of speculation in that it offers profit possibilities in return for the assumption of risk. We have, it is true, repeatedly argued against the acceptance of an admitted risk to principal without the presence of a compensating chance for profit. In so doing, however, we have not advocated speculation in place of investment, but only intelligent speculation in preference to obviously unsound and ill-advised forms of investment. We are convinced (as stated in the Introduction) that the public generally will derive far better results from fixed-value investments, if selected with exceeding care, than from speculative operations, even though these may be aided by considerable education in financial matters. It may well be that the results of investment will prove disappointing; but if so, the results of speculation would have been disastrous.

We are thus back around the circle to the recognition that fixed-value investment serves a useful purpose. It is quite clear also that periodic reexamination of investment holdings is necessary to reduce the risk of loss. What principles and practical methods can be followed in such supervision?

Principles and Problems of Systematic Supervision; Switching.—It is generally understood that the investor should examine his holdings at intervals to see whether all of them may still be regarded as entirely safe; and that if the soundness of any issue has become questionable, he should exchange it for a better one. In making such a "switch" the investor must be prepared to accept a moderate loss on the holding he sells out, which loss he must charge against his aggregate investment income.

In the early years of systematic investment supervision, this policy worked out extremely well. Seasoned securities of the high-grade type tended to cling rather tenaciously to their established price levels and frequently failed to reflect a progressive deterioration of their intrinsic position until some time after this impairment was discoverable by analysis. It was possible, therefore, for the alert investor to sell out such holdings to some heedless and unsuspecting victim, who was attracted by the reputation of the issue and the slight discount at which it was obtainable in comparison with other issues of its class. The impersonal character of the securities market relieves this procedure of any ethical stigma, and it is considered merely as establishing a proper premium for shrewdness and a deserved penalty for lack of care.

Increased Sensitivity of Security Prices.—In more recent years, however, investment issues have lost what may have been called their "price inertia," and their quotations have come to reflect promptly any materially adverse development. This fact creates a serious difficulty in the way of effective switching to maintain investment quality. By the time that any real impairment of security is manifest, the issue may have fallen in price not only to a speculative level but to a level even lower than the decline in earnings would seem to justify. (One reason for this excessive price decline is that an unfavorable apparent *trend* has come to influence prices even more severely than the absolute earnings figures.) The owner's natural reluctance to accept a large loss is reinforced by the reasonable belief that he would be selling the issue at an unduly low price, and he is likely to find himself compelled almost unavoidably to maintain a speculative position with respect to that security.

Exceptional Margins of Safety as Insurance against Doubt.— The only effective means of meeting this difficulty lies in following counsels of perfection in making the original investment. The degree of safety enjoyed by the issue, as shown by quantitative measures, must be so far in *excess* of the minimum standards that a large shrinkage can be suffered before its position need be called into question. Such a policy should reduce to a very small figure the proportion of holdings about which the investor will subsequently find himself in doubt. It would also permit him to make his exchanges when the showing of the issue is still comparatively strong and while, therefore, there is a better chance that the market price will have been maintained.

Example and Conclusion.—As a concrete example, let us assume that the investor buys an issue such as the General Baking Company 5½s (described in Chap. XI) which have earned their interest an average of 20 times, as compared with the minimum requirement of three times. If a decline in profits should reduce the coverage to four times, he might prefer to switch into some other issue (if one can be found) which is earning its interest eight to ten times. On these assumptions he would have a fair chance of obtaining a full price for the General Baking issue, since it would still be making an impressive exhibit. But if the influence of the downward trend of earnings has depressed the quotation to a large discount, then he could decide to retain the issue rather than accept an appreciable loss. In so doing he would have the great advantage of being able to feel that the safety of investment was still not in any real danger.

Such a policy of demanding very high safety margins would obviously prove especially beneficial if a period of acute depression and market unsettlement should supervene. It is not practicable, however, to recommend this as a standard practice for all investors, because the supply of such strongly buttressed issues is too limited; and because, further, it is contrary to human nature for investors to take *extreme* precautions against future collapse when current conditions make for optimism.[1]

Policy in Depression.—Assuming that the investor has exercised merely reasonable caution in the choice of his fixed-value holdings, how will he fare and what policy should he follow in a period of depression? If the depression is a moderate one, his investments should be only mildly affected marketwise and still less in their intrinsic position. If conditions should approximate those of 1930–1933 he could not hope to escape a severe shrinkage in the quotations and considerable uneasiness over the safety of his holdings. But any reasoned policy of fixed-value investment requires the assumption that disturbances of the 1930–1933 amplitude are nonrecurring in their nature and need not be specifically guarded against in the future. If the 1921–1922 experience is accepted instead as typical of the "recurrent severe depression," a carefully selected investment list should give a reasonably good account of itself in such a period. The investor should not be stampeded into selling out holdings with a strong past record because of a current decline in earnings. He is likely, however, to pay more attention than usual to the question of improving the quality of his securities, and in many cases it should be possible to gain some benefits through carefully considered switches.

---

[1] We must caution the reader, however, against assuming that very large coverage of interest charges is, in itself, a complete assurance of safety. An operating loss eliminates the margin of safety, however high it may have been. Hence, inherent stability is an essential requirement, as we emphasize in our Studebaker example given in Chap. II.

"Bargain" Issues.—Where an issue has suffered impairment in both safety and its market quotation, the owner would like to find some means of exchanging into a better security selling at no higher price. In the next chapter we shall consider what possibilities exist for the purchase of safe securities at substantial discounts from par. While such "bargain" opportunities must clearly be exceptional, it is possible to locate them by means of careful but open-minded examination of large numbers of bonds and preferred stocks. In most cases issues of this character will not be well known and actively traded in. Consequently the typical investor is likely to hesitate to make *direct* purchases of this kind, for no matter how impressive the statistical exhibit may be, he will fear that the low price is a warning of some hidden weakness. He will be more likely to acquire such securities in exchange for others which he already owns and which are in a doubtful position, since such an exchange can scarcely result in increasing his risk and most probably will mean an improvement.

Sources of Investment Advice and Supervision.—Supervision of securities involves the question of who should do it as well as how to do it. Investors have the choice of various agencies for this purpose, of which the more important are the following:

1. The investor himself.
2. His commercial bank.
3. An investment banking (or underwriting) house.
4. A New York Stock Exchange firm.
5. The advisory department of a large trust company.
6. Independent investment counsel or supervisory service.

The last two agencies charge fees for their service, while the three preceding supply advice and information gratis.

Advice from Commercial Bankers.—The investor should not be his own sole consultant unless he has training and experience sufficient to qualify him to advise others professionally. In most cases he should at least supplement his own judgment by conference with others. The practice of consulting one's bank about investments is wide-spread, and it is undeniably of great benefit, especially to the smaller investor. If followed consistently it would afford almost complete protection against the hypnotic wiles of the high-pressure stock salesman and his worthless "blue sky" flotations. It is doubtful, however, whether the commercial banker is the most suitable adviser to an investor of means. While his judgment is usually sound, his knowledge of securities is likely to be somewhat superficial; and he cannot be expected to spare the time necessary for a thoroughgoing analysis of his clients' holdings and problems.

Advice from Investment Banking Houses.—There are objections of another kind to the advisory service of an investment banking house. An institution with securities of its own to sell cannot be looked to for entirely impartial guidance. However ethical its aims may be, the compelling force of self-interest is bound to affect its judgment. This is particularly true when the advice is supplied by a bond salesman whose livelihood depends upon persuading his customers to buy the securities that his firm has "on its shelves." It is true that the reputable underwriting houses consider themselves as bound in some degree by a fiduciary responsibility toward their clients. The endeavor to give them sound advice and to sell them suitable securities arises not only from the dictates of good business practice, but more compellingly from the obligations of a professional code of ethics.

Nevertheless, the sale of securities is not a profession but a business, and is necessarily carried on as such. While in the typical transaction it is to the advantage of the seller to give the buyer full value and satisfaction, conditions may arise in which their interests are in serious conflict. Hence it is impracticable, and in a sense unfair, to require investment banking houses to act as impartial advisers to buyers of securities; and, broadly speaking, it is unwise for the investor to rely primarily upon the advice of sellers of securities.

Advice from New York Stock Exchange Firms.—The investment departments of the large Stock Exchange firms present a somewhat different picture. While they also have a pecuniary interest in the transactions of their customers, their advice is much more likely to be painstaking and thoroughly impartial. Stock Exchange houses do not ordinarily own securities for sale. While at times they participate in selling syndicates, which carry larger allowances than the ordinary market commission, their interest in pushing such individual issues is less vital than that of the underwriting houses who actually own them. At bottom, the investment business or bond department of a Stock Exchange firm is perhaps more important to them as a badge of respectability than for the profits it yields. Attacks made upon them as agencies of speculation may be answered in part by pointing to the necessary services which they render to conservative investors. Consequently, the investor who consults a large Stock Exchange firm regarding a small bond purchase is likely to receive time and attention out of all proportion to the commission involved. Admittedly this practice is found profitable in the end, as a cold business proposition, because a certain proportion of the bond customers later develop into active stock traders. On behalf of the Stock Exchange houses it should be said that they make no effort to persuade their bond clients to speculate in stocks; but the atmosphere of a brokerage office is perhaps not without its seductive influence.

Advice from Investment Counsel.—While the idea of giving investment advice on a fee basis is not a new one, it has only recently developed into an important financial activity. The work is now being done by special departments of large trust companies, by a division of the statistical services, and by private firms designating themselves as investment counsel or investment consultants. The advantage of such agencies is that they can be entirely impartial, having no interest in the sale of any securities or in any commissions on their client's transactions. The chief disadvantage is the cost of the service, which averages about $\frac{1}{2}\%$ per annum on the principal involved. As applied strictly to investment funds this charge would amount to about one-tenth of the annual income, which must be considered substantial.

In order to make their fees appear less burdensome, some of the private investment consultants endeavor to forecast the general course of the bond market and to advise their clients as to when to buy or sell. It is doubtful whether trading in bonds, to catch the market swings, can be carried on successfully by the investor. If the course of the bond market can be predicted it should be possible to predict that of the stock market as well, and there would be undoubted technical advantages in trading in stocks rather than in bonds. We are sceptical of the ability of any paid agency to provide reliable forecasts of the market action of either bonds or stocks. Furthermore we are convinced that any combined effort to advise upon the choice of individual high-grade investments and upon the course of bond *prices* is fundamentally illogical and confusing. Much as the investor would like to be able to buy at just the right time and to sell out when prices are about to fall, experience shows that he is not likely to be brilliantly successful in such efforts, and that by injecting the trading element into his investment operations he will disrupt the income return on his capital and inevitably shift his interest into speculative directions.

It is not clear as yet whether advice on a fee basis will work out satisfactorily in the field of standard high-grade investments, because of their relatively small income return. In the purely speculative field the objection to paying for advice is that if the adviser *knew* whereof he spoke he would not need to bother with a consultant's duties. It may be that the profession of adviser on securities will find its most practicable field in the intermediate region, where the adviser will deal with problems arising from depreciated investments, and where he will propose advantageous exchanges and recommend bargain issues selling considerably below their intrinsic value.

PART III

SENIOR SECURITIES WITH SPECULATIVE FEATURES

CHAPTER XXII

SENIOR ISSUES AT BARGAIN LEVELS. PRIVILEGED ISSUES

We come now to the second major division of our revised classification of securities, namely bonds and preferred stocks presumed by the buyer to be subject to substantial change in principal value. In our introductory discussion (Chap. V) we subdivided this group under two heads: those issues which are speculative because of inadequate safety, and those which are speculative because they possess a conversion or similar privilege which makes possible substantial variations in market price.

AN INTERMEDIATE TYPE

There is, however, a third kind of bond or preferred stock, which may either be placed in this general division or may be considered more or less as a connecting link between the fixed-value and the speculative senior issues. We refer to those senior securities which the purchaser may regard as safe investments, but which nevertheless sell at so low a price as to offer a chance of considerable enhancement in market value. Many readers are likely to doubt the existence of such an alluring combination, and hold to the common view that a low price for a bond or preferred stock invariably points to a lack of safety, the cause of which may be either evident or concealed. Admitting that in the great majority of instances this observation is correct, the fact remains that market prices are not an infallible measure of value, and hence that there will always be a small minority of cases in which the price may be speculative but the security measures up to strict investment standards.

UNDERVALUED ISSUES

Bonds	Low price in 1932–1933	Yield at low price, %	Number of times fixed charges were earned		Ratio of market value of common and preferred to bonds at par
			Average 1927–1931	Minimum 1927–1931	
American Ice Co. Debenture 5s/1953	52	10.95	9.47	7.11	1.57
Brooklyn City R.R. Co. 1st 5s/1941	50	15.50	2.44†	2.15	0.38
Brooklyn Union Elevated R.R. 1st 5s/1950	60	9.79	1.88‡	1.81	‡
Chesapeake & Ohio R.R. Refunding 4½s/1995	60	7.55	3.97	3.45	0.40
Investors Equity Co. Debenture 5s/1948 (Assumed by Tri-Continental Corp.)	55	10.65	10.62‖	9.21	4.40¶
Owens-Illinois Glass Co. Debenture 5s/1939	65*	13.20	11.72"	9.00	3.36
Shell Pipe Line Corp. Debenture 5s/1952	56½	10.10	9.65((	6.83	))
Tobacco Products Corp. (N. J.) 6½s/2022	73¼	8.85	14.22**	10.12**	10.19**
Union Pacific R.R. Co. Debenture 4½s/1967	58	8.16	3.40	2.72	0.34

Preferred Stocks	Low price in 1932–1933	Yield at low price, %	Number of times pfd. dividends (and bond interest) were earned		Ratio of market value of common to preferred (and bonds)
			Average 1927–1931	Minimum 1927–1931	
Electric Auto-Lite Co. 7 % Cum. Pfd.	61	11.48	20.33	8.14	2.98
General Cigar Co. 7 % Cum. Pfd.	75	9.33	5.02	4.16	1.44
General Railway Signal Co. 6 % Cum. Pfd.	65	9.23	14.70	8.68	2.97
Gold Dust Corp. $6 Cum. Pfd.	70	8.57	8.35††	7.12	4.25
MacAndrews & Forbes Co. 6 % Cum. Pfd.	57¼	10.43	8.94	6.20	2.87
United States Gypsum Co. 7 % Cum. Pfd.	84¾	8.25	9.93	6.43	2.29

* Over-the-counter price prior to listing.

† Based on exhibit of Brooklyn and Queens Transit Corporation which has assumed the bonds. Data cover five-year period ended June 30, 1932.

‡ Based on exhibit of New York Rapid Transit Corporation which has assumed the bonds. Data cover five-year period ended June 30, 1931.

§ Not computed. All common stock owned by Brooklyn-Manhattan Transit Corporation and subsidiaries.

|| Based on two full years since organization (1930 and 1931). Earnings coverage based on net earnings of Tri-Continental Corporation, exclusive of profits on sale of securities in 1930 and losses thereon in 1931. Assumed net earnings of 5 % on the cash assets and net market value of Investors Equity portfolio were added to the Tri-Continental earnings for 1930 and 1931.

¶ Based on the exhibit of Tri-Continental Corporation which assumed the bonds, and giving effect to additional shares issued in the absorption of Investors Equity Company by Tri-Continental Corporation. Giving effect to this merger the net equity per $1,000 bond in cash and marketable securities at market as of Dec. 31, 1931, was $6,896.47.

" Based on consolidated income account as reported to the New York Stock Exchange in application to list the new bonds, dated May 2, 1932. Charges on consolidated funded debt, including the new bonds and other obligations assumed in the merger with Illinois Pacific Coast Company, were used in computing this ratio.

((Average for four years, 1928–1931, inclusive. Five-year average of Shell Union Oil Corporation which guarantees principal and interest on this issue was 2.89 times. Minimum coverage on latter basis was deficit 3.22 times in 1931.

)) Not computed. All common stock owned by Shell Union Oil Corporation. Stock-value ratio for latter on date of low price for this issue was 0.37.

** Analyzed as a bond of the American Tobacco Company, which guarantees principal and interest on this issue under the terms of a lease.

†† Number of times interest, subsidiary preferred dividends and parent company preferred dividends were earned. Based on 1929–1931, inclusive. Comparable figures for 1927–1928 not obtainable.

The untrained investor would no doubt do well to avoid seeking these "bargain issues," because his search is likely to lead him astray into the acceptance of unsound securities. But to the experienced and capable analyst this field offers opportunities greater than its restricted size would indicate. If he has mastered the technique of the selection of sound fixed-value investments, he can apply his skill most effectively to the discovery of safe issues selling at low prices. This work requires the painstaking examination of literally hundreds of securities in order to find a mere handful of this desirable character, and they must then be subjected to especially critical study in order to detect concealed elements of weakness.

During periods of severe market dislocation, such as occurred in 1931–1933, the proportion of such bargain opportunities will increase. Quite an impressive number of securities will be found in the low-price range which nevertheless are making a strong statistical exhibit even under the conditions of maximum depression. We append, on page 238, a group of listed bonds and preferred stocks which met all our quantitative tests of safety for 1931 as well as in previous years, which did not appear subject to any persuasive objections of the qualitative kind, and which in 1932 sold at abnormally low levels for an investment issue.

Some of these bonds did not meet our stock-value test on the basis of the very low stock prices obtaining in 1932; but considering the abnormal market conditions then obtaining and the fact that the stock-value ratio is a secondary test, we do not believe that this fact would have disqualified the issues as sound investments. Because of the weaker contractual position of preferred stocks we have included only such issues as met both tests fully.

In concluding chapters of this book devoted to "Discrepancies between Price and Value," we shall discuss in some detail the typical conditions under which such bargain opportunities are likely to arise among senior issues. This knowledge will be found a useful aid in the process of locating such attractive offerings. Moreover, an adequate explanation of the low price, if for reasons not related to the intrinsic value, will reassure the analyst against the fear of unknown perils lurking in the situation and give him more confidence in his judgment.

SENIOR ISSUES WITH SPECULATIVE PRIVILEGES

In addition to enjoying a prior claim for a fixed amount of principal and income, a bond or preferred stock may also be given the right to share in benefits accruing to the common stock. These privileges are of three kinds, designated as follows:

1. Convertible—conferring the right to exchange the senior issue for common stock on stipulated terms.
2. Participating—under which additional income may be paid to the senior security holder, dependent usually upon the amount of common dividends declared.
3. Subscription—by which holders of the bond or preferred stock may purchase common shares, at prices, in amounts, and during periods, stipulated.[1]

Since the conversion privilege is the most familiar of the three, we shall frequently use the term "convertible issues" to refer to privileged issues in general.

Such Issues Attractive in Form.—By means of any one of these three provisions a senior security can be given virtually all the profit possibilities that attach to the common stock of the enterprise. Such issues must therefore be considered as the most attractive of all in point of *form*, since they permit the combination of maximum safety with the chance of unlimited appreciation in value. A bond which meets all the requirements

[1] There is still a fourth type of profit-sharing arrangement, of less importance than the three just described, which made its first appearance in the 1928–1929 bull market. This is the so-called "optional" bond or preferred stock. The option consists of taking interest or dividend payments in a fixed amount of common stock (*i.e.*, at a fixed price per share) in lieu of cash.

For example, Commercial Investment Trust $6 Convertible Preference, Optional Series of 1929, gave the holder the option to take his dividend at the annual rate of one-thirteenth share of common instead of $6 in cash. This was equivalent to a price of $78 per share for the common, which meant that the option would be valuable whenever the stock was selling above 78. Similarly, Warner Brothers Pictures, Inc., Optional 6% Convertible Debentures, due 1939, issued in 1929, gave the owner the option to take his interest payments at the annual rate of one share of common stock instead of $60 in cash.

It may be said that this optional arrangement is a modified form of conversion privilege, under which the interest or dividend amounts are made separately convertible into common stock. In most, possibly all, of these issues, the principal is convertible as well. The separate convertibility of the income payments adds somewhat, but not a great deal, to the attractiveness of the privilege.

of a sound investment and in addition possesses an interesting conversion privilege would undoubtedly constitute a highly desirable purchase.

Their Investment Record Unenviable: Reasons.—Despite this impressive argument in favor of privileged senior issues as a *form* of investment, we must recognize that actual experience with this class has not been generally satisfactory. For this discrepancy between promise and performance, reasons of two different kinds may be advanced.

The first is that only a small fraction of the privileged issues have actually met the rigorous requirements of a sound investment. The conversion feature has most often been offered to compensate for inadequate security.[1] This weakness was most pronounced during the period of greatest vogue for convertible issues, between 1926 and 1929.[2] During these years it was broadly true that the strongly entrenched industrial enterprises raised money through sales of common stock, while the weaker— or weakly capitalized—undertakings resorted to privileged senior securities.

The second reason is related to the conditions under which profit may accrue from the conversion privilege. While there is indeed no upper limit to the price which a convertible bond may reach, there is a very real limitation on the amount of profit which the holder may realize while still maintaining an *investment position*. After a privileged issue has advanced with the common stock, its price soon becomes dependent *in both directions* upon changes in the stock quotation; and to that extent the continued holding of the senior issue becomes a *speculative* operation. An example will make this clear:

Let us assume the purchase of a high-grade 5% bond at par, convertible into two shares of common for each $100 bond

---

[1] The *Report of the Industrial Securities Committee of the Investment Bankers Association of America* for 1927 quotes, presumably with approval, a suggestion that since a certain percentage of the senior securities of moderate-sized industrial companies "are liable to show substantial losses over a period of five or ten years," investors therein should be given a participation in future earnings through a conversion or other privilege to compensate for this risk. See *Proceedings of the Sixteenth Annual Convention of the Investment Bankers Association of America*, 1927, pp. 144–145.

[2] So far as the writers are aware no comprehensive compilation of the dollar volume of privileged issues has been made and regularly maintained. Circumstantial evidence of the increase in the volume of this type of financing is given in Note 33 of the Appendix.

(*i.e.*, convertible into common stock at 50). The common stock is selling at 45 when the bond is bought.

First stage: (*a*) If the stock declines to 35, the bond may remain close to par. This illustrates the pronounced technical advantage of a convertible issue over the common stock. (*b*) If the stock advances to 55, the price of the bond will probably rise to 115 or more. (Its "immediate conversion value" would be 110, but a premium would be justified because of its advantage over the stock.) This illustrates the undoubted speculative possibilities of such a convertible issue.

Second stage: The stock advances further to 65. The conversion value of the bond is now 130, and it will sell at that figure, or slightly higher. At this point the original purchaser is faced with a problem. Within wide limits, the future price of his bond depends entirely upon the course of the common stock. In order to seek a larger profit he must risk the loss of the profit in hand, which in fact constitutes a substantial part of the present market value of his security. (A drop in the price of the common could readily induce a decline in the bond from 130 to 100.) If he elects to hold the issue, he places himself to a considerable degree in the position of the stockholders, and this similarity increases rapidly as the price advances further. If, for example, he is still holding the bond at a level say of 180 (90 for the stock) he has for all practical purposes assumed the status and risks of a stockholder.

Unlimited Profit in Such Issues Identified with Stockholder's Position.—The unlimited profit possibilities of a privileged issue are thus in an important sense illusory. They must be identified not with the ownership of a bond or preferred stock, but with the assumption of a common stockholder's position—which any holder of a nonconvertible may effect by exchanging from his bond into a stock. Practically speaking, the range of profit possibilities for a convertible issue, *while still maintaining the advantage of an investment holding*, must usually be limited to somewhere between 25 and 35% of its face value. For this reason original purchasers of privileged issues do not ordinarily hold them for more than a small fraction of the maximum market gains scored by the most successful among them, and consequently they do not actually realize these very large possible profits. Thus the profits taken may not offset the losses occasioned by unsound commitments in this field.

Examples of Attractive Issues.—The two objections just discussed must considerably temper our enthusiasm for privileged senior issues as a class, but they by no means destroy their inherent advantages nor the possibilities of exploiting them with reasonable success. Although *most* new convertible offerings may be inadequately secured, there are fairly frequent exceptions to the rule, and these exceptions should be of prime interest to the alert investor. We append three leading examples of such opportunities, taken from the utility, the railroad, and the industrial field.

1. *American Telephone and Telegraph Company Convertible 6s, Due* 1925.—These notes were originally offered to shareholders in August 1918 at 94 to yield 7.10% to maturity. The statistical exhibit of the company gave every assurance that the notes were a sound investment. They were convertible from August 1, 1920, to August 1, 1925, into common stock of the company, par for par, upon the payment of $6 a share for stock thus obtained, with the usual adjustments of accrued interest and dividends.

The notes could be purchased on August 31, 1918 at 94⅞ when the stock was selling at 98½. At these prices, the notes were quoted almost on a parity with the stock, in other words, a very slight advance in the price of the stock would create a profit for the holder of the convertible note. In October 1920 when the conversion privilege was operative, the bonds could be bought at 95¼, while the stock was selling at 97½. As the stock gradually advanced during the next five years, the market value of the convertible notes increased correspondingly, reaching 136½ in 1925.

2. *Chesapeake and Ohio Railway Company Convertible 5s, Due* 1946.—These bonds were originally offered to shareholders in June 1916. They were convertible into common stock at 75 until April 1, 1920; at 80 from the latter date until April 1, 1923; at 90 from the latter date until April 1, 1926; and at 100 from the latter date until April 1, 1936.

Late in 1924 they could have been bought on a parity basis (that is, without payment of a premium for the conversion privilege) at prices close to par. Specifically, they sold on November 28, 1924, at 101 when the stock sold at 91. At that time the company's earnings were showing continued improvement and indicated that the bonds were adequately secured. (Fixed charges were covered twice in 1924.) The value of the conversion privilege was shown by the fact that the stock sold at 131 in the next year, making the bonds worth 145.

3. *Rand Kardex Bureau, Inc.*, 5½s, *Due* 1931.—These bonds were originally offered in December 1925 at 99½. They carried stock-purchase warrants (detachable after January 1, 1927) entitling the holder to purchase 22½ shares of Class *A* common at $40 a share during 1926; at $42.50 per share during 1927; at $45 a share during 1928; at $47.50 a share during 1929; and at $50 a share during 1930. (The Class *A* stock was in reality a participating preferred issue.) The bonds could be turned in at par in payment for the stock purchased under the warrants, a provision which virtually made the bonds convertible into the stock.

The bonds appeared to be adequately secured. The previous exhibit (based on the earnings of the predecessor companies) showed the following coverage for the interest on the new bond issue:

Year	Number of Times Interest Covered
1921	1.7
1922	2.3
1923	6.7
1924	7.2
1925 (9 mo.)	12.2

Net quick assets exceed twice the face value of the bond issue.

When the bonds were offered to the public the Class *A* stock was quoted at about 42, indicating an immediate value for the stock-purchase warrants. The following year the stock advanced to 53, and the bonds to 130½. In 1927 (when Rand Kardex merged with Remington Typewriter) the stock advanced to 76 and the bonds to 190.

Example of an Unattractive Issue.—By way of contrast with these examples we shall supply an illustration of a superficially attractive but basically unsound convertible offering, such as characterized the 1928–1929 period.

National Trade Journals, Inc., 6% *Convertible Notes, Due* 1938.—The company was organized in February 1928 to acquire and publish about a dozen trade journals. In November 1928 it sold $2,800,000 of the above notes at 97½. The notes were initially convertible into 27 shares of common stock (at $37.03 per share) until November 1, 1930; into 25 shares (at $40 a share) from the latter date until November 1, 1932; and at prices which progressively increased to $52.63 a share during the last two years of the life of the bonds.

These bonds could have been purchased at the time of issuance and for several months thereafter at prices only slightly above their parity value as compared with the market value of the equivalent stock. Specifically, they could have been bought at 97½ on November 30, when the stock sold at 34⅛, which meant that the stock needed to advance only two points to assure a profit on conversion.

However, at no time did the bonds appear to be adequately secured, despite the attractive picture presented in the offering circular. The circular exhibited "estimated" earnings of the predecessor enterprise for the 3½ years preceding which averaged 4.16 times the charges on the bond issue. But close to half of these estimated earnings were expected to be derived from economies predicted to result from the consolidation in the way of reduction of salaries, etc. The conservative investor would not be justified in taking these "earnings" for granted, particularly in a hazardous and competitive business of this type, with a relatively small amount of tangible assets.

Eliminating the estimated "earnings" mentioned in the preceding paragraph the exhibit at the time of issuance and thereafter was as follows:

Year	Price range of bonds	Price range of stock	Prevailing conversion price	Times interest earned	Earned per share on common
1925	....	....	....	1.73*	$0.78*
1926	....	....	....	2.52*	1.84*
1927	....	....	....	2.80*	2.20*
1928	100 −97½	35⅞−30	$37.03	1.69†	1.95
1929	99 −50	34⅝− 5	37.03	1.86†	1.04
1930	42 −10	6⅜− ½	37.03−$40	0.09†	1.68(d)
1931	10½− 5	1	40.00	Receivership	

* Estimated as indicated above. Per-share figures are taken after estimating federal taxes.

† Actual earnings for last ten months of 1928 and succeeding calendar years.

Receivers were appointed in June 1931. The properties were sold in August of that year and bondholders later received about 8½ cents on the dollar.

Principle Derived.—From these contrasting instances an investment principle may be developed which should afford a valuable guide to the selection of privileged senior issues. The principle is as follows: *A privileged senior issue, selling approximately at or above face value, must meet the requirements either of a straight fixed-value investment, or of a straight common-stock speculation, and it must be bought with one or the other qualification clearly in view.*

The alternative given supplies two different approaches to the purchase of a privileged security. It may be bought as a sound investment with an *incidental* chance of profit through an enhancement of principal; or it may be bought *primarily* as an attractive form of speculation in the common stock. Generally speaking, there should be no middle ground. The *investor* interested in safety of principal should not abate his requirements in return for a conversion privilege; the *speculator* should not be attracted to an enterprise of mediocre promise because of the pseudo-security provided by the bond contract.

Our opposition to any compromise between the purely investment and the admittedly speculative attitude is based primarily on subjective grounds. Where an intermediate stand is taken the result is usually confusion, clouded thinking, and self-deception. The investor who relaxes his safety requirements to obtain a profit-sharing privilege is frequently not prepared, financially or mentally, for the inevitable loss if fortune should frown on the venture. The speculator who wants to reduce his risk by operating in convertible issues is likely to find his primary interest divided between the enterprise itself and the terms of the privilege; and he will probably be uncertain in his own mind as to whether he is at bottom a stockholder or a bondholder. (Privileged issues *selling at substantial discounts from par* are not in general subject to this principle, since they belong to the second category of speculative senior securities to be considered later.)

Reverting to our examples, it will be seen at once that the American Telephone and Telegraph Convertible 6s could properly have been purchased as an investment without any regard to the conversion feature. The strong possibility that this privilege would be of value made the note almost uniquely attractive at the time of issuance. Somewhat similar statements could be made with respect to the Chesapeake and Ohio and the Rand

Kardex bonds. Any of these three securities should also have been attractive to a speculator who was persuaded that the related common stock was due for an advance in price.

On the other hand the National Trade Journals debentures could not have passed stringent qualitative and quantitative tests of safety. Hence they should properly have been of interest only to a person who had full confidence in the future value of the stock. It is undoubtedly true, however, that most of the buying of this issue was not motivated by the primary desire to invest or speculate in the National Trade Journals common stock; but it was based rather on the attractive terms of the conversion privilege and on the feeling that the issue was "fairly safe" as a bond investment. It is precisely this compromise between true investment and true speculation that we disapprove, chiefly because the purchaser has no clear-cut idea of the purpose of his commitment or of the risk that he is incurring.

Rules Regarding Retention or Sale.—Having stated a basic principle to guide the *selection* of privileged issues, we ask next what rules can be established regarding their subsequent retention or sale. Convertibles bought primarily as a form of commitment in the common stock may be held for a larger profit than those acquired from the investment standpoint. If a bond of the former class advances from 100 to 150 the large premium need not in itself be a controlling reason for selling out; the owner must be guided rather by his views as to whether the common stock has advanced enough to justify taking his profit. But when the purchase is made primarily as a safe bond investment, then the limitation on the amount of profit which can conservatively be waited for comes directly into play. For the reasons explained in detail above, the conservative buyer of privileged issues will not ordinarily hold them for more than a 25 to 35% advance. This means that a really successful investment operation in the convertible field does not cover a long period of time. Hence such issues should be bought with the *possibility* of long-term holding in mind, but with the *hope* that the potential profit will be realized fairly soon.

The foregoing discussion leads to the statement of another investment rule, *viz.:*

In the typical case, a convertible bond should not be converted by the investor.

It is true that the object of the privilege is to bring about such conversion when it seems advantageous. If the price of the bond advances substantially, its current yield will shrink to an unattractive figure, and there is ordinarily a substantial gain in income to be realized through the exchange into stock. Nevertheless when the investor does exchange his bond into the stock, he abandons the priority and the unqualified claim to principal and interest upon which the purchase was originally premised. If after the conversion is made things should go badly, his shares may decline in value far below the original cost of his bond and he will lose not only his profit but part of his principal as well.

Moreover he is running the risk of transforming himself— generally, as well as in the specific instances—from a bond investor into a stock speculator. It must be recognized that there is something insidious about even a good convertible bond; it can easily prove a costly snare to the unwary. To avoid this danger the investor must cling determinedly to a conservative viewpoint. When the price of his bond has passed out of the investment range, he must sell it; most important of all, he must not consider his judgment impugned if the bond subsequently rises to a much higher level. The market behavior of the issue, once it has entered the speculative range, is no more the investor's affair than the price gyrations of any speculative stock about which he knows nothing.

If the course of action here recommended is followed by investors generally, the conversion of bonds would be brought about only through their purchase for this specific purpose by persons who have decided independently to acquire the shares either for speculation or supposed investment.[1] The arguments against the investor's converting convertible issues apply with equal force against his exercising stock-purchase warrants attached to bonds bought for investment purposes.

A continued policy of investment in privileged issues would, under favorable conditions, require rather frequent taking of profits and replacement by new securities not selling at an excessive premium. More concretely, a bond bought at 100 would be sold, say, at 125 and be replaced by another good con-

---

[1] In actual practice, conversions often result also from arbitrage operations involving the purchase of the bond and the simultaneous sale of the stock at a price slightly higher than the "conversion parity."

vertible issue purchasable at about par. It is not likely that satisfactory opportunities of this kind will be continuously available, nor that the investor would have the means of locating all those that are available. There is a fair possibility that a larger number of really attractive convertibles will be found in the years following 1933 than in the period preceding. From 1926 to 1929 there was a tremendous outpouring of privileged issues, but in the great majority of cases their investment quality was poor. In 1931–1933 the very small investment demand then in evidence was concentrated upon the idea of safety to such an extent that conversion privileges practically disappeared from new bond offerings. If the pendulum swings back toward a middle point, we may again find that participating features will at times be employed to facilitate the sale of sound bond offerings; and opportunities similar to the American Telephone and Telegraph 6% bond issue of 1918 may again be available to the careful and shrewd investor.

CHAPTER XXIII

TECHNICAL CHARACTERISTICS OF PRIVILEGED SENIOR SECURITIES

In the preceding chapter privileged senior issues were considered in their relationship to the broader principles of investment and speculation. To arrive at an adequate knowledge of this group of securities from their practical side, a more intensive discussion of their characteristics is now in order. Such a study may conveniently be carried on from three successive viewpoints: (1) considerations common to all three types of privilege—conversion, participation, and subscription (*i.e.*, "warrant"); (2) the relative merits of each type, as compared with the others; (3) technical aspects of each type, considered by itself.[1]

CONSIDERATIONS GENERALLY APPLICABLE TO PRIVILEGED ISSUES

The attractiveness of a profit-sharing feature depends upon two major, but entirely unrelated factors: (1) the *terms* of the arrangement; and (2) the *prospects* of there being profits to share. To use a simple illustration:

Company A	Company B
6% bond selling at 100	6% bond selling at 100
Convertible into stock at 50	Convertible into stock at 33⅓
(*i.e.*, two shares of stock for	(*i.e.*, three shares of stock for
a $100 bond)	a $100 bond)
Stock selling at 30	Stock selling at 30

Terms of the Privilege vs. Prospects for the Enterprise.—The *terms* of the conversion privilege are evidently more attractive in the case of Bond *B*; for the stock need advance only a little more than 3 points to assure a profit, while Stock *A* must advance over 20 points to make conversion profitable. Nevertheless, it

---

[1] This subject is treated at what may appear to be disproportionate length because of the growing importance of privileged issues and the absence of thoroughgoing discussion thereof in the standard descriptive textbooks.

is quite possible that Bond *A* may turn out to be the more advantageous purchase. For conceivably Stock *B* may fail to advance at all while Stock *A* may double or triple in price.

As between the two factors, it is undoubtedly true that it is more profitable to select the right *company* than to select the issue with the most desirable *terms*. There is certainly no mathematical basis on which the attractiveness of the enterprise may be offset against the terms of the privilege and a balance struck between these two entirely dissociated elements of value. But in analyzing privileged issues of the investment grade, the *terms* of the privilege must receive the greater attention, not because they are more important, but because they can be more definitely dealt with. It may seem a comparatively easy matter to determine that one enterprise is more promising than another. But it is by no means so easy to establish that one common stock at a given price is clearly preferable to another stock at its current price.

Reverting to our example, if it were quite certain, or even reasonably probable, that Stock *A* is more likely to advance to 50 than Stock *B* to advance to 33, then both issues would not be quoted at 30. Stock *A*, of course, would be selling higher. The point we make is that the market price in general *reflects already* any superiority which one enterprise has demonstrated over another. The investor who prefers Bond *A* because he expects its related stock to rise a great deal faster than Stock *B*, is exercising independent judgment in a field where certainty is lacking and where mistakes are necessarily frequent. For this reason we doubt whether a successful policy of buying privileged issues *from the investment approach* can be based primarily upon the purchaser's view regarding the future expansion of the profits of the enterprise. (In stating this point we are merely repeating a principle previously laid down in the field of fixed-value investment.)

Where the speculative approach is followed, *i.e.*, where the issue is bought primarily as a desirable method of acquiring an interest in the stock, it would be quite logical, of course, to assign dominant weight to the buyer's judgment as to the future of the company.

Three Important Elements. 1. *Extent of the Privilege.*—In examining the *terms* of a profit-sharing privilege, three component elements are seen to enter. These are:

1. The *extent* of the profit-sharing or speculative interest per dollar of investment.

2. The *closeness* of the privilege to a realizable profit at the time of purchase.

3. The *duration* of the privilege.

The amount of speculative interest attaching to a convertible or warrant-bearing senior security is equal to the current market value of the number of shares of stock covered by the privilege. Other things being equal, the larger the amount of the speculative interest per dollar of investment, the more attractive the privilege.

Examples: Rand Kardex 5½s, previously described, carried warrants to buy 22½ shares of Class *A* stock initially at 40. Current price of Class *A* stock was 42. The "speculative interest" amounted to 22½ × 42, or $945 per $1,000 bond.

Reliable Stores Corporation 6s, offered in 1927, carried warrants to buy only 5 shares of common stock initially at 10. Current price of the common was 12. Hence the "speculative interest" amounted to 5 × 12, or only $60 per $1,000 bond.

International Rubber Products Company 7s offered an extraordinary example of a large speculative interest attaching to a bond. As a result of peculiar provisions surrounding their issuance in 1922, each $1,000 note was convertible into 100 shares of stock and also carried the right to purchase 400 additional shares at 10. When the stock sold at 10 in 1925, the speculative interest per $1,000 note amounted to 500 × 10, or $5,000. If the notes were then selling, say, at 120, the speculative interest would have equalled 417% of the bond investment— or 70 *times* as great as in the case of the Reliable Stores offering.

The practical importance of the amount of speculative interest can be illustrated by the following comparison, covering the three examples above given.

Item	Reliable Stores 6s	Rand Kardex 5½s	Intercontinental Rubber 7s
Number of shares covered by each $1,000 bond........	5	22½	500
Base price.................	$10.00	$ 40.00	$ 10.00
Increase in value of bond when stock advances:			
25% above base price....	12.50	225.00	1,250.00
50% above base price....	25.00	450.00	2,500.00
100% above base price....	50.00	900.00	5,000.00

In the case of convertible bonds the speculative interest always amounts to 100% of the bond at par when the stock sells at the conversion price. Hence in these issues our first and second component elements express the same fact. If a bond selling at par is convertible into stock at 50, and if the stock sells at 30, then the speculative interest amounts to 60% of the commitment; which is the same thing as saying that the current price of the stock is 60% of that needed before conversion would be profitable. Stock-purchase-warrant issues disclose no such fixed relationship between the amount of the speculative interest and the proximity of this interest to a realizable profit. In the case of the Reliable Stores 6s, the speculative interest was very small, but it showed an actual profit at the time of issuance, since the stock was selling *above* the subscription price.

Significance of Call on Large Number of Shares at Low Price.—It may be said parenthetically that a speculative interest in a large number of shares selling at a low price is technically more attractive than one in a smaller number of shares selling at a high price. This is because low-priced shares are apt to fluctuate over a wider range *percentagewise* than higher priced stocks. Hence if a bond is both well secured and convertible into many shares at a low price, it will have an excellent chance for very large profit without being subject to the offsetting risk of greater loss through a speculative dip in the price of the stock.

For example, as a matter of *form* of privilege, the Ohio Copper Company 7s, due 1931, convertible into 1,000 shares of stock selling at $1, had better possibilities than the Atchison, Topeka and Santa Fe Convertible 4½s, due 1948, convertible into 6 shares of common, selling at 166⅔, although in each case the amount of speculative interest equalled $1,000 per bond. As it turned out, Ohio Copper stock advanced from less than $1 a share in 1928 to 4⅞ in 1929, making the bond worth close to 500% of par. It would have required a rise in the price of Atchison from 166 to 800 to yield the same profit on the convertible 4½s, but the highest price reached in 1929 was under 300.

In the case of participating issues, the extent of the profit-sharing interest would ordinarily be considered in terms of the amount of extra income which may conceivably be obtained as a result of the privilege. A limited extra payment (*e.g.*, Bayuk Cigars, Inc., 7% Preferred, which may receive not more than

1% additional) is of course less attractive than an unlimited participation (*e.g.*, White Rock Mineral Springs Company 5% Second Preferred, which received a total of 26¼% in 1930).

2 *and* 3. *Closeness and Duration of the Privilege.*—The implications of the second and third factors in valuing a privilege are readily apparent. A privilege having a long period to run is in that respect more desirable than one expiring in a short time. The nearer the current price of the stock to the level at which conversion or subscription becomes profitable, the more attractive does the privilege become. In the case of a participation feature, it is similarly desirable that the current dividends or earnings on the common stock should be close to the figure at which the extra distribution on the senior issue commences.

By "conversion price" is meant the price of the common stock equivalent to a price of 100 for the convertible issue. If a preferred stock is convertible into 1⅔ as many shares of common, the conversion price of the common is therefore 60. The term "conversion parity" or "conversion level," may be used to designate that price of the common which is equivalent to a given quotation for the convertible issue, or *vice versa*. It can be found by multiplying the price of the convertible issue by the conversion price of the common. If the preferred stock just mentioned is selling at 90, the conversion parity of the common becomes 60 × 90% = 54. This means that to a buyer of the preferred at 90 an advance in the common above 54 will create a realizable profit.

COMPARATIVE MERITS OF THE THREE TYPES OF PRIVILEGES

From the theoretical standpoint, a participating feature—unlimited in time and possible amount—is the most desirable type of profit-sharing privilege. This arrangement enables the investor to derive the specific benefit of participation in profits (*viz.*, increased income), without modifying his original position as a senior-security holder. These benefits may be received over a long period of years. By contrast, a conversion privilege can result in higher income only through actual exchange into the stock and consequent surrender of the senior position. Its real advantage consists therefore, only of the opportunity to make a profit through the sale of the convertible issue at the right time. Similarly the benefits from a subscription privilege may conservatively be realized only through sale of the warrants (or by the subscription to and prompt sale of the stock).   If the common stock is purchased and held for permanent income, the operation involves the risking of additional money on a basis entirely different from the original purchase of the senior issue.

Example of Advantage of Unlimited Participation Privilege.— An excellent practical example of the theoretical advantages attaching to a well-entrenched participating security is afforded by Westinghouse Electric and Manufacturing Company Preferred. This issue is entitled to cumulative prior dividends of $3.50 per annum (7% on $50 par), and in addition participates equally per share with the common in any dividends paid on the latter in excess of $3.50.   As far back as 1917, Westinghouse Preferred could have been bought at 52½, representing an attractive straight investment with additional possibilities through its participating feature.   In the ensuing 15 years to 1932 a total of about $7 per share was disbursed in extra dividends above the basic 7%.   In the meantime there was an opportunity to sell out at a large profit (the high price being 284 in 1929), which corresponded to the enhancement possibilities of a convertible or subscription-warrant issue.   If the stock was not sold, the profit was naturally lost in the ensuing market decline.   But the investor's original position remained unimpaired, for at the low point of 1932 the issue was still paying the 7% dividend and selling at 52½ per share—although the common had passed its dividend and had fallen to 15⅝.

In this instance the investor was able to participate in the surplus profits of the common stock in good years while maintaining his preferred position, so that when the bad years came, he lost only his temporary *profit*.   Had the issue been convertible instead of participating, the investor could have received the higher dividends only through converting, and would later have found the dividend omitted on his common shares and their value fallen far below his original investment.

Participating Issues at Disadvantage, Marketwise.—While from the standpoint of long-pull-investment holding, participating issues are theoretically the most desirable, they may behave somewhat less satisfactorily in a major market upswing than do convertible or subscription-warrant issues.   During such a period a participating senior security may regularly sell below its proper comparative price.   In the case of Westinghouse Preferred for example, its price during 1929 was usually from

5 to 10 points lower than that of the common, although its intrinsic value per share could not be less than that of the junior stock.[1]

The reason for this phenomenon is as follows: The price of the common stock is made largely by speculators interested chiefly in quick profits, to secure which they need an active market. The preferred stock, being closely held, is relatively inactive. Consequently the speculators are willing to pay several points more for the inferior common issue simply because it can be bought and sold more readily, and because other speculators are likely to be willing to pay more for it also.

The same anomaly arises in the case of closely held common stocks with voting power, compared with the more active nonvoting issue of the same company. American Tobacco *B* (nonvoting) for example, has for years sold higher than the original (voting) stock. A similar situation has been presented by the two common issues of Bethlehem Steel, Pan American Petroleum, Liggett and Myers, and others.[2] The paradoxical principle holds true for the securities market generally that *in the absence of a special demand* relative scarcity is likely to make for a lower rather than a higher price.

In cases such as Westinghouse and American Tobacco the proper corporate policy would be to extend to the holder of the intrinsically more valuable issue the privilege of exchanging it for the more active but intrinsically inferior issue. The White Rock company actually took this step. While the holders of the participating preferred might make a mistake in accepting such an offer, they cannot object to its being made to them, and the common stockholders may gain but cannot lose through its acceptance.

Relative Price Behavior of Convertible and Warrant-bearing Issues.—From the standpoint of price behavior under favorable

[1] A much greater price discrepancy of this kind existed in the case of White Rock Mineral Springs Participating Preferred and common during 1929 and 1930. Because of this market situation, holders of nearly all the participating preferred shares accepted an offer to exchange into common stock, although this meant no gain in income and the loss of their senior position.

[2] The persistently wide spread between the market prices for R. J. Reynolds Tobacco Company Common and Class *B* stocks rests on the special circumstance that officers and employees of the company who own the common stock enjoy certain profit-sharing benefits not accorded to holders of the Class *B* stock.

market conditions the best results are obtained by holders of senior securities with detachable stock-purchase warrants. To illustrate this point we shall compare certain price relationships shown in 1929 between four privileged issues and the corresponding common stocks. The issues are as follows:

1. Mohawk Hudson Power Corporation 7% Second Preferred, carrying warrants to buy 2 shares of common at 50 for each share of preferred.

2. White Sewing Machine Corporation 6% Debentures, due 1936, carrying warrants to buy 2½ shares of common stock for each $100 bond.

3. Central States Electric Corporation 6% Preferred, convertible into common stock at $118 per share.

4. Independent Oil and Gas Company Debentures 6s, due 1939, convertible into common stock at $32 per share.

Senior issue	Market price of common	Conversion or subscription price of common	Price of senior issue	Realizable value of senior issue based on privilege (conversion or subscription parity)	Amount by which senior issue sold above parity ("premium"), points
Mohawk Hudson 2d Pfd...........	52½	50	163*	105	58
White Sewing Machine 6s..........	39	40	123½†	97½	26
Central States Electric Pfd........	116	118	97	98	−1
Independent Oil & Gas 6s..........	31	32	105	97	8

* Consisting of 107 for the preferred stock, ex-warrants, plus 56 for the warrants.
† Consisting of 98½ for the bonds, ex-warrants, plus 25 for the warrants.

The above table shows in striking fashion that in speculative markets issues with purchase warrants have a tendency to sell at large premiums in relation to the common-stock price, and that these premiums are much greater than in the case of similarly situated convertible issues.

Advantage of Separability of Speculative Component.—This advantage of subscription-warrant issues is due largely to the fact that their speculative component (*i.e.*, the subscription warrant itself) can be entirely separated from their investment component (*i.e.*, the bond or preferred stock ex-warrants). Speculators are always looking for a chance to make large profits on a small cash commitment. This is a distinguishing characteristic of stock option warrants, as will be shown in detail in our later discussion of these instruments. In an advancing market, therefore, speculators bid for the warrants attached to these privileged issues, and hence they sell separately at a substantial price even though they may have no immediate exercisable value. These speculators greatly prefer buying the option warrants to buying a corresponding *convertible bond,* because the latter requires a much larger cash investment per share of common stock involved.[1] It follows, therefore, that the separate market values of the bond plus the option warrant (which combine to make the price of the bond "with warrants") may considerably exceed the single quotation for a closely similar convertible issue.

Second Advantage of Warrant-bearing Issues.—Subscription-warrant issues have a second point of superiority, in respect to callable provisions. A right reserved by the corporation to redeem an issue prior to maturity must in general be considered as a disadvantage to the holder; for presumably it will be exercised only when it is to the benefit of the issuer to do so, which means usually that the security would otherwise sell for more than the call price.[2] A callable provision, unless at a very high premium, might entirely vitiate the value of a participating privilege. For with such a provision there would be danger of redemption as soon as the company grew prosperous enough to place the issue in line for extra distributions.[3] In some cases participating issues which are callable are made convertible as well, in order to give them a chance to benefit from any large advance in the market price of the common which may have taken place up to the time of call. (See for examples: National Distillers Products Corporation $2.50 Cumulative Participating Convertible Pre-

---

[1] Note that the Independent Oil and Gas bonds represented a commitment of $35 per share of common, while the White Sewing Machine warrants involved a commitment of only $10 per share of common. But the former meant *ownership* of either a fixed claim or a share of stock, while the latter meant only the *right to buy* a share of stock at a price above the market.

[2] In some cases a callable feature works out to the advantage of the holder, by facilitating new financing which involves the redemption of the old issue at a price above the previous market. But the same result could be obtained, if there were no right to call, by an offer to "buy in" the security. This was done in the case of United States Steel Corporation 5s, due 1951, which were not callable but were bought in at 110.

[3] Dewing cites the case of Union Pacific Railroad—Oregon Short Line Participating 4s, issued in 1903, which were secured by the pledge of Northern Securities Company stock. The bondholders had the right to participate in any dividends in excess of 4% declared on the deposited collateral. The bonds were called at 102½ just at the time when participating distributions seemed likely to occur. See Arthur S. Dewing, *Financial Policy of Corporations,* p. 175n., New York, 1926.

ferred; Kelsey-Hayes Wheel Company $1.50 Participating Convertible Class *A* stock.) Participating *bonds* are generally limited in their right to participate in surplus earnings and are commonly callable. (See White Sewing Machine Corporation Participating Debenture 6s, due 1940; United Steel Works Corporation Participating 6½s, Series *A*, due 1947; neither of these issues is convertible.) Sometimes participating issues are protected against loss of the privilege through redemption by setting the call price at a very high figure. Something of this sort was apparently attempted in the case of San Francisco Toll-Bridge Company Participating 7s, due 1942, which were callable at 120 through November 1, 1933, and at lower prices thereafter.

Another device to prevent vitiating the participating privilege through redemption is to make the issue callable at a price which may be directly dependent upon the value of the participating privilege. For example, Siemens and Halske Participating Debentures, due in 2930, are callable after April 1, 1942, *at the average market price for the issue during the six months preceding notice of redemption, but at not less than the original issue price* (which was over 230% of the par value). The Kreuger and Toll 5% Participating Debentures had similar provisions.

Even in the case of a convertible issue a callable feature is technically a serious drawback because it may operate to reduce the duration of the privilege. Conceivably a convertible bond may be called just when the privilege is about to acquire real value.[1]

In the case of issues with stock-purchase warrants, the subscription privilege almost invariably runs its full time even though the senior issue itself may be called prior to maturity. If the warrant is detachable, it simply continues its separate existence until its own expiration date. Frequently, the subscription privilege is made "nondetachable," *i.e.*, it can be exercised only by presentation of the senior security. But even in these instances, if the issue should be redeemed prior to the expiration

---

[1] This danger was avoided in the case of Atchison, Topeka and Santa Fe Railway Convertible 4½s, due 1948, by permitting the issue to be called only after the conversion privilege expires in 1938. Another protective device recently employed is to give the holder of a convertible issue a stock-purchase warrant, at the time the issue is redeemed, entitling the holder to buy the number of shares of common stock which would have been received upon conversion if the senior issue had not been redeemed. See Freeport Texas Company 6% Cumulative Convertible Preferred, issued in January 1933.

of the purchase-option period it is customary to give the holder a separate warrant running for the balance of the time originally provided.

Example: Prior to January 1, 1934, United Aircraft and Transport Corporation had outstanding 150,000 shares of 6% Cumulative Preferred stock. These shares carried nondetachable warrants for one share of common stock at $30 a share for each two shares of preferred stock held. The subscription privilege was to run to November 1, 1938, and was protected by a provision for the issuance of a detached warrant evidencing the same privilege per share in case the preferred stock was redeemed prior to November 1, 1938. Some of the preferred stock was called for redemption on January 1, 1933, and detached warrants were accordingly issued to the holders thereof. (A year later the remainder of the issue was called and additional warrants issued.)

Third Advantage of Warrant-bearing Issues.—Subscription-warrant issues have still a third advantage over other privileged securities, and this is in a practical sense probably the most important of all. Let us consider what courses of conduct are open to holders of each type in the favorable event that the company prospers, that a high dividend is paid on the common, and that the common sells at a high price.

1. Holder of a participating issue:
   a. May sell at a profit.
   b. May hold and receive participating income.

2. Holder of a convertible issue:
   a. May sell at a profit.
   b. May hold, but will receive no benefit from high common dividend.
   c. May convert to secure larger income, but sacrificing his senior position.

3. Holder of an issue with stock-purchase warrants:
   a. May sell at a profit.
   b. May hold, but will receive no benefit from high common dividend.
   c. May subscribe to common to receive high dividend. He may invest new capital or he may sell or apply his security ex-warrants to provide funds to pay for the common. In either case he undertakes the risks of a common stockholder in order to receive the high dividend income.
   d. He may dispose of his warrant at a cash profit and retain his original security, ex-warrants. (The warrant may be sold directly, or he may subscribe to the stock and immediately sell it at the current indicated profit.)

The fourth option above listed is peculiar to a subscription-warrant issue and has no counterpart in convertible or participating securities. It permits the holder to cash his profit from the speculative component of the issue, and still maintain his original *investment* position. Since the typical buyer of a privileged senior issue should be interested primarily in making a sound investment—with a secondary opportunity to profit from the privilege—this fourth optional course of conduct may prove a great convenience. He is not under the necessity of selling the entire commitment, as he would be if he owned a convertible, which would then require him to find some new medium for the funds involved. The reluctance to sell one good thing and buy another, which characterizes the typical investor, is one of the reasons that holders of high-priced convertibles are prone to convert them rather than to dispose of them. In the case of participating issues also, the owner can protect his *principal* profit only by selling out and thus creating a reinvestment problem.

Example: The theoretical and practical advantage of subscription-warrant issues in this respect may be illustrated in the case of Commercial Investment Trust Corporation 6½% Preferred. This was issued in 1925, and carried warrants to buy common stock at an initial price of $80 per share. In 1929 the warrants sold as high as $69.50 per share of preferred. The holder of this issue was therefore enabled to sell out its speculative component at a high price and to retain his original preferred-stock commitment, which maintained an investment status throughout the depression until it was finally called for redemption at 110 on April 1, 1933. At the time of the redemption call the common stock was selling at the equivalent of about $50 per old share. If the preferred stock had been convertible, instead of carrying warrants, many of the holders would undoubtedly have been led to convert and to retain the common shares. Instead of netting a large profit they would have been faced with a substantial loss.

Summary.—To summarize this section, it may be said that, for *long-pull holding*, a sound participating issue represents the best form of profit-sharing privilege. From the standpoint of maximum *price advance* under favorable market conditions, a senior issue with detachable stock-purchase warrants is likely to show the best results. Furthermore, subscription-warrant issues as a class have definite advantages in that the privilege is ordinarily not subject to curtailment through early redemption of the security; and they permit the realization of a speculative profit while retaining the original investment position.

CHAPTER XXIV

TECHNICAL ASPECTS OF CONVERTIBLE ISSUES

The third division of the subject of privileged issues relates to technical aspects of each type, separately considered. We shall first discuss convertible issues.

The effective terms of a conversion privilege are frequently subject to change during the life of the issue. These changes are of two kinds: (1) a decrease in the conversion price, to protect the holder against "dilution"; and (2) an increase in the conversion price (in accordance usually with a "sliding-scale" arrangement) for the benefit of the company.

Dilution, and Antidilution Clauses.—The value of a common stock is said to be diluted if there is an increase in the number of shares without a corresponding increase in assets and earning power. Dilution may arise through split-ups, stock dividends, offers of subscription rights at a low price, and issuance of stock for property or services at a low valuation per share. The standard "antidilution" provisions of a convertible issue endeavor to reduce the conversion price proportionately to any decrease in the per-share value arising through any act of dilution.

The method may be expressed in a formula, as follows: Let C be the conversion price, O be the number of shares now outstanding, N be the number of new shares to be issued, and P be the price at which they are to be issued.

Then

$$C' \text{ (the new conversion price)} = \frac{CO + NP}{O + N}$$

The application of this formula to Chesapeake Corporation Convertible Collateral 5s, due 1947, is given in the Appendix, Note 34. A simpler example of an antidilution adjustment is afforded by the Central States Electric Corporation 6% Convertible Preferred previously referred to (page 258). After its issuance in 1928, the common stock received successive stock dividends of 100 and 200%. The conversion price was accordingly first cut in half (from $118 to $59 per share) and then again reduced by two-thirds (to $19.66 per share).

A much less frequent provision merely reduces the conversion price to any lower figure at which new shares may be issued. This is, of course, more favorable to the holder of the convertible issue.[1]

Protection against Dilution Not Complete.—While practically all convertibles now have antidilution provisions, exceptions arise now and then.[2]   As a matter of course, a prospective buyer should make certain that such protection exists for the issue he is considering.

It should be borne in mind that the effect of these provisions is to preserve only the principal or par value of the privileged issue against dilution. If a convertible is selling considerably above par, the *premium* will still be subject to impairment through additional stock issues or a special dividend. A simple illustration will make this clear.

A bond is convertible into stock, par for par. The usual antidilution clauses are present. Both bond and stock are selling at 200.

Stockholders are given the right to buy new stock, share for share, at par ($100). These rights will be worth $50 per share, and the new stock (or the old stock "ex-rights") will be worth 150. No change will be made in the conversion basis, because the new stock is not issued below the old conversion price. However, the effect of offering these rights must be to compel immediate conversion of the bonds, since otherwise they would lose 25% of their value. As the stock will be worth only 150 "ex-rights," instead of 200, the value of the unconverted bonds would drop proportionately.

The above discussion indicates that when a large premium or market profit is created for a privileged issue, the situation is vulnerable to sudden change. While prompt action will always prevent loss through such changes, their effect is always to terminate the effective life of the privilege.[3]   The same result

---

[1] See Appendix, Note 35, for example (Consolidated Textile Corporation 7s, due 1923).

[2] See Appendix, Note 36, for example (American Telephone and Telegraph Company Convertible 4½s, due 1933).

[3] To guard against this form of dilution, holders of convertible issues are sometimes given the right to subscribe to any new offerings of common stock on the same basis as if they owned the amount of common shares into which their holdings are convertible. See the indentures securing New York, New Haven and Hartford Railroad Company Convertible Debenture

will follow, of course, from the calling of a privileged issue for redemption at a price below its then conversion value.

Where the number of shares is *reduced* through recapitalization, it is customary to *increase* the conversion price proportionately. Such recapitalization measures include increases in par value; "reverse split-ups" (*i.e.*, issuance of one no-par share in place of, say, five old shares); and exchanges of the old stock for fewer new shares through consolidation with another company.[1]

Sliding Scales Designed to Accelerate Conversion.—The provisions just discussed are intended to maintain equitably the original basis of conversion in the event of subsequent capitalization changes. On the other hand, a "sliding-scale" arrangement is intended definitely to reduce the value of the privilege as time goes on. The underlying purpose is to accelerate conversion, in other words, to curtail the effective duration and hence the real value of the option. Obviously, any diminution of the worth of the privilege to its recipients must correspondingly benefit the donors of the privilege, who are the company's common shareholders.

The more usual terms of a sliding scale prescribe a series of increases in the conversion price in successive periods of time. A more recent variation makes the conversion price increase as soon as a certain portion of the issue has been exchanged.

Examples: American Telephone and Telegraph Company Ten-year Debenture 4½s, due 1939, issued in 1929, were made convertible into common at $180 per share during 1930, at $190 per share during 1931 and 1932, and at $200 per share during 1933 to 1937, inclusive. These prices were later reduced through the issuance of additional stock at $100, in accordance with the standard antidilution provision.

Anaconda Copper Mining Company Debenture 7s, due 1938, were issued in the amount of $50,000,000. The first $10,000,000 presented were convertible into common stock at $53 per share; the second $10,000,000 were convertible at $56; the third at $59; the fourth at $62, and the final lot at $65.

Sliding Scale Based on Time Intervals.—The former type of sliding scale, based on time intervals, is a readily understandable

6s, due 1948, and Commercial Investment Trust Corporation Convertible Debenture 5½s, due 1949.

[1] See Appendix, Note 37, for example of Dodge Brothers, Inc., Convertible Debenture 6s.

method of reducing the liberality of a conversion privilege.   Its effect can be shown in the case of Porto Rican-American Tobacco Company 6s, due 1942.   These were convertible into pledged Congress Cigar Company, Inc., stock at $80 per share prior to January 2, 1929; at $85 during the next three years; and at $90 thereafter.   During 1928 the highest price reached by Congress Cigar was 87¼, which was only a moderate premium above the conversion price.   Nevertheless a number of holders were induced to convert before the year-end, because of the impending rise in the conversion basis.   These conversions proved very ill-advised, since the price of the common fell to 43 in 1929, against a low of 89 for the bonds.   In this instance, the adverse change in the conversion basis not only meant a smaller potential profit for those who delayed conversion until after 1928, but also involved a risk of serious loss through inducing conversion at the wrong time.

Sliding Scale Based on Extent Privilege Is Exercised.—The second method, however, based on the quantities converted, is not so simple in its implications.   Since it gives the first lot of bonds converted an advantage over the next, it evidently provides a *competitive* stimulus to early conversion.   By so doing it creates a conflict in the minds of the holder between the desire to retain his senior position and the fear of losing the more favorable basis of conversion through prior action by other bondholders.   This fear of being forestalled should ordinarily result in large-scale conversions as soon as the stock advances moderately above the initial conversion price, *i.e.*, as soon as the bond is worth slightly more than the original cost.

Example: The sequence of events normally to be expected is shown fairly well by the market action of Engineers Public Service Company $5 Convertible Preferred in 1928 and 1929.

This issue (320,000 shares) was convertible into common on the following basis:

The first ⅛ of the issue at $47.62 per share of common (2.1 shares of common for 1 preferred).

The next ⅛ of the issue at $52.63 per share of common (1.9 shares of common for 1 preferred).

The next ⅛ of the issue at $58.82 per share of common (1.7 shares of common for 1 preferred).

The next ⅛ of the issue at $62.50 per share of common (1.6 shares of common for 1 preferred).

The second ½ of the issue at $66.67 per share of common (1.5 shares of common for 1 preferred).

The market prices of the $5 preferred and the common at various dates are shown as follows:

ENGINEERS PUBLIC SERVICE COMPANY $5 CONVERTIBLE PREFERRED

Date	Price range of common on that day	Prevailing conversion price	Price range of preferred on that day
8/16/28	37–37	47.62	97– 98
8/28/28	41–42	47.62	98– 98
9/19/28	42–44	47.62	98– 98
10/ 1/28	47–50	47.62	101–103
10/15/28	46–47	47.62	97– 99
10/16/28	45–47	52.63	95– 96
11/ 2/28	45–46	52.63	91– 92
11/28/28	47–48	52.63	92– 92
12/28/28	48–48	52.63	92– 92
1/ 5/29	50–51	52.63	94– 94
1/ 8/29	50–52	58.82	91– 92
1/31/29	59–60	58.82	99–105
2/ 7/29	54–56	62.50	94– 94
3/ 6/29	52–54	62.50	94– 96
4/ 8/29	49–49	62.50	91– 91
5/ 8/29	51–52	62.50	92– 92
6/14/29	55–60	62.50	95– 96
6/29/29	61–62	62.50	99– 99
7/12/29	61–63	62.50	98– 99
7/23/29	68–71	62.50	107–112
8/ 5/29	76–80	62.50	120–123
8/21/29	72–73	66.67	112–112
9/ 4/29	72–73	66.67	112–112
10/ 3/29	68–72	66.67	108–109
10/30/29	32–45	66.67	90– 90
11/13/29	32–36	66.67	80– 80
11/21/29	39–42	66.67	90– 90
12/23/29	37–38	66.67	92– 92

According to this analysis the price of the senior issue should oscillate over a relatively narrow range while the common stock is advancing and while successive blocks of bonds or preferred are being converted. When the last block is reached, the *competitive* element disappears and the bond or preferred stock is now in the position of an ordinary convertible, free to advance indefinitely with the stock.

It should be pointed out that issues with such a sliding-scale provision do not always follow this theoretical behavior pattern.

The Anaconda Copper Company Convertible 7s, for example, actually sold at a high premium (30%) in 1928, before the first block was exhausted. This seems to have been one of the anomalous incidents of the highly speculative atmosphere at the time. From the standpoint of critical analysis, a convertible of this type must be considered as having very limited possibilities of enhancement until the common stock approaches the last and highest conversion price.[1]

The sliding-scale privilege on a "block" basis belongs to the objectionable category of devices which tend to mislead the holder of securities as to the real nature and value of what he owns. The competitive pressure to take advantage of a limited opportunity introduces an element of compulsion into the exercise of the conversion right which is directly opposed to that freedom of choice for a reasonable time which is the essential merit of such a privilege. There seems no reason why investment bankers should inject so confusing and contradictory a feature into a security issue. Sound practice would dictate its complete abandonment, or in any event the studious avoidance of such issues by intelligent investors.

Issues Convertible into Preferred Stock.—Many bond issues were formerly made convertible into preferred stock. Ordinarily some increase in income was offered to make the provision appear attractive. (For examples, see Missouri-Kansas-Texas Railroad Company Adjustment 5s, due 1967, convertible prior to January 1, 1932 into $7 preferred stock; Central States Electric Corporation Debenture 5s, due 1948, convertible into $6 preferred stock; G. R. Kinney Company Secured 7½s, due 1936, convertible into $8 preferred stock; American Electric Power Corporation 6s, due 1957, convertible into $7 preferred stock.)

There have been instances in which a fair-sized profit has been realized through such a conversion right; but the upper limitation on the market value of the ordinary preferred stock is likely to keep down the maximum benefits from such a privilege to a modest figure. Moreover, since developments in recent years have made preferred stocks in general appear far less desirable

[1] In some cases (*e.g.*, Porto Rican-American 6s, already mentioned, and International Paper and Power Company First Preferred) the conversion privilege ceases entirely after a certain fraction of the issue has been converted. This maintains the competitive factor throughout the life of the privilege and in theory should prevent it from ever having any substantial value.

than formerly, the right to convert, say, from a 6% bond into a 7% preferred is likely to constitute more of a danger to the unwary than an inducement to the alert investor. If the latter is looking for convertibles, he should canvass the market thoroughly and endeavor to find a suitably secured issue convertible into common stock. In a few cases where bonds are convertible into preferred stock the latter is in turn convertible into common, or participates therewith, and this double arrangement may be equivalent to convertibility of the bond into common stock. For example, International Hydro-Electric System 6s, due 1944, are convertible into Class *A* stock, which is in reality a participating second preferred.

There are also bond issues convertible into either preferred or common, or into a combination of certain amounts of each.[1] While any individual issue of this sort may turn out well, in general it may be said that complicated provisions of this sort should be avoided (both by issuing companies and by security buyers) because they tend to create confusion.

Bonds Convertible at the Option of the Company.—The unending flood of variations in the terms of conversion and other privileges which developed during the 1920s made it difficult for the untrained investor to distinguish between the attractive, the merely harmless, and the positively harmful. Hence he proved an easy victim to unsound financing practices which in former times might have stood out as questionable because of their departure from the standard. As an example of this sort we cite the various Associated Gas and Electric Company "Convertible Obligations" which were made convertible by their terms into preferred or Class *A* stock *at the option of the company.* Such a contraption was nothing more than a preferred stock masquerading as a bond. If the purchasers were entirely aware of this fact and were willing to invest in the preferred stock, they would presumably have no cause to complain. But it goes without saying that an artifice of this kind lends itself far too readily to concealment and possible misrepresentation.[2]

[1] See, for example, the Chicago, Milwaukee, St. Paul and Pacific Railroad Company Convertible Adjustment Mortgage 5s, Series *A*, due January 1, 2000, which are convertible into five shares of the preferred and five shares of common. For other examples see p. 623 in the Appendix.

[2] These anomalous securities were variously entitled "investment certificates," "convertible debenture certificates," "interest-bearing allotment certificates," and "convertible obligations." In 1932 the company com-

Bonds Convertible into Other Bonds.—Some bonds are convertible into other bonds. The usual case is that of a short-term issue, the holder of which is given the right to exchange into a long-term bond of the same company. Frequently the long-term bond is deposited as collateral security for the note. (For example, Interborough Rapid Transit Company 7s, due 1932, were secured by deposit of $1,736 of the same company's First and Refunding 5s, due 1966, for each $1,000 note, and they were also convertible into the deposited collateral, the final rate being $1,000 of 5s for $900 of 7% notes.) The holder thus has an option either to demand repayment at an early date or to make a long-term commitment in the enterprise. In practice, this amounts merely to the chance of a moderate profit at or before maturity, in the event that the company prospers, or interest rates fall, or both.

Unlike the case of a bond convertible into a preferred stock, there is usually a *reduction* in the coupon rate when a short-term note is converted into a long-term bond. The reason is that short-term notes are ordinarily issued when interest rates, either in general or for the specific company, are regarded as abnormally high, so that the company is unwilling to incur so steep a rate for a long-term bond. It is thus expected that when normal conditions return, long-term bonds can be floated at a much lower rate; and hence the right to exchange the note for a long-term bond, even on a basis involving some reduction in income, may prove to be valuable.[1]

---

pelled the conversion of the large majority of them, but the holder was given an option (in addition to those already granted by the terms of the issues) of converting into equally anomalous "Convertible Obligations, Series *A* and *B*, due 2022," which are likewise convertible into stock at the option of the company. The company was deterred from compelling the conversion of some $17,000,000 "5½% Investment Certificates" after November 15, 1933, by a provision in the indenture for that issue prohibiting the exercise of the company's option in case dividends on the $5.50 Dividend Series Preferred were in arrears (no dividends having been paid thereon since June 15, 1932).

It is interesting to note that the Pennsylvania Securities Commission prohibited the sale of these "Convertible Obligations" in December 1932 because of their objectionable provisions. The company resisted the Commission's order in the Federal District Court of Philadelphia, but later dropped its suit (see 135 *Chronicle* 4383, 4559; 136 *Chronicle* 326, 1011).

[1] See the following issues taken from the 1920–1921 period: Shawinigan Water and Power Company 7½% Gold Notes, issued in 1920 and due in

Convertible Bonds with an Original Market Value in Excess of Par.—One of the extraordinary developments of the 1928–1929 financial pyrotechnics was the offering of convertible issues with an original market value greatly in excess of par. This is illustrated by Atchison, Topeka and Santa Fe Railway Company Convertible 4½s, due 1948, and by American Telephone and Telegraph Company Convertible 4½s, due 1939. Initial trading in the former on the New York Curb Market (on a "when issued" basis) in November 1928 was around 125, and initial trading in the latter on the New York Stock Exchange (on a "when issued" basis) on May 1, 1929, was at 142. Obviously such investments represented primarily a commitment in the common stock, since they were immediately subject to the danger of a substantial loss of principal value if the stock declined. Furthermore the income return was entirely too low to come under our definition of investment. It would undoubtedly have been more straightforward financing to have offered additional common stock directly instead of such an equivocal "bond." Apparently the convertible bond form was employed to permit certain institutions, prohibited by law from buying stocks, to make a "bond investment" equivalent to a common-stock purchase. It is to be hoped that so questionable a financial practice will not again be resorted to in future market booms.

A Technical Feature of Some Convertible Issues.—A technical feature of the American Telephone and Telegraph convertible issue deserves mention. The bonds were made convertible at 180; but instead of presenting $180 of bonds to obtain a share of stock, the holder might present $100 of bonds and $80 in cash. The effect of such an option is to make the bond more valuable whenever the stock sells above 180 (*i.e.*, whenever the conversion value of the bond exceeds 100). This is illustrated as follows:

1926, convertible into First and Refunding 6s, Series *B*, due 1950, which were pledged as security; San Joaquin Light and Power Corporation Convertible Collateral Trust 8s, issued in 1920 and due in 1935, convertible into the pledged Series *C* First and Refunding 6s, due 1950; Great Western Power Company of California Convertible Gold 8s, issued in 1920 and due in 1930, convertible into pledged First and Refunding 7s, Series *B*, due in 1950.

Another type of bond-for-bond conversion is represented by Dawson Railway and Coal 5s, due 1951, which are convertible into El Paso and Southwestern Railroad Company First 5s, due 1965 (the parent company and a subsidiary of the Southern Pacific). Such examples are rare and do not invite generalization.

If the stock sells at 360, a straight conversion basis of 180 would make the bond worth 200. But by the provision accepting $80 per share in cash, the value of the bond becomes 360 − 80 = 280.

This arrangement may be characterized as a combination of a conversion privilege at 180 with a stock purchase right at 100.

Delayed Conversion Privilege.—The privilege of converting is sometimes not operative immediately upon issuance of the obligation.

Examples: This was true, for example, of Brooklyn Union Gas Company Convertible 5½s, discussed in Note 36 of the Appendix. Although they were issued in December 1925 the right to convert did not accrue until January 1, 1929. Similarly, New York, New Haven and Hartford Railroad Company Convertible Debenture 6s, due in 1948, although issued in 1907 were not convertible until January 15, 1923; Chesapeake Corporation Convertible 5s, due 1947, were issued in 1927 but did not become convertible until May 15, 1932.

More commonly the suspension of the conversion privilege does not last so long as these examples indicate, but in any event this practice introduces an additional factor of uncertainty and tends to render the privilege less valuable than it would be otherwise. This feature may account in part for the spread, indicated in Note 36 of the Appendix, which existed during 1926, 1927, and the early part of 1928 between the Brooklyn Union Gas Company 5½s and the related common stock.

CHAPTER XXV

SENIOR SECURITIES WITH WARRANTS.
PARTICIPATING ISSUES. SWITCHING AND HEDGING

Nearly all the variations found in convertible issues have their counterpart in the terms of subscription warrants. The purchase price of the stock is ordinarily subject to change, up or down, corresponding to the standard provisions for adjusting a conversion price.

Example: White Eagle Oil and Refining Company Debenture 5½s, due 1937, were offered in March 1927 and carried warrants entitling the holder to subscribe on or before March 15, 1932, to ten shares of the capital stock of the company at the following prices:

$32 per share to and including March 15, 1928, and thereafter at
$34 per share to and including March 15, 1929, and thereafter at
$36 per share to and including March 15, 1930, and thereafter at
$38 per share to and including March 15, 1931, and thereafter at
$40 per share to and including March 15, 1932.

On January 27, 1930, the Standard Oil Company of New York acquired the White Eagle properties by assuming the liabilities of the latter company and exchanging 8½ shares of Standard Oil of New York for each 10 shares of White Eagle. In accordance with the terms of the indenture protecting the warrants against dilution and providing for readjustment of the subscription price in the case of a sale of the properties or merger of the company, the warrants thereafter entitled the holder to subscribe to 8½ shares of Standard Oil of New York (now Socony-Vacuum Corporation) at $42.35 per share to and including March 15, 1930; at $44.71 for the next year; and at $47.06 for the following year.

Sliding Scales of Both Types.—Sliding-scale arrangements of both types are also encountered in option-warrant issues.

Examples: Interstate Department Stores, Inc., 7% Preferred, issued in 1928, carried nondetachable warrants entitling the holder to purchase common stock, share for share, at the following prices:

$37 per share up to January 31, 1929.
$42 per share up to January 31, 1931.
$47 per share up to January 31, 1933.

Central States Electric Corporation Optional 5½% Debentures, due 1954, carried detachable warrants entitling the holder to buy, on or before September 15, 1934, 10 shares of common stock for each $1,000 bond, at the following prices:

$89 per share for the first 25% of the warrants exercised.
$94 per share for the next 25% of the warrants exercised.
$99 per share for the next 25% of the warrants exercised.
$104 per share for the last 25% of the warrants exercised.

As with convertibles, a sliding scale based on the "block" principle detracts greatly from the value of the privilege until the last block, *i.e.*, the highest price, is reached, at which time it becomes an ordinary purchase option.

Methods of Payment.—Stock-purchase warrants attached to bonds or preferred stocks frequently provide that payment for the common stock may be made either in cash as by turning in the senior security itself at par. Such an arrangement may prove directly equivalent to a conversion privilege. For example, each share of American and Foreign Power Second Preferred was issued with warrants to buy 4 shares of common at $25 per share. Instead of paying cash, the holder can tender preferred stock at a value of $100 per share. If he does so, he is actually *converting* his preferred stock with warrants into common.

Similarly, the Rand Kardex 5½% bonds, described in Chap. XXII, could be tendered at par, in lieu of cash, upon exercising the warrants. Since the warrants attached to a $1,000 bond called for payment of $900 (22½ shares at 40) the owner of a $1,000 bond making payment in this fashion would have a $100 bond remaining. These provisions were thus equivalent to convertibility of 90% of each bond into common.

Advantage of Option to Pay Cash.—The option to pay cash instead of turning in the senior issue must be considered an advantage over a straight conversion privilege—first, because the bond or preferred, "ex-warrants," may be worth more than par, thus increasing the profit; and secondly, because, as previously explained, the holder may be glad to retain his investment while realizing a cash profit on its speculative component; and thirdly, because the warrant is likely to sell separately at a greater premium over its realizable value than a pure convertible. All these advantages are illustrated by the Mohawk Hudson Power Corporation Second Preferred with warrants as shown in the table on page 258. This stock was tenderable at par, in lieu of cash, upon exercise of the warrants, thus having rights equivalent to convertibility; but the warrant arrangement proved far more profitable than an equivalent conversion privilege.

Detachability.—Stock-purchase warrants are either detachable, or nondetachable, or nondetachable for a certain period and detachable thereafter. A detachable warrant may be exercised upon presentation of the warrant alone. Hence it may be sold separately from the issue of which it originally formed a part. A nondetachable warrant or right may be exercised only in conjunction with the senior issue, *i.e.*, the bond or preferred stock must be physically presented at the time of making payment for the common shares. Hence such warrants may not be dealt in separately. For example, the warrants attached to Montecatini 7s, due 1937, and those accompanying the Fiat Debenture 7s, due 1946, were detachable immediately after issuance. Those attached to Loews, Inc., $6.50 Preferred, offered in December 1927 were not detachable until July 1, 1928; and the warrants attached to the Loews, Inc., 6% Debentures, due 1941, were not detachable until October 1, 1926, six months after their issuance. On the other hand, the warrants attached to Crown-Zellerbach Corporation Debenture 6s, due 1940, and to Interstate Department Stores, Inc., 7% Preferred were not detachable during the life of the warrant, unless the senior issue to which they were attached were called for redemption.

In an active stock market, separate option warrants are popular with speculators (as pointed out before), and they sell at considerable premiums above their immediately realizable value. Other things being equal, therefore, an issue with detachable warrants will sell higher than one with a nonseparable right. In view of this fact it may be asked why all subscription warrants are not made immediately detachable, to give the holder the benefit of their superior market appeal. The reason for making a warrant nondetachable is that the company as well as the underwriters of the issue wish to avoid the establishment of an unduly low price for its bonds ex-warrants. Such a low price is likely to follow if large purchases of the bond with warrants are made by out-and-out speculators. For these holders, having no interest in the bond as such, are likely to detach the warrant and sell the bond ex-warrants for whatever it will bring. Selling pressure from this source, coupled with the absence of any steady demand for the issue due to lack of "seasoning," may result in so low a price as to constitute an apparent reflection upon the corporation's credit, which is evidently undesirable.

The compromise arrangement—which makes the warrant detachable only after an interval—is based upon the assumption that after the security has had time to become fairly well known in the investment world, a proper price may more readily be established for the issue ex-warrants, even in the face of sales by those who have profited from the warrants.

When once these subscription warrants were made detachable from the related senior issue, they were bound to assume an existence and characteristics of their own. From a mere appendage of bond financing they developed into an independent form of security and a major vehicle of speculation during the madness of 1928–1929. It is an amazing fact that the option warrants created by one company, American and Foreign Power, reached an indicated market value in 1929 of over a billion dollars, a figure which exceeded the market value of *all* the railroad common stocks of the United States listed on the New York Stock Exchange in July 1932, less than three years later.

It will be necessary, therefore, to consider in a later chapter the characteristics of stock-purchase warrants, viewed as an independent speculative medium. At that time we shall discuss the relationships between the prices of such warrants and of the preferred and common shares of the same corporations.

Participating Issues.—Most of the traits of this type of privilege have already been brought out in the preceding comparison with the other forms. A distinction may be made between two kinds of participation. The more usual arrangement depends upon the dividend paid upon the common; less frequently, the profit sharing is determined by the earnings without reference to the dividend rate.

Examples: Westinghouse Electric and Manufacturing Company Preferred, already described, is a standard example of the first type; Geo. A. Fuller Company First and Second Preferred and Celanese Corporation of America Participating First Preferred are instances of the second. The Fuller issues are each entitled to cumulative dividends of $6 per share and to additional contingent dividends of not exceeding $3 per share annually. The amount of the participating dividends is determined by taking 6% and $3\frac{1}{2}$%, respectively, of the net earnings after deducting regular preferred dividends and sinking fund. Celanese Participating Preferred is entitled to 10% of earnings available after First Preferred dividends, without limitation.

Preferred shares constitute the great bulk of participating issues; participating bonds are rare and likely to deviate widely in other respects from the standard bond pattern. The Kreuger and Toll Participating Debentures, for example, while nominally a bond, were in essence a nonvoting common stock. The Green Bay and Western Railway (Participating) Debentures, Series *A* and Series *B*, are in reality preferred and common stocks respectively. Spanish River Pulp and Paper Mills, Ltd., First 6s, due 1931 and redeemed in 1928, are one of the few examples of an investment-type bond with a participating privilege.[1] Siemens and Halske A. G. (a German enterprise) issued a series of Participating Debentures, due 2930, carrying interest equal to the rate of dividend paid upon the common stock, but not less than 6%.

Participating preferred stocks originally had a standard pattern, exemplified by Westinghouse Electric and Manufacturing Company Preferred. The order of payment is first a fixed preference to the senior shares, then a similar amount on the common shares, and finally an equal participation, share for share, in additional dividends. This pattern arose from the common-law right of all classes of stock to share equally in earnings and assets, except as otherwise provided by agreement. Other examples of this arrangement are Chicago, Milwaukee, St. Paul and Pacific Railroad Company Preferred, Wabash Railway Company 5% Preferred *A* and Consolidated Film Industries, Inc., Preferred.

In recent years, however, a wide diversity of participating arrangements have made their appearance, so that there is now no standard pattern. A number of variations are described in our classification of security forms in Note 3 of the Appendix.

---

[1] See Appendix, Note 38, for details concerning this issue.

Participating issues require two kinds of calculation, one showing the number of times the fixed interest or dividend is earned, and the other showing the amount available for distribution under the participation privilege.

Example:

CELANESE CORPORATION OF AMERICA, 1933

Net for dividends..........................	$5,454,000
Prior preferred dividend ($7)..............	800,000
First participating preferred dividend at $7 rate...................................	1,011,000
First participating preferred: additional participation............................	364,000
Balance for common......................	3,279,000
Prior preferred dividends earned...........	6.8 times
Prior preferred and participating preferred ($7) dividends earned..................	3.0 times
Earned for participating preferred: participating basis..........................	$9.52 per share

Privileged Issues Compared with the Related Common Stocks. In our previous discussion of the merits of privileged issues as a class it was pointed out that they sometimes offer a very attractive combination of security and chance for profit. More frequently, a decision may be reached that the privileged senior security is preferable to the common stock of the enterprise. Since a conclusion of this kind is based on *comparative* elements only, it is likely to involve smaller risks of error than one which asserts the *absolute* attractiveness of an issue. An example will make this clear.

Example: Keith-Albee-Orpheum Corporation $7 Cumulative Preferred Stock was convertible into three shares of common. In 1928 the preferred received its full dividend but none was paid on the common. A switch from the common to the preferred could frequently have been effected at or near to parity during that year, as is indicated by the following actual price-relationships on the dates indicated:

Date, 1928	Keith-Albee-Orpheum Pfd.	Keith-Albee-Orpheum Com.
Aug. 8............	84⅝	27
Sept. 13..........	99¾	33
Oct. 9............	95	31¾
Nov. 3............	102	34

Such a switch was advisable on the ground that the holder of the common had everything to gain and little or nothing to lose thereby. The preferred was yielding about 7% or better on the above prices, and the dividend had been covered with a fair margin. The common, on the other hand, was yielding nothing. In the case of a rise in the price of the common the preferred could be counted to go along with it percentagewise, which it did later. If, on the other hand, the common were to decline (which it did subsequently) the preferred could be counted upon to decline less, percentagewise.

Meanwhile, Radio-Keith-Orpheum Corporation was organized in October 1928 as a holding company for Keith-Albee-Orpheum and other similar enterprises. R-K-O offered its Class *A* common stock, share for share, for Keith-Albee-Orpheum common and the Keith-Albee-Orpheum $7 Preferred became convertible into three shares of either Keith-Albee-Orpheum common or R-K-O *A*. For some months thereafter it was frequently possible for the holders of R-K-O *A* to switch into Keith-Albee-Orpheum Preferred at prices very close to parity.

"Parity," "Premium," and "Discount."—When the price of a convertible bond or preferred is exactly equivalent, on an exchange basis, to the current price of the common stock, the two issues are said to be selling at a *parity*.[1] When the price of the senior issue is above parity it is said to be selling at a *premium*, and the difference between its price and conversion parity is called the amount of the premium, or the "spread." Conversely if the price of the convertible is below parity, the difference is sometimes called the *discount*.[2]

---

[1] This should not be confused with *par*, which means simply the face value of the security in question. "Par," when applied to the *price* of a stock, always means $100 per share and has no reference to the real par value of the share which may be quite different.

[2] If the senior issue may be promptly exchanged for the common, a discount results in creating an *arbitrage* opportunity. This is a chance to make a profit (usually small) without risk of loss by: (1) simultaneously buying the senior issue and selling the common stock; (2) immediately converting the senior issue into the common stock; and (3) delivering the common stock against the sale, thus completing the transaction. Arbitraging of this "open-and-shut" kind is done rather extensively in active, rising markets; but the opportunities are usually monopolized by brokers specializing in such operations. Other forms of intersecurity arbitrage operations arise from reorganizations, mergers, stock split-ups, rights to buy new stocks, *etc.* For detailed discussion see Meyer H. Weinstein, *Arbitrage in Securities*,

A Fruitful Field for Dependable Analysis.—The Radio-Keith-Orpheum illustration is one of the infrequent examples of an absolutely dependable conclusion arrived at by security analysis. Holders of R-K-O *A* could not possibly lose by exchanging into Keith-Albee convertible preferred and they had excellent prospects, which in fact were realized, of deriving substantial benefits in the form of both increased income and greater market value. In this respect, privileged issues offer a fruitful field for the more scientific application of the technique of analysis. The above example is typical also of the price relationships created by an active and advancing market. When there is a senior issue convertible into common, the concentration of speculative interest in the latter often results in establishing a price level closely equivalent to (and sometimes even higher than) the price of the senior issue, to which the public pays little attention.

A Second Example.—As a second illustration of this interesting phenomenon, we cite the price history of Consolidated Textile Corporation Convertible 7s and related stock during the market rise and fall of 1919–1921. The company was organized in September 1919 as a consolidation of a half-dozen well-established textile manufacturing concerns. Its stock was listed on the New York Stock Exchange and rose from a low price of 31¾ in November 1919 to a high of 46¼ in April 1920. In the latter month $3,000,000 of 7% convertible notes, due in 1923, were offered to the public at 98½. These notes have been described in Note 35 of the Appendix. The unusual protective provision against dilution of the conversion privilege gave them a special attractiveness. Consolidated earnings of the predecessor companies during the preceding three years had averaged nine times the charges on the notes and showed a favorable upward trend. By their terms the notes were convertible (beginning May 1, 1920) into 22 shares of common stock, on which dividends were currently being paid at the rate of $3 a share per annum.

During the month of April 1920 the holder of stock in the company could have sold out and shifted into notes representing an equivalent amount of stock without paying more than a nominal sum for the privilege. For example, on April 23rd the

Harper & Bros., 1931. In the older sense, the term "arbitrage" applied to simultaneous purchases and sales of the same security in different markets (*e.g.*, New York and London), and to similar operations involving foreign exchange.

notes could have been bought at 98½ (equivalent to 44¾ for the stock) and the stock could have been sold at 44. The difference of $17 on 22 shares of stock was a small price to pay for the superior advantages of the notes as an alternative commitment. The notes yielded an income of $70 per year, whereas the dividends on the stock at the current rate were $66 on 22 shares. Moreover, if the stock were to continue its rise in market price the notes could be counted to go along correspondingly, whereas if the stock were to decline in price the notes could undoubtedly be expected to hold up better in price than the stock.

The sequel furnishes a clear demonstration of the wisdom of such a shift. The company's earnings suffered a setback during the 1920–1921 depression. The common stock declined rapidly and shareholders were offered new stock in November 1920 at only $21 a share. This resulted in a lowering of the conversion price from $45.45 to $21, a factor which again gave the notes an attractive profit possibility and caused them to rise somewhat in price. (They had continuously sold fairly close to par.) Early in 1921 the dividend on the stock was suspended and it declined to a low price of 12¾. Meanwhile, however, the bonds continued to sell at 95 or above; and during the second half of the year, when the stock was reaching its lowest levels, the bonds enjoyed an appreciation in price prior to their being called at 102½ on October 1, 1921.

Conclusion from Foregoing.—It is clear that a convertible issue selling on a parity with the common is preferable thereto, except when its price is so far above an investment level that it has become merely a form of commitment in the common stock. (Brooklyn Union Gas Company Convertible 5½s, due 1936, are an example of the latter type of situation. The bonds, convertible into 20 shares of common from January 1, 1929, sold at 147 or higher during the years 1927–1932, inclusive, and sold at 489 in 1929.) It is generally worthwhile to pay some moderate premium in order to obtain the superior safety of the senior issue. This is certainly true when the convertible yields a higher income return than the common, and it holds good to some extent even if the income yield is lower. The Keith-Albee-Orpheum preferred paid a 7% dividend while the Radio-Keith-Orpheum *A* received nothing, so that here the preferred was undoubtedly more attractive than the common at a price, say, five points above parity (*e.g.* 110 for the preferred against 35 for the common).

Switching.—As a practical rule therefore, holders of common stocks who wish to retain their interest in the company should always exchange into a convertible senior issue of the enterprise, whenever it sells both at an investment level on its own account and also close to parity on a conversion basis. Just how large a premium a common stockholder should be willing to pay in making such an exchange, is a matter of individual judgment. Because of his confidence in the future of his company, he is usually unwilling to pay anything substantial for insurance against a decline in value. But experience shows that he would be wise to give up somewhat more than he thinks is necessary in order to secure the strategic advantages that even a fairly sound convertible issue possesses over a common stock.[1]

Hedging.—These advantages of a strong convertible issue over a common stock become manifest when the market declines. The price of the senior issue will ordinarily suffer less severely than the common, so that a good-sized spread may thereby be established, instead of the near-parity previously existing. In the Consolidated Textile example the bond actually advanced (first because of the favorable change in the conversion basis and then because of redemption), while the common declined more than 60%. This possibility suggests a special form of market operation, known as "hedging," in which the operator buys the convertible and sells the common stock short against it, at an approximate parity.[2] In the event of a protracted *rise*, he can convert the senior issue and thus close out the transaction at only a slight loss, consisting of the original spread plus carrying

[1] The same reasoning holds true when both issues are confessedly speculative.

Example: Western Maryland Railroad Preferred is convertible into common share for share. It sold no higher than the common during the greater part of 1928–1933. Yet if any one was willing to own the common he should have switched into the preferred, which had all the possibilities of the common *plus* its senior position. Early in 1934 the preferred sold at a fair premium above the common—23 against 17.

[2] "Hedging" in commodities is a superficially similar but basically different type of operation. Generally speaking, its purpose is to protect a normal manufacturing or distributing profit against the chance of speculative loss through commodity price changes. A miller, having bought wheat which he will sell as flour some months later, will sell wheat futures as a "hedge" against the possibility of a decline in wheat destroying his profit margin. When the flour is disposed of, he covers (buys back) the wheat sold as protection. Most commodity hedging is thus designed as a safeguard, while security hedging is usually intended to yield direct profits.

expenses. But if the market declines substantially, he can "undo" the operation at a considerable profit, by selling out the senior issue and buying back the common.

A practical illustration of a hedging operation is afforded by Keith-Albee-Orpheum Convertible Preferred and Radio-Keith-Orpheum *A* as follows, the hedge being established on March, 1, 1929, and the positions reversed or "undone" on March 26, 1929:

1. Sold (short) 300 R-K-O *A* @ 39⅞ on March
   1, 1929.............................. $11,962.50
   Less commission ($45) and tax ($12)..... 57.00

   Proceeds of short sale.............................. $11,905.50
   Bought 300 R-K-O *A* @ 29 on March 26,
   1929.............................. $ 8,700.00
   Plus commission on this purchase........ 45.00

   Cost of cover................................... $ 8,745.00

   Profit on short sale............................. $ 3,160.50

2. Bought on March 1, 1929, 100 Keith-Albee-
   Orpheum Pfd. @ 120................. $12,000.00
   Plus commission ($25).................. 25.00

   Cost of long stock................................. $12,025.00
   Sold 100 Keith-Albee Orpheum Pfd. on
   March 26, 1929 @ 98................. $ 9,800.00
   Less commission ($20) and tax ($4)....... 24.00

   Proceeds of long stock.................. $ 9,776.00
   Plus dividend received on long stock...... 175.00
   (Preferred sold ex-div. on March 19,
   1929)

   Net proceeds from sale of long stock and
   dividends thereon.............................. $ 9,951.00

   Loss on long stock................................. $ 2,074.00

3. Profit on short sale....................... $ 3,160.50
   Loss on long stock...................... 2,074.00

   Net profit on hedge................... $ 1,086.50

The profit indicated was about 9% on the capital tied up in the transaction, and since it covered a period of 26 days the profit was at the rate of over 100% per year. Since there was no chance of loss on the transaction, a considerable part of the cost of the preferred stock could properly have been borrowed, thus largely increasing the percentage of profit on the capital supplied by the operator. With favorable surrounding conditions, operations of this kind offer a chance for large gains against a small maximum loss. They are particularly suitable as a form of protection against other financial commitments, for they yield their profit in a declining market when other holdings are likely to show losses.

Some Technical Aspects of Hedging.—Hedging has numerous technical aspects, however, which make it less simple and "foolproof" than our brief description would indicate. An exhaustive discussion of hedging would fall outside the scope of this volume, and for this reason we shall merely list below certain elements which the experienced hedger will take into account in embarking upon such operations:

1. Ability to borrow stock sold and to maintain short position indefinitely.[1]

2. Original cost of establishing position, including spread and commissions.

3. Cost of maintaining the position, including interest charges on long holdings, dividends on short stock, possible premiums payable for borrowing stock, and stamp taxes in connection with reborrowings of stock—less offsets in the forms of dividends or interest receivable on long securities and possible interest credit on short position.

4. Amount of profit at which operation will probably be closed out if opportunity offers. Relationship between this maximum profit and probable maximum loss, consisting of (2) plus (3).

It should be borne in mind in these, as in all other operations in securities, that the potential profit to be taken into account is not the maximum figure which might conceivably be reached in the market, but merely the highest figure which the operator is likely to wait for before he closes out his position. Once a given profit is taken, the additional profit which might have been realized subsequently becomes of merely academic interest.

An Intermediate Form of Hedging.—An intermediate form of hedging consists of purchasing a convertible issue and selling only part of the related common shares, say, one-half of the

[1] At this writing the proposed Stock Exchange Control Act may result in preventing short selling, and thus eliminate hedging operations of the kind under discussion.

amount receivable upon conversion. On this basis a profit may be realized in the event of either a substantial advance or a substantial decline in the common stock. This is probably the most scientific method of hedging, since it requires no opinion as to the future course of prices. An ideal situation of this kind would neet the following two requirements:

1. A strongly entrenched senior issue which can be relied on to maintain a price close to par even if the common should drop precipitately. A good convertible bond, maturing in a short time, is an ideal *type* for this purpose.

2. A common stock in which the speculative interest is large and which is therefore subject to wide fluctuations in either direction.

An example of this form of hedge is supplied by operations carried on in 1918–1919 in Pierce Oil 6s, due in 1920, and the company's common stock.[1]

The advantages possessed by convertibles, along the lines just described, are shared also by participating and purchase-warrant issues. The latter types of privileged securities may, of course, be used as media for hedging operations. Similarly, it may be found most desirable to switch from common stocks into such issues. The Rand Kardex 5½s, described on page 245, were not only an attractive direct commitment at the time of issuance, but they were certainly a desirable substitute for the Class *A* stock. Furthermore they offered an interesting hedging opportunity. In like manner, persons committed to a permanent investment in Westinghouse Electric and Manufacturing Company would certainly have been wise to switch from the common stock into the participating preferred when the latter sold at a lower price than the common in 1929 or 1930. In this case, however, a hedging operation between the preferred and common would have involved special hazards, because the senior issue was not convertible into the junior shares.

[1] This operation is analyzed in the Appendix, Note 39.

CHAPTER XXVI

SENIOR SECURITIES OF QUESTIONABLE SAFETY

At the low point of the 1932 securities market the safety of at least 80% of all corporate bonds and preferred stocks was open to some appreciable degree of doubt.[1] Even prior to the 1929 crash the number of speculative senior securities was very large, and it must inevitably be still larger for some years to come. The financial world is faced, therefore, with the unpleasant fact that a considerable proportion of American securities now belong to what may be called a misfit category. A low-grade bond or preferred stock constitutes a relatively unpopular form of commitment. The investor must not buy them, and the speculator generally prefers to devote his attention to common stocks. There seems to be much logic to the view that if one decides to speculate he should choose a thoroughly speculative medium, and not subject himself to the upper limitations of market value and income return, or to the possibility of confusion between speculation and investment, which attach to the lower priced bonds and preferred stocks.

Limitation of Profit on Low-priced Bonds Not a Real Drawback.—But however impressive may be the objection to these nondescript securities, the fact remains that they exist in enormous quantities, that they are owned by innumerable security holders, and that hence they must be taken seriously into account in any survey of security analysis. It is reasonable to conclude that the large supply of such issues, coupled with the lack of a natural demand for them, will make for a level of prices below their intrinsic value. Even if an inherent unattractiveness in the *form* of such securities be admitted, this may be more than offset by the attractive *price* at which they may be purchased. Furthermore, the limitations of principal profit in the case of a low-priced bond, as compared with a common stock, may be of only minor practical importance, because the profit actually *realized* by the common-stock buyer is ordinarily no greater than

[1] See Appendix, Note 40, for data on bond prices in 1932.

286

that obtainable from a speculative senior security. If, for example, we are considering a 5% bond selling at 35, its maximum possible price appreciation is about 70 points, or 200%. The average common-stock purchase at 35 cannot be held for a greater profit than this without a dangerous surrender to "bull-market psychology."

Two Viewpoints with Respect to Speculative Bonds.—There are two directly opposite angles from which a speculative bond may be viewed. It may be considered in its relation to investment standards and yields, in which case the leading question is whether the low price and higher income return will compensate for the concession made in the safety factor. Or it may be thought of in terms of a common-stock commitment, in which event the contrary question arises, *viz.:* "Does the *smaller* risk of loss involved in this low-priced bond, as compared with a common stock, compensate for the smaller possibilities of profit?" The nearer a bond comes to meeting investment requirements—and the closer it sells to an investment price—the more likely are those interested to regard it from the investment viewpoint. The opposite approach is evidently suggested in the case of a bond in default or selling at an extremely low price. We are faced here with the familiar difficulty of classification arising from the absence of definite lines of demarcation. Some issues can always be found reflecting any conceivable status in the gamut between complete worthlessness and absolute safety.

Common-stock Approach Preferable.—We believe, however, that the sounder and more fruitful approach to the field of speculative senior securities lies from the direction of common stocks. This will carry with it a more thorough appreciation of the risk involved and therefore a greater insistence upon either reasonable assurance of safety, or upon especially attractive possibilities of profit, or both. It induces also—among intelligent security buyers at least—a more intensive examination of the corporate picture than would ordinarily be made in viewing a security from the investment angle.

Such an approach would be distinctly unfavorable to the purchase of slightly substandard bonds selling at moderate discounts from par. These, together with high-coupon bonds of second grade, belong in the category of "business men's investments" which we considered and decided against in Chap. VII. It may be objected that a general adoption of this attitude would result in wide and sudden fluctuations in the price of many issues. Assuming that a 5% bond deserves to sell at par as long as it meets strict investment standards, then as soon as it falls slightly below these standards, its price would suffer a precipitous decline, say to 70; and conversely, a slight improvement in its exhibit would warrant its jumping suddenly back to par. Apparently there would be no justification for intermediate quotations between 70 and 100.

The real situation is not so simple as this, however. Differences of opinion may properly exist in the minds of investors as to whether a given issue is adequately secured, particularly since the standards are qualitative and personal as well as arithmetical and objective. The range between 70 and 100 may therefore logically reflect a greater or lesser agreement concerning the safety of the issue. This would mean that an investor would be justified in buying such a bond, say at 85, if his own considered judgment regarded it as sound; although he would recognize that there was doubt on this score in the minds of other investors which would account for its appreciable discount from a prime investment price. According to this view, the levels between 70 and 100, approximately, may be designated as the range of "subjective variations" in the status of the issue.

The field of speculative values proper would therefore commence somewhere near the 70 level (for bonds with a coupon rate of 5% or larger) and would offer maximum possibilities of appreciation of at least 50% of the cost. In making such commitments, it is recommended that the same general attitude be taken as in the careful purchase of a common stock; in other words, that the income account and the balance sheet be submitted to the same intensive analysis and that the same effort be made to evaluate future possibilities—favorable and unfavorable.

Important Distinctions between Common Stocks and Speculative Senior Issues.—We shall not seek, therefore, to set up standards of selection for speculative senior issues in any sense corresponding to the quantitative tests applicable to fixed-value securities. On the other hand, while they should preferably be considered in their relationship to the common-stock approach and technique, it is necessary to appreciate certain rather important points of difference that exist between common stocks as a class and speculative senior issues.

Low-priced Bonds Associated with Corporate Weakness.—The limitation on the profit possibilities of senior securities has already been referred to. Its significance varies with the individual case, but in general we do not consider it a controlling disadvantage. A more emphatic objection is made against low-priced bonds and preferred stocks on the ground that they are associated with corporate weakness, retrogression, or depression. Obviously the enterprise behind such a security is not highly successful; and furthermore, it must have been following a downward course, since the issue originally sold at a much higher level. In 1928 and 1929 this consideration was enough to condemn all such issues absolutely in the eyes of the general public. Businesses were divided into two groups: those which were successful and progressing, and those which were on the downgrade or making no headway. The common shares of the first group were desirable no matter how high the price; but no security belonging to the second group was attractive irrespective of how low it sold.

This concept of permanently strong and permanently weak corporations has been pretty well dissipated by the subsequent depression, and we are back to the older realization that Time brings unpredictable changes in the fortunes of business undertakings. The fact that the low price of a bond or preferred stock results from a decline in earnings need not signify that the company's outlook is hopeless and that there is nothing ahead but still poorer results. With the return of reasonable prosperity, many of the enterprises which fared badly in 1931 to 1933 will assuredly regain a good part of their former earning power. Consequently there is just as much reason to expect substantial recoveries in the quotations of depressed senior securities as in the price of common stocks generally.

Many Undervalued in Relation to Their Status and Contractual Position.—We have already mentioned that the unpopularity of speculative senior securities tends to make them sell at lower prices than common stocks, in relation to their intrinsic value. From the standpoint of the intelligent buyer this must be considered a point in their favor. With respect to their intrinsic position, speculative bonds—and, to a lesser degree, preferred stocks—derive important advantages from their contractual rights. The fixed obligation to pay bond interest will usually result in the continuation of such payments as long as they are in any way possible. If we assume that a fairly large proportion of a group of carefully selected low-priced bonds will escape default, the income received on the group as a whole over a period of time will undoubtedly far exceed the dividend return on similarly priced common stocks.

Preferred shares occupy an immeasurably weaker position in this regard, but even here the provisions transferring voting control to the senior shares in the event of suspension of dividends will be found in some cases to impel their continuance. Where the cash resources are ample, the desire to maintain an unbroken record and to avoid accumulations will frequently result in paying preferred dividends even though poor earnings have depressed the market price.

Examples: Century Ribbon Mills, Inc., failed to earn its 7% preferred dividend in six years out of the seven from 1926 to 1932, inclusive, and the price repeatedly declined to about 50. Yet the preferred dividend was continued without interruption during this entire period, while the common received nothing. Similarly, a purchaser of Universal Pictures Company First Preferred at about 30 in 1929 would have received the 8% dividend during three years of depression before the payment was finally suspended.

Contrasting Importance of Contractual Terms in Speculation and Investment.—The reader should appreciate the distinction between the *investment* and the *speculative* qualities of preferred stocks in this matter of dividend continuance. From the investment standpoint, *i.e.*, the dependability of the dividend, the absence of an enforceable claim is a disadvantage as compared with bonds. From the speculative standpoint, *i.e.*, the possibility of dividends being continued under unfavorable conditions, preferred stocks have certain semicontractual claims to consideration by the directors which undoubtedly give them an advantage over common stocks.

Bearing of Working Capital on Safety of Speculative Senior Issues.—A large working capital, which has been characteristic of even nonprosperous industrials for some years past, is much more directly advantageous to the senior securities than to the common stock. Not only does it make possible the continuance of interest or preferred-dividend payments, but it has an important bearing also on the retirement of the principal, either at maturity, or by sinking-fund operations, or by voluntary repurchase. Sinking-fund provisions, for bonds as well as preferred stocks, contribute to the improvement of both the market quotation and the intrinsic position of the issue. This advantage is not found in the case of common stocks.

Examples: Francis H. Leggett Company, manufacturers and wholesalers of food products, issued $2,000,000 of 7% preferred stock carrying a sinking-fund provision which retired 3% of the issue annually. By June 30, 1932, the amount outstanding had been reduced to $608,500 and because of the small balance remaining, the issue was called for redemption at 110, in the depth of the depression. Similarly, the Century Ribbon Mills Preferred was reduced from $2,000,000 to $1,000,000 between 1922 and 1932; and the amount of Universal Pictures First Preferred outstanding was reduced from $3,000,000 to $1,786,400 between 1924 and October 29, 1932.

Importance of Large Net-quick-asset Coverage.—Where a low-priced bond is covered several times over by net quick assets, it presents a special type of opportunity, because experience shows that the chances of repayment are good, even though the earnings may be poor or irregular.

Examples: Electric Refrigeration Corporation (Kelvinator) 6s, due 1936, sold at 66 in November 1929 when the net quick assets of the company according to its latest statement amounted to $6,008,900 for the $2,528,500 of bonds outstanding. It is true that the company had operated at a deficit in 1927 and 1928, but fixed charges were earned nearly nine times in the year ended September 30, 1929, and the net quick assets were nearly four times the market value of the bond issue. The bonds recovered to a price close to par in 1930 and were redeemed at 105 in 1931. Similarly, Electric Refrigeration Building Corporation First 6s, due 1936, which were in effect guaranteed by Kelvinator Corporation under a lease, sold at 70 in July 1932 when the net current assets of the parent company amounted to about six times the $1,073,000 of bonds outstanding and over eight times the total market value of the issue. The bonds were called at 101½ in 1933.

Other examples which may be cited in this connection are Murray Corporation First 6½s, due 1934, which sold at 68 in 1932 (because of current operating deficits) although the company had net current assets of over 2½ times the par value of the issue and nearly four times their market value at that price;

Sidney Blumenthal and Company 7% Notes, due 1936, which
sold at 70 in 1926 when the company had net quick assets of
twice the par value of the issue and nearly three times the total
market value thereof (they were called at 103 in 1930); Belding,
Heminway Company 6s, due 1936, which sold at 67 in 1930 when
the company had net current assets of nearly three times the par
value of the issue and over four times its market value. In the
latter case drastic liquidation of inventories occurred in 1930 and
1931, proceeds from which were used to retire about 80% of the
bond issue through purchases in the market. The balance of the
issue was called for payment at 101 early in 1934.

In the typical case of this kind the chance of profit will exceed
the chance of loss and the probable amount of profit will exceed
the probable amount of loss. It may well be that the risk
involved in each individual case is still so considerable as to
preclude us from applying the term investment to such a commitment. Nevertheless, we suggest that if the insurance
principle of diversification of risk be followed by making a
number of such commitments at the same time, the net result
should be sufficiently dependable to warrant our calling the
group purchase an *investment operation.* This was one of the
possibilities envisaged in our broadened definition of investment
as given in Chap. IV.

Limitations upon Importance of Quick-asset Position.—It is
clear that considerable weight attaches to the working-capital
exhibit in selecting speculative bonds. This importance must
not be exaggerated, however, to the point of assuming that
whenever a bond is fully covered by net quick assets its safety
is thereby assured. The current assets shown in any balance
sheet may be greatly reduced by subsequent operating losses;
more important still, the stated values frequently prove entirely
undependable in the event of receivership.[1]

Of the many examples of this point which can be given, we
shall mention R. Hoe and Company 7% Notes and Ajax Rubber
Company First 8s. Although these obligations were covered
by net working capital in 1929, they subsequently sold as low as
2 cents on the dollar. (See also our discussion of Willys-Overland
Company First 6½s and Berkey and Gay Furniture Company
First 6s in Note 32 of the Appendix.)

[1] The comparative reliability of the various components in the quick-
assets figure (cash assets, receivables, inventories) will receive detailed
treatment in a discussion of balance-sheet analysis in Part VI.

We must distinguish, therefore, between the mere fact that the working capital, as reported, covers the funded debt, and the more significant fact that it exceeds the bond issue many times over. The former statement is always interesting, but by no means conclusive. If added to other favorable factors, such as a good earnings coverage in normal years, and a generally satisfactory qualitative showing, it might make the issue quite attractive, but preferably as part of a group-purchase in the field.

EXAMPLES OF LOW-PRICED INDUSTRIAL BONDS COVERED BY NET CURRENT ASSETS, 1932

Name of issue	Due	Low price 1932	Date of balance sheet	Net current assets*	Funded debt at par*	Normal interest coverage	
						Period	Times earned†
American Seating 6s.....	1936	17	Sept. 1932	$ 3,826	$ 3,056	1924–1930	5.2
Crucible Steel 5s........	1940	39	June 1932	16,163	13,250	1924–1930	9.4
McKesson & Robbins 5⅛s.................	1950	25	June 1932	42,885	20,848	1925–1930	4.1
Marion Steam Shovel 6s.	1947	21	June 1932	4,598	2,417	1922–1930	3.9
National Acme 6s.......	1942	54	Dec. 1931	4,327	1,963	1922–1930	5.5

* 000 omitted.
† Coverage for 1931 charges, adjusted where necessary.

Speculative Preferred Stocks. *Stages in Their Price History.—* Speculative preferred stocks are more subject than speculative bonds to manipulative activity, so that from time to time such preferred shares are overvalued in the market in the same way as common stocks. We thus have three possible stages in the price history of a preferred issue, in each of which the market quotation tends to be out of line with the value:

1. The first stage is that of original issuance, when investors are persuaded to buy the offering at a full investment price not justified by its intrinsic merit.

2. In the second stage the lack of investment merit has become manifest and the price drops to a speculative level. During this period the decline is likely to be overdone, for reasons previously discussed.

3. A third stage sometimes appears in which the issue is manipulated upward by the methods applied to common stocks. On such occasions certain factors of questionable importance—such as the amount of dividend accumulations—are overemphasized, in order to persuade the speculative public to buy the issue at an inflated price.

An example of this third or manipulative stage will be given a little later.

The Rule of "Maximum Valuation for Senior Issues."—Both as a safeguard against being led astray by the propaganda which is characteristic of the third stage, and also as a general guide in dealing with speculative senior issues, the following principle of security analysis is presented, which we shall call "the rule of maximum valuation for senior issues."

A senior issue cannot be worth, intrinsically, any more than a common stock would be worth if it occupied the position of that senior issue.

This statement may be understood more readily by means of an example.

Company X and Company Y have the same value. Company X has 80,000 shares of preferred and 200,000 shares of common. Company Y has only 80,000 shares of common and no preferred. Then our principle states that a share of Company X preferred cannot be worth more than a share of Company Y common. This is true because Company Y common represents the same value that lies behind *both* the preferred and common of Company X.

Instead of comparing two equivalent companies such as X and Y, we may assume that Company X is recapitalized so that the old common is eliminated and the preferred becomes the sole stock issue, *i.e.*, the new common stock. (To coin a term, we may call such an assumed change the "commonizing" of a preferred stock.) Then our principle merely states the obvious fact that the value of such a hypothetical common stock cannot be *less* than the value of the preferred stock it replaces, because it is equivalent to the preferred *plus* the old common. The same idea may be applied to a speculative bond, followed either by common stock only or by both preferred and common. If the bond is "commonized," *i.e.*, if it is assumed to be turned into a common stock, with the old stock issues eliminated, then the value of the new common stock thus created cannot be less than the present value of the bond.

This relationship must hold true regardless of how high the coupon or dividend rate, the par value, or the redemption price of the senior issue may be; and, particularly, regardless of what amount of unpaid interest or dividends may have accumulated. For if we had a preferred stock with accumulations of $1,000 per share, the value of the issue could be no greater than if it were a common stock (without dividend accumulations) representing complete ownership of the business. The unpaid dividends cannot create any additional value for the company's securities in the aggregate, they merely affect the division of the total value between the preferred and the common.

Excessive Emphasis Placed on Amount of Accrued Dividends.— While a very small amount of analysis will show the above statements to be almost self-evident truths, the public fails to observe the simplest rules of logic when once it is in a gambling mood. Hence preferred shares with large dividend accruals lend themselves readily to market manipulation in which the accumulations are made the basis for a large advance in the price of both the preferred and common. An excellent example of such a performance was provided by American Zinc, Lead and Smelting Company shares in 1928.

American Zinc preferred stock was created in 1916 as a stock dividend on the common, the transaction thus amounting to a split-up of old common into preferred and new common. The preferred was given a stated par of $25, but had all the attributes of a $100-par stock ($6 cumulative dividends, redemption and liquidating value of $100). This arrangement was evidently a device to permit carrying the preferred issue in the balance sheet as a much smaller liability than it actually represented. Between 1920 and 1927, the company reported continuous deficits (except for a negligible profit in 1922); preferred dividends were suspended in 1921, and by 1928 about $40 per share had accumulated.

In 1928 the company benefited moderately from the prevailing prosperity and barely earned $6 per share on the preferred. However, the company's issues were subjected to manipulation which advanced the price of the preferred from 35 in 1927 to 118 in 1928, while the common rose even more spectacularly from 6 to 57. These advances were accompanied by rumors of a plan to pay off the accumulated dividends—exactly how, not being stated. Naturally enough, this development failed to materialize.

The irrationality of the gambling spirit is well shown here by the absurd acceptance of unpaid preferred dividends *as a source of value for both the preferred and the common.* The speculative argument in behalf of the common stock ran as follows: "The accumulated preferred dividends are going to be paid off. This will be good for the common. Therefore let us buy the common." According to this topsy-turvy reasoning, if there were no unpaid preferred dividends ahead of the common it would be less attractive (even at the same price), because there would then be in prospect no wonderful plan for clearing up the accumulations.

We may use the American Zinc example to demonstrate the practical application of our "rule of maximum valuation for senior issues." Was American Zinc Preferred too high at 118 in 1928? Assuming the preferred stockholders owned the company completely, this would then mean a price of 118 for a *common* stock earning $6 per share in 1928 after eight years of deficits. Even in the hectic days of 1928 speculators would not have been at all attracted to such a common stock at that price, so that the application of our rule should have prevented the purchase of the preferred stock at its inflated value.

The quotation of 57 reached by American Zinc common was evidently the height of absurdity, since it represented the following valuation for the company:

Preferred stock, 80,000 sh. @ 118.........	$ 9,440,000
Common stock, 200,000 sh. @ 57..........	11,400,000
Total valuation......................	$20,840,000
Earnings, 1928.......................	481,000
Average earnings, 1920–1927............	*188,000(d)*

In order to equal the above valuation for the American Zinc Company the hypothetical common stock (80,000 shares basis) would have had to sell at $260 *per share*, earning a bare $6 and paying no dividend. This figure indicates the extent to which the heedless public was led astray in this case by the exploitation of unpaid dividends.

American Hide and Leather Company offers another, but less striking, example of this point. In no year between 1922 and 1928 inclusive, did the company earn more than $4.41 on the preferred, and the average profits were very small. Yet in each of these seven years, the preferred stock sold as high as 66 or higher. This recurring strength was based largely on the speculative appeal of the enormous accumulated preferred dividends which grew from about $120 to $175 per share during this period.

Applying our rule, we may consider American Hide and Leather Preferred as representing complete ownership of the business, which to all intents and purposes it did. We should then have a common stock which had paid no dividends for many years and with average earnings at best (using the 1922–1927 period) of barely $2 per share. Evidently a price of above 65 for such a common stock would be far too high. Consequently this price was excessive for American Hide and Leather Preferred, nor could the existence of accumulated dividends, however large, affect this conclusion in the slightest.

Variation in Capital Structure Affects Total Market Value of Securities.—From the foregoing discussion it might be inferred that the value of a single capital-stock issue must always be equivalent to the combined values of any preferred and common-stock issues into which it might be split. In a theoretical sense this is entirely true; but in practice it may not be true at all, because a division of capitalization into senior securities and common stock may have a real advantage over a single common-stock issue. This subject will receive extended treatment under the heading of "Capitalization Structure" in Chap. XL.

The distinction between the idea just suggested and our "rule of maximum valuation" may be clarified as follows:

1. Assume Company X = Company Y
2. Company X has preferred (P) and common (C); Company Y has common only (C')
3. Then it *would appear* that

Value of P + value of C = value of C'

since each side of the equation represents equal things, namely the total value of each company.

But this apparent relationship may not hold good in practice because the preferred-and-common capitalization method may have real advantages over a single common-stock issue.

On the other hand, our "rule of maximum valuation" merely states that the value of P alone cannot exceed value of C'. This should hold true in practice as well as in theory, except in so far as manipulative or heedlessly speculative activity brushes aside all rational considerations.

Our rule is stated in negative form and is therefore essentially negative in its application. It is most useful in detecting instances where preferred stocks or bonds are *not worth* their market price. To apply it positively it would be necessary, first to arrive at a value for the preferred on a "commonized" basis (*i.e.*, representing complete ownership of the business) and then to determine what deduction from this value should be made to reflect the part of the ownership fairly ascribable to the existing common stock. At times this approach will be found useful in establishing the fact that a given senior issue is worth more than its market price. But such a procedure brings us far outside the range of mathematical formulas and into the difficult and indefinite field of common-stock valuation, with which we have next to deal.

PART IV

THEORY OF COMMON-STOCK INVESTMENT. THE DIVIDEND FACTOR

CHAPTER XXVII

THE THEORY OF COMMON-STOCK INVESTMENT

In our introductory discussion we set forth the difficulties inherent in efforts to apply the analytical technique to speculative situations. Since the speculative factors bulk particularly large in common stocks, it follows that analysis of such issues is likely to prove inconclusive and unsatisfactory, and even where it appears to be conclusive, there is danger that it may be misleading. At this point it is necessary to consider the function of common-stock analysis in greater detail. We must begin with three realistic premises. The first is that common stocks are of basic importance in our financial scheme and of fascinating interest to many people; the second is that owners and buyers of common stocks are generally anxious to arrive at an intelligent idea of their value; and thirdly, even when the underlying motive of purchase is mere speculative greed, human nature desires to conceal this unlovely impulse behind a screen of apparent logic and good sense. To adapt the aphorism of Voltaire, it may be said that if there were no such thing as common-stock analysis it would be necessary to counterfeit it.

Broad Merits of Common-stock Analysis.—We are thus led to the question: "To what extent is common-stock analysis a valid and truly valuable exercise, and to what extent is it an empty but indispensable ceremony attending the wagering of money on the future of business and of the stock market?" We shall ultimately find the answer to run somewhat as follows: "As far as the *typical* common stock is concerned—an issue picked at random from the list—an analysis, however elaborate, is unlikely to yield a dependable conclusion as to its attractiveness or its real value. But in individual cases, the exhibit may be such as to permit reasonably confident conclusions to be drawn from the processes of analysis." It would follow that analysis is of positive or scientific value only in the case of the exceptional common stock, and that for common stocks in general, it must be regarded either as a somewhat questionable aid to speculative judgment, or as a highly illusory method of aiming at values which defy calculation, and which must somehow be calculated none the less.

Perhaps the most effective way of clarifying the subject is through the historical approach. Such a survey will throw light not only upon the changing status of common-stock analysis but also upon a closely related subject of major importance, namely the theory of common-stock investment. We shall encounter at first a set of old established and seemingly logical principles for common-stock investment. Through the advent of new conditions, we shall find the validity of these principles impaired. Their insufficiency will give rise to an entirely different concept of common-stock selection, the so-called "new-era theory," which beneath its superficial plausibility will hold possibilities of untold mischief in store. With the prewar theory obsolete and the new-era theory exploded, we must finally make the attempt to establish a new set of logically sound and reasonably dependable principles of common-stock investment.

History of Common-stock Analysis.—Turning first to the history of common-stock *analysis*, we shall find that two conflicting factors have been at work during the past 30 years. On the one hand there has been an increase in the *investment prestige* of common stocks as a class, due chiefly to the expanding number which have shown substantial earnings, continued dividends, and a strong financial condition. Accompanying this progress was a considerable advance in the frequency and adequacy of corporate statements, thus supplying the public and the securities analyst with a wealth of statistical data. Finally, an impressive theory was constructed asserting the preeminence of common stocks as long-term investments. But at the time that the interest in common stocks reached its height, in the period between 1927 and 1929, the *basis of valuation* employed by the stock-buying public departed more and more from the factual approach and technique of security analysis, and concerned itself increasingly with the elements of potentiality and prophecy.

Moreover, the heightened instability in the affairs of industrial companies and groups of enterprises, which has undermined the investment quality of bonds in general, has of course been still more hostile to the maintenance of true investment quality in common stocks.

Analysis Vitiated by Two Types of Instability.—The extent to which common-stock analysis has been vitiated by these two developments, (1) the instability of tangibles, and (2) the dominant importance of intangibles—may be better realized by a contrast of specific common stocks prior to 1920 and in more recent times. Let us consider four typical examples: Pennsylvania Railroad, Atchison, Topeka and Santa Fe Railway, National Biscuit, and American Can.

PENNSYLVANIA RAILROAD COMPANY

Year	Range for stock	Earned per share	Paid per share
1904	70–56	$4.63	$3.00
1905	74–66	4.98	3.00
1906	74–61	5.83	3.25
1907	71–52	5.32	3.50
1908	68–52	4.46	3.00
1909	76–63	4.37	3.00
1910	69–61	4.60	3.00
1911	65–59	4.14	3.00
1912	63–60	4.64	3.00
1913	62–53	4.20	3.00
1923	48–41	5.16	3.00
1924	50–42	3.82	3.00
1925	55–43	6.23	3.00
1926	57–49	6.77	3.125
1927	68–57	6.83	3.50
1928	77–62	7.34	3.50
1929	110–73	8.82	3.875
1930	87–53	5.28	4.00
1931	64–16	1.48	3.25
1932	23– 7	1.03	0.50

American Can was a typical example of a prewar speculative stock. It was speculative for three good and sufficient reasons: (1) it paid no dividend; (2) its earnings were small and irregular; (3) the issue was "watered," *i.e.*, a substantial part of its stated value represented no actual investment in the business. By contrast, Pennsylvania, Atchison, and National Biscuit were regarded as investment common stocks—also for three good and sufficient reasons: (1) they showed a satisfactory record of continued dividends; (2) the earnings were reasonably stable and averaged substantially in excess of the dividends paid; and (3) each dollar of stock was backed by a dollar or more of actual investment in the business.

ATCHISON, TOPEKA AND SANTA FE RAILWAY COMPANY

Year	Range of stock	Earned per share	Paid per share
1904	89– 64	$ 9.47*	$ 4.00
1905	93– 78	5.92*	4.00
1906	111– 85	12.31*	4.50
1907	108– 66	15.02*	6.00
1908	101– 66	7.74*	5.00
1909	125– 98	12.10*	5.50
1910	124– 91	8.89*	6.00
1911	117–100	9.30*	6.00
1912	112–103	8.19*	6.00
1913	106– 90	8.62*	6.00
1923	105– 94	15.48	6.00
1924	121– 97	15.47	6.00
1925	141–116	17.19	7.00
1926	172–122	23.42	7.00
1927	200–162	18.74	10.00
1928	204–183	18.09	10.00
1929	299–195	22.69	10.00
1930	243–168	12.86	10.00
1931	203– 79	6.96	9.00
1932	94– 18	0.55	2.50

* Fiscal years ended June 30.

If we study the range of market price of these issues during the decade preceding the War (or the 1909–1918 period for National Biscuit), we note that American Can fluctuated widely from year to year in the fashion regularly associated with speculative media, but that Pennsylvania, Atchison, and National Biscuit showed much narrower variations, and evidently tended to oscillate about a base price (*i.e.*, 97 for Atchison, 64 for Pennsylvania, and 120 for National Biscuit) which seemed to represent a well-defined view of their investment or intrinsic value.

Prewar Conception of Investment in Common Stocks.—Hence the prewar relationship between analysis and investment on the one hand and price changes and speculation on the other may be set forth as follows: Investment in common stocks was confined to those showing stable dividends and fairly stable earnings; and such issues in turn were expected to maintain a fairly stable

NATIONAL BISCUIT COMPANY

Year	Range for stock	Earned per share	Paid per share
1909	120– 97	$ 7.67*	$ 5.75
1910	120– 100	9.86*	6.00
1911	144– 117	10.05*	8.75
1912	161– 114	9.59*	7.00
1913	130– 104	11.73*	7.00
1914	139– 120	9.52*	7.00
1915	132– 116	8.20*	7.00
1916	131– 118	9.72*	7.00
1917	123– 80	9.87†	7.00
1918	111– 90	11.63	7.00
	(old basis)‡	(old basis)‡	(old basis)‡
1923	370– 266	$35.42	$21.00
1924	541– 352	38.15	28.00
1925	553– 455	40.53	28.00
1926	714– 518	44.24	35.00
1927	1309– 663	49.77	42.00
1928	1367–1117	51.17	49.00
1929	1657– 980	57.40	52.50
1930	1628–1148	59.68	56.00
1931	1466– 637	50.05	49.00
1932	820– 354	42.70	49.00

* Earnings for the year ended Jan. 31 of the following year.
† Eleven months ending Dec. 31, 1917.
‡ Stock was split 4 for 1 in 1922, followed by a 75 % stock dividend. In 1930 it was again split 2½ for 1. Published figures applicable to new stock were one-seventh of those given above for 1923–1929. Likewise the above figures for 1930–1932 are 17½ times the published figures for those years.

market level. The function of analysis was primarily to search for elements of *weakness* in the picture. If the earnings were not properly stated; if the balance sheet revealed a poor current position, or the funded debt was growing too rapidly; if the physical plant was not properly maintained; if dangerous new competition was threatening, or if the company was losing ground in the industry; if the management was deteriorating or was likely to change for the worse; if there was reason to fear for the future of the industry as a whole—any of these defects, or some other one, might be sufficient to condemn the issue from the standpoint of the cautious investor.

On the positive side, analysis was concerned with finding those issues which met all the requirements of investment and *in*

AMERICAN CAN COMPANY

Year	Range for stock	Earned per share	Paid per share
1904	....	$ 0.51*	0
1905	....	*1.39(d)*†	0
1906	....	*1.30(d)*‡	0
1907	8– 3	*0.57(d)*	0
1908	10– 4	*0.44(d)*	0
1909	15– 8	*0.32(d)*	0
1910	14– 7	*0.15(d)*	0
1911	13– 9	0.07	0
1912	47– 11	8.86	0
1913	47– 21	5.21	0
1923	108– 74	19.64	$ 5.00
1924	164– 96	20.51	6.00
1925	297–158	32.75	7.00
	(old basis)§	(old basis)§	(old basis)§
1926	379–233	26.34	13.25
1927	466–262	24.66	12.00
1928	705–423	41.16	12.00
1929	1107–516	48.12	30.00
1930	940–628	48.48	30.00
1931	779–349	30.66	30.00
1932	443–178	19.56	24.00

* Fiscal year ended Mar. 31, 1905.
† Nine months ended Dec. 31, 1905.
‡ Excluding fire losses of 58 cents a share.
§ Stock was split 6 for 1 in 1926. Published figures applicable to new stock were one-sixth of those given for 1926–1932.

addition offered the best chance for future enhancement. The process was largely a matter of comparing similar issues in the investment class, *e.g.*, the group of dividend-paying Northwestern railroads. Chief emphasis would be laid upon the relative showing for past years, in particular the average earnings in relation to price, and the stability and the trend of earnings. To a lesser extent, the analyst sought to look into the future and to select the industries or the individual companies which were likely to show the most rapid growth.

Speculation Characterized by Emphasis on Future Prospects.—In the prewar period it was the well-considered view that when *prime emphasis* was laid upon what was expected of the future, instead of what had been accomplished in the past, a speculative attitude was thereby taken. Speculation, in its etymology, meant looking forward; investment was allied to "vested interests,"—to property rights and values taking root in the *past*. The future was uncertain, therefore speculative; the past was known, therefore the source of safety. Let us consider a buyer of American Can common in 1910. He may have bought it believing that its price was going to advance or be "put up"; or that its earnings were going to increase; or that it was soon going to pay a dividend; or possibly that it was destined to develop into one of the country's strongest industrials. From the prewar standpoint, while one of these reasons may have been more intelligent or creditable than another, each of them constituted a *speculative* motive for the purchase.

Technique of Investing in Common Stocks Resembled That for Bonds.—Evidently there was a close similarity between the technique of investing in common stocks and that of investing in bonds. The common-stock investor, also, wanted a stable business and one showing an adequate margin of earnings over dividend requirements. Naturally he had to content himself with a smaller margin of safety than he would demand of a bond, a disadvantage which was offset by a larger income return (6% was standard on a good common stock compared with 4½% on a high-grade bond), by the chance of an increased dividend if the business continued to prosper, and—generally of least importance in his eyes—by the possibility of a profit. A common-stock investor was likely to consider himself as in no very different position from that of a purchaser of second-grade bonds; essentially his venture amounted to sacrificing a certain *degree* of safety in return for larger income. The Pennsylvania and Atchison examples during the 1904–1913 decade, will supply specific confirmation of the above description.

Buying Common Stocks Viewed as Taking a Share in a Business. Another useful approach to the attitude of the prewar common-stock investor is from the standpoint of taking an interest in a private business. The typical common-stock investor was a business man, and it seemed sensible to him to value any corporate enterprise in much the same manner as he would value his own business. This meant that he gave at least as much attention to the asset values behind the shares as he did to their earnings records. It is essential to bear in the mind that a private business has always been valued primarily on the basis of the "net worth" as shown by its statement. A man contemplating the purchase of a partnership or stock interest in a private undertaking will always start with the value of that interest as shown "on the books," *i.e.*, the balance sheet, and will then consider whether the record and prospects are good enough to make such a commitment attractive. An interest in a private business may of course be sold for more or less than its proportionate asset value; but the book value is still invariably the starting point of the calculation, and the deal is finally made and viewed in terms of the premium or discount from book value involved.

Broadly speaking, the same attitude was formerly taken in an investment purchase of a marketable common stock. The first point of departure was the par value, presumably representing the amount of cash or property originally paid into the business; the second basal figure was the book value, representing the par value plus a ratable interest in the accumulated surplus. Hence in considering a common stock, investors asked themselves: "Is this issue a desirable purchase at the premium above book value, or the discount below book value, represented by the market price?" "Watered stock" was repeatedly inveighed against as a deception practised upon the stock-buying public, who were misled by a fictitious statement of the asset values existing behind the shares. Hence one of the protective functions of security analysis was to discover whether the value of the fixed assets, as stated on the balance sheet of a company, fairly represented the actual cost or reasonable worth of the properties.

Investment in Common Stocks Based on Threefold Concept.—We thus see that investment in common stocks was formerly based upon the threefold concept of: (1) a suitable and established dividend return; (2) a stable and adequate earnings record; and (3) a satisfactory backing of tangible assets. Each of these three elements could be made the subject of careful analytical study viewing the issue both by itself and in comparison with others of its class. Common-stock commitments motivated by any other viewpoint were characterized as speculative, and it was not expected that they should be justified by a serious analysis.

THE NEW-ERA THEORY

During the postwar period, and particularly during the latter stage of the bull market culminating in 1929, the public acquired a completely different attitude towards the investment merits of common stocks. Two of the three elements above stated lost nearly all of their significance and the third, the earnings record, took on an entirely novel complexion. The new theory or principle may be summed up in the sentence: "The value of a common stock depends entirely upon what it will earn in the future."

From this dictum the following corollaries were drawn:

1. That the dividend rate should have slight bearing upon the value.

2. That since no relationship apparently existed between assets and earning power, the asset value was entirely devoid of importance.

3. That past earnings were significant only to the extent that they indicated what *changes* in the earnings were likely to take place in the future.

This complete revolution in the philosophy of common-stock investment took place virtually without realization by the stock-buying public and with only the most superficial recognition by financial observers. An effort must be made to reach a thorough comprehension of what this changed viewpoint really signifies. To do so we must consider it from three angles, its causes, its consequences, and its logical validity.

Causes for This Changed Viewpoint.—Why did the *investing* public turn its attention from dividends, from asset values, and from earnings, to transfer it almost exclusively to the earnings *trend*, *i.e.*, to the *changes* in earnings expected in the future? The answer was, first, that the records of the past were proving an undependable guide to investment; and secondly, that the rewards offered by the future had become irresistibly alluring.

The new-era concepts had their root first of all in the obsolescence of the old-established standards. During the last generation the tempo of economic change has been speeded up to such a degree that the fact of being *long established* has ceased to be, as once it was, a warranty of *stability*. Corporations enjoying decade-long prosperity have been precipitated into insolvency within a few years. Other enterprises, which had been small or unsuccessful or in doubtful repute, have just as quickly acquired dominant size, impressive earnings, and the highest rating. The major group upon which investment interest was chiefly concentrated, viz., the railroads, failed signally to participate in the expansion of national wealth and income, and showed repeated signs of definite retrogression. The street railways, another important medium of investment prior to 1914, rapidly lost the greater portion of their value as the result of the development of new transportation agencies. The electric and gas companies followed an irregular course during this period, since they were harmed rather than helped by the war and postwar inflation, and their impressive growth is a relatively recent phenomenon. The history of industrial companies was a hodge-podge of violent changes, in which the benefits of prosperity were so unequally and so impermanently distributed as to bring about the most unexpected failures alongside of the most dazzling successes.

In the face of all this instability it was inevitable that the threefold basis of common-stock investment should prove totally inadequate. Past earnings and dividends could no longer be considered, in themselves, an index of future earnings and dividends. Furthermore, these future earnings showed no tendency whatever to be controlled by the amount of the actual investment in the business—the asset values—but instead depended entirely upon a favorable industrial position and upon capable or fortunate managerial policies. In numerous cases of receivership, the current assets dwindled and the fixed assets proved almost worthless. Because of this absence of any connection between both assets and earnings, and between assets and realizable values in bankruptcy, less and less attention came to be paid either by financial writers or by the general public to the formerly important question of "net worth," or "book value," and it may be said that by 1929 book value had practically disappeared as an element in determining the attractiveness of a security issue. It is a significant confirmation of this point that "watered stock," once so burning an issue, is now a forgotten phrase.

Attention Shifted to the Trend of Earnings.—Thus the prewar approach to investment, based upon past records and tangible facts, became outworn and was discarded. Could anything be put in its place?   A new conception was given central importance —that of *trend of earnings*.   The past was important only in so far as it showed the direction in which the future could be expected to move.   A continuous increase in profits proved that the company was on the upgrade and promised still better results in the future than had been accomplished to date.   Conversely, if the earnings had declined, or even remained stationary during a prosperous period, the future must be thought unpromising and the issue was certainly to be avoided.

The Common-stocks-as-long-term-investments Doctrine.— Along with this idea as to what constituted the basis for common-stock selection, there emerged a companion theory that common stocks represented the most profitable and therefore the most desirable media for long-term investment.   This gospel was based upon a certain amount of research, showing that diversified lists of common stocks had regularly increased in value over stated intervals of time for many years past.   The figures indicated that such diversified common-stock holdings yielded both a higher income return and a greater principal profit than purchases of standard bonds.

The combination of these two ideas supplied the "investment theory" upon which the 1927–1929 stock market proceeded. Amplifying the principle stated on page 307, the theory ran as follows:

1. "The value of a common stock depends on what it can earn in the future."
2. "Good common stocks will prove sound and profitable investments."
3. "Good common stocks are those which have shown a rising trend of earnings."

These statements sound innocent and plausible.   Yet they concealed two theoretical weaknesses which could and did result in untold mischief.   The first of these defects was that they abolished the fundamental distinctions between investment and speculation.   The second was that they ignored the *price* of a stock in determining whether it was a desirable purchase.

New-era Investment Equivalent to Prewar Speculation.—A moment's thought will show that "new-era investment," as practiced by the representative investment trusts, was almost identical with speculation as popularly defined in preboom days. Such "investment" meant buying common stocks instead of bonds, emphasizing enhancement of principal instead of income, and stressing the changes of the future instead of the facts of the established past. It would not be inaccurate to state that new-era investment was simply old-style speculation confined to common stocks with a satisfactory trend of earnings. The impressive new concept underlying the greatest stock-market boom in history appears to be no more than a thinly disguised version of the old cynical epigram: "Investment is successful speculation."

Stocks Regarded as Attractive Irrespective of Their Prices.— The notion that the desirability of a common stock was entirely independent of its price seems incredibly absurd. Yet the new-era theory led directly to this thesis. If a public-utility stock was selling at 35 times its *maximum* recorded earnings, instead of 10 times its *average* earnings, which was the preboom standard, the conclusion to be drawn was not that the stock was now too high but merely that the standard of value had been raised. Instead of judging the market price by established standards of value, the new era based its standards of value upon the market price. Hence all upper limits disappeared, not only upon the price at which a stock *could* sell, but even upon the price at which it would *deserve* to sell. This fantastic reasoning actually led to the purchase for investment at $100 per share of common stocks earning $2.50 per share. The identical reasoning would support the purchase of these same shares at $200, at $1,000, or at any conceivable price.

An alluring corollary of this principle was that making money in the stock market was now the easiest thing in the world. It was only necessary to buy "good" stocks, regardless of price, and then to let nature take her upward course. The results of such a doctrine could not fail to be tragic. Countless people asked themselves, "Why work for a living when a fortune can be made in Wall Street without working?" The ensuing migration from business into the financial district resembled the famous gold rush to the Klondike, with the not unimportant difference that there really was gold in the Klondike.

Investment Trusts Adopted This New Doctrine.—An ironical sidelight is thrown on this 1928–1929 theory by the practice of the investment trusts. These were formed for the purpose of giving the untrained public the benefit of expert administration of its funds—a plausible idea, and one which had been working well in England. The earliest American investment trusts laid considerable emphasis upon certain time-tried principles of successful investment, which they were much better qualified to follow than the typical individual. The most important of these principles were:

1. To buy in times of depression and low prices, and to sell out in times of prosperity and high prices.
2. To diversify holdings in many fields and probably in many countries.
3. To discover and acquire undervalued individual securities as the result of comprehensive and expert statistical investigations.

The rapidity and completeness with which these traditional principles disappeared from investment-trust technique is one of the many marvels of the period. The idea of buying in times of depression was obviously inapplicable. It suffered from the fatal weakness that investment trusts could be organized only in good times, so that they were virtually compelled to make their initial commitments in bull markets. The idea of world-wide geographical distribution had never exerted a powerful appeal upon the provincially minded Americans (who possibly were right in this respect); and with things going so much better here than abroad this principle was dropped by common consent.

Analysis Abandoned by Investment Trusts.—But most para-doxical was the early abandonment of research and analysis in guiding investment-trust policies. However, since these financial institutions owed their existence to the new-era phi-losophy, it was natural and perhaps only just that they should adhere closely to it. Under its canons investment had now become so beautifully simple that research was unnecessary and statistical data a mere incumbrance. The investment process consisted merely of finding prominent companies with a rising trend of earnings, and then buying their shares regardless of price. Hence the sound policy was to buy only what every one else was buying—a select list of highly popular and exceedingly expensive issues, appropriately known as the "blue chips." The original idea of searching for the undervalued and neglected issues dropped completely out of sight. Investment trusts actually boasted that their portfolios consisted exclusively of the active and standard (*i.e.*, the most popular and highest priced) common stocks. With but slight exaggeration, it might be asserted that under this convenient technique of investment, the affairs of a ten-million-dollar investment trust could be administered by the intelligence, the training, and the actual labors of a single thirty-dollar-a-week clerk.

The man in the street, having been urged to entrust his funds to the superior skill of investment experts—for substantial compensation—was soon reassuringly told that the trusts would be careful to buy nothing except what the man in the street was buying himself.

The Justification Offered.—Irrationality could go no further; yet it is important to note that mass speculation can flourish only in an atmosphere of illogic and unreality. The self-deception of the mass speculator must, however, have its elements of justification. This is usually some generalized statement, sound enough within its proper field, but twisted to fit the speculative mania. In real-estate booms, the "reasoning" is usually based upon the inherent permanence and growth of land values. In the new-era bull market, the "rational" basis was the record of long-term improvement shown by diversified common-stock holdings.

A Sound Premise Used to Support an Unsound Conclusion.—There was, however, a radical fallacy involved in the new-era application of this historical fact. This should be apparent from even a superficial examination of the data contained in the small and rather sketchy volume from which the new-era theory may be said to have sprung. The book is entitled *Common Stocks as Long-Term Investments*, by Edgar Lawrence Smith, published in 1924. Common stocks were shown to have a tendency to increase in value with the years, for the simple reason that they earned more than they paid out in dividends, and thus the reinvested earnings added to their worth. In a representative case, the company would earn an average of 9%, pay 6% in dividends, and add 3% to surplus. With good management and reasonable luck the fair value of the stock would increase with its book value, at the annual rate of 3% *compounded*. This was, of course, a theoretical rather than a standard pattern; but the numerous instances of results poorer than "normal" might be offset by examples of more rapid growth.

The attractiveness of common stocks for the long pull thus lay essentially in the fact that they earned more than the bond-interest rate upon their cost. This would be true, typically, of a stock earning $10 and selling at 100. But as soon as the price was advanced to a much higher price in relation to earnings, this advantage disappeared, *and with it disappeared the entire theoretical basis for investment purchases of common stocks.* When investors paid $200 per share for a stock earning $10, they were buying an earning power no greater than the bond-interest rate, without the extra protection afforded by a prior claim. Hence in using the past performances of common stocks as the reason for paying prices 20 to 40 times their earnings, the new-era exponents were starting with a sound premise and twisting it into a woefully unsound conclusion.

In fact their rush to take advantage of the inherent attractiveness of common stocks itself produced conditions entirely different from those which had given rise to this attractiveness and upon which it basically depended, *viz.,* the fact that earnings had averaged some 10% on market price. As we have seen, Edgar Lawrence Smith plausibly explained the growth of common-stock values as arising from the building up of asset values through the reinvestment of surplus earnings. Paradoxically enough, the new-era theory which exploited this finding refused to accord the slightest importance to the asset values behind the stocks it favored. Furthermore, the validity of Mr. Smith's conclusions rested necessarily upon the assumption that common stocks could be counted on to behave in the future about as they had in the past. Yet the new-era theory threw out of account the past earnings of corporations except in so far as they were regarded as pointing to a *trend* for the future.

Examples Showing Emphasis on Trend of Earnings.—Take three companies with the following exhibits:

EARNINGS PER SHARE

Year	Company A (Electric Power & Light)	Company B (Bangor & Aroostook R.R.)	Company C (Chicago Yellow Cab)
1925	$1.01	$6.22	$5.52
1926	1.45	8.69	5.60
1927	2.09	8.41	4.54
1928	2.37	6.94	4.58
1929	2.98	8.30	4.47
5-year average.........	$1.98	$7.71	$4.94
High price, 1929.......	86⅝	90⅜	35

The 1929 high prices for these three companies show that the new-era attitude was enthusiastically favorable to Company *A*, unimpressed by Company *B*, and definitely hostile to Company *C*. The market considered Company *A* shares worth more than twice as much as Company *C* shares, although the latter earned 50% more per share than Company *A* in 1929 and its average earnings were 150% greater.

Average versus Trend of Earnings.—These relationships between price and earnings in 1929 show definitely that the past exhibit was no longer a measure of normal earning power but merely a weathervane to show which way the winds of profit were blowing. That the *average earnings* had ceased to be a dependable measure of future earnings must indeed be admitted, because of the greater instability of the typical business to which we have previously alluded. But it did not follow at all that the *trend of earnings* must therefore be a more dependable guide than the *average;* and even if it were more dependable it would not necessarily provide a safe basis, entirely by itself, for investment.

The accepted assumption that because earnings have moved in a certain direction for some years past they will continue to move in that direction, is fundamentally no different from the discarded assumption that because earnings averaged a certain amount in the past they will continue to average about that amount in the future. It may well be that the earnings *trend* offers a more dependable clue to the future than does the earnings average. But at best such an indication of future results is far from certain, and, more important still, there is no method of establishing a logical relationship between trend and price.[1] This means that the value placed upon a satisfactory trend must be wholly arbitrary, and hence speculative, and hence inevitably subject to exaggeration and later collapse.

Danger in Projecting Trends into the Future.—There are several reasons why we cannot be sure that a trend of profits shown in the past will continue in the future. In the broad economic sense, there is the law of diminishing returns and of increasing

---

[1] The new-era investment theory was conspicuously reticent on the mathematical side. The relationship between price and earnings, or price and trend of earnings was anything that the market pleased to make it (note the price of Electric Power and Light compared with its earnings record given on p. 313). If an attempt were to be made to give a mathematical expression to the underlying idea of valuation, it might be said that it was based on the *derivative* of the earnings, stated in terms of Time.

competition which must finally flatten out any sharply upward curve of growth. There is also the flow and ebb of the business cycle, from which the particular danger arises that the earnings curve will look most impressive on the very eve of a serious setback. Considering the 1927–1929 period we observe that since the trend-of-earnings theory was at bottom only a pretext to excuse rank speculation under the guise of "investment," the profit-mad public was quite willing to accept the flimsiest evidence of the existence of a favorable trend. Rising earnings for a period of five, or four, or even three years only, were regarded as an assurance of uninterrupted future growth and a warrant for projecting the curve of profits indefinitely upward.

Example: The prevalent heedlessness on this score was most evident in connection with the numerous common-stock flotations during this period. The craze for a showing of rising profits resulted in the promotion of many industrial enterprises which had been favored by temporary good fortune and were just approaching, or had already reached, the peak of their prosperity. A typical example of this practice is found in the offering of preferred and common stock of Schletter and Zander, Inc., a manufacturer of hosiery (name changed later to Signature Hosiery Company). The company was organized in 1929, to succeed a company organized in 1922, and the financing was effected by the sale of 44,810 shares of $3.50 convertible preferred shares at $50 a share and 261,349 voting-trust certificates for common stock at $26 per share. The offering circular presented the following exhibit of earnings from the constituent properties:

Year	Net after federal taxes	Per share of preferred	Per share of common
1925	$ 172,058	$ 3.84	$0.06
1926	339,920	7.58	0.70
1927	563,856	12.58	1.56
1928	1,021,308	22.79	3.31

The subsequent record was as follows:

Year	Net after federal taxes	Per share of preferred	Per share of common
1929	812,136	18.13	2.51
1930	*179,875(d)*	*4.01(d)*	*1.81(d)*

In 1931 liquidation of the company's assets was begun and a total of $17 per share in liquidating dividends on the preferred have been paid up to the end of 1933.   (Assets then remaining for liquidation were negligible.)

This example illustrates one of the paradoxes of financial history, *viz.*, that at the very period when the increasing instability of individual companies had made the purchase of common stocks far more precarious than before, the gospel of common stocks as safe and satisfactory investments was preached to and avidly accepted by the American public.

CHAPTER XXVIII

A PROPOSED CANON OF COMMON-STOCK INVESTMENT

Our extended discussion of the theory of common-stock investment has thus far led only to negative conclusions. The older approach, centering upon the conception of a stable average earning power, appears to have been vitiated by the increasing instability of the typical business. As for the new-era view, which turned upon the earnings trend as the sole criterion of value, whatever truth may lurk in this generalization, its blind adoption as a basis for common-stock purchases, without calculation or restraint, was certain to end in an appalling debacle. Is there anything at all left, then, of the idea of sound investment in common stocks?

A careful review of the preceding criticism will show that it need not be so destructive to the notion of investment in common stocks as a first impression would suggest. The instability of individual companies may conceivably be offset by means of thoroughgoing diversification. Moreover, the trend of earnings, while most dangerous as a *sole* basis for selection, may prove a useful *indication* of investment merit. If this approach is a sound one, there may be formulated an acceptable canon of common-stock investment, containing the following elements:

1. Investment is conceived as a *group* operation, in which diversification of risk is depended upon to yield a favorable average result.

2. The individual issues are selected by means of qualitative and quantitative tests corresponding to those employed in the choice of fixed-value investments.

3. A greater effort is made, than in the case of bond selection, to determine the future outlook of the issues considered.

Whether a policy of common-stock acquisition based upon the above principles deserves the title of investment, is undoubtedly open to debate. The importance of the question, and the lack of well-defined and authoritative views thereon, compel us to weigh here the leading arguments for and against this proposition.

Basic Conditions.—May the ownership of a carefully selected diversified group of common stocks, purchased at reasonable prices, be characterized as a sound investment policy? The affirmative answer depends on the assumption that certain basic and long-established elements in this country's economic experience may still be counted upon. These are: (1) that our national wealth and earning power will increase; (2) that such increase will reflect itself in the increased resources and profits of our important corporations, and (3) that such increases will in the main take place through the normal process of investment of new capital and reinvestment of undistributed earnings. The third assumption signifies that a broad causal connection exists between accumulating surplus and future earning power, so that common-stock selection is not a matter purely of chance or guesswork, but should be governed by an analysis of past records in relation to current market prices.

If these fundamental conditions still obtain, then common stocks with suitable exhibits should on the whole present the same favorable opportunities in the future as they have for generations past. The cardinal defect of instability may not be regarded, therefore, as menacing the long-range development of common stocks as a whole. It does indeed exert a powerful temporary effect upon all business through the variations of the economic cycle, and it has permanently adverse effects upon individual enterprises and single industries. But of these two dangers, the latter may be offset in part by careful selection and chiefly by wide diversification; the former may be guarded against by unvarying insistence upon the reasonableness of the price paid for each purchase.

Purchase Price Must Have Rational Basis.—This criterion of *reasonableness* is vital to all investment methods, and particularly to any theory of investing in common stocks. The absence of this controlling test constituted the fatal weakness of the new-era doctrine. The price paid for a common-stock investment must be justified by a conservative valuation based upon the past and current record. Paradoxically, this would be true even if it were admitted that the past record afforded no indication whatever of the future. For without these tests based on the past, no upper limit whatever could be set for stock values and investment must soon be transformed—as it was—into unbridled speculation. Hence even an illogical measure of value is better than no quantitative limits at all.

Past Record of Primary Importance.—But while we have just admitted for the sake of argument that a company's past record supplies no worthwhile guide to its future, such an extreme assumption is far from warranted. Behind the undoubted vicissitudes and instability of individual companies there remains the fact that by and large a good past record offers better promise for the future than does a poor one. If a hundred enterprises are taken which had average profits of $6 per share over the past ten years, and another hundred with average profits of only $1 per share, there is every reason to expect the first group to report larger aggregate profits than the second over the *next* ten years. The basic reason therefor is that future earnings are not determined entirely by luck or by competitive managerial skill, but that capital, experience, reputation, trade contacts, and all the other factors which contributed to past earning power, are bound to exert a considerable influence upon the future. This is certainly true if, as Edgar Lawrence Smith's studies have suggested, the growth of common-stock values as a whole is related chiefly to the upbuilding of *net worth* through the reinvestment of surplus earnings as well as the raising of new capital. It should also be pointed out that unless the past record has a bearing upon the future value of *stock* issues, it could have no proper bearing either upon the safety of a *bond* issue, and hence no logical basis could exist for bond investment.

These arguments merely assert that, by and large, strong companies are safer than weak ones, and that enterprises with good records are preferable to those with poor records. With every allowance for the surprises that Time holds in store, the validity of this generalization can scarcely be questioned. It is true that any single enterprise may rise from failure to success, or fall from success into failure; but on a *group* basis, the strongly entrenched companies are almost certain to fare better than those with poor past earnings and unsatisfactory balance-sheet positions.

Approximations to Insurance Principles and Practice.—The trend of the foregoing discussion leads to the conclusion that the principles of common-stock investment may be closely likened to the operations of insurance companies. Underwriters charge a premium which will compensate for the risk assumed in each case, measured by actuarial experience. At times considerable emphasis is placed upon the so-called "moral hazard," which involves consideration of other than factual data. Any individual risk may result in a loss far exceeding the premium received, but the average result is depended upon to yield a business profit. Similarly, the common-stock investor endeavors to obtain full value for each individual risk he assumes, basing this value upon the company's statistical exhibit. Corresponding to the underwriter's concern with the non-quantitative moral risk, the investor seeks to form a satisfactory judgment as to the future prospects of the enterprise considered. Finally, he will rely, as does the insurance company, upon diversification to average out the effects of unforeseeable future developments upon individual commitments.

Purchase of a Single Common Stock Not an Investment.—This conception of common-stock investment is similar to the pre-boom viewpoint, in its emphasis upon analysis of the past record to justify the price paid. It is influenced also by the new-era philosophy to the extent that it attaches more significance than formerly to the future indications, among which the earnings trend must of course be included. With respect to diversification, while this had always been regarded as a desirable element in investment, our formulation goes much farther in that it holds diversification to be an integral part of all standard common-stock-investment operations. In our view, the purchase of a single common stock can no more constitute an investment than the issuance of a single policy on a life or a building can properly constitute an insurance underwriting.[1]

Group Purchases May Constitute an Investment Operation.— Group purchases of carefully selected common stocks at attractive prices will very probably fall under our original definition of investment as given in Chap IV. ("An investment operation is one which, upon thorough analysis, promises safety of principal and a satisfactory return.") A satisfactory current dividend return should ordinarily be required for such an operation, but it is not an absolutely essential factor. If values exist far in excess of the price paid, and if they are being added to as time goes on, the investor may look to future dividends or even to future enhancement of market value, instead of to the current dividend

---

[1] An exception to this rule should be made in the comparatively rare cases when a common stock meets the safety requirements of a sound bond. Such an individual common-stock purchase would deserve the title of investment, since investment quality depends upon the *position* of the issue and not upon its contractual rights. This exceptional type of common-stock situation will be discussed in our later chapter on "The Significance of Current Asset Value."

rate, to justify his commitment from the standpoint of "satisfactory return."

Of more importance is the question whether such carefully selected group purchases actually do "promise safety of principal." This must remain partly a matter of the definition of "safety." We may avoid controversy on this point by defining common-stock investment in *subjective* terms. It would then embrace purchases made with the *intention* and the *reasonable expectation* of securing safety of principal. This conception would correspond with the familiar view of a common-stock "investor" as one interested in the intrinsic value of these issues, as distinguished from the "speculator" who is concerned only or chiefly with their market fluctuations. In the following chapters on Common-Stock Analysis we shall regularly use the term "investor" in this accepted sense.

Price an Integral Part of Every Investment Decision.—But the fact that the investor is interested in intrinsic values does not mean that market prices have no significance for him. An issue is attractive only if the indicated value amply justifies the price paid; hence the price is an integral part of any investment decision. This is true not only at the time of purchase but throughout the period of subsequent ownership. While the expectation may be to hold the issue indefinitely for income and enhancement in value, it will often prove desirable from the investment standpoint to dispose of it if it should cease to be attractive—either because its quality has deteriorated or because the price has risen to a level not justified by the demonstrable value. The common-stock investor must, therefore, necessarily concern himself with prices and price changes. He is unavoidably a participant in the stock market. In theory, of course, he need only follow the old and trite policy of buying when prices are low and selling out when prices are high. No advice could be easier to give or more difficult to follow. High prices and low prices are not marked with distinguishing signals, like red and green traffic lights. Not only are "high" and "low" always relative terms, but in Wall Street their meaning is mainly retrospective.

Disturbing Influence of Market Fluctuations.—The wider the fluctuations of the market, and the longer they persist in one direction, the more difficult it is to preserve the investment viewpoint in dealing with common stocks. The attention is bound to be diverted from the investment question, which is whether the price is attractive or unattractive in relation to value, to the speculative question whether the market is near its low or its high point.

This difficulty was so overshadowing in the years between 1927 and 1933 that common-stock investment virtually ceased to have any sound practical significance during that period. If an investor had sold out his common stocks early in 1927, because prices had outstripped values, he was almost certain to regret his action during the ensuing two years of further spectacular advances. Similarly those who hailed the crash of 1929 as an opportunity to buy common stocks at reasonable prices were to be confronted by appalling market losses as a result of the subsequent protracted decline.

It should be evident that a stock-market pattern such as that of 1927–1933 is fatal to the whole idea of rational investment in common stocks. On the other hand, the market history of Pennsylvania Railroad and Atchison during 1904–1913, and National Biscuit during 1909–1918, given on pages 301 to 303, shows fluctuations of far smaller amplitude. These would have permitted an investment policy which either ignored the fluctuations altogether, or else exploited them intelligently by buying at low levels and selling out when optimism was in the ascendency. It is for the future to tell whether the stock market will adhere to the pyrotechnic pattern of 1927–1933 or return to the lesser variations of the preceding years. Only in the latter event can common-stock investment establish itself on firm ground. Since we have repeatedly stressed the abnormal character of recent financial history, we are of course committed to the view that future markets will resemble those of 1904–1913 far more than those of 1927–1933. Hence we are not pessimistic as to the possibility of establishing a sound policy of common-stock investment.

The Danger of Speculative Contagion in Common-stock Investment.—We doubt, however, whether many individuals are qualified by nature to follow consistently such an investment policy without deviating into the primrose path of market speculation. The chief reason for this hazard is that the distinctions between common-stock investment and common-stock speculation are too intangible to hold human nature in check. When investment is confined to bonds (and high-grade preferred stocks) its attributes are well-defined and obvious. The bond buyer readily places himself in an entirely different category from the speculator in common stocks, whose losses and gains are alike remote from the former's interests. But when the investor employs the same medium as the speculator, the line of demarcation between one approach and the other is one of mental attitude only, and hence is relatively insecure. It is not likely to keep him immune from speculative contagion, especially when this is rampant in the very issues in which he has made his investment. Prior to 1926, a fairly definite separation could be made between investment common stocks and speculative common stocks. The former fluctuated over a much narrower range percentage-wise, since their prices were determined largely by their established dividend rate. This is shown strikingly by the market record of Atchison between 1916 and 1925, given below.

ATCHISON, TOPEKA AND SANTA FE RAILWAY COMPANY

Year	Range for common stock	Earned per share	Paid per share
1916	1u9–100	$14.74	$6
1917	108– 75	14.50	6
1918	100– 81	10.59*	6
1919	104– 81	15.41*	6
1920	90– 76	12.54*	6
1921	94– 76	14.69†	6
1922	109– 92	12.41	6
1923	105– 94	15.48	6
1924	121– 97	15.47	6
1925	141–116	17.19	7

* Results for these years based on actual operations. Results of federal operation were: 1918—$9.98; 1919—$16.55; 1920—$13.98.

† Includes nonrecurrent income. Excluding the latter the figure for 1921 would have been $11.29.

Hence the issues which the common-stock investor dealt in served to set him apart from the speculative public and make it easier for him to maintain his conservative viewpoint. The new era was marked by a concentration of *speculative* interest on those issues which had formerly deserved an *investment rating*. This made for an extraordinary confusion in the mental processes of the entire financial community, and the straightening out of this confusion may be a matter of many years. In consequence it is likely to be far more difficult in the future than it was prior to 1927 for the typical common-stock investor to hold aloof from speculative influences.

Investment Trusts May Solve This Problem.—A possible solution of this vexing problem may come through a sounder development of the investment-trust idea. It is true that to date the American investment trust has proved a melancholy fiasco. But this has been due to the insanity of the period in which they were created here, and not to intrinsic defects in the previously established theory of investment-trust procedure. This established procedure is in fact ideally suited to the canons of common-stock investment as developed in this chapter. As compared with the individual, an investment trust has better facilities for statistical analysis and investigation, larger funds with which to achieve suitable diversification, and presumably superior training and judgment on the part of its management. Most important of all, perhaps, is the fact that an investment trust may be compelled by its charter or other contractual engagements to follow unremittingly conservative policies of common-stock purchase and disposal. Even when the sirens of speculation prove too alluring for mere human nature to withstand, the investment-trust managers may find their salvation, Ulysses-like, in the fetters of a legal obligation.

Analytical Technique Essential to Intelligent Common-stock Investment.—Our survey of common-stock investment has led us to a very tentative and qualified endorsement of such a program. But the popular interest in common stocks is certain to be far less tentative and qualified. Since a successful policy of common-stock investment will yield much greater rewards than come from the purchase of sound bonds, a large portion of the public will constantly venture into this field. They will be actuated by a legitimate desire for profit, and by a natural but frequently excessive optimism and self-confidence. They will properly seek to arm themselves with an adequate knowledge of financial practice and with the tools and technique necessary for an intelligent analysis of corporate statements.

Such information and equipment for the common-stock investor form the subject-matter of the following chapters.

CHAPTER XXIX

THE DIVIDEND FACTOR IN COMMON-STOCK ANALYSIS

A natural classification of the elements entering into the valuation of a common stock would be under the three headings:

1. The dividend rate and record.
2. Income-account factors (earning power).
3. Balance-sheet factors (asset value).

The dividend rate is a simple fact and requires no analysis; but its exact significance is exceedingly difficult to appraise. From one point of view the dividend rate is all-important; but from another and equally valid standpoint it must be considered an accidental and minor factor. A basic confusion has grown up in the minds of managements and stockholders alike as to what constitutes a proper dividend policy. The result has been to create a definite conflict between two aspects of common-stock ownership: the one being the possession of a marketable security, and the other being the assumption of a partnership interest in a business. Let us consider the matter in detail from this twofold approach.

Dividend Return as a Factor in Common-stock Investment.— Until recent years the dividend return was the overshadowing factor in common-stock investment. This point of view was based on simple logic. The prime purpose of a business corporation is to pay dividends to its owners. A successful company is one which can pay dividends regularly and presumably increase the rate as time goes on. Since the idea of investment is closely bound up with that of dependable income, it follows that investment in common stocks would ordinarily be confined to those with a well-established dividend. It would follow also that the *price* paid for an investment common stock would be determined chiefly by the amount of the dividend.

We have seen that the traditional common-stock investor sought to place himself as nearly as possible in the position of an investor in a bond or a preferred stock. He aimed primarily at a steady income return which in general would be both somewhat larger and somewhat less certain than that provided by good senior securities. Excellent illustrations of the effect of this attitude upon the price of common stocks are afforded by the records of the earnings, dividends, and annual price variations of American Sugar Refining between 1907 and 1913, presented herewith, and of the Atchison, Topeka and Santa Fe Railway between 1916 and 1925, which were given on page 323.

AMERICAN SUGAR REFINING COMPANY

Year	Range for stock	Earned per share	Paid per share
1907	138– 93	$10.22	$7.00
1908	138– 99	7.45	7.00
1909	136–115	14.20	7.00
1910	128–112	5.38	7.00
1911	123–113	18.92	7.00
1912	134–114	5.34	7.00
1913	118–100	0.02(d)	7.00

The market range of both issues is surprisingly narrow, considering the continuous gyrations of the stock market generally during that period. The most striking feature of the exhibit is the slight influence exercised both by the irregular earnings of American Sugar and by the exceptionally well-maintained and increasing earning power on the part of Atchison. It is clear that the price of American Sugar was dominated throughout by its $7 rate and that of Atchison by its $6 rate, even though the earnings records would apparently have justified an entirely different range of relative market values.

Established Principle of Withholding Dividends.—We have, therefore, on the one hand an ingrained and powerfully motivated tradition which centers investment interest upon the present and past dividend rate. But on the other hand we have an equally authoritative and well-established principle of *corporate management* which subordinates the current dividend to the future welfare of the company and its shareholders. It is considered proper managerial policy to withhold current earnings from stockholders, for the sake of any of the following advantages:

1. To strengthen the financial (working-capital) position.
2. To increase productive capacity.
3. To eliminate an original overcapitalization.

When a management withholds and reinvests profits, thus building up an accumulated surplus, it claims confidently to be acting for the best interests of the shareholders. For by this policy the continuance of the established dividend rate is undoubtedly better assured, and furthermore a gradual but continuous increase in the regular payment is thereby made possible. The rank and file of stockholders will give such policies their support, either because they are individually convinced that this procedure redounds to their advantage, or because they accept uncritically the authority of the managements and bankers who recommend it.

But this approval by stockholders of what is called a "conservative dividend policy" has about it a peculiar element of the perfunctory and even the reluctant. The typical investor would most certainly prefer to have his dividend today and let tomorrow take care of itself. No instances are on record in which the withholding of dividends for the sake of future profits has been hailed with such enthusiasm as to advance the price of the stock. The direct opposite has invariably been true. *Given two companies in the same general position and with the same earning power, the one paying the larger dividend will always sell at the higher price.*

Policy of Withholding Dividends Questionable.—This is an arresting fact, and it should serve to call into question the traditional theory of corporate finance that the smaller the percentage of earnings paid out in dividends, the better for the company and its stockholders. While investors have been taught to pay lip service to this theory, their instincts—and perhaps their better judgment—are in revolt against it. If we try to bring a fresh and critical viewpoint to bear upon this subject, we shall find that weighty objections may be leveled against the accepted dividend policy of American corporations.

Examining this policy more closely, we see that it rests upon two quite distinct assumptions. The first is that it is advantageous to the stockholders to leave a substantial part of the annual earnings in the business; the second is that it is desirable to maintain a steady dividend rate in the face of fluctuations in profits. As to the second point, there would be no question at all, provided the dividend *stability* is achieved without too great sacrifice in the *amount* of the dividend. Assume that the earnings vary between $5 and $15 annually over a period of years,

averaging $10.   No doubt the stockholder's advantage would be
best served by maintaining a stable dividend rate of $8, sometimes
drawing upon the surplus to maintain it, but on the average
increasing the surplus at the rate of $2 per share annually.

This would be an ideal arrangement.   But in practice it is
rarely followed.   We find that stability of dividends is usually
accomplished by the simple expedient of paying out a *small part*
of the average earnings.   By a *reductio ad absurdum* it is clear
that any company which earned $10 per share on the average
could readily stabilize its dividend at $1.   The question arises
very properly whether the shareholders might not prefer a much
larger aggregate dividend, even with some irregularity.   This
point is well illustrated in the case of Atchison.

The Case of Atchison.—Atchison maintained its dividend at
the annual rate of $6 for the 15 years between 1910 and 1924.
During this time the average earnings were in excess of $12 per
share, so that the stability was attained by withholding over
half the earnings from the stockholders.   Eventually this policy
bore fruit in an advance of the dividend to $10, which rate was
paid between 1927 and 1931, and was accompanied by a rise
in the market price to nearly $300 per share in 1929.   Less than
a year after the last payment at the $10 rate (in December 1931)
the dividend was omitted entirely (in June 1932).   Viewed
critically, the stability of the Atchison dividend between 1910
and 1924 must be considered as of dubious benefit to the stockholders.   During its continuance they received an unduly small
return in relation to the earnings; when the rate was finally
advanced, the importance attached to such a move promoted
excessive speculation in the shares; finally, the reinvestment of
the enormous sums out of earnings failed to protect the shareholders from a complete loss of income in 1932.   Allowance
must be made, of course, for the unprecedented character of
the depression in 1932.   But the fact remains that the actual
operating losses in dollars per share up to the passing of the
dividend were entirely insignificant in comparison with the
surplus accumulated out of the profits of previous years.

United States Steel, Another Example.—The Atchison case
illustrates the two major objections to what is characterized
and generally approved of as a "conservative dividend policy."
The first objection is that stockholders receive *both currently
and ultimately* too low a return in relation to the earnings of their

property; the second is that the "saving up of profits for a rainy day" often fails to safeguard even the moderate dividend rate when the rainy day actually arrives. A similarly striking example of the ineffectiveness of a large accumulated surplus is shown by the country's leading industrial, United States Steel.

The following figures tell a remarkable story:

Profits available for the common stock,
    1901–1930 . . . . $2,344,000,000
Dividends paid:
    Cash   891,000,000
    Stock   203,000,000
Undistributed earnings   1,250,000,000
Loss after preferred dividends Jan. 1, 1931–
    June 30, 1932   59,000,000
Common dividend passed June 30, 1932.

A year and a half of declining business was sufficient to outweigh the beneficial influence of 30 years of continuous reinvestment of profits.

The Merits of "Plowing-back" Earnings.—These examples serve to direct our critical attention to the other assumption on which American dividend policies are based, *viz.*, that it is advantageous to the stockholders if a large portion of the annual earnings are retained in the business. This may well be true; but in determining its truth a number of factors must be considered which are usually left out of account. The customary reasoning on this point may be stated in the form of a syllogism, as follows:

Major premise—Whatever benefits the company benefits the stockholders.

Minor premise—A company is benefited if its earnings are retained rather than paid out in dividends.

Conclusion—Stockholders are benefited by the withholding of corporate earnings.

The weakness of the above reasoning rests oт course in the major premise. Whatever benefits a business benefits its owners, *provided* the benefit is not conferred upon the corporation at the *expense* of the stockholders. Taking money away from the stockholders and presenting it to the company will undoubtedly strengthen the enterprise, but whether it is to the owners' advantage is an entirely different question. It is customary to commend managements for "plowing earnings back into the property"; but in measuring the benefits from such a policy, the time element is usually left out of account. It stands to reason that if a business paid out only a small part of its earnings in dividends, the value of the stock should increase over a period of years; but it is by no means so certain that this increase will compensate the stockholders for the dividends withheld from them, *particularly if interest on these amounts is compounded.*

An inductive study would undoubtedly show that the earning power of corporations does not in general expand proportionately with increases in accumulated surplus. *Assuming that the reported earnings were actually available for distribution,* then stockholders in general would certainly fare better in dollars and cents if they drew out practically all of these earnings in dividends. An unconscious realization of this fact has much to do with the tendency of common stocks paying liberal dividends to sell higher than others with the same earning power but paying out only a small part thereof.

Dividend Policies Arbitrary and Sometimes Selfishly Determined.—One of the obstacles in the way of an intelligent understanding by stockholders of the dividend question is the accepted notion that the determination of dividend policies is entirely a managerial function, in the same way as the general running of the business. This is legally true, and the courts will not interfere with the dividend action or inaction except upon an exceedingly convincing showing of unfairness. But if stockholders' opinion were properly informed, it would insist upon curtailing the despotic powers given the directorate over the dividend policy. Experience shows that these unrestricted powers are likely to be abused, and for various reasons. Boards of directors usually consist largely of executive officers and their friends. The officers are naturally desirous of retaining as much cash as possible in the treasury, in order to simplify their financial problems; they are also inclined to expand the business persistently for the sake of personal aggrandizement and to secure higher salaries. This is a leading cause of the unwise increase of manufacturing facilities which has proved recurrently one of the chief unsettling factors in our economic situation.

The discretionary power over the dividend policy may also be abused in more sinister fashion, sometimes to permit the acquisition of shares at an unduly low price, at other times to facilitate unloading at a high quotation. The heavy surtaxes imposed upon large incomes frequently make it undesirable from the standpoint of the large stockholders that earnings be paid out in dividends. Hence dividend policies may be determined at times from the standpoint of the taxable status of the large stockholders who control the directorate. This is particularly true in cases where these dominant stockholders receive substantial salaries as executives. In such cases they are perfectly willing to leave their share of the earnings in the corporate treasury, since the latter is under their control and since by so doing they retain control over the earnings accruing to the other stockholders as well.

Arbitrary Control of Dividend Policy Complicates Analysis of Common Stocks.—Viewing American corporate dividend policies as a whole, it cannot be said that the virtually unlimited power given the management on this score has redounded to the benefit of the stockholders. In entirely too many cases the right to pay out or withhold earnings at will is exercised in an unintelligent or inequitable manner. Dividend policies are often so arbitrarily managed as to introduce an additional uncertainty in the analysis of a common stock. Besides the difficulty of judging the earning power, there is the second difficulty of predicting what part of the earnings the directors will see fit to disburse in dividends.

It is important to note that this feature is peculiar to American corporate finance and has no close counterpart in the other important countries. The typical English, French, or German company pays out practically all the earnings of each year, except those carried to reserves.[1] Hence they do not build up large profit-and-loss surpluses, such as are common in the United States. Capital for expansion purposes is provided abroad not out of undistributed earnings but through the sale of additional stock. To some extent, perhaps, the reserve accounts shown in foreign balance sheets will serve the same purpose as an American surplus account, but these reserve accounts rarely attain a comparable magnitude.

Plowing Back Due to Watered Stock.—The American theory of "plowing back" earnings appears to have grown out of the stock-watering practices of prewar days. Many of our large industrial companies made their initial appearance with no tangible assets behind their common shares and with inadequate protection for their preferred issues. Hence it was natural that the management should seek to make good these deficiencies

[1] See Appendix, Note 41, for discussion and examples.

out of subsequent earnings.   This was particularly true because
additional stock could not be sold at its par value, and it was
difficult therefore to obtain new capital for expansion except
through undistributed profits.[1]

Examples: Concrete examples of the relation between overcapitalization and dividend policies are afforded by the outstanding cases of Woolworth and United States Steel Corporation.

In the original sale of F. W. Woolworth Company shares to the
public, made in 1911, the company issued preferred stock to
represent all the tangible assets and common stock to represent
the good-will.   The balance sheet accordingly carried a good-will
item of $50,000,000 among the assets, offsetting a corresponding
liability for 500,000 shares of common, par $100.[2]   As Wool-
worth prospered, a large surplus was built up out of earnings and
amounts were charged against this surplus to reduce the good-will
account, until finally it was written down to $1.[3]

In the case of United States Steel Corporation, the original
capitalization exceeded tangible assets by an amount not less
than the common stock at par, *viz.*, $508,000,000.   This "water"
in the balance sheet was not shown as a good-will item, as in the
case of Woolworth, but was concealed by an overvaluation of the
fixed assets (*i.e.*, of the "Property Investment Accounts").
Through various accounting methods, however, the management
applied earnings from operations to the writing off of these
intangible or fictitious assets.   By the end of 1929 a total of
$508,000,000—equal to the entire original common-stock issue—
had been taken from earnings or surplus and deducted from the
property account.

Some of the accounting policies above referred to will be discussed again, with respect to their influence on investment values,
in our chapters on Analysis of the Income Account and Balance-
sheet Analysis.   From the dividend standpoint it is clear that in
both of these examples the decision to retain large amounts of

---

[1] The no-par-value device is largely a postwar development.

[2] This was for many years a standard scheme for financing of industrial
companies.   It was followed by Sears Roebuck, Cluett Peabody, National
Cloak and Suit, and others.

[3] It should be noted that when the good-will of Woolworth was originally
listed in the balance sheet at $50,000,000, its actual value (as measured by
the market price of the shares) was only some $20,000,000.   But when the
good-will was written down to $1, in 1925, its real value was apparently
many times $50,000,000.

earnings, instead of paying them out to the stockholders, was due in part to the desire to eliminate intangible items from the asset accounts.

Conclusions from the Foregoing.—From the foregoing discussion certain conclusions may be drawn. These bear, first, on the very practical question of what significance should be accorded the dividend rate as compared with the reported earnings; and, secondly, upon the more theoretical but exceedingly important question of what dividend policies should be considered as most desirable from the standpoint of the stockholders' interest.

Experience would confirm the established verdict of the stock market that a dollar of earnings is worth more to the stockholder if paid him in dividends than when carried to surplus. The common-stock investor should ordinarily require both an adequate earning power and an adequate dividend. If the dividend is disproportionately small, an investment purchase will be justified only on an exceptionally impressive showing of earnings (or by a very special situation with respect to liquid assets). On the other hand, of course, an extra-liberal dividend policy cannot compensate for inadequate earnings, since with such a showing the dividend rate must necessarily be undependable.

To aid in developing these ideas quantitatively, we submit the following definitions:

The *dividend rate* is the amount of annual dividends paid per share, expressed either in dollars or as a percentage of a $100 par value. (If the par value is less than $100, it is inadvisable to refer to the dividend rate as a percentage figure since this may lead to confusion.)

The *earnings rate* is the amount of annual earnings per share, expressed either in dollars or as a percentage of a $100 par value.

The *dividend ratio, dividend return* or *dividend yield,* is the ratio of the dividend paid to the market price (*e.g.,* a stock paying $6 annually and selling at 120 has a dividend ratio of 5%).

The *earnings ratio, earnings return* or *earnings yield* is the ratio of the annual earnings to the market price (*e.g.,* a stock earning $6 and selling at 50 shows an earnings yield of 12%).[1]

Let us assume that a common stock, *A*, earning $10 and paying $7 should sell at 100. This is a 10% earnings ratio and a 7%

---

[1] The term *earnings basis* has the same meaning as *earnings ratio.* However, the term *dividend basis* is ambiguous, since it is used sometimes to denote the rate and sometimes the ratio.

dividend return.   Then a similar common stock, *B*, earning $10 but paying only $6, should sell lower than 100.   Its price evidently should be somewhere between 85.71 (representing a 7% dividend yield) and 100 (representing a 10% earnings yield). In general the price should tend to be established nearer to the lower limit than to the upper limit.   A fair approximation of the proper relative price would be about 90, at which level the dividend yield is 6⅔% and the earnings ratio is 11.1%.   If the investor makes a small concession in dividend yield below the standard, he is entitled to demand a more than corresponding increase in the earning power above standard.

In the opposite case a similar stock, *C*, may earn $10 but pay $8.   Here the investor is justified in paying some premium above 100 because of the larger dividend.   The upper limit, of course, would be 114³⁄₇ at which price the dividend ratio would be the standard 7%, but the earnings ratio would be only 8¾%. Here again the proper price should be closer to the lower than to the upper limit, say 105, at which figure the dividend yield would be 7.62% and the earnings ratio 9.52%.

Suggested Principle for Dividend Payments.—While these figures are arbitrarily taken, they correspond fairly well with the actualities of investment values under reasonably normal conditions in the stock market.   The dividend *rate* is seen to be important, apart from the earnings, not only because the investor naturally wants cash income from his capital, but also because the earnings which are *not* paid out in dividends have a tendency to lose part of their effective value for the stockholder. Because of this fact American shareholders would do well to adopt a different attitude than is now prevalent with respect to corporate dividend policies.   We should suggest the following principle as a desirable modification of the traditional viewpoint:

Principle: Stockholders are entitled to receive the earnings on their capital except to the extent they decide to reinvest them in the business.   The management should retain or reinvest earnings only with the specific approval of the stockholders. Such "earnings" as must be retained to protect the company's position are not true earnings at all.   They should not be reported as profits but should be deducted in the income statement as necessary reserves, with an adequate explanation thereof.   *A compulsory surplus is an imaginary surplus.*

Were this principle to be generally accepted, the withholding of earnings would not be taken as a matter of course and of arbitrary determination by the management, but it would require justification corresponding to that now expected in the case of changes in capitalization and in the sale of additional stock. The result would be to subject dividend policies to greater scrutiny and more intelligent criticism than they now receive, thus imposing a salutary check upon the tendency of managements to expand unwisely and to accumulate excessive working capital.[1]

If it should become the standard policy to disburse the major portion of each year's earnings (as is done abroad), then the rate of dividend will vary with business conditions. This would apparently introduce an added factor of instability into stock values. But the objection to the present practice is that it *fails* to produce the stable dividend rate which is its avowed purpose and the justification for the sacrifice it imposes. Hence instead of a dependable dividend which mitigates the uncertainty of earnings we have a frequently arbitrary and unaccountable dividend policy which aggravates the earnings hazard. The sensible remedy would be to transfer to the stockholder the task of averaging out his own annual income return. Since the common-stock investor must form some fairly satisfactory opinion of average earning power, which transcends the annual fluctuations, he may as readily accustom himself to forming a similar idea of average *income*. As in fact the two ideas are substantially identical, dividend fluctuations *of this kind* would not make matters more difficult for the common-stock investor. In the end such fluctuations will work out more to his advantage than the present method of attempting, usually unsuccessfully, to stabilize the dividend by large additions to the surplus account.

[1] The suggested procedure under the British Companies Act of 1929 requires that dividend payments be approved by the shareholders at their annual meeting, but prohibits the approval of a rate greater than that recommended by the directors. Despite the latter proviso, the mere fact that the dividend policy is submitted to the stockholders for their specific approval or criticism carries an exceedingly valuable reminder to the management of its responsibilities, and to the owners of their rights, on this important question.

Although this procedure is not required by the Companies Act in all cases, it is generally followed in England. See Companies Act of 1929, Sections 6–10; Table A to the Companies Act of 1929, pars. 89–93; *Palmer's Company Law*, pp. 222–224, 13th ed., 1929.

On the former basis, the stockholder's average income would probably be considerably larger.

A Paradox.—While we have concluded that the payment of a liberal portion of the earnings in dividends adds definitely to the attractiveness of a common stock, it must be recognized that this conclusion involves a curious paradox. Value is increased by taking away value. The more the stockholder subtracts in dividends from the capital and surplus fund, the larger value he places upon what is left. It is like the famous legend of the Sybilline Books, except that here the price of the remainder is *increased* because part has been taken away.

This point is well illustrated by a comparison of Atchison and Union Pacific—two railroads of similar standing—over the ten-year period between January 1, 1915 and December 31, 1924.

Item	Per share of common	
	Union Pacific	Atchison
Earned, 10 years 1915–1924................	$142.00	$137
Net adjustments in surplus account.........	dr.　1.50*	cr.　13
Total available for stockholders.............	$140.50	$150
Dividends paid...........................	$ 97.50	$ 60
Increase in market price..................	33.00	25
Total realizable by stockholders.............	$130.50	$ 85
Increase in earnings, 1924 over 1914.........	9%†	109%†
Increase in book value, 1924 over 1914......	25%	70%
Increase in dividend rate, 1924 over 1914.....	25%	none
Increase in market price, 1924 over 1914.....	28%	27%
Market price, Dec. 31, 1914................	116	93
Market price, Dec. 31, 1924................	149	118
Earnings, year ended June 30, 1914.........	$ 13.10	$ 7.40
Earnings, calendar year 1924...............	14.30	15.45

* Excluding about $7 per share transferred from reserves to surplus.
† Calendar year 1924 compared with year ended June 30, 1914.

It is to be noted that because Atchison failed to increase its dividend the market price of the shares failed to reflect adequately the large increase both in earning power and in book value. The more liberal dividend policy of Union Pacific produced the opposite result.

This anomaly of the stock market is explained in good part by the underlying conflict of the two prevailing ideas regarding dividends which we have discussed in this chapter. In the following brief summary of the situation we endeavor to indicate the relation between the theoretical and the practical aspects of the dividend question.

Summary.—1. In some cases the stockholders derive positive benefits from an ultraconservative dividend policy, *i.e.*, through much larger eventual earnings and dividends. In such instances the market's judgment proves to be wrong in penalizing the shares because of their small dividend. The price of these shares should be higher rather than lower on account of the fact that profits have been added to surplus instead of having been paid out in dividends.

2. Far more frequently, however, the stockholders derive much greater benefits from dividend payments than from additions to surplus. This happens because either: (a) the reinvested profits fail to add proportionately to the earning power, or (b) they are not true "profits" at all but reserves which *had* to be retained merely to protect the business. In this majority of cases the market's disposition to emphasize the dividend and to ignore the additions to surplus turns out to be sound.

3. The confusion of thought arises from the fact that the stockholder votes in accordance with the first premise and invests on the basis of the second. If the stockholders asserted themselves intelligently, this paradox would tend to disappear. For in that case the withholding of a large percentage of the earnings would become an exceptional practice, subject to close scrutiny by the stockholders and presumably approved by them from a considered conviction that such retention would be beneficial to the owners of the shares. Such a ceremonious endorsement of a low dividend rate would probably and properly dispel the stock market's scepticism on this point and permit the price to reflect the earnings which are accumulating as well as those which were paid out.

The foregoing discussion may appear to conflict with the suggestion, advanced in the previous chapter, that long-term increases in common-stock values are often due to the reinvestment of undistributed profits. We must distinguish here between the two lines of argument. Taking our standard case of a company earning $10 per share and paying dividends of $7, we have pointed out that the repeated annual additions of $3 per share to surplus should serve to increase the value of the stock over a period of years. This may very well be true, and at the same time the rate of increase in value may be substantially less than 3% per annum compounded. If we take the reverse case, *viz.*, $3 paid in dividends and $7 added to surplus, the distinction is clearer. Undoubtedly the large addition to surplus will expand the value of the stock, but quite probably also this value will fail to increase at the annual rate of 7% compounded. Hence the argument against reinvesting large proportions of the yearly earnings would remain perfectly valid. Our criticism is advanced against the latter type of policy, *e.g.*, the retention of 70% of the earnings, and not against the normal reinvestment of some 30% of the profits.

CHAPTER XXX

STOCK DIVIDENDS

Distributions made in the form of stock instead of cash are of two kinds, which may be called *extraordinary* and *periodic*. An extraordinary stock dividend may be defined as one which capitalizes part of the accumulated surplus of past years, *i.e.*, it transfers a substantial amount from the accumulated surplus to stated capital and gives the stockholders additional shares to represent the funds thus transferred.

A periodic stock dividend may be defined as one which capitalizes part of only the *current year's* earnings. Hence it is almost always of relatively small size. It is called periodic because such dividends are usually repeated over a number of years in accordance with an established policy.

EXTRAORDINARY STOCK DIVIDENDS

Extraordinary stock dividends are legal and legitimate, but by and large they produce unfortunate effects. The only reason for such a dividend which is at once sound and practical is that it will adjust the market price of the shares to a more convenient level. Widespread public interest and an active market are desirable attributes of a common stock, and these are diminished when the normal price range has advanced to such a high figure as, say, $300 or $400 per share. Hence an increase in the number of shares and the reduction in value of each share, by means of a large stock dividend, would be a logical step to take.

Example: In 1917 Bethlehem Steel stock was selling above $500 per share. A stock dividend of 200% was paid (and additional shares were sold at par) bringing the market price down to about 150.

Split-ups.—Exactly the same result may be obtained by reducing the par value of the shares, such a move being referred to familiarly as a "split-up." During the recent bull market, reductions in par value were much more frequent than large stock dividends on stocks with par value because the rise in market price had so far outstripped the accumulated surplus

339

that a distribution of the latter would have been insufficient for the purpose.

Example: In 1926 General Electric stock was selling at 360. Four new shares of no-par value were given for each old share of $100 par value, thus reducing the market price to about 90. To have effected the same result by a 300% stock dividend would have required the transfer of 540 millions from surplus to capital, but the surplus was then only 100 millions. A similar situation existed in 1930 when General Electric shares were again split four for one.

In the case of Woolworth, the original common issue of 500,000 shares was increased to 9,750,000 shares by the following steps, involving both stock dividends and split-ups.

	Total Shares Outstanding
1920: Stock dividend of 30%, reducing the price from about 140 to about 110.	650,000
1924: Par value cut from $100 to $25, reducing the price from about 320 to about 80.	2,600,000
1927: Stock dividend of 50%, reducing the price from about 180 to about 120.	3,900,000
1929: Par value cut from $25 to $10, reducing the price from about 225 to about 90.	9,750,000

American Can combined both devices at one time in 1926. It reduced the par value from $100 to $25 and also paid a stock dividend of 50%. Hence six shares were issued for one, and the price was reduced from about 300 to about 50.

Stock Splits and Stock Dividends in No-par Stock.—In the case of common stocks of no-par value, a split-up or a stock dividend leads to exactly the same results, and to all practical purposes they are indistinguishable. While a stock dividend requires the transfer of a certain sum on the books from surplus to capital, the infinite latitude in accounting permitted by no-par stock may make this transfer a purely nominal affair.

Examples: Central States Electric Corporation paid a 900% stock dividend in 1926, increasing the number of shares (no par) from 109,000 to 1,090,000. The old stock had a book value of about $44 per share at the end of 1925, but the new stock was charged against surplus at the rate of only $1 per share.

Similarly in 1929, the Coca-Cola Company paid a 100% stock dividend in Class *A* stock without par value. This stock was booked at $5 per share (lower than the stated value of the common) despite the fact that the Class *A* stock has all of the characteristics of a $50-par, 6% preferred issue, except formal designation of such a par figure. (See also the accounting by this company of its 100% dividend payable in common stock in 1927, and also our discussion of its treatment of repurchases of Class *A* shares in Chap. XLII.)

Objections to Extraordinary Stock Dividends and Split-ups.— Extraordinary stock dividends and stock split-ups are both open to the serious objection that their declaration exercises an undue influence upon market prices, and hence that they afford an avenue for manipulation and for unfair profits by insiders. It is obvious that in theory a large stock dividend gives the stockholder nothing that he did not own before. His two pieces of paper now represent the same ownership formerly expressed by one piece of paper. This reasoning led the United States Supreme Court to decide that stock dividends are not income and consequently not subject to income tax. In practice, however, a stock dividend may readily be given exceptional speculative importance. For stock speculation is largely a matter of *A* trying to decide what *B*, *C*, and *D* are likely to think—with *B*, *C*, and *D* trying to do the same. Hence a stock dividend, even if it has no real significance of any kind, can and does serve as a stimulus to that *mutual attempt at taking advantage of each other* which often lies at the bottom of speculators' activities.

An Investment Element of Importance.—The essentially illusive character of large stock dividends would be more evident were it not for the fact that an investment element of real importance may also enter into the picture. The payment of an extraordinary stock dividend is usually the forerunner of an increase in the regular *cash* dividend rate. Since investors are legitimately interested in the cash dividend, they must necessarily be interested also in any stock dividend, for this may have a bearing upon the probable cash dividend. This serves to confuse the issue and to make less obtrusive the purely manipulative aspects of stock-dividend declarations.

The dividend history of a successful industrial corporation frequently discloses the following sequence:

1. A protracted period of small dividends in relation to earnings, with the upbuilding of a huge surplus.

2. The sudden payment of a large stock dividend.
3. An immediate increase in the regular cash dividend payments.[1]

No policy could be more conducive to the confusion of investment and speculative attitudes, or lend itself more easily to the taking of unfair advantage by those in control.

PERIODIC STOCK DIVIDENDS

This policy represents a great advance in basic soundness over the haphazard and often inequitable practices which we have been discussing. Such practices involve first the large accumulation of undistributed earnings in the surplus account and secondly the ultimate capitalization thereof through stock dividends at arbitrary times and in arbitrary amounts. Assuming that in many cases it may be desirable to retain a good part of each year's earnings in the business, then the interests of the stockholders would be best served by giving them currently a tangible evidence of their ownership of these reinvested profits.

If an enterprise regularly earns $12 per share and pays out only $5 in cash, the stockholders would benefit greatly by receiving *each year* a stock dividend representing a good part of the $7 added to their company's resources. In theory, of course, the additional stock certificate gives him nothing that he would not own without it; in other words, without a stock dividend his old certificate would still fully represent the ownership of the added $7 per share. But in actuality the payment of periodic stock dividend produces important advantages. Among them are the following:

1. The stockholder can sell the stock-dividend certificate, so that at his option he can have either cash or more stock to represent the reinvested earnings. Without a stock dividend he might in theory accomplish the same end by selling a small part of the shares represented by his old certificate, but in practice this is difficult to calculate and inconvenient in execution.

[1] For example, American Can issued six shares for one in 1926 through a four-for-one split and a 50% stock dividend. The dividend rate was $7 per share on the old stock, but a $2 rate was immediately inaugurated on the new stock, which was equivalent to $12 a share on the old. The rate on the new stock was stepped up to $5 a share in 1929. Likewise National Biscuit paid a $7 dividend annually from 1912 through 1922, although it earned substantially in excess of that figure. The stock was split seven for one in 1922 through issuing four new shares for each old share, followed by a 75% stock dividend. Dividends on the new shares were inaugurated at $3 a share, equivalent to $21 a share on the old.

2. He is likely to receive larger cash dividends as a result of such a policy, because the established cash rate will usually be continued on the increased number of shares. For example, if a company earning $12 pays out $5 in cash and 5% in stock, in the next year it will most probably pay $5 in cash on the new capitalization, equivalent to $5.25 on the previous holdings. Without the stock dividend, it would probably continue the $5 rate unchanged.[1]

3. By adding the reinvested profits to the stated capital (instead of to surplus) the management is placed under a direct obligation to earn money and pay dividends on these added resources. No such accountability exists with respect to the profit and loss surplus. The stock-dividend procedure will serve not only as a challenge to the efficiency of the management but also as a proper test of the wisdom of reinvesting the sums involved.

4. Issues paying periodic stock dividends enjoy a higher market value than similar common stock not paying such dividends.

Variations in the Practice of Periodic Stock-dividend Payment. The practice of disbursing periodic stock dividends developed fairly rapidly from about 1923 until the subsequent depression. Three variations of the idea were resorted to:

1. The standard method was to pay a stock dividend in addition to the regular cash dividend. These stock dividends were paid either monthly,[2] quarterly,[3] semiannually,[4] or annually.[5]

2. Sometimes a periodic stock dividend was offered in *lieu* of the regular cash dividend. This took the form of an option

---

[1] For examples of this sequence see: Cities Service Company, which paid 6% in cash and 6% in stock between March 1, 1925 and June 1, 1932; Sears, Roebuck and Company which paid $2.50 a share in cash and 4% in stock (annual rates) from the middle of 1928 through the first quarter of 1931; Auburn Automobile Company which paid $1 in cash and 2% in stock (quarterly) from January 1928 to July 1931; R. H. Macy Company, Inc., which during 1928–1932 paid annual stock dividends of 5% along with increasing cash dividends.

[2] Cities Service Company, from July 1, 1929 to June 1, 1932; Gas and Electric Securities Company between 1926 and 1931.

[3] Sears, Roebuck and Company between 1928 and 1931; Auburn Automobile Company between 1928 and 1931; Federal Light and Traction Company between 1925 and 1932.

[4] American Water Works and Electric Company between 1927 and 1930; American Gas and Electric Company between 1914 and 1932, with additional sporadic stock dividends; American Power and Light Company between 1923 and 1931, with extras in stock.

[5] Continental Can Company in 1924 and 1925; R. H. Macy Company, Inc., between 1928 and 1932; Truscon Steel Company between 1926 and 1931; General Electric Company between 1922 and 1925 (5% in special stock).

to the stockholder to take a certain amount of either cash or stock.

Example: The Seagrave Corporation paid a dividend quarterly at the annual rate of either $1.20 in cash or 10% in stock between 1925 and 1929, inclusive.[1]

3. In a few cases stock dividends only were paid, with no cash disbursement or option. The most prominent exponent of periodic stock dividends, the North American Company, followed this procedure by paying dividends of 2½% in stock, quarterly, between 1923 and 1933, in which year the payment was reduced to 2% quarterly. (In February 1934 the stock dividend was reduced to 1% quarterly and supplemented by a cash dividend of 12½ cents.)

Objectionable Feature of Periodic Stock Dividends.—Nearly every financial practice is open to abuse, and periodic stock dividends have proved no exception. The objectionable feature in this case has been to establish a regular stock-dividend rate exceeding in *market value* the amount of the earnings carried to surplus. This practice makes the issue appear unduly attractive to the unintelligent buyer, who is deceived by the high cash value of the current payments in stock. It requires some insight into corporate accounting methods to realize the true significance of such stock-dividend payments.

Let us use the outstanding North American Company case as an illustration. As we have stated, this company paid continuous stock dividends on the common shares at the rate of 10% annually for ten years. During most of this period the 10% stock dividend represented a payment of only $1 per share, as far as its books were concerned. This followed from the fact that prior to 1927 the par value of the stock was $10, and that after the shares were made no-par they were still given a "stated value" on the books at $10 per share. Hence 10% of either the par or the stated value amounted to only $1 per share. But from the investor's viewpoint he was receiving dividends worth much more than $1 per share, because the market price of North American common far exceeded its par or stated value.

The facts will appear from the table shown on page 345.

It will be noted that beginning with the third quarterly payment in 1931, the amount charged against earnings for the stock

---

[1] Compare this arrangement with the optional dividend or interest payments on preferred stocks and bonds, mentioned on p. 59.

dividend was advanced from $1 to $1.468 per share annually. This followed a request from the New York Stock Exchange that the charge against earnings or earned surplus covering the stock dividends reflect the interest of the new shares in the capital surplus as well as in the stated capital. Even after this change was made, however, there remained a wide discrepancy between the amount at which the dividends were valued on the books and the value given these dividends by the stock market, and presumably by the stockholders, until the quotation suffered a further severe decline.

Year	Earnings per share*	Range of market price	Value of the 10% stock dividend	
			Per company's books	To the stockholders (average market value)
1932	$2.01	43–14	$1.47	$ 2.85
1931	3.41	90–26	1.24†	5.80
1930	4.53	133–57	1.00	9.50
1929	5.03	187–67	1.00	12.70
1928	4.68	97–56	1.00	7.65
1927	4.06	65–46	1.00	5.55
1926	4.05	67–42	1.00	5.45
1925	3.74	75–41	1.00	5.80
1924	3.32	45–22	1.00	3.35
1923	3.59	24–18	1.00	2.10

* Based on the average number of shares outstanding during the year.
† First two quarterly dividends in 1931 were booked at $1 and last two at $1 to capital stock and 46.8 cents to capital surplus.

Danger of Vicious Circle Developing.—An arrangement of this kind is likely to develop into a vicious circle. The higher the market price, the greater the apparent value of the stock dividends, which in turn will seem to justify a still higher market price. (With a 10% stock dividend the dividend return obviously remains at 10% regardless of how high the market price may climb.) Such a result is deceptive and supplies an unwholesome impetus to riotous speculation as well as to thoughtless investment. In effect it is the opposite of the practice followed many years ago by such companies as American Can and National Biscuit, when the market price was kept far below the true value of the shares by an unduly "conservative" dividend policy. It is fully as objectionable, of course, to pursue a policy calculated to create a market price higher than that warranted by the earnings and other value factors. Such an unjustified price must necessarily be of temporary duration and is likely to result (as does all improper accounting) in giving the initiated an unfair advantage over the investing public.[1]

Historical Development.—From the historical standpoint it is interesting to note that the North American Company began its stock-dividend policy at about the same time that the first protagonist of the idea had decided to abandon it. This was the American Light and Traction Company, which during 1910–1919 had paid dividends at the annual rate of both $10 in cash and 10% in stock. During 1916 when the stock sold at about 400, the stockholders were receiving dividends with a realizable value of some $50 annually, although the earnings were only about $25 per share. Such a dividend policy could be permanently successful only if the company could continuously reinvest in its business ever-increasing amounts of profits, upon which in turn it could realize 20% annually. The law of diminishing returns (and the voracious growth of compound interest) would clearly outlaw such a possibility. In the depression of 1920–1921 American Light and Traction found it necessary to reduce its dividend rate sharply. The market quotation fell below 80, an astounding decline for an *investment stock* during that period. (The price range of Atchison during the years 1916–1921 was between 109 and 76.) This experience led the directors to give up the periodic stock-dividend idea in 1925, at the very time when it was coming into general favor among other public-utility holding companies.

Example of Vicious Pyramiding on Stock Dividends.—During the boom years periodic stock dividends were made the medium of an especially vicious pyramiding of reported profits. An operating company would pay out stock dividends with a market value more than its current earnings, and in turn an investment

---

[1] The North American Company has an excellent reputation and its policy was clearly not devised with any such sinister purpose in view. The company has been at pains to justify its stock-dividend payments in communications to its shareholders. The arguments center, however, on the advantages of reinvesting earnings and on the propriety of issuing additional common shares to represent these added resources. The discrepancy between the book value and the market value of these stock dividends, and the misconceptions which might arise therefrom, were hardly touched upon. It was particularly unfortunate that a company of high standing should have adopted this questionable practice, since its example was all too readily followed and exploited by other enterprises less scrupulously managed.

trust or holding company would report these stock dividends as income in an amount equal to the market value. For example, Central States Electric Corporation, which is a large holder of North American Company common stock, reported a total income in 1928 (exclusive of profits on the sale of securities) of $7,188,178. Of this sum, $6,396,225 was represented by stock of North American received during the year and taken on the recipient's books *at the market value* for North American immediately following the date of record for each quarterly dividend. The average price at which these stock dividends were taken on the books as income was $74 per share, or $7.40 for the 10% dividend, in a year in which North American earned $4.68 per share on the average number of shares outstanding. Nevertheless, the stock market capitalized these artificial earnings of Central States Electric Corporation to arrive at its valuation of that company's shares.

Market Price of Shares Should Be Recognized in Stock-dividend Payments.—The New York Stock Exchange finally adopted a new listing requirement under which corporations agree not to take into their income accounts stock dividends received, at a valuation greater than the amount at which such stock dividends were charged "against earnings, earned surplus or undivided profits by the issuing company in relation thereto."

While this regulation was properly conceived, it does not go to the heart of the matter. The abuses of the periodic stock-dividend procedure may readily be prevented by the simple rule that stock dividends *at market value* must not exceed the earnings available for dividends. Declarations may be made in the following form: "A stock dividend of 5% is hereby declared. The approximate market value of this dividend is about $6 per share, and it represents the capitalization of $7 per share retained in the business out of current earnings of $10 per share."

A procedure similar to this was followed by Famous Players Lasky Corporation in 1926 and 1927. An extra dividend of $2 per share was declared to be paid in no-par stock at a price to be fixed by the directors. The price was set at $100 per share, which was not far from the market price. Such a declaration does not imply that the directors guarantee or endorse the fairness of the current market price, but merely that they recognize the current market price as an objective fact.

Advantages of Stock Dividends Payable in Preferred Stock.—
Dividends may be paid in preferred stock instead of common
stock. The chief exemplar of this method is General Electric
Company, which distributed extra dividends of 5% annually
between 1922 and 1925, in addition to the regular payment of
$8 in cash. These extra dividends were paid in 6% special
stock, par value $10, which was in reality a preferred stock. A
similar procedure was followed by S. H. Kress Company and
by Hartman Corporation. The theoretical advantage of this
method is that the amount of the dividend paid is clearly fixed
at the effective par value[1] of the preferred shares issued, thus
obviating the complication presented by differences between
book value and market value. Where the company has no
senior securities, or only a small amount, the issuance of preferred stock to represent reinvested earnings will not weaken
the capital structure.

In the case of S. H. Kress, the 1931–1933 depression reduced
its working-capital requirements (due to lower volume of sales)
so that it found it advisable to use its surplus cash to retire part
of the preferred stock. This may be said to represent the ideal
arrangement from the stockholder's standpoint in dealing with
undistributed earnings. The two steps involved are as follows:

1. In prosperous years earnings are retained for expansion or added
working capital, but the stockholders receive preferred shares periodically
to represent a portion thereof.
2. If a subsequent decline in business shows that the additional capital
is no longer needed, it is paid out to the stockholders through the redemption
of their preferred shares.

Unfortunately the Kress example appears to be unique in
American financial history. Furthermore its impressiveness is
reduced by the fact that the dividends paid in the form of
preferred stock represented only 20% of the surplus earnings
during the period. It does, however, supply a concrete indication

---

[1] If payment is made in a *convertible preferred stock* the danger of overvaluation is, of course, not fully eliminated. For example, Columbia Gas and
Electric Corporation during 1932 paid $1.125 to common stockholders in
5% Convertible Preference Stock (par $100) which was convertible into
common in the ratio of one share of preference to five shares of common.
The preference stock sold as high as 108 during 1932 and 138 in 1933, or at
equivalents substantially in excess of the earnings of the company on its
common stock during those years.

of the type of dividend policies which stockholders should demand from their managements.

The Foregoing Summarized.—These various policies, which we have discussed at length in this and the preceding chapter, may be summed up in the following three statements:[1]

1. Withholding and reinvestment of a substantial part of the earnings must be clearly justified to the stockholders on the grounds of concrete benefits therefrom exceeding the value of the cash if paid to the stockholders. Such withholding should be specifically approved by the stockholders.

2. If retention of profits is in any sense a matter of *necessity* rather than *choice*, the stockholders should be advised of this fact, and the amounts involved should be designated as "reserves" instead of as "surplus profits."

3. Earnings voluntarily retained in the business should be capitalized in good part by the periodic issuance of additional stock, with current market value not exceeding such reinvested earnings. If the additional capital is subsequently found no longer to be needed in the business, it should be distributed to the shareholders against the retirement of the stock previously issued to represent it.

[1] For some interesting legal aspects of the power to declare or withhold dividends see A. A. Berle, and G. C. Means, *The Modern Corporation and Private Property*, pp. 260–263, New York, 1932.

PART V

ANALYSIS OF THE INCOME ACCOUNT
THE EARNINGS FACTOR IN COMMON-STOCK VALUATION

CHAPTER XXXI

ANALYSIS OF THE INCOME ACCOUNT

In our historical discussion of the theory of investment in common stocks we traced the transfer of emphasis from the net worth of an enterprise to its capitalized earning power. While there are sound and compelling reasons behind this development, it is none the less one which has removed much of the firm ground that formerly lay—or seemed to lie—beneath investment analysis, and has subjected it to a multiplicity of added hazards. When an investor was able to take very much the same attitude in valuing shares of stock as in valuing his own business, he was dealing with concepts familiar to his individual experience and matured judgment. Given sufficient information, he was not likely to go far astray, except perhaps in his estimate of future earning power. The interrelations of balance sheet and income statement gave him a double check on intrinsic values, which corresponded to the formulas of banks or credit agencies in appraising the eligibility of the enterprise for credit.

Disadvantages of Sole Emphasis on Earning Power.—Now that common-stock values have come to depend exclusively upon the earnings exhibit, a gulf has been created between the concepts of private business and the guiding rules of investment. When the business man lays down his own statement and picks up the report of a large corporation, he apparently enters a new and entirely different world of values. For certainly he does not appraise his own business solely on the basis of its recent operating results without reference to its financial resources. When in his capacity as investor or speculator the business man elects to pay no attention whatever to corporate balance sheets, he is placing himself at a serious disadvantage in several different respects: In the first place, he is embracing a *new* set of ideas which are alien to his everyday business experience. In the second place, instead of the twofold test of value afforded by both earnings and assets, he is relying upon a *single* and therefore less dependable criterion. In the third place, these earnings statements on which he relies exclusively are subject to more rapid and radical *changes* than those which occur in balance sheets. Hence an exaggerated degree of instability is introduced into his concept of stock values. In the fourth place, the earnings statements are far more subject to *misleading* presentation and mistaken inferences than is the typical balance sheet when scrutinized by an investor of experience.

Warning against Sole Reliance upon Earnings Exhibit.—In approaching the analysis of earnings statements we must, therefore, utter an emphatic warning against exclusive preoccupation with this factor in dealing with investment values. With due recognition of the greatly restricted importance of the asset picture, it must nevertheless be asserted that a company's resources still have some significance and require some attention. This is particularly true, as will be seen later on, because the meaning of any income statement cannot properly be understood except with reference to the balance sheet at the beginning and the end of the period.

Simplified Statement of Wall Street's Method of Appraising Common Stocks.—Viewing the subject from another angle, we may say that the Wall-Street method of appraising common stocks has been simplified to the following standard formula:

1. Find out what the stock is earning. (This usually means the earnings per share as shown in the last report.)
2. Multiply these per-share earnings by some suitable "coefficient of quality" which will reflect:
    a. The dividend rate and record.
    b. The standing of the company—its size, reputation, financial position, and prospects.
    c. The type of business (*e.g.*, an electric light company will sell at a higher multiple of earnings than a baking company).
    d. The temper of the general market. (Bull-market multipliers are larger than those used in bear markets.)

The foregoing may be summarized in the following formula:

$$Price = earnings\ per\ share \times quality\ coefficient.[1]$$

---

[1] Where there are no earnings or where the amount is recognized as being far below "normal," Wall Street is reluctantly compelled to apply what is at bottom a more rational method of valuation, *i.e.*, one ascribing greater

The result of this procedure is that in most cases the "earnings per share" have attained a weight in determining value which is equivalent to the *weight of all the other factors taken together.* The truth of this is evident if it be remembered that the "quality coefficient" is itself largely determined by the *earnings trend*, which in turn is taken from the stated earnings over a period.

Earnings Not Only Fluctuate but Are Subject to Arbitrary Determination.—But these earnings per share, on which the entire edifice of value has come to be built, are not only highly fluctuating but are subject also in extraordinary degree to arbitrary determination and manipulation. It will be illuminating if we summarize at this point the various devices, legitimate and otherwise, by which the per-share earnings may *at the choice of those in control* be made to appear either larger or smaller.

1. By allocating items to surplus instead of to income, or *vice versa.*
2. By over- or understating amortization and other reserve charges.
3. By varying the capital structure, as between senior securities and common stock. (Such moves are decided upon by managements and ratified by the stockholders as a matter of course.)
4. By the use made of large capital funds not employed in the conduct of the business.

Significance of the Foregoing to the Analyst.—These intricacies of corporate accounting and financial policies undoubtedly provide a broad field for the activities of the securities analyst. There are unbounded opportunities for shrewd detective work, for critical comparisons, for discovering and pointing out a state of affairs quite different from that indicated by the publicized "per-share earnings."

That this work may be of exceeding value cannot be denied. In a number of cases it will lead to a convincing conclusion that the market price is far out of line with intrinsic or comparative worth, and hence to profitable action based upon this sound foundation. But it is necessary to caution the analyst against overconfidence in the practical utility of his findings. It is always good to know the truth but it may not always be wise to act upon it, particularly in Wall Street. And it must always be remembered that the truth which the analyst uncovers is first of all not the *whole* truth and, secondly, not the *immutable* truth. The result of his study is only a *more nearly correct*

---

weight to *average* earning power, working capital, *etc.* But this is the *exceptional* procedure.

version of the past. His information may have lost its relevance by the time he acquires it, or in any event by the time the market place is finally ready to respond to it.

With full allowance for these pitfalls, it goes without saying, none the less, that security analysis must devote thoroughgoing study to corporate income accounts. It will aid our exposition if we classify this study under three headings, *viz.:*

1. The accounting aspect.
   Leading question: What are tne true earnings for the period studied?

2. The business aspect.
   Leading question: What indications does the earnings record carry as to the *future* earning power of the company?

3. The aspect of investment finance.
   Leading question: What elements in the earnings exhibit must be taken into account, and what standards followed, in endeavoring to arrive at a reasonable *valuation* of the shares?

CRITICISM AND RESTATEMENT OF THE INCOME ACCOUNT

If an income statement is to be informing in any true sense, it must at least present a fair and undistorted picture of the year's operating results. Direct misstatement of the figures in the case of publicly owned companies is a rare occurrence. The Ivar Kreuger frauds, revealed in 1932, partook of this character; but these were quite unique in the baldness as well as in the extent of the deception. The statements of most important companies are audited by independent public accountants and their reports are reasonably dependable within the rather limited sphere of accounting accuracy. But from the standpoint of common-stock analysis these audited statements may require critical interpretation and adjustment, especially with respect to three important elements:

1. Nonrecurrent profits and losses.
2. Operations of subsidiaries or affiliates.
3. Reserves.

General Observations on the Income Account.—Accounting procedure allows considerable leeway to the management in the method of treating nonrecurrent items. It is a standard and proper rule that transactions applicable to past years should be excluded from current income and entered as a charge or credit direct to the surplus account. Yet there are many kinds of entries which may technically be considered part of the current year's results, but which are none the less of a special and non-recurrent nature. Accounting rules permit the management to decide whether to show these operations as part of the *income* or to report them as adjustments of *surplus*. Following are a number of examples of entries of this type:

1. Profit or loss on sale of fixed assets.
2. Profit or loss on sale of marketable securities.
3. Discount or premium on retirement of capital obligations.
4. Proceeds of life insurance policies.
5. Tax refunds and interest thereon.
6. Gain or loss as result of litigation.
7. Extraordinary write-downs of inventory.
8. Extraordinary write-downs of receivables.
9. Cost of maintaining nonoperating properties.

Wide variations will be found in corporate practice respecting items such as the above. Under each heading examples may be given of either inclusion in or exclusion from the income account. Which is the better accounting procedure in some of these cases may be a rather controversial question; but as far as the analyst is concerned, his object requires that all these items be segregated from the *ordinary operating results* of the year. For what the investor chiefly wants to learn from an annual report is the *indicated earning power* under the given set of conditions, *i.e.*, what the company might be expected to earn year after year if the business conditions prevailing during the period were to continue unchanged.

The analyst must endeavor also to adjust the reported earnings so as to reflect as accurately as possible the company's interest in results of controlled or affiliated companies. In most cases consolidated reports are made, so that such adjustments are unnecessary. But there are many instances in which the statements are incomplete or misleading because either: (1) they fail to reflect any part of the profits or losses of important subsidiaries; or (2) they include as income dividends from subsidiaries which are substantially less or greater than the current earnings of the controlled enterprises.

The third aspect of the income account to which the analyst must give critical attention is the matter of reserves for depreciation and other amortization, and reserves for future losses and other contingencies. These reserves are subject in good part to arbitrary determination by the management. Hence they may readily be overstated or understated, in which case the final figure of reported earnings will be correspondingly distorted. With respect to amortization charges another and more subtle element enters which may at times be of considerable importance, and that is the fact that the deductions from income, as calculated by the management based on the book cost of the property, may not properly reflect the amortization which the *individual investor* should charge against his own commitment in the enterprise.

Nonrecurrent Items: Profits or Losses from Sale of Fixed Assets.—We shall proceed to a more detailed discussion of these three types of adjustment of the reported income account, beginning with the subject of nonrecurrent items. Profits or losses from the sale of fixed assets belong quite obviously to this category; and they should be excluded from the year's result in order to gain an idea of the "indicated earning power" based on the assumed continuance of the then existing business conditions. Approved accounting practice recommends that profit on sales of capital assets be shown only as a credit to the surplus account. In numerous instances, however, such profits are reported by the company as part of its current net income, creating a distorted picture of the earnings for the period.

Examples: A glaring example of this practice is presented by the report of the Manhattan Electrical Supply Company for 1926. This showed earnings of $882,000 or $10.25 per share, which was regarded as a very favorable exhibit. But a subsequent application to list additional shares on the New York Stock Exchange revealed that out of this $882,000 reported as earned, no less than $586,700 had been realized through the sale of the company's battery business. Hence the earnings from ordinary operations were only $295,300, or about $3.40 per share. The inclusion of this special profit in income was particularly objectionable because in the very same year the company had charged to surplus extraordinary *losses* amounting to $544,000. Obviously the special losses belonged to the same category as the special profits and the two items should have been grouped together. The effect of including the one in income and charging the other to surplus was misleading in the highest degree. Still more discreditable was the failure to make any clear reference to the profit from the battery sale either in the income account itself or in the extended remarks which accompanied it in the annual report.[1]

During 1931 the United States Steel Corporation reported "special income" of some $19,300,000, the greater part of which was due to "profit on sale of fixed property"—understood to be certain public-utility holdings in Gary, Indiana. This item was included in the year's earnings and resulted in a final "net income" of $13,000,000. But since this credit was definitely of a nonrecurring nature, the analyst would be compelled to eliminate it from his consideration of the 1931 operating results, which would accordingly register a *loss* of $6,300,000 before preferred dividends. United States Steel's accounting method in 1931 is at variance with its previous policy, as shown by its treatment of the large sums received in the form of income-tax refunds in the three preceding years. These receipts were not reported as current income but were credited directly to surplus.

Profits from Sale of Marketable Securities.—Profits realized by a business corporation from the sale of marketable securities are also of a special character and must be separated from the ordinary operating results.

Examples: The report of National Transit Company, a former Standard Oil subsidiary, for the 1928 year illustrates the distorting effect due to the inclusion in the income account of profits from this source. The method of presenting the story to the stockholders is also open to serious criticism. The consolidated income account for 1927 and 1928 was stated in approximately the following terms:

Item	1927	1928
Operating revenues	$3,432,000	$3,419,000
Dividends, interest, and miscellaneous income	463,000	370,000
Total revenues	$3,895,000	$3,789,000
"Operating expenses, including depreciation and profit and loss direct items" (in 1928 "including profits from sale of securities")	3,264,000	2,599,000
Net income	$ 631,000	$1,190,000
(Earned per share)	($1.24)	($2.34)

[1] The president's remarks contained only the following in respect to this transaction: "After several years of unprofitable experience in the battery business the directors arranged a sale of same on satisfactory terms."

The increase in the earnings per share appeared quite impressive. But a study of the detailed figures of the parent company alone as submitted to the Interstate Commerce Commission, would have revealed that $560,000 of the 1928 income was due to its profits from the sale of securities. This happens to be almost exactly equal to the increase in consolidated net earnings over the previous year. Allowing on the one hand for income tax and other offsets against these special profits, but on the other hand for probable additional profits from the sale of securities by the manufacturing subsidiary, it seems likely that all or nearly all of the apparent improvement in earnings for 1928 was due to nonoperating items. Such gains must clearly be eliminated from any comparison or calculation of *earning power*. The form of statement resorted to by National Transit, in which such profits are applied to *reduce operating expenses*, is bizarre to say the least.

The sale by the New York, Chicago and St. Louis Railroad Company, through a subsidiary, of its holdings of Pere Marquette stock in 1929 gave rise later to an even more extraordinary form of bookkeeping manipulation. We shall describe these transactions in connection with our treatment of items involving nonconsolidated subsidiaries. During 1931, F. W. Woolworth Company included in its income a profit of nearly $10,000,000 on the sale of a part interest in its British subsidiary. The effect of this inclusion was to make the per-share earnings appear larger than in any previous year, when in fact they had experienced a recession. It is somewhat surprising to note that in the same year the company charged against *surplus* an additional tax accrual of $2,000,000 which seemed to be closely related to the special profit included in *income*.

Reduction in the market value of securities should be considered as a nonrecurring item in the same way as losses from the sale of such securities. The same would be true of shrinkage in the value of foreign exchange. In most cases corporations charge such write-downs, when made, against surplus. The General Motors report for 1931 included both such adjustments, totalling $20,575,000 as deductions from *income*, but was careful to designate them as "extraordinary and nonrecurring losses."

Methods Used by Investment Trusts in Reporting Sale of Marketable Securities.—Investment-trust statements raise special questions with respect to the treatment of profits or losses realized from the sale of securities and changes in security values.   Prior to 1930 most of these companies reported profits from the sale of securities as part of their regular income; but they showed the appreciation on *unsold* securities in the form of a memorandum or footnote to the balance sheet.   But when large losses were taken in 1930 and subsequently, they were shown in most cases not in the income account, but as charges against capital, surplus, or reserves.   The *unrealized* depreciation was still recorded by most companies in the form of an explanatory comment on the balance sheet, which continued to carry the securities owned at original cost.   A minority of investment trusts reduced the carrying price of their portfolio to the market by means of charges against capital and surplus.

It may logically be contended that since dealing in securities is an integral part of the investment-trust business, the results from sales and even the changes in portfolio values should be regarded as ordinary rather than extraordinary elements in the year's report.   Certainly a study confined to the interest and dividend receipts less expenses would prove of negligible value. If *any* useful results can be expected from an analysis of investment-trust exhibits, such analysis must clearly be based on the three items: investment income, profits or losses on the sale of securities, and changes in market values.   But the reader must bear in mind the basic fact that security analysis, however intelligently conducted, cannot be more dependable than the conditions with which it deals.   No study of investment-trust reports during the pyrotechnic years of 1928 to 1933 could possibly shed light upon the future results to be expected from these enterprises.   The value of such a study would have to be confined to certain comparisons between one company and another.   Even here it would be difficult to distinguish confidently between superior management and luckier guesses on the market.

Similar Problem in the Case of Banks and Insurance Companies. The same problem is involved in analyzing the results shown by insurance companies and by banks.   Public interest in insurance securities is concentrated largely upon the shares of fire insurance companies.   These enterprises represent a combination of the insurance business and the investment-trust business.   They have available for investment their capital funds, plus substantial amounts received as premiums paid in advance.   Generally speaking, only a small portion of these funds are subject to legal restrictions as regards investment, and the balance is handled in much the same way as the resources of the investment trusts. The underwriting business as such has rarely proved highly profitable. More frequently than not it shows a deficit, which is offset, however, by interest and dividend income. The profits or losses shown on security operations, including changes in their market value, exert a predominant influence upon the public's attitude toward fire-insurance-company stocks. The same has been true of bank stocks to a smaller, but none the less significant, degree. The tremendous overspeculation in these issues during the late 1920s was stimulated largely by the participation of the banks, directly or through affiliates, in the fabulous profits made in the securities markets.

The fact that the operations of financial institutions generally —such as investment trusts, banks, and insurance companies— must necessarily reflect changes in security values, makes their shares a dangerous medium for widespread public dealings. Since in these enterprises an increase in security values may be held to be part of the year's profits, there is an inevitable tendency to regard the gains made in good times as part of the "earning power," and to value the shares accordingly. This results of course in an absurd overvaluation, to be followed by collapse and a correspondingly excessive depreciation. Such violent fluctuations are particularly harmful in the case of financial institutions because they may affect public confidence. It is true also that rampant speculation (called "investment") in bank and insurance-company stocks leads to the ill-advised launching of new enterprises, to the unwise expansion of old ones, and to a general relaxation of established standards of conservatism and even of probity.

The securities analyst, in discharging his function of investment counsellor, should do his best to discourage the purchase of stocks of banking and insurance institutions by the ordinary small investor. Prior to the boom of the 1920s such securities were owned almost exclusively by those having or commanding large financial experience and matured judgment. These qualities are needed to avoid the special danger of misjudging values in this field by reason of the dependence of their reported earnings upon fluctuations in security prices.

Herein lies also a paradoxical difficulty of the investment-trust movement. Given a proper technique of management, these organizations may well prove a logical vehicle for the placing of small investor's funds. But considered as a marketable security dealt in by small investors, the investment-trust stock itself is a dangerously volatile instrument. Apparently this troublesome factor can be held in check only by educating or effectively cautioning the general public on the interpretation of investment-trust reports. The prospects of this being done are none too bright.

Profits through Repurchase of Senior Securities at a Discount.—At times a substantial profit is realized by corporations through the repurchase of their own senior securities at less than par value. The inclusion of such gains in current income is certainly a misleading practice, first, because they are obviously nonrecurring and, secondly, because they are at best a questionable sort of profit, since they are made at the expense of the company's own security holders.

Example: A peculiar example of this accounting practice was furnished as long ago as 1915 by Utah Securities Corporation, a holding company controlling Utah Power and Light Company. The following income account illustrates this point:

YEAR ENDED MARCH 31, 1915

Earnings of Utah Securities Corporation including surplus of subsidiaries accruing to it	$ 771,299
Expenses and taxes	30,288
Net earnings	$ 741,011
Profit on redemption of 6% notes	1,309,657
Income from all sources accruing to Utah Securities Corporation	$2,050,668
Deduct interest charges on 6% notes	1,063,009
Combined net income for the year	$ 987,659

The above income account shows that the chief "earnings" of Utah Securities were derived from the repurchase of its own obligations at a discount. Had it not been for this extraordinary item the company would have failed to cover its interest charges.

The widespread repurchases of senior securities at a substantial discount constituted one of the unique features of the 1931–1933 depression years. It was made possible by the disproportion that existed between the strong cash positions and the poor earnings of many enterprises. Because of the latter influence, the senior securities sold at low prices; and because of the former, the issuing companies were able to buy them back in large amounts. This practice was most in evidence among the investment trusts.

Example: The International Securities Corporation of America, to use an outstanding example, repurchased in the fiscal year ending November 30, 1932, no less than $12,684,000 of its 5% bonds, representing nearly half of the issue. The average price paid was about 55 and the operation showed a profit of about $6,000,000 which served to offset the shrinkage in the value of the investment portfolio.

A contrary result appears when senior securities are retired at a cost exceeding the face or stated value. When this premium involves a large amount it is always charged against surplus and not against current income.

Examples: As prominent illustrations of this practice, we cite the charge of $40,600,000 against surplus made by United States Steel Corporation in 1929, in connection with the retirement at 110 of $307,000,000 of its own and subsidiaries' bonds; also the charge of $9,600,000 made against surplus in 1927 by Goodyear Tire and Rubber Company, growing out of the retirement at a premium of various bond and preferred-stock issues and their replacement by new securities bearing lower coupon and dividend rates. From the analyst's standpoint, either profit or expense in such special transactions involving the company's own securities should be regarded as nonrecurring and excluded from the operating results.

A Comprehensive Example.—American Machine and Metals, Inc. (successor to Manhattan Electrical Supply Company mentioned earlier in this chapter) included in its current *income* for 1932 a profit realized from the repurchase of its own bonds at a discount. Because the reports for 1931 and 1932 illustrate to an unusual degree the arbitrary nature of much corporate accounting, we reproduce herewith in full the income account and the appended capital and surplus adjustments.

We find again in 1932, as in 1926, the highly objectionable practice of including extraordinary profits in income while charging special losses to surplus. It does not make much difference that in the later year the nature of the special profit—gain through repurchase of bonds at less than par—is disclosed in the report. Stockholders and stock buyers for the most part pay attention only to the final figure of earnings per share, as presented by the company; nor are they likely to inquire carefully into the manner in which it is determined.   The significance of some of the charges made by this company against surplus in 1932 will be taken up later under the appropriate headings.

REPORT OF AMERICAN MACHINE AND METALS, INC., FOR 1931 AND 1932

Item	1932	1931
Income account:		
Net before depreciation and interest..	Loss $ 136,885	Profit $101,534
Add profit on bonds repurchased.....	174,278	270,701
Profit, including bonds repurchased..	37,393	372,236
Depreciation.....................	87,918	184,562
Bond interest.....................	119,273	140,658
Final net profit or loss..............	Loss 169,789	Profit 47,015
Charges against capital, capital surplus and earned surplus:		
Deferred moving expense and mine development..................	111,014	
Provision for losses on:		
Doubtful notes, interest thereon, and claims..................	600,000	
Inventories.....................	385,000	
Investments....................	54,999	
Liquidation of subsidiary.........	39,298	
Depletion of ore reserves...........	28,406	32,515
Write-down of fixed assets (net).....	557,578	
Reduction of ore reserves and mineral rights......................	681,742	
Federal tax refund, etc..............	cr. 7,198	cr. 12,269
Total charges not shown in income account......................	$2,450,839	$ 20,246
Result shown in income account.......	dr. 169,798	cr. 47,015
Received from sale of additional stock ..	cr. 44,000	
Combined change in capital and surplus	dr. $2,576,637	cr. $ 26,769

Other Nonrecurrent Items.—The remaining group of nonrecurrent profit items is not important enough to merit detailed discussion.   In most cases it is of minor consequence whether they appear as part of the year's earnings or are credited to surplus where they properly belong.

Examples: Borg Warner Corporation included the sum of $443,330, proceeds of life insurance policies, in its income for 1931. (On the other hand Alfred Decker and Cohn, Inc., receiving a similar payment of $442,289 in 1930, more soundly credited it direct to surplus under the heading of "Special Revenue.") Bendix Aviation Corporation reported as income for the year 1929 the sum of $901,282 received in settlement of a patent suit; and again in 1931 it included in current earnings an amount $242,656 paid to it as back royalties collected through litigation. The 1932 earnings of Gulf Oil Corporation included the sum of $5,512,000 representing the value of oil previously in litigation. By means of this item, designated as nonrecurrent, it was able to turn a loss of $2,768,000 into a profit of $2,743,000. While tax refunds are regularly shown as credits to surplus only, the accumulated interest received thereon sometimes appears as part of the income account, *e.g.*, $2,000,000 reported by E. I. du Pont de Nemours and Company in 1926, and an unstated but apparently much larger sum included in the earnings of United States Steel for 1930.

CHAPTER XXXII

EXTRAORDINARY LOSSES AND OTHER SPECIAL ITEMS IN THE INCOME ACCOUNT

The question of nonrecurrent losses is likely to create peculiar difficulties in the analysis of income accounts. To what extent should write-downs of inventories and receivables be regarded as extraordinary deductions not fairly chargeable against the year's operating results? In the disastrous year 1932 such charge-offs were made by nearly every business. The accounting methods used showed wide divergences; but the majority of companies spared their income accounts as much as possible and subtracted these losses from surplus. Was this proper?

The analyst must approach a year like 1932 in the spirit of common sense rather than of narrow exactitude. No doubt the efforts of many concerns to avoid too harrowing an income exhibit resulted in some highly questionable uses of the surplus account. But in the typical case there seems to be little point in any endeavor to arrive at a precise restatement of the year's operating results. For since prevailing business conditions were thoroughly abnormal, and themselves presumably *nonrecurring*, there is but slight advantage in knowing the exact amount of the operating losses as distinguished from the extraordinary losses. Such a figure would have no value as a guide to future results; and it is doubtful whether its use as a component of a long-term average would be especially instructive. The real significance of most 1932 reports is to be looked for in the balance sheet, particularly in the extent to which working capital was maintained or impaired, and also in certain contradictory effects following from the drastic write-downs of fixed assets. Both of these factors will be considered later.

Manufactured Earnings.—An examination of the wholesale charges made against surplus in 1932 by American Machine and Metals, detailed on page 362, suggests the possibility that *excessive* provision for losses may have been made in that year with the intention of benefiting future income accounts. If the receivables and inventories were written down to an unduly low figure on December 31, 1932, this artificially low "cost price" would give rise to a correspondingly inflated profit in the following years. This point may be made clear by the use of hypothetical figures as follows:

> Assume fair value of inventory and receivables on
> Dec. 31, 1932 to be........................ $2,000,000
> Assume profit for 1933 based on such fair value.. 200,000
> But assume that, by special and excessive charges
> to surplus, the inventory and receivables had
> been written down to...................... 1,600,000
> Then the amounts realized therefrom will show a
> correspondingly greater profit for 1933, which
> might mean *reported* earnings for 1933 of.... 600,000
> This would be three times the proper figure.

The above example illustrates a whole set of practices which constitute perhaps the most vicious type of accounting manipulation. They consist, in brief, of taking sums out of surplus (or even capital) and then reporting these same sums as income. The charge to surplus goes unnoticed; the credit to income may have a determining influence upon the market price of the securities of the company. We shall later point out that the "conservative" writing down of the property account has precisely this result, in that it permits a decreased depreciation charge and hence an increase in the apparent earnings. The dangers inherent in accounting methods of this sort are the more serious because they are so little realized by the public, so difficult to detect even by the expert analyst, and so impervious to legislative or stock-exchange correction.

The basing of common-stock values on reported per-share earnings has made it much easier for managements to exercise an arbitrary and unwholesome control over the price level of their shares. While it should be emphasized that the overwhelming majority of managements are honest, it must be emphasized also that loose or "purposive" accounting is a highly contagious disease.

Losses on Inventories.—Except under quite extraordinary general business conditions write-downs of inventories to market value are regarded as properly chargeable against the year's income and not against surplus. When a company departs from standard practice in this respect, an analyst must take this point carefully into account, particularly in comparing the published results with those of other companies. A good illustration of this rule is afforded by a comparison of the reports submitted by United States Rubber Company and by Goodyear Tire and Rubber Company for the years 1925–1927, during which time rubber prices were subject to wide fluctuations.

In these three years Goodyear charged against *earnings* a total of $11,500,000 as reserves against decline of raw-material prices. Of this amount one-half was used to absorb actual losses sustained and the other half was carried forward into 1928 (and eventually used up in 1930).

United States Rubber during this period charged a total of $20,446,000 for inventory reserves and write-downs, all of which was absorbed by actual losses taken. But the form of annual statement, as submitted to the stockholders, excluded these deductions from income and made them appear as special adjustments of surplus. (In 1927, moreover, the inventory loss of $8,910,000 was apparently offset by a special credit of $8,000,000 from the transfer of *past* earnings of the crude-rubber producing subsidiary.)

The result of these divergent bases of reporting annual income was that the per-share earnings of the two companies, as compiled by the statistical manuals, made an entirely misleading comparative exhibit. The following per share earnings are taken from Poor's *Manual* for 1928:

Year	U.S. Rubber	Goodyear
1925	$14.92	$9.45
1926	10.54	3.79
1927	1.26	9.02
3-year average...................	$ 8.91	$7.42

For proper comparative purposes the statements must manifestly be considered on an identical basis, or as close thereto as possible. Such a comparison might be made by three possible methods, *viz.:*

1. As reported by United States Rubber, *i.e.*, excluding inventory reserves and losses from the current income account.

2. As reported by Goodyear, *i.e.*, reducing the earnings of the period of high prices for crude rubber by a reserve for future losses, and using this reserve to absorb the later shrinkage.

3. Eliminating such reserves, as an arbitrary effort of the management to level out the earnings. On this basis the inventory losses would be deducted

from the results of the year in which they were actually sustained. (The Standard Statistics Company's analysis of Goodyear includes a revision of the reported earnings in conformity with this approach.)

We have then, for comparative purposes, three statements of the per-share earnings for the period:

Year	1. Omitting adjustments of inventory		2. Allowing for inventory adjustments, as made by the companies		3. Excluding *reserves* and charging losses to the year in which decline occurred	
	U.S. Rubber	Goodyear	U.S. Rubber	Goodyear	U.S. Rubber	Goodyear
1925	$14.92	$18.48	$11.21	$9.45	$14.92	$18.48
1926	10.54	3.79	0.00	3.79	*14.71(d)*	*2.53(d)*
1927	1.26*	13.24	*9.73(d)**	9.02	1.26*	13.24
3-year average	$ 8.91	$12.17	$ 0.49	$7.42	$ 0.49	$ 9.73

* Excluding credit for profits made prior to 1926 by United States Rubber Plantations, Inc.

The range of market prices for the two common issues during this period suggests that the accounting methods followed by U.S. Rubber served rather effectively to obscure the unsatisfactory nature of its results for these years.

Year	U.S. Rubber common		Goodyear common	
	High	Low	High	Low
1925	97	33	50	25
1926	88	50	40	27
1927	67	37	69	29
Average of highs and lows..	62		40	

Idle-plant Expense.—The cost of carrying nonoperating properties is almost always charged against income. Many statements for 1932 earmarked substantial deductions under this heading.

Examples: Youngstown Sheet and Tube Company, which reported a charge of $2,759,000 for "Maintenance Expense, Insurance and Taxes of Plants, Mines, and Other Properties that were Idle." Stewart Warner Corporation followed the exceptional policy of charging against *surplus* in 1932, instead of income, the sum of $309,000 for "Depreciation of Plant Facilities not used in current year's production."

The analyst may properly consider idle-plant expense as belonging to a somewhat different category from ordinary charges against income. In theory, at least, these expenses should be of a temporary and therefore nonrecurring type. Presumably the management can terminate these losses at any time by disposing of or abandoning the property. Because, for the time being, the company elects to spend money to carry these assets along in the expectation that future value will justify the outlay, it does not seem logical to consider these assets as equivalent to a permanent liability, *i.e.*, as a permanent drag upon the company's earning power, which makes the stock worth considerably less than it would be if these "assets" did not exist.

Example: The practical implications of this point are illustrated by the case of New York Transit Company, a carrier of oil by pipe line. In 1926, owing to new competitive conditions, it lost all the business formerly carried by its principal line, which thereupon became "idle plant." The depreciation, taxes, and other expenses of this property were so heavy as to absorb the earnings of the company's other profitable assets (consisting of a smaller pipe line and high-grade-bond investments). This created an apparent net loss and caused the dividend to be passed. The price of the stock accordingly declined to a figure far less than the company's holdings of cash and marketable securities alone. In this uncritical appraisal by the stock market, the idle asset was considered equivalent to a serious and permanent liability.

In 1928, however, the directors determined to put an end to these heavy carrying charges and succeeded in selling the unused pipe line for a substantial sum of money. Thereafter, the stockholders received special cash distributions aggregating $72 per share (nearly twice the average market price for 1926 and 1927), and they still retained ownership of a profitable business which resumed regular dividends. Even if no money had been realized from the idle property, its mere abandonment would have led to a considerable increase in the value of the shares.

This is an impressive, if somewhat extreme, example of the practical utility of security analysis in detecting discrepancies between intrinsic value and market price. It is customary to refer with great respect to the "bloodless verdict of the market place," as though it represented invariably the composite judgment of countless shrewd, informed, and calculating minds. Very frequently, however, these appraisals are based on mob psychology, on faulty reasoning, and on the most superficial examination of inadequate information. The analyst, on his side, is usually unable to apply his technique effectively to correcting or taking advantage of these popular errors, for the reason that surrounding conditions change so rapidly that his own conclusions may become inapplicable before he can profit by them. But in the exceptional case, as illustrated by our last example, the facts and the logic of the case may be sharply enough defined to warrant a high degree of confidence in the practical value of his analysis.

Deferred Charges.—A business sometimes incurs expenses which may fairly be considered as applicable to a number of years following rather than to the single twelve-month period in which the outlay was made. Under this heading might be included the following:

Organization expense (legal fees, etc.).
Moving expenses.
Development expenses (for new products or processes; also for opening up a mine, etc.).
Discount on obligations sold.

Under approved accounting methods such costs are spread over an appropriate period of years. The amount involved is entered upon the balance sheet as a Deferred Charge, which is written off by annual charges against earnings. In the case of bond-discount the period is fixed by the life of the issue; mine development expenses are similarly prorated on the basis of the tonnage mined. For most other items the number of years must be arbitrarily taken, five years being a customary figure.

In order to relieve the reported earnings of these annual deductions it has become common practice to write off such "expense applicable to future years" by a single charge against surplus. In theory this practice is improper, because it results in the understatement of operating expenses for a succeeding period of years and hence in the exaggeration of the net income. If, to take a simple example, the president's salary were paid for ten years in advance and the entire outlay charged against surplus as a "special expense," it is clear that the profits of the ensuing period would thereby be overstated. There is the danger also that expenses of a character frequently repeated— e.g., advertising campaigns, or cost of developing new automobile models—might be omitted from the income account by designating them as deferred charges and then writing them off against surplus.

Ordinarily the amounts involved in such accounting transactions are not large enough to warrant the analyst in making an issue of them. Security analysis is a severely practical activity, and it must not linger over matters which are not likely to affect the ultimate judgment. At times, however, these items may assume appreciable importance.

Examples: The Kraft Cheese Company for example, during some years prior to 1927 carried a substantial part of its advertising outlays as a deferred charge to be absorbed in the operations of subsequent years. In 1926 it spent about $1,000,000 for advertising and charged only one-half of this amount against current income. But in the same year the balance of this expenditure was deducted from surplus, and furthermore an additional $480,000 was similarly written off against surplus to cancel the balance carried forward from prior years as a deferred charge. By this means the company was able to report to its stockholders the sum of $1,071,000 as earned for 1926. But when in the following year it applied to list additional shares, it found it necessary to adopt a less questionable basis of reporting its income to the New York Stock Exchange, so that its profit for 1926 was restated to read $461,296, instead of $1,071,000.

The 1932 report of International Telephone and Telegraph Co. showed various charges *against surplus* aggregating $35,817,000, which included the following: "Write-off of certain deferred charges which have today no tangible value although originally set up to be amortized over a period of years in accordance with accepted accounting principles, $4,655,696."

Hudson Motor Car Company charged against surplus instead of income the following items (among others) during 1930–1931.

1930.	Special adjustment of tools and materials due to development of new models	$2,266,000
1931.	Reserve for special tools	2,000,000
	Rearrangement of plant equipment	633,000
	Special advertising	1,400,000

In 1933 Gold Dust Corporation appropriated out of surplus the sum of $2,000,000 as a reserve for the "net cost of the introduction and exploitation of new products." About one third of this amount was expended during 1933 and the balance carried forward into 1934.

The effect of these accounting practices is to relieve the reported earnings of expenditures which most companies charge currently thereagainst, and which in any event should be charged against earnings in installments over a short period of years.

Amortization of Bond Discount.—Bonds are usually floated by corporations at a price to net the treasury less than par. The discount suffered is part of the cost of borrowing the money, *i.e.*, part of the interest burden, and it should be amortized over the life of the bond issue by an annual charge against earnings, included with the statement of interest paid. It was formerly considered "conservative" to write off such bond discounts by a single charge against surplus, in order not to show so intangible an item among the assets on the balance sheet. More recently these write-offs against surplus have become popular for the opposite reason, *viz.*, to eliminate future annual deductions from earnings, and in that way to make the shares more "valuable."

Example: Associated Gas and Electric Company charged against surplus in 1932 the sum of $5,892,000 for "debt discount and expense" written off.[1]

[1] The New York Stock Exchange has inveighed vigorously against this practice. Witness the following from an address in February 1933 by J. M. B. Hoxsey, Executive Assistant to the Committee on Stock List: "The advance write-off (of Bond Discount) should not occur at all, either against Capital Surplus or against earned surplus. It distorts future income accounts and conceals from investors highly pertinent information. I cannot conceive a case justifying a charge against capital surplus for this reason. If it must be made, it should really be a charge against current income. I feel that an auditor should refuse to certify an account, even with qualification, where an entry of this sort has been made against capital surplus; certainly unless he secures from the corporation a written statement that, independent of whether the auditor retains his employment or not, all future annual statements will show as a footnote the fact that such Bond Discount and Expense has been written off, together with the amount of the charge which would fall against the year in question had this not been done."

CHAPTER XXXIII

MISLEADING ARTIFICES IN THE INCOME ACCOUNT. EARNINGS OF SUBSIDIARIES

Flagrant Example of Padded Income Account.—On comparatively rare occasions, managements resort to padding their income account by including items in earnings which have no real existence. Perhaps the most flagrant instance of this kind which has come to our knowledge occurred in the 1929–1930 reports of Park and Tilford, Inc., an enterprise with shares listed on the New York Stock Exchange. For these years the company reported net income as follows:

> 1929—$1,001,130 = $4.72 per share.
> 1930— 124,563 = 0.57 per share.

An examination of the balance sheets discloses that during these two years the item of Good-will and Trade-marks was written up successively from $1,000,000 to $1,600,000 and then to $2,000,000, and these increases *deducted* from the expenses for the period. The extraordinary character of the bookkeeping employed will be apparent from a study of the condensed balance sheets as of three dates, shown on page 373.

These figures show a reduction of $1,600,000 in net current assets in fifteen months, or $1,000,000 more than the cash dividends paid. This shrinkage was concealed by a $1,000,000 write-up of Good-will and Trade-marks. No statement relating to these amazing entries was vouchsafed to the stockholders in the annual reports, or to the New York Stock Exchange in subsequent listing applications. In answer to an individual inquiry, however, the company stated that these additions to Good-will and Trade-marks represented expenditures for advertising and other sales efforts to develop the business of Tintex Company, Inc., a subsidiary.[1]

The charging of current advertising expense to the good-will account is inadmissible under all canons of sound accounting.

[1] In the 1930 report the wording in the balance sheet was changed from "Good-will and Trade-marks" to "Tintex Good-will and Trade-marks."

PARK AND TILFORD, INC.

Balance Sheet	Sept. 30, 1929	Dec. 31, 1929	Dec. 31, 1930
Fixed assets	$1,250,000	$1,250,000	$1,250,000
Deferred charges	132,000	163,000	32,000
Good-will and Trade-marks	1,000,000	1,600,000	2,000,000
Net current assets	4,797,000	4,080,000	3,154,000
Bonds and mortgages	2,195,000	2,195,000	2,095,000
Capital and surplus	4,984,000	4,898,000	4,341,000
Total	$7,179,000	$7,093,000	$6,436,000

Adjusted earnings	First 9 months 1929	Last 3 months 1929	Year 1929	Year 1930
Earnings for stock as reported	$929,000	$72,000	$1,001,000	$125,000
Cash dividends paid	463,000	158,000	621,000	453,000
Charges against surplus				229,000
Added to capital and surplus	466,000	decrease 86,000	380,000	decrease 557,000
Earnings for stock as corrected (excluding increase in intangibles and deducting charges to surplus)	929,000	528,000(d)	401,000	504,000(d)

To do so without any disclosure to the stockholders is still more discreditable. It is difficult to believe, moreover, that the sum of $600,000 could have been expended for this purpose by Park and Tilford in the *three months* between September 30 and December 31, 1929. The entry appears therefore to have included a recrediting to *current* income of expenditures made in a *previous period*, and to that extent the results for the fourth quarter of 1929 may have been flagrantly distorted. Needless to say, no accountants' certificate accompanied the annual statements of this enterprise.[1]

Checks upon the Reliability of Published Earnings Statements.—The Park and Tilford case illustrates the necessity of relating an analysis of income accounts to an examination of the appurtenant balance sheets. It suggests also a further check upon the reliability of published earnings statements, *viz.*, by the amount of federal income tax accrued. Since the repeal of the Excess Profits Tax in 1921, corporation income taxes have been levied at the following rates:

Year	Rate, %	Year	Rate, %
1922	12½	1928	12
1923	12½	1929	11
1924	12½	1930	12
1925	13	1931	12
1926	13½	1932	13¾*
1927	13½	1933	13¾*

* 14½ on consolidated returns.

If the amount of the federal tax accrual for the year is stated, it is possible readily to calculate the income on which it was based and to compare this taxable profit with the earnings

[1] In April 1933 Park and Tilford common stock was changed from a no-par to a $1-par issue. This step was accompanied by a reduction of the stated capital from $3,278,330 to only $218,722 and permitted the transfer of the difference, $3,059,608, from capital stock to capital surplus. Judging from the procedure of other companies in this respect, this move seems likely to be followed by the writing down of the "Tintex Good-will and Trade-marks" account to $1, and the charging of the $2,000,000 present valuation thereof against the capital surplus. In this way the expenditures for advertising, etc., may prove finally to have been disposed of by a charge not against earnings but against *stated capital*. This subject is discussed at some length in Chap. XXXV.

reported to the stockholders. Certain differences may properly exist, due chiefly to the fact that interest received on federal, state, and municipal obligations and dividends from domestic corporations are not included in the taxable income, though they of course belong in the net income reported to shareholders. But if, after making such adjustments as are known to be required, a wide discrepancy exists between the tax paid and the reported income, the analyst is justified in viewing the latter with suspicion and in seeking further light from the company.

The Park and Tilford figures analyzed from this viewpoint supply the following suggestive results:

Period	Federal income tax accrued	Net income before federal tax	
		A. As indicated by the tax accrued	B. As reported to the stockholders
5 mo. to Dec. 1925	$36,881	$284,000	$ 297,000
1926	66,624	494,000	533,000
1927	51,319	380,000	792,000
1928	79,852	665,000	1,315,000
1929	81,623*	742,000	1,076,000

* Including $6,623 additional paid in 1931.

The close correspondence of the tax accrual with the reported income during the earlier period makes the later discrepancy appear the more striking. These figures eloquently cast suspicion upon the truthfulness of the reports made to the stockholders during 1927–1929, at which time considerable manipulation was apparently going on in the shares.

This and other examples discussed herein point strongly to the need for independent audits of corporate statements by certified public accountants. It may be suggested also that annual reports should include a detailed reconcilement of the net earnings reported to the shareholders with the net income upon which the federal tax is paid.

Another Extraordinary Case of Manipulated Accounting.—An accounting vagary fully as extraordinary as that of Park and Tilford, though exercising a smaller influence on the reported earnings, was indulged in by United Cigar Stores Company of America, from 1924–1927. The "theory" behind the entries was explained by the company for the first time in May 1927 in a listing application which contained the following paragraphs:[1]

The Company owns several hundred long-term leaseholds on business buildings in the principal cities of the United States, which up until May, 1924, were not set up on the books. Accordingly, at that time they were appraised by the Company and Messrs. F. W. Lafrentz and Company, certified public accountants of New York City, in excess of $20,000,000.

The Board of Directors have, since that time, authorized every three months the setting up among the assets of the Company a portion of this valuation and the capitalization thereof, in the form of dividends, payable in Common Stock at par on the Common Stock on the quarterly basis of 1¼% on the Common Stock issued and outstanding.

The entire capital surplus created in this manner has been absorbed by the issuance of Common Stock at par for an equal amount and accordingly is not a part of the existing surplus of the Company. No cash dividends have been declared out of such capital surplus so created.

The present estimated value of such leaseholds, using the same basis of appraisal as in 1924, is more than twice the present value shown on the books of the Company.

The effect of the inclusion of "Appreciation of Leaseholds" in earnings is shown herewith:

Year	Net earnings as reported	Earned per share of common ($25-par basis)	Market range ($25-par basis)	Amount of "Leasehold Appreciation" included in earnings	Earned per share of common excluding lease appreciation
1924	$6,697,000	$4.69	64–43	$1,248,000	$3.77
1925	8,813,000	5.95	116–60	1,295,000*	5.05
1926	9,855,000	5.02	110–83	2,302,000	3.81
1927	9,952,000†	4.63	100–81	2,437,000	3.43

* The 5% stock dividend paid in 1925 amounted to $1,737,770. There is an unexplained difference between the two figures, which in the other years are identical.

† Excluding refund of federal taxes of $229,017 applicable to prior years.

In passing judgment on the inclusion of leasehold appreciation in the current earnings of United Cigar Stores, a number of considerations might well be borne in mind.

1. Leaseholds are essentially as much a liability as they are an asset. They are an obligation to pay rent for premises occupied. Ironically enough, these very leaseholds of United Cigar Stores eventually plunged it into bankruptcy.

[1] See application to list 6% Cumulative Preferred Stock of United Cigar Stores Company of America on the New York Stock Exchange, dated May 18, 1927 (Application #A-7552).

2. Assuming leaseholds may acquire a capital value to the occupant, such value is highly intangible, and it is contrary to accounting principles to mark up above actual cost the value of such intangibles in a balance sheet.

3. If the value of any capital asset is to be marked up, such enhancement must be credited to Capital Surplus. By no stretch of the imagination can it be considered as *income*.

4. The $20,000,000 appreciation of the United Cigar Stores leases took place prior to May 1924, but it was *treated as income in subsequent years*. There was thus no connection between the $2,437,000 appreciation included in the profits of 1927 and the operations or developments of that year.

5. If the leaseholds had really increased in value, the effect should be visible in *larger earnings* realized from these favorable locations. Any other recognition given this enhancement would mean counting the same value twice. In fact, however, allowing for extensions of the business financed by additional capitalization, the per-share earnings of United Cigar Stores showed no advancing trend.

6. Whatever value is given to leaseholds must be amortized over the life of the lease. If the United Cigar Stores investors were paying a high price for the shares because of earnings produced by these valuable leases, then they should *deduct* from earnings an allowance to write off this capital value by the time it disappears through the expiration of the leases.[1] The United Cigar Stores Company continued to amortize its leaseholds on the basis of *original cost*, which apparently was practically nothing.

The surprising truth of the matter, therefore, is that the effect of the appreciation of leasehold values—if it had occurred—should have been to *reduce* the subsequent operating profits by an increased amortization charge.

7. The padding of the United Cigar Stores income for 1924–1927 was made the more reprehensible by the failure to reveal the facts clearly in the annual reports to shareholders.[2] Disclosure of the essential facts to the New York Stock Exchange was made nearly three years after the practice was initiated. It may have been compelled by legal considerations growing out of the sale to the public at that time of a new issue of preferred stock, underwritten by large financial institutions. The following year the policy of including leasehold appreciation in earnings was discontinued.

These accounting maneuvers of United Cigar Stores may be fairly described, therefore, as the *unexplained* inclusion in current earnings of an *imaginary* appreciation of an *intangible* asset—the asset being in reality a *liability*, the enhancement being related to a *previous* period, and the proper effect of the appreciation, if it had occurred, being to *reduce* the subsequent earnings by virtue of higher amortization charges.

[1] This subject is treated fully in a succeeding chapter.

[2] The reports stated the "Net Profit for the year, including Enhancement of Leasehold Values" (giving amount of the latter), but no indication was afforded that this enhancement was arbitrarily computed and had taken place in previous years.

The federal-income-tax check, described in the Park and Tilford example, will also give interesting results if applied to United Cigar Stores:

Year	Federal tax reserve	Income before tax		
		A. Indicated by tax reserve	B. Reported to stockholders	C. Reported to stockholders less leasehold appreciation
1924	$700,000	$5,600,000	$ 7,397,000	$6,149,000
1925	825,000	6,346,000	9,638,000	8,343,000
1926	900,000	6,667,000	10,755,000	8,453,000
1927	900,000	6,667,000	10,852,000*	8,415,000*
1928	700,000	5,833,000	9,053,000	9,053,000
1929	13,000	118,000	3,132,000†	3,132,000†
1930	none	none	1,552,000	1,552,000

* Eliminating tax refund of $229,000 evidently applicable to prior years.
† This is also reported as $2,947,000, after an adjustment.

Moral Drawn from Foregoing Examples.—A moral of considerable practical utility may be drawn from the United Cigar Stores example. When an enterprise pursues questionable accounting policies, *all* its securities must be shunned by the investor, no matter how safe or attractive some of them may appear. This is well illustrated by United Cigar Stores Preferred, which made an exceedingly impressive statistical showing for many successive years but which later proved all but worthless. Investors confronted with the strange bookkeeping above detailed might have reasoned that the issue was still perfectly sound, because, when the overstatement of earnings was corrected, the margin of safety remained more than ample. Such reasoning is fallacious. You cannot make a quantitative deduction to allow for an unscrupulous management; the only way to deal with such situations is to avoid them.

Fictitious Value Placed on Stock Dividends Received.—From 1922 on, most of the United Cigar Stores common shares were held by Tobacco Products Corporation, an enterprise controlled by the same interests. This was an important company, the market value of its shares averaging more than $100,000,000 in 1926 and 1927. The accounting practice of Tobacco Products introduced still another way of padding the income account, *viz.*, by placing a fictitious valuation upon stock dividends received.

For the year 1926 the company's earnings statement read as follows:

Net income	$10,790,000
Income tax	400,000
Class A dividend	3,136,000
Balance for common stock	7,254,000
Earned per share	11
Market range for common	117–95

Detailed information regarding the company's affairs during that period has never been published (the New York Stock Exchange having been unaccountably willing to list new shares on submission of an extremely sketchy exhibit). Sufficient information is available, however, to indicate that the net income was made up substantially as follows:

Rental received from lease of assets to American Tobacco Co	$ 2,500,000
Cash dividends on United Cigar Stores common (80% of total paid)	2,950,000
Stock dividends on United Cigar Stores common (par value $1,840,000), less expenses	5,340,000
	$10,790,000

It is to be noted that Tobacco Products must have valued the stock dividends received from United Cigar Stores at about three times their face value, *i.e.*, at three times the value at which United Cigar charged them against surplus. Presumably the basis of this valuation by Tobacco Products was the market price of United Cigar Stores shares, which price was easily manipulated due to the small amount of stock not owned by Tobacco Products.

When a holding company takes into its income account stock dividends received at a higher value than that assigned them by the subsidiary which pays them, we have a particularly dangerous form of pyramiding of earnings. The New York Stock Exchange, beginning in 1929, has made stringent regulations forbidding this practice. (The point was discussed in Chap. XXX.) In the case of Tobacco Products the device was especially objectionable because the stock dividend was issued in the first instance to represent a fictitious element of earnings, *i.e.*, the appreciation of leasehold values. By unscrupulous exploitation of the holding-company mechanism these imaginary profits were effectively multiplied by three.

On a consolidated earnings basis, the report of Tobacco Products for 1926 would read as follows:

American Tobacco Co. lease income, less income tax, etc..................................	$2,100,000
80% of earnings on United Cigar Stores common	5,828,000*
	$7,928,000
Class *A* dividend...........................	3,136,000
	$4,792,000
Balance for common........................	$4,792,000
Earned per share...........................	$ 7.27

* Excluding leasehold appreciation.

The reported earnings for Tobacco Products common given as $11 per share are seen to have been overstated by about 50%.

It may be stated as a Wall-Street maxim that where manipulation of accounts is found there will be found also stock juggling in some form or other. Familiarity with the methods of questionable finance should assist the analyst, and perhaps even the public, in detecting such practices when they are perpetrated.[1]

SUBSIDIARY COMPANIES AND CONSOLIDATED REPORTS

This title introduces our second general type of adjustment of reported earnings. When an enterprise controls one or more important subsidiaries, a *consolidated* income account is necessary to supply a true picture of the year's operations. Figures showing the parent company's results only are incomplete and may be quite misleading. As previously remarked, they may either understate the earnings by not showing all the current profits made by the subsidiaries, or they may overstate the earnings by failure to deduct subsidiaries' losses or by including dividends from subsidiaries in excess of their actual income for the year.

Nonconsolidated Profits.—Prior to the separation of the Reading Company's railroad and coal properties in 1923, the reported earnings on Reading stock were consistently understated because they did not include the profits of the coal subsidiary,

---

[1] To avoid an implication of inconsistency, because of our favorable comments on Tobacco Products Corporation 6½s, due 2022, in a previous chapter, we must point out that a complete change of management took place in this situation during 1930.

which paid no dividend to its parent company. Until January 1923 the Consolidated Gas Company of New York similarly reported only that part of its subsidiaries' earnings which were received by it in the form of dividends. Since the New York Edison Company and other electric subsidiaries were more important than the gas company itself, this form of statement was obviously far from adequate. To obtain an accurate picture of the earning power of Consolidated Gas, the analyst would have found it necessary to construct a consolidated income account from the detailed information submitted, not to the stockholders, but to the Public Service Commission of New York State.

Nonconsolidated Losses.—The opposite situation, *viz.*, failure to deduct the losses of subsidiaries, is illustrated by the report of Warren Brothers Company for 1923. The statement of this road-building enterprise showed record earnings of $724,000, equal after preferred dividends to $11.44 per share of common. These earnings, however, failed to allow for losses sustained by two subsidiaries, the results of which were not available to the stockholders until the company listed its shares in 1925. A corrected income account for 1923 would have read as follows:

Earnings reported by Warren Brothers Co....................$724,000

Deduct:

72% of $754,000 loss of Warren Construction Co..	$543,000	
100% of $80,000 loss of Southern Roads Co......	80,000	623,000

Consolidated net profit.................................. $101,000
Preferred dividends (of Warren Bros. Co. only)............ 151,000

Balance for common.................................... $ 50,000(d)

This deficit is without allowance for federal taxes. The company's report for 1924 showed, significantly enough, that no taxes had been paid for the year 1923 despite the earnings of $724,000 reported to the stockholders.[1]

Special Dividends Paid by Subsidiaries.—When earnings of nonconsolidated subsidiaries are allowed to accumulate in their surplus accounts, they may be used later to bolster up the results

---

[1] The stock of Warren Brothers advanced materially after 1925. Hence knowledge that the true earnings of the company for 1923 had been much less than reported would have been harmful rather than helpful from the practical standpoint. This illustrates once again our point that analysis is not a dependable tool in dealing with *speculative* situations.

of a poor year by means of a large special dividend paid over to the parent company.

Examples: Such dividends, amounting to $11,000,000, were taken by the Erie Railroad Company in 1922 from the Pennsylvania Coal Company and Hillside Coal and Iron Company. The Northern Pacific Railway Company similarly eked out its depleted earnings in 1930 and 1931 by means of large sums taken as special dividends from the Chicago, Burlington and Quincy Railroad Company, the Northern Express Company and the Northwestern Improvement Company, the last being a real-estate, coal, and iron-ore subsidiary. The 1931 earnings of the New York, Chicago, and St. Louis Railroad Company included a back dividend of some $1,600,000 on its holdings of Wheeling and Lake Erie Railway Company Prior Preferred Stock, only a part of which was earned in that year by the Wheeling road.

This device of concealing a subsidiary's profits in good years and drawing upon them in bad ones may seem quite praiseworthy as a method of stabilizing the reported earning power. But such benevolent deceptions are frowned upon by enlightened opinion, as illustrated by the more recent regulations of the New York Stock Exchange which insist upon full disclosure of subsidiaries' earnings. It is the duty of managements to disclose the truth and the whole truth about the results of each period; it is the function of the stockholders to deduce the "normal earning power" of their company by averaging out the earnings of prosperity and depression. Manipulation of the reported earnings by the management even for the desirable purpose of maintaining them on an even keel is objectionable none the less because it may too readily lead to manipulation for more sinister reasons.

Revision of Income Account Should Be Made to Allow for Such Items.—It is clear from the foregoing discussion that whenever necessary and possible the reported earnings should be corrected to reflect the current results of all subsidiaries as well as of the parent company. A subsidiary is ordinarily defined as a company controlled through ownership of a majority of the voting stock.

Example: The Louisville and Nashville Railroad Company, for example, is regarded as a subsidiary of the Atlantic Coast Line, which holds 51% of the former's capital stock. Hence a competent analysis of the earning power of Atlantic Coast Line would make due allowance for its share of the losses or the undistributed profits of the Louisville and Nashville. This procedure would not be customary however, in connection with the Union Pacific's substantial holdings of Illinois Central and of other railroad shares, because these blocks are not large enough to constitute control.

But there is no good reason to insist too rigidly upon full majority ownership as the criterion of parent-subsidiary relationship.

Example: Northern Pacific and Great Northern each own 48.6% of the Chicago, Burlington and Quincy's stock. This degree of control is large enough to warrant treating the Burlington as a subsidiary of each of the two Northern roads, and to justify an analysis upon a consolidated basis, *i.e.*, including the proper proportion of the Burlington's results in the exhibits of both Northern Pacific and Great Northern. It might be said, therefore, that whenever one enterprise owns an *important interest* in another, an analysis of the former company may properly take into account its full share of the current results of the latter.

Method of Revision Illustrated.—The method of working out this form of revision of reported earnings may be illustrated by the case of E. I. du Pont de Nemours and Company. In 1932 du Pont reported earnings of $19,770,000 for its common stock, amounting to $1.81 per share on 10,872,000 shares. These earnings included $12,500,000 received as dividends on its holdings of some 10,000,000 shares of General Motors common stock, representing about 23% of the total outstanding. General Motors, however, not only failed to earn its common dividend in 1932 but showed a loss of 21 cents per share after preferred dividends. On a combined basis, therefore, the reported earnings of du Pont common for 1932 would have to be reduced by about $14,600,000, representing its share of the loss of General Motors after dividends. (This corresponds to the reduction in the book

Year	du Pont earnings per share	Adjustments to reflect du Pont's interest in operating results of General Motors	Earnings per share of du Pont as adjusted
1932	$1.81	−$1.35	$0.46
1931	4.30	− 0.51	3.79
1930	4.52	+ 0.04	4.56
1929	6.99	+ 1.12	8.11

value of its holdings of General Motors for the year.) The same calculation made for the four years 1929–1932 will give the corrected results as shown in table on page 383.

The report of General Motors Corporation for 1931 is worthy of appreciative attention because it includes a supplementary calculation of the kind suggested in this and the previous chapter *i.e.*, exclusive of special and nonrecurring profits or losses, and inclusive of General Motors' interest in the results of nonconsolidated subsidiaries. The report contains the following statement of per-share earnings for 1931 and 1930:

EARNINGS PER SHARE, INCLUDING THE EQUITY IN UNDIVIDED PROFITS OR LOSSES OF NONCONSOLIDATED SUBSIDIARIES

Year	Including nonrecurrent items	Excluding nonrecurrent items
1931	$2.01	$2.43
1930	3.25	3.04

Suggested Procedure for Statistical Agencies.—While this procedure may seem to complicate a report, it is in fact a salutary antidote against the oversimplification of common-stock analysis which resulted from exclusive preoccupation with the single figure of per-share earnings. The statistical manuals and agencies have naturally come to feature the per-share earnings in their analyses of corporations. They might, however, perform a more useful service if they omitted a calculation of the per-share earnings in all cases where the company's reports appear to contain irregularities or complications in any of the following directions:

1. By reason of nonrecurrent items included in income, or because of charges to surplus which might properly belong in the income account.

2. Because current results of subsidiaries are not accurately reflected in the parent company's statements.

3. Because the depreciation and other amortization changes are irregularly computed.

Distorted Earnings through Parent-subsidiary Relationships.—Examples are available of the use of the parent-subsidiary relationship to produce astonishing distortions in the reported income. We shall give two illustrations taken from the railroad field. These instances are the more impressive because the stringent accounting regulations of the Interstate Commerce Commission might be expected to prevent any misrepresentation of earnings.

Examples: In 1925, Western Pacific Railroad *Corporation* paid dividends of $7.56 upon its preferred stock and $5 upon its common stock. Its income account showed earnings slightly exceeding the dividends paid. These earnings consisted almost entirely of dividends aggregating $4,450,000 received from its operating subsidiary, the Western Pacific Railroad *Company.* The year's earnings of the railroad, itself, however, were only $2,450,000. Furthermore its accumulated surplus was insufficient to permit the larger dividend which the parent company desired to report as its income for the year. To achieve this end, the parent company went to the extraordinary lengths of *donating* the sum of $1,500,000 to the operating company, and it immediately took the same money back as a *dividend* from its subsidiary. The donation it charged against its *surplus;* the receipt of the same money as dividends it reported as *earnings.* In this devious fashion it was able to report $5 "earned" upon its common stock, when in fact the applicable earnings were only about $2 per share.

In support of our previous statement that bad accounting practices are contagious, we may point out that the Western Pacific example of 1925 was followed by the New York, Chicago, and St. Louis Railroad Company ("Nickel Plate") in 1930 and 1931. The details are briefly as follows:

In 1929 Nickel Plate sold, through a subsidiary, its holdings of Pere Marquette stock to Chesapeake and Ohio, which was under the same control. A profit of $10,665,000 was realized on this sale, which gain was properly credited to surplus. In 1930 Nickel Plate needed to increase its income; whereupon it took the $10,665,000 profit out of its surplus, returned it to the subsidiary's treasury and then took $3,000,000 thereof in the form of a "dividend" from this subsidiary, which it included in its 1930 *income.* A similar dividend of $2,100,000 was included in the income account for 1931.

These extraordinary devices may have been resorted to for what was considered the necessary purpose of establishing a net income large enough to keep the company's bonds legal for trust-fund investments. The result, however, was the same as that from all other misleading accounting practices, *viz.,* to lead the public astray and to give those "on the inside" an unfair advantage.

CHAPTER XXXIV

THE RELATION OF DEPRECIATION AND SIMILAR CHARGES TO EARNING POWER

A critical analysis of an income account must pay particular attention to the amounts deducted for depreciation and kindred charges. These items differ from ordinary operating expenses in that they do not signify a current and corresponding outlay of cash. They represent the estimated shrinkage in the value of the fixed or capital assets, due to wearing out, to using up, or to their approaching extinction for whatever cause. The important charges of this character may be classified as follows:

1. Depreciation (and obsolescence), replacements, renewals or retirements.
2. Depletion or exhaustion.
3. Amortization of leaseholds and leasehold improvements.
4. Amortization of patents.

All these items may properly be embraced under the title "amortization," but we shall sometimes refer to them generically as "depreciation items," or simply as "depreciation," because the latter is a more familiar term.

Leading Questions Relative to Depreciation.—The accounting theory which governs depreciation charges is simple enough. If a capital asset has a limited life, provision must be made to write off the cost of that asset by charges against earnings distributed over the period of its life. Corporate reports, however, do not always follow accepted accounting methods in computing their depreciation charges; and furthermore on frequent occasions the allowance which may be justified from a technical accounting standpoint may fail to meet the situation properly from the investment standpoint. The analyst should ask three questions with respect to the depreciation feature of an income account:

1. Have the amortization charges been *deducted* from the earnings as here reported?
2. Are the *rates* employed reasonable, as judged by standard accounting practice?
3. Does the cost or *base* to which these rates are applied reflect reasonably well the fair value of these assets to the investor?

Depreciation Charges Represent an Expense of Operation.— Earnings are sometimes reported to the public without deducting the amortization charges which have actually been made on the books of the company. In our discussion of bond analysis we pointed out that this practice has become distressingly prevalent in the offering of public-utility bonds to the public, and we insisted that the bond buyer reject any presentation which omitted so important a deduction from earnings. The argument is often made that depreciation charges may properly be ignored because they are mere bookkeeping entries and do not represent a real outlay of cash. This is a highly inaccurate statement of the case. Depreciation is not a mere bookkeeping conception, because for the most part it registers an actual diminution of capital values, for which adequate provision must be made if creditors or owners are to avoid deceiving themselves.

Moreover, in the majority of cases the depreciation charges are consumed or offset over a period of time by even larger cash expenditures made for replacements or extensions. More often than not, therefore, depreciation charges are eventually found to be related to actual cash outlays, and turn out to be as truly an expense of the business as wages or rents. Minority cases are fairly numerous in which a good part of the depreciation reserve remains unexpended over a long period of time. In these instances a reduction of the annual charges may sometimes be justified in the investor's calculations, as we shall later explain. The broad principle remains, however, that an adequate depreciation allowance is essential in arriving at a fair statement of earnings.

Leading Example of Omission of Such Charges.—While depreciation charges have frequently been omitted in bond-circular computations, it is a comparatively rare occurrence to find them similarly excluded from the annual reports to stockholders. The practice of the Cities Service Company prior to 1930 in this respect is fortunately quite exceptional, but it deserves specific mention because of the magnitude of the enterprise and its enormous number of small stockholders. This company carried on for many years a continuous sale of its shares to the public, through its fiscal agents, and in its sales literature and annual reports it maintained the invariable policy of stating earnings without deduction of depreciation and depletion charges. The two lines of business in which it was engaged—public utility and oil—normally require the setting aside of a substantial portion of gross earnings for amortization of property. The following condensed income account for a typical year will show how thoroughly misleading were the statements on which Cities Service securities were continuously and aggressively sold to the public.

Item	Year 1925	Year 1931 (appended for comparison)
Gross earnings	$127,108,000	$177,047,000
Net before depreciation and depletion	40,249,000	63,107,000
Interest and preferred dividends	26,628,000	44,942,000
Minority interest	2,124,000	2,757,000
Balance for common stocks and reserves	11,497,000	15,408,000
Reserves for depreciation, etc..	12,700,000(est.)*	18,063,000 (actual)
Actual balance for common....	Deficit	Deficit
Earned per share, as reported..	$3.05	$0.42
Earned per share, actual	Nil	Nil
Average market price of common	39	13

* This estimate of 10 % of gross for depreciation, etc., appears reasonable in the light of: (1) the actual reserves for 1925 set up by the larger subsidiaries in their separate reports; (2) the reserves for the entire system disclosed in 1930 and subsequent years; and (3) the practice followed by representative companies in the same fields.

Criticism of the Justification Offered.—The Cities Service Company has defended its failure to allow for amortization in its statements on the ground that its property was increasing so rapidly in value that no real depreciation was taking place at all. But whatever increases in value were taking place should have reflected themselves in the earning power *after deduction* of normal depreciation charges. The fact that year after year there seem to have been no substantial earnings available for the common stock, after such deductions are allowed for, would strongly gainsay the claim of increasing value. In any case it is a complete logical fallacy, equivalent to "counting the same trick twice," to take *double credit* in the income statement for enhanced property values, *i.e.*, first in the increased net profits therefrom, and secondly in the waiving of the customary amortization charges on the ground that there is no net shrinkage in capital value.[1] This misconception, or misrepresentation, is identical with that of United Cigar Stores, previously discussed, whereby portions of a purported increase in leasehold values were added each year to net earnings—although the effect of such an increased value should already have shown itself in larger profits from these favorably located stores.

Another Misleading Practice.—The Cities Service accounting methods illustrate another practice which is perhaps even more dangerous than the frank omission of depreciation from the reported earnings. This is the device of charging a part of the year's amortization to income and the balance direct to surplus. A fairly careful investor will know enough to reject an earnings statement in which it is announced that no allowance for depreciation has been made. But where the figures are given "after depreciation" he is not so likely to make further inquiry on this point, so that an understatement of these charges may do more real damage than their complete and admitted exclusion.

It should be clear that except under quite abnormal conditions whatever depreciation accrues during a year is charged against the earnings and should not be handled as an "adjustment of surplus." For some years many of the Cities Service subsidiaries deducted the major portion of the annual depreciation allowance from surplus, thus showing a correspondingly larger "earning power" in their income accounts. This practice was apparently related to a provision in Cities Service Power and Light Company bond indentures requiring certain minimum deductions for maintenance and depreciation combined. These requirements, set up ostensibly for the protection of the investor, were actually used to mislead him. The charges made by the subsidiaries were normally much greater than the indenture minima, and the actual allowances should, of course, have been reported in the income statements. But the "indenture requirements" afforded a convenient pretext for drastically reducing the charges against

---

[1] The Middle States Oil Corporation used the same practice and arguments in what amounted to a fraudulent campaign of market manipulation and dishonest reports, during which the shares were advanced to 71¾ in 1919 and then were practically wiped out through receivership (1924) and reorganization (1929). Curiously enough, the company actually overpaid large amounts of income taxes to make its inflated earnings appear more plausible.

income and deducting the balance from the surplus account, where the investor was not likely to notice it.[1]

It is unfortunate that something resembling this practice has been resorted to at times by conservatively managed companies. Note the following in the reports of the Detroit Edison Company for 1931 and 1930.

Item	1931	1930
Gross...............................	$49,233,000	$53,707,000
Net before depreciation...............	21,421,000	24,041,000
Depreciation.......................	4,000,000	6,900,000
(Per cent of gross).................	8.1%	12.8%
Fixed charges.......................	5,992,000	6,024,000
Balance for common..................	11,429,000	11,117,000
Earned per share....................	$8.98	$8.75
Additional depreciation charged to surplus.............................	1,500,000	
Earned per share after charge to surplus	$7.80	$8.75

Although Detroit Edison's depreciation charges have been unusually liberal by comparison with the average for the industry, the accounting method employed for 1931 might well be criticized for two reasons. In the first place its effect, if not its purpose, was to disguise the actual decline in earnings from the previous year. Secondly, because of the high reputation of the company, this device was likely to be imitated by other enterprises, and thus it might furnish an unwholesome stimulus to the new practice of overstating earnings by the transfer of charges to the surplus account.

Another Example of Tricky Accounting.—An extraordinary example of tricky accounting is displayed by Iowa Public Service Company. For 1929 this company reported a property account of $25,200,000, gross earnings of $4,200,000, and a depreciation charge of only $78,000. The inadequacy of this figure is patent. In succeeding years the depreciation allowance was gradually increased, reaching $220,000 in 1932, which was still a somewhat subnormal figure. In 1932, the company made formal confession of the insufficiency of its past depreciation charges, by the following unique procedure:

[1] See also Chap. XII for a discussion of this practice as it affects the stated coverage for bond interest.

1. It reduced the stated value of its common stock by $1,587,000 and transferred this sum to capital surplus.

2. It immediately used up this capital surplus by charging against it $1,500,000 for additional depreciation and $87,000 for contingencies.

In this case we see a good part of the necessary depreciation charge excluded from the income statements over a period of years, and finally allowed for by reducing the amount at which the common stock is valued. An incidental effect of this mischievous accounting was to permit the parent company (American Electric Power Corporation) to take out in dividends a sum exceeding the true earnings and the initial surplus combined, to the serious prejudice of the bondholders and the first preferred stockholders.

Double Accounting Policies on Depreciation.—An issue of overshadowing importance in the accounting practices of public-utility systems is revealed by a study of the 53-page prospectus issued by American Water Works and Electric Company in connection with the sale of its ten-year convertible 5% bonds in February 1934. The prospectus disclosed for the first time that the company had regularly been employing two different bases for calculating property amortization charges, one for income-tax purposes and the other, *at a much lower rate*, in its reports to its security holders. The significance of this situation, in its relation both to the earning power of the common stock and the soundness of the senior issues, should be manifest from the following figures:

Year	Charge for "renewals, replacements, etc." reported to the public	Charge for "depreciation" reported to the government	Balance for common stock reported to the public	
			Total	Per share
1930	$4,105,000	$6,781,000	$5,377,000	$3.10
1931	3,095,000	7,089,000	5,117,000	2.80
1932	2,747,000	7,023,000	2,396,000	1.42
1933 (11 mo.)	2,654,000	6,384,000	1,870,000	1.07

Year	Balance for common stock on basis of depreciation taken in the tax returns	Per share
1930	$2,701,000	$1,60
1931	1,123,000	0.70
1932	1,880,000(d)	nil
1933 (11 mo.)	1,860,000(d)	nil

Immediately the question arises: "Can both of these methods of calculating reserves properly be used at the same time? If not, which is the right one?" The prospectus contains a long statement on the subject, sufficiently technical and dexterous to allay the suspicions of nearly all the few who might read it. But a careful study of what is said must induce grave doubts in the mind of the analyst as to the propriety of the figures submitted to the stockholders. The much larger deductions taken in the tax returns are based upon the standard theory of depreciation and upon standard tables of estimated life of the depreciable assets. The long discussion in the prospectus yields no evidence that the "Reserve for Renewals, etc.," set up in the reports to the public, is fully adequate to take care of the loss of value of the capital assets by reason of wear and tear and obsolescence.

It is claimed that the practices followed are customary in the industry, and that the rates charged are determined in some cases by agreement with state or municipal authorities and generally upon consultation between engineers and officers of the company. Various considerations are taken into account, including the annual earnings available in each year. But on the vital question of *adequacy* the entire statement is most unconvincing.

It is significant that the auditors certifying the consolidated accounts specifically exclude the matter of reserves for renewals, etc., from the scope of their examination and certificate. They do say that the actions taken by the company with respect to such reserves "are not an attempt to measure depreciation on the basis of estimated life." It is also significant that the annual charges for renewals and replacements have averaged less than 1% of the property account and less than 1% of the capitalization of the system. Finally we must note the sharp reduction in these charges during 1931–1933, the cut being much heavier percentagewise than the decline in gross or net revenue.

A Serious Problem for the Investor.—Disclosures such as these must induce gravely disturbing questions in the analysis of public-utility securities. The amount of amortization charged has so vital a bearing on the reported earnings in this field that the knowledge that there are two entirely different bases for calculating these charges must, at the very least, create considerable misgivings with respect to the published figures. The indications are very strong that the size of the net profits of utility companies is largely susceptible to arbitrary fashioning by the managements, according as they adopt one policy or the other with respect to depreciation allowances. Hence the wide differences which appear in the exhibits of different systems —to which we shall refer a little later—would seem to be occasioned not so much by divergences in the *character* of the enterprise as by purely voluntary decisions by those in control. In the individual case of American Water Works and Electric Company, just discussed, the conservatively minded analyst or investor could scarcely be expected to accept the reported figures as trustworthy indications of the true and comparable earning power of the enterprise.[1]

Further Study of the American Water Works Prospectus.—We may digress for a moment from the subject of depreciation allowances to pursue the disclosures of the American Water Works and Electric prospectus a little further, for they include a number of matters relevant to the technique of security analysis. The implications of the income-tax figures are quite arresting, as is shown by the following:

[1] The whole subject of annual allowances for depreciation or renewals is a thorny and controversial one. A study published by Professor Henry E. Riggs in 1922 (*Depreciation of Public Utility Properties*, McGraw-Hill Book Company, Inc.) tends to support the less conservative retirement-reserve method as against the straight-line depreciation method. The implications in Professor William Z. Ripley's discussion (*Main Street and Wall Street*, pp. 172–175, 333–336) are rather in the opposite direction. There is also official disagreement on the subject, the Interstate Commerce Commission and the Bureau of Internal Revenue favoring the straight-line depreciation method for railroad and income-tax accounting, and the National Association of Railroad and Utilities Commissioners favoring the retirement-reserve method by prescribing it in its uniform classification of accounts. The former method is also prescribed by the uniform system of accounts for electric and gas corporations promulgated by the Public Service Commission of the State of New York. In general the electric light and gas companies favor Retirement Accounting and the telephone companies favor Depreciation Accounting. For a readable discussion of this controversy see Eliot Jones and Truman C. Bigham, *Principles of Public Utilities*, Chap. X, especially pp. 478–494, The Macmillan Company, 1931.

Assuming that the allowances made by American Water Works and Electric Company in its published reports are adequate, as they may be, the company must still appear vulnerable from the standpoint of its income-tax accruals, and, furthermore, important adjustments would be needed in comparing its figures with those of other companies using a more liberal basis for depreciation.

Year	Federal income tax paid or accrued	Rate of tax, %	Net earnings as indicated by tax†	Net earnings as reported†	Difference approximately
1930	$1,501,000	12	$12,500,000	$13,694,000	$1,194,000
1931	472,000	12	3,932,000	12,423,000	8,490,000
1932	103,000	14½*	713,000	9,345,000	8,630,000
1933 (11 mo.)	588,000	14½*	4,068,000	8,792,000	4,720,000

* Consolidated return basis.
† Before deducting subsidiaries' preferred dividends.

DIFFERENCES DISCLOSED IN PROSPECTUS BETWEEN INCOME-TAX FIGURES
AND REPORTED NET EARNINGS

Year	Excess of charge for depreciation	Excess of charge for amortization of bond discount	Loss from abandonment, etc. of electric railway properties	Total differences disclosed
1930	$2,676,000	$260,000	$ 480,000	$3,416,000
1931	3,994,000	280,000	4,128,000	8,402,000
1932	4,276,000	290,000	3,411,000*	7,977,000
1933 (11 mo.)	3,730,000	253,000	1,366,000*	5,349,000

* Includes $1,118,000 in 1932 and $100,000 in 1933 for loss on investments in Wheeling Traction Company, etc.

The net income of the system for tax purposes is evidently quite a different matter from the net income as published. In addition to the significant disparities in the depreciation charges, large write-offs were made for the abandonment of electric-railway properties, which were claimed as capital losses on the income-tax returns. Such losses, being of nonrecurring nature, are properly separable from the year's operating results. But the astonishing fact in the situation is that these charges, apparently exceeding $9,000,000 during the four-year period, appear *nowhere at all* in the company's statements. Not only are they excluded from the income account, but no trace can be found of them in the balance sheets—either in the property accounts, or in the surplus, or in any of the reserves. By a strange contrivance, these losses were taken on the books of the subsidiaries in such a way as not to be reflected in the *consolidated* balance sheets. (Apparently the value of other properties was marked up on the books to offset the amounts written off.)

A Minimum Adjustment of the Published Earnings.—In endeavoring to evaluate these discrepancies in the American Water Works and Electric statement, we must recognize that at least a minimum reduction is to be made in the reported earnings, *i.e.*, the amount by which the income tax would be increased if the company actually earned the sums it reported. For in gauging the *future earning power* on the basis of the past record, it must be assumed that the company will have to pay income taxes commensurate with what it earns. If the reported figures are true profits, it is unthinkable that the company could continuously persuade the government to accept taxes on a much smaller sum. Hence the *minimum* adjustment of the earnings for the period, required if we accept the company's own figures of its earnings and adjust its income tax thereto, would be as follows:

| Year | Tax reported | Tax required by reported earnings | Difference | | Maximum earnings per share of common as adjusted |
			Amount	Per share of common	
1930	$1,501,000	$1,644,000	$ 143,000	$0.08	$3.02
1931	472,000	1,490,000	1,018,000	0.61	2.19
1932	103,000	1,380,000	1,277,000	0.70	0.72
1933 (11 mo.)	588,000	1,280,000	692,000	0.39	0.68

If, however, the depreciation charges on which the tax return is based are correct, the reduction in earnings would be far greater than the minimum above computed.

The American Water Works and Electric prospectus has a certain historical significance as the first issued for a major bond offering under the Securities Act of 1933. The importance of the additional information regarding the company's affairs therein elicited may be considered as a practical vindication of the stringent requirements of the Act as regards disclosure of facts. But it also brings home the inherent complexity of present-day accounting, and therefore of present-day investment problems. The need for competent and impartial analysis would seem correspondingly more impressive.

Reporting of Depletion Charges.—Depletion represents the using up of capital assets by turning them into products for sale. It applies to companies producing metals, oil and gas, sulphur, timber, etc. As the holdings, or reserves, of these products are exhausted their value must gradually be written off through charges against earnings. In the case of the older mining companies (including particularly the copper and sulphur producers) the depletion charges are determined by certain technical requirements of the federal income tax law, which rest upon the amount and value of the reserves as they were supposed to exist on March 1, 1913. Because of the artificial base used in these computations, many companies have omitted the depletion charge from their reports to stockholders.

Independent Calculation by Investor Necessary.—As we shall show later, the investor in a mining concern must ordinarily compute his own depletion allowance, based upon the *amount which he has paid* for his share of the mining property. Consequently a depletion charge based either on the company's original book cost or on the special figure set up for income-tax purposes would be confusing rather than helpful. The omission of the depletion charge of mining companies is not to be criticized, therefore; but the stockholder in such enterprises must be well aware of the fact in studying their reports. Furthermore, in any comparison of mining companies a proper distinction must be drawn between those which do and those which do not deduct their depletion charges in reporting their earnings. Following are some examples of companies which pursue one or the other policy:

Companies Which Report Earnings without Deduction for Depletion:	Companies Which Report Earnings after Deduction for Depletion:
Alaska Juneau Gold Mining Co.	Cerro de Pasco Copper Corp.
Anaconda Copper Mining Co.	Granby Consolidated Mining etc., Co. (copper)
Dome Mines, Ltd. (gold)	Homestake Mining Co. (gold)
Kennecott Copper Corp.	International Nickel Co. of Canada, Ltd.
Noranda Mines, Ltd. (copper and gold)	Patino Mines, etc. (tin)
Texas Gulf Sulphur Co.	Phelps Dodge Corp. (copper)
	St. Joseph Lead Co.

Depletion in the Oil Industry.—In the oil industry depletion charges are more closely related to the actual cost of doing business than in the case of mining enterprises. The latter ordinarily invest in a single property or group of properties, the cost of which is then written off over a fairly long period of years. But the typical large oil producer normally spends substantial sums each year on new leases and new wells. These additional holdings are needed to make up for the shrinkage of reserves through production. The depletion charge corresponds in some measure, therefore, to a current cash outlay for the purpose of maintaining reserves and production. New wells may yield as high as 80% of their total output during the first year. Hence nearly all the cost of such "flush production" must be written off in a single fiscal period; and most of the "earnings" from this source are in reality a return of the capital expended thereon. If the investment is not written off rapidly through depletion and other charges, the profit and the value of the property account will both be grossly overstated. In the case of an oil company actively engaged in development work, the various headings under which write-offs must be made include the following:

1. *Depreciation* of tangible assets.

2. *Depletion* of oil and gas reserves, based upon the cost of the leases.

3. *Unprofitable leases* written off. Part of the acquisitions and exploration will always prove totally valueless and must be charged against the revenue from the productive leases.

4. *Intangible drilling costs.* It is established practice to charge all drilling expenses against operations, except for the tangible property erected, which is written off through regular depreciation charges.

Example: The case of Marland Oil in 1926 illustrates the extent to which reported earnings of oil companies are dependent upon the accounting policies with respect to amortization. This company spent large sums annually on new leases and wells to maintain its rate of production. Prior to 1926 it charged the so-called "intangible drilling costs" to capital account and then wrote them off against earnings through an annual amortization charge. In 1926 Marland adopted the more conservative policy of charging off all these "intangible costs" currently against earnings. The effect on profits is shown in the table on page 398.[1]

From the above discussion and illustration it is evident that oil company profits can be made largely a matter of oil company bookkeeping. It is particularly important, for this reason,

[1] In 1932 Consolidated Oil Corporation made the contrary change, *i.e.*, it began to charge intangible drilling costs to capital instead of to earnings

that those interested in oil-company securities make sure that the management follows standard and conservative accounting policies, and that all deductions made on its books are shown in the earnings statements which it submits.[1]

MARLAND OIL CO.

Item	1925	1926	1927
Gross earnings and miscellaneous income	$73,231,000	$87,360,000	$58,980,000
Net before reserves	24,495,000	30,303,000	9,808,000
Amortization charges	9,696,000	18,612,000	17,499,000
Balance for stock	14,799,000	11,691,000	7,691,000(d)

Adequacy of Amortization Rates in Use: 1. As Shown by Listing Statements.—The second general consideration in regard to amortization charges is whether the *rates* employed are suitable, judged by standard accounting practice. It is not to be expected that the investor or even the securities analyst can deal exhaustively with so technical a problem. There are, however, two lines of inquiry which may be followed without undue difficulty and which at times will yield useful results. The first is to consult the listing applications if the securities are dealt in on the New York Stock Exchange (or the prospectus, if the securities have been issued under the Securities Act of 1933) in order to see in what terms the company has stated its depreciation policy.

Examples: If standard methods are followed, they are likely to be announced in somewhat the following manner:

(From listing application of Electric Storage Battery Company, dated December 17, 1928.)

---

as formerly. As a result of this less conservative policy it was able to show a small net profit for the year, instead of an undoubted loss under its former method of accounting.

[1] The item of Film Exhaustion and Amortization shown in the income accounts of moving-picture producers bears a close resemblance to the charges for oil depletion. The cost of making a picture is written off against earnings in accordance with an actuarial schedule under which over 90% thereof is absorbed during the first year after release. Since both a new picture and a new oil well must return their cost within a short time after completion, these charges for "amortization" are really direct *operating costs* in disguise.

The policy of this Company in regard to depreciation . . . is as follows: On buildings the term of life is twenty to thirty-three years, depending upon the character of construction. Machinery, tools and fixtures are written off at the rate of one to ten years, depending upon the character of the equipment. Office furniture and fixtures are written off in ten years. On all depreciable properties rates are determined by actual experience and engineers' estimates as to the productive life of the equipment. In respect to depreciation of current assets, a reserve is set aside to cover probable loss from bad debts.

(From the listing application of Midland Steel Products Company, dated February 11, 1930.)

The following are the rates of depreciation used:

	Rate of depreciation per year, %
Buildings	2
Grounds, driveways, and walks	2
Machinery	7
Furniture and fixtures	10
Railroad sidings	2
Automobiles and trucks	25

Tools and dies—amortized over life of job when number of units required can be determined, otherwise written off at close of each fiscal year.

These rates have been used by the Company for several years, being standard practice in the industry.

The rates are based upon the estimated life of the respective property involved. Thus, with respect to buildings, the cost is depreciated, over 50 years; grounds, driveways, and walks, over 50 years; machinery over 14 years; furniture and fixtures, over 10 years; railroad sidings, over 50 years. No residual value at the expiration of said periods is considered in determining the rates used.

In contrast to these statements indicating adherence to a standard policy, we may point to the listing application of such prominent companies as American Sugar Refining, American Car and Foundry, Baldwin Locomotive Works, and American Can.

The American Sugar Refining Company's listing application, dated December 6, 1923 contained the following statement:

The Company maintains a very liberal policy as to depreciation as shown by the annual profit and loss statement of past years. The value of its properties is at all times fully maintained by the making of all needful and proper repairs thereto and renewals and replacements thereof.

This declaration sounds reassuring, but it is far too indefinite to satisfy the analyst. The actual depreciation charges, as shown in the following record, disclose an unusually arbitrary and erratic policy.

ANNUAL CHARGES BY AMERICAN SUGAR REFINING COMPANY FOR DEPRECIATION

Year	Charged to income	Charged to surplus
1916–1920	$2,000,000	None
1921	None	None
1922–1923	1,000,000	None
1924	None	None
1925	1,000,000	None
1926	1,000,000	$2,000,000
1927	1,000,000	1,000,000
1928	1,250,000	500,000
1929	1,000,000	500,000
1930	1,000,000	542,631
1931	1,000,000	None
1932	1,000,000	None

The additional charges to surplus made in the years 1926–1930, inclusive, appear to strengthen our contention that American Sugar's depreciation allowances have been both arbitrary and inadequate.

The American Car and Foundry's application, dated April 2, 1925 contains the following:

The Company has no depreciation account as such. However, its equivalent is found in the policy and the practice of the Company to maintain at all times its plants and properties in first class physical condition and in a high state of efficiency by repairing, renewing and replacing equipment and buildings as their physical conditions may require, and by replacing facilities with those of more modern type, when such action results in more economical production. This procedure amply covers depreciation and obsolescence and the cost is charged to Operating Expenses.

Here again a sceptical attitude on the part of the analyst is "amply" warranted. The same is true in respect of American Can which managed—inexplicably—to avoid all reference to its depreciation policy in its listing application dated February 26, 1926; although it did mention that the company had spent approximately $50,000,000 on extensions and improvement of properties since February 1907 and that "during this period properties have been depreciated by at least $20,000,000."

Baldwin Locomotive Works, in its listing application dated October 3, 1929, makes the following rather astonishing statement on depreciation:

The amount of the depreciation upon plant and equipment as determined by the Federal Government for the five years 1924 to 1928 inclusive has totaled $5,112,258.09 which has been deducted either from income or surplus as follows:

Year	From income	From surplus	Total depreciation
1924	$ 600,000	none	$ 600,000.00
1925	none	none	none
1926	none	none	none
1927	1,000,000	$2,637,881.01	3,637,881.01
1928	600,000	274,377.08	874,377.08
	$2,200,000	$2,912,258.09	$5,112,258.09

It is expected that in future years the amount of depreciation based upon the estimated useful life of depreciable properties as determined by the Federal Government, allowed by the Commissioner of Taxes as a proper deduction from income and agreed to by our engineers, will govern the amount to be used by the Works in its calculation of depreciation.

EARNINGS PER SHARE OF COMMON

Year	As reported	As corrected for annual depreciation charge of $1,022,000
1924	$ 0.40(d)	$ 2.51(d)
1925	6.02(d)	11.13(d)
1926	22.42	17.31
1927	5.21	5.10
1928	5.34(d)	7.45(d)
5-year average.......	$ 3.33	$ .06

COMPARATIVE DEPRECIATION—RETIREMENT ALLOWANCES OF PUBLIC UTILITIES, 1930

Company	Gross (000 omitted)	Depreciation D or retirement reserve R (000 omitted)	Ratio of D or R to gross, %	Maintenance	Ratio of maintenance to gross, %	Ratio of year's depreciation or retirement reserve to average property account, %
Kansas City Power & Light Co.	$ 14,504	$ 2,036D	14.0	$?		3.2
Pacific Lighting Corp.	48,838	6,784D	13.9	?		3.1
Detroit Edison Co.	53,707	6,900R	12.8	3,199	6.0	3.3
Southern California Edison Co.	41,129	5,014D	12.2	1,180	2.9	1.5
Pacific Gas & Electric Co.	76,578	8,866D	11.6	3,796	5.0	1.7
North American Co.	133,751	14,274D	10.7	?		2.0
Engineers Public Service Co.	53,042	4,906R	9.2	3,446	6.5	1.7
American Gas & Electric Co.	68,601	5,898D*	8.6	?		1.4
International Hydro-Electric System	46,414	3,970D	8.6	3,321	7.2	1.0
Public Service Corp. of N. J.	138,162	11,904D	8.6	12,881	9.3	1.9
Columbia Gas & Electric Corp.	96,130	8,138R†	8.5	?		1.4
Commonwealth Edison Co.	84,004	7,109R	8.5	?		2.4
Electric Power & Light Corp.	75,048	6,165R†	8.2	?		0.8
Utilities Power & Light Corp.	52,416	4,256D	8.1	3,613	6.9	1.3
Duquesne Light Co.	28,676	2,294R	8.0	1,410	4.9	1.4
Northern States Power Co. (Del.)	33,272	2,560R	7.7	1,778	5.3	1.1
American Water Works & Electric Co.	54,067	4,105R	7.6	4,252	7.9	1.4
United Gas Improvement Co.	108,374	8,040R	7.4	5,586	5.2	1.7
Consolidated Gas, etc., of Baltimore	28,582	2,075R	7.3	1,389	4.9	1.2
National Power & Light Co.	80,376	5,901R	7.3	?		0.9
Commonwealth & Southern Corp.	137,752	9,548R	6.9	?		2.0
Detroit City Gas Co.	18,446	1,271D	6.8	1,199	6.5	1.8
Public Service Co. of Northern Ill.	35,405	2,400R	6.8	2,013	5.7	1.6
Peoples Gas Light & Coke Co.	39,881	2,584R	6.5	?		0.9
American Power & Light Co.	87,088	5,556R	6.4	?		1.3
Consolidated Gas Co. (N. Y.)	238,758	15,033R	6.3	17,047	7.1	1.1
Illinois Power & Light Corp.	37,123	2,239D	6.0	3,628	9.7	0.7
Niagara Hudson Power Corp.	78,834	4,753R	6.0	?		0.7
Associated Gas & Electric Co.	84,219	4,849R	5.8	?		0.7
Penna. Power & Light Co.	31,006	1,500R	4.8	2,464	7.9	0.5
American & Foreign Power Co.	78,656	3,437R	4.4	?		0.7
Midland United Co.	46,289	1,843R	4.0	?		0.7
Brooklyn Union Gas Co.	25,698	669R	2.6	2,034	7.9	0.6

* Subsidiaries also charged $615,000 against surplus for special depreciation in 1930. Including this item the total charge for depreciation was 9.5 % of gross. Note that in 1933 the depreciation charge against earnings was $7,698,000 on gross earnings of $57,000,000.

† This figure included depletion.

Evidently the income statements of Baldwin for this period were anything but accurate. The average annual earnings per share of common stock for 1924–1928, as reported to the stockholders, were strikingly higher than the correct figure, as shown at the bottom of page 401.

2. **As Shown by Comparison with Similar Enterprises.**—The failure of numerous important companies to follow standard policies on depreciation lends significance to the second line of inquiry on this point, which proceeds through a comparative analysis of two or more companies in the same field. It is not to be expected that such a study will reveal a definite "normal" depreciation rate for the industry—in terms of either the property account or gross receipts—to which nearly all the enterprises will closely adhere. But something approaching a modal range is likely to be disclosed within which the policy of most units will fall. A wide departure from this range shown by an individual concern would then supply grounds for suspecting that the depreciation charges are being either skimped or padded, as the case may be. More careful inquiry would, of course, be in order before reaching a final conclusion on this point.

Public-utility Depreciation Reserves.—An example of this type of study is presented herewith, covering a number of the more important public-utility systems for 1930. The depreciation ratios used in that year were fairly representative of the normal policy of each company.

These figures would seem to bear out the observation, made in our discussion of the American Water Works and Electric Company prospectus above, that public-utility-depreciation charges are subject in large measure to arbitrary determination by the management. They would also confirm the implication that companies using the designation "Reserve for Renewals, Replacements, or Retirements" are likely to be following a less conservative policy than those charging a Reserve for Depreciation. A comparison of the ten companies at the head of the table with the ten at the foot yields the significant figures shown in the table at top of page 404.

It is clear that the depreciation or retirement charges made by public-utility companies require careful scrutiny by the analyst and investor. Allowances for any variations in the basis used must be made in any comparative analysis. Furthermore, the safety of senior securities or the attractiveness of a common stock should not be considered as established unless the question of the adequacy of the amortization charges has been raised and answered in the affirmative. In this connection the amount of the maintenance expenditures should also be scrutinized.

Group	Number using term:		Average rate of allowance to:	
	Depreciation	Retirements, etc.	Gross, %	Property account, %
First ten companies........	8	2	11.02	2.08
Last ten companies........	1	9	5.28	0.88

Comparisons of Two Companies.—When the analyst knows that a company's depreciation policy differs from the standard, there is special reason to check the adequacy of the allowance. Comparison with a single company in the same field may yield significant results, as is shown by the following data respecting American Can, American Sugar, and American Car and Foundry.

Company	Average property account (net) 1928–1932	Average depreciation charge 1928–1932	% of depreciation charge to property account
American Sugar Refining......	$ 60,665,000	$1,050,000*	1.73†
National Sugar Refining.......	19,250,000‡	922,000‡	4.79‡
American Can...............	133,628,000	2,000,000	1.50
Continental Can.............	42,582,000	1,988,000	4.67
American Car and Foundry....	72,000,000	1,186,000§	1.65
American Steel Foundries......	31,000,000	1,136,000	3.66

* Exclusive of depreciation charged to surplus. Including the latter, this figure would be $1,358,500.
† Including depreciation charged to surplus this figure would be 2.24 %.
‡ Based on the four years 1929–1932, inclusive. Figure for 1928 unavailable.
§ Estimated at one-half of the expenditures for renewals and repairs. In the case of United States Steel for the period 1901–1933, the charge for depreciation averaged about 40 % of the total allowances for both maintenance and depreciation.

Both comparatively and absolutely the depreciation allowances made by American Sugar, American Can, and American Car and Foundry appear to have been inadequate.

Depreciation Charges Often an Issue in Mergers.—Comparative depreciation charges at times become quite an issue in determining the fairness of proposed terms of consolidation.

Example: In 1924 a merger plan was announced embracing the Chesapeake and Ohio, Hocking Valley, Pere Marquette, "Nickel Plate," and Erie railroads. Some Chesapeake and Ohio stockholders dissented, and they convinced the Interstate Commerce Commission that the terms of the consolidation were highly unfair to their road. Among other matters they pointed out that the earnings of Chesapeake and Ohio in the preceding three years had in reality been much higher than stated, due to the unusually heavy charges made against them for depreciation and retirement of equipment.[1] A similar objection was made in connection with the projected merger of Bethlehem Steel and Youngstown Sheet and Tube in 1929, which plan was also defeated. Some figures on these two steel producers are given herewith.

1928	Bethlehem Steel	Youngstown Sheet & Tube
Property account, Dec. 31, 1927.........	$673,000,000	$204,000,000
Sales................................	295,000,000	141,000,00
Depreciation, depletion, and obsolescence	13,658,000	8,321,000
Ratio: depreciation to property account..	2.03%	4.08%
Ratio: depreciation to sales.............	4.63%	5.90%

Failure to State Depreciation Charges Inexcusable.—A small number of important companies report earnings after depreciation but fail to state the amount so deducted. Some idea of their policy may generally be gained by scrutinizing the accumulated depreciation reserves given in the balance sheet over a period of years. But there is no convincing excuse for failure to state the amount of the depreciation charge; if no satisfactory indications on this point are available, the conservatively minded should resolve any doubt *against* the company.

A Case of Excessive Depreciation Charges Concealed by Accounting Methods.—It is possible, though not likely, that in

[1] Large expenditures made by Chesapeake and Ohio upon its equipment in 1926–1928 and charged to operating expense were later claimed by the Interstate Commerce Commission to represent *capital* outlays. In 1933 this controversy was taken into the courts, and the Interstate Commerce Commission was sustained.

such instances the management may actually be charging off an excessive amount for depreciation. A classic example of such a practice is the record of National Biscuit Company for many years prior to 1922. During this time the company was constantly adding to the number of its factories, but its property account failed to show any appreciable increase, except in the single year 1920. The reports to stockholders were supremely ambiguous on the matter of depreciation charges,[1] but according to the financial manuals the company's policy was as follows: "Depreciation is $300,000 per annum, and all items of replacement and building alterations are charged direct to operating expense."

NATIONAL BISCUIT COMPANY

Year ended	Earnings for common stock	Net plant account at end of year
Jan. 31, 1911.......	$ 2,883,000	$53,159,000
1912.......	2,937,000	53,464,000
1913.......	2,803,000	53,740,000
1914.......	3,432,000	54,777,000
1915.......	2,784,000	54,886,000
1916.......	2,393,000	55,207,000
1917.......	2,843,000	55,484,000
Dec. 31, 1917.......	2,886,000 (11 mo.)	53,231,000
1918.......	3,400,000	52,678,000
1919.......	3,614,000	53,955,000
1920.......	3,807,000	57,788,000
1921.......	3,941,000	57,925,000
1922.......	9,289,000	61,700,000
1923.......	10,357,000	64,400,000
1924.......	11,145,000	67,292,000
1925.......	11,845,000	69,745,000

It is difficult to avoid the conclusion, however, that the capital investments in additional plants were actually being charged against the profits, and that the real earnings were in all probability much larger than those reported to the public. Coincident with the issuance of seven shares of stock for one,

[1] Prior to 1919, the company's balance sheet each year stated its fixed assets "Less Depreciation Account—$300,000." Evidently this was the deduction for the current year, and not the amount accumulated.

and the tripling of the cash-dividend rate in 1922, this policy of understating earnings was terminated. The result was a sudden doubling of the apparent earning power, accompanied by an equally sudden expansion in the plant account. The contrast between the two periods is shown forcibly in the table on page 406.

CHAPTER XXXV

AMORTIZATION CHARGES FROM THE INVESTOR'S STANDPOINT

The third general question about depreciation charges relates to the property values against which they are applied. This point is undoubtedly of greater practical importance than the other two, because it involves, not the failure of the corporation to follow permitted accounting methods—which is a phenomenon of increasing rarity—but rather the failure of such permitted accounting to reflect the real situation confronting the investor in the company. Developments since 1929 have tended to make such divergence much more frequent and more significant than previously.

Problem Indicated by Hypothetical Example.—The point at issue may be more readily comprehensible by the use at the outset of a simplified and therefore hypothetical example.

Let us assume that companies *A*, *B*, and *C* are all engaged in the trucking business. Each has a single truck; each is capitalized at 100 shares of stock, no par; and each earns $2,000 per annum before depreciation.

Company *A* paid $10,000 for its truck.
Company *B* paid $5,000 for its truck.
Company *C* paid $10,000 for its truck but followed "an ultra conservative policy" and wrote its value down to $1.

Assume that *B*'s acquisition of a cheaper truck was a stroke of luck and that in fact the management of the three companies are equally capable and their general situation identical.

The accountants give these trucks a depreciable life of four years. On this basis the income accounts of the three corporations are as follows:

Item	Company A	Company B	Company C
Net before depreciation........	$2,000	$2,000	$2,000
Depreciation (at 25%).........	2,500	1,250	0
Balance for common stock......	*500(d)*	750	2,000
Earned per share.............	0	$7.50	$20

Typical Market Appraisals.—According to these audited statements, A is losing money, B is earning 15% on its capital, and C is doing very well indeed. An "investor," steeped in the recent wisdom of stock-exchange valuations, would consider the shares of Company A practically worthless—$5 per share, perhaps, being a generous appraisal. On the other hand he might value the shares of B and C at about ten times the earnings, which would produce $75 per share for B stock and no less than $200 per share for C stock. Such a procedure would result in the following total valuations for the three enterprises:

Company A..................................	$ 500
Company B..................................	7,500
Company C..................................	20,000

The absurdity of these valuations should be too patent for argument. Nevertheless they represent merely a faithful application of current accounting methods and the established Wall Street reasoning. The results are, first, that a company with a less valuable asset is *for that very reason* declared to be worth more than a company with a more valuable asset; and secondly that by the single gesture of writing down its assets to zero, a company has been able to increase enormously the market price of its shares.

Irrationality of These Valuations Disclosed by the Balance Sheet.—The irrationality of these conclusions would be even more glaring if the balance sheets are examined. Assume that the companies have been in business three years and (for simplicity) that they started with no working capital. Company A, having lost money steadily, has of course paid no dividends; Company B has paid out two-thirds of its earnings, *i.e.* $5 per share annually; and Company C has paid out three-fourths of its profits or $15 per share. The balance sheets would then read as shown in the table at the top of page 410.

Although Company A has a profit-and-loss deficit, it has accumulated the largest amount of cash, presumably "earmarked" as a depreciation fund. Company C, which has shown the largest earnings, has by far the smallest cash holdings. The suggested market value of $5 per share for Company A would amount to only *one-twelfth* of its cash; while the price of $200 for Company C shares would equal more than *twelve times* the cash behind them.

Item	Company *A*	Company *B*	Company *C*
Assets:			
Truck.....................	$10,000	$5,000	$ 1
Cash.....................	6,000	4,500	1,500
Total.....................	$16,000	$9,500	$1,501
Liabilities:			
Capital stock...............	$10,000	$5,000	$ 1
Depreciation reserve.........	7,500	3,750	
Profit and loss..............	*1,500(d)*	750	1,500
Total.....................	$16,000	$9,500	$1,501

A More Rational Approach.—These are the Alice-in-Wonderland results to which the accepted logic of the stock market would lead us. Let us now ask a more sensible question, *viz.,* "How would a *business man* determine the *reasonable value* of these three enterprises?" Common sense would tell him immediately that all three business as such, independent of their assets, are of equal value. As a practical business matter he would be inclined to place a somewhat higher valuation on the more expensive vehicle owned by Companies *A* and *C* than upon the cheaper truck of Company *B*. Nor is there the slightest doubt that this business man will give full weight to the relative cash holdings of each company.

His reasoning would therefore run somewhat as follows: Each business is worth, in the first instance, the amount of its cash plus the fair market value of its truck. Something might properly be paid also for the good-will, because the earnings on the average capital required for the business, after allowing for *necessary depreciation,* would be quite substantial This good-will value would be the same for all three enterprises.

Item	Company *A*	Company *B*	Company *C*
Cash........................	$6,000	$4,500	$1,500
Truck (estimated).............	1,500	1,000	1,500
Good-will (estimated)..........	2,000	2,000	2,000
Total value.................	$9,500	$7,500	$5,000

What is the relation of the *companies' depreciation charges* to these valuations? The answer is that the charge made by

Company *B* might well be accepted as relevant because it *corresponds fairly well with the conditions of the business.* Partly by coincidence, this fact results in making the business-man's valuation of Company *B* identical with that reached by the Wall Street method. But in the case of Company *A* and Company *C*, the depreciation charges made by the managements are entirely out of line with the realities of the business. In the one case they have been made far too high because of the excessive cost of the fixed assets. Such an error of judgment should be corrected by writing down the property account (and the capital account) to a fair going-value, against which a businesslike depreciation charge will accrue. In the case of Company *C* the assets have been deliberately undervalued for the purpose of suppressing a depreciation charge which *must* be allowed for out of earnings because the owner's investment is actually depreciating. If the business man or the investor is *going to pay anything* for the truck (or for the business itself which requires a truck) he cannot avoid allowing for depreciation on the amount so paid by merely *making believe* that there is no such investment.

Practical Application of Foregoing Reasoning.—Let us consider now how the above reasoning may be applied to actual situations which confront the security buyer.

Examples: As an initial example, we shall present the exhibit of the Eureka Pipe Line Company for the three years 1924–1926.

Year	Gross revenues	Net before depreciation	Depreciation	Balance for stock
1924	$1,999,000	$300,000	$314,000	$14,000(d)
1925	2,102,000	541,000	498,000	43,000
1926	1,982,000	486,000	500,000	14,000(d)
3-year average........	2,028,000	442,000	437,000	5,000
Per share of common (on 50,000 shares) ..	....	$8.84	$8.74	$0.10

The final column would imply that during the three years under review there was practically no earning power for the shares, so that presumably the stock would have no value on a going-concern basis. But would such a conclusion be justified from a business standpoint? The question will turn, as in our hypothetical examples, upon the correctness of the depreciation charges. The following data will throw additional light upon this aspect of the Eureka Pipe Line's record (figures in thousands):

Year	Depreciation charged for year	Actually expended for plant replacements, etc.	Depreciation charge unspent	Earnings after depreciation	Total cash available from year's operations	Dividend paid	Added to net quick assets
1924	$314	$ 61	$253	$14(d)	$239	$350	−$111
1925	498	cr. 30	528	43	571	200	371
1926	500	239	261	14(d)	247	200	47
3-year average.........	437	90	347	5	352	250	102

We find that the expenditures on property account averaged only $90,000 per annum, so that there was available *in actual cash* the sum of $352,000 per annum to be added to working capital or used for dividends (which were charged against previously accumulated surplus). It is clear that this business had been a producer of cash income for the owners; and for that reason it had substantial going-concern value, although the high depreciation charges made it appear that there was none.

How to Determine the Proper Depreciation Charge.—In this case, therefore, as in our hypothetical example, the investor or the analyst must reject the company's basis for depreciation and endeavor to establish some other basis more consonant with the actual conditions of the business. How can the proper charge be determined? The answer was given without difficulty for the trucking companies, because we knew just what depreciation had to be allowed for in order to maintain these enterprises in operation. But in practice such exact knowledge is hardly ever available. We do not know how long the Eureka Pipe Line's fixed assets will last nor how much it would cost to replace them. The best we can do is to formulate some rough estimates based on the discoverable facts. The only virtue of these estimates may be that they are in all probability closer to the mark than the company's figures, which we realize are untenable.

Concept of "Expended Depreciation."—Taking a business attitude towards the Eureka Pipe Line's exhibit, it is evident at the start that the depreciation allowance should be *not less* than the average expenditures made on the property. The primary reason for reducing the company's depreciation charges is because they do not properly reflect the cash available from operations. The expenditures on property account, including new fixed assets, represent in effect the portion of the depreciation reserve which is not available in cash, and that portion should hence be considered as the minimum amount of depreciation which must be allowed for in conducting the business. We may call this item the *Expended Depreciation Charge.* (If the increase in the property account exceeds the year's depreciation, then all of the latter must be considered as "expended.") In the case of Eureka Pipe Line, such expenditures averaged $90,000 for the three years 1924–1926 and a closely similar amount over a much longer period.

Long-term Depreciation a Form of Obsolescence.—The second question is what amount should be provided as a reserve to take care of the eventual wearing out of the entire property—in other words, for the major replacements which may have to be made at some distant date. This is the leading function of the depreciation charge in most theoretical discussions of the subject; and our trucking company examples were based on a simple application of this idea (the total fixed-asset account having to be replaced at the end of four years). But we must recognize that in practice such complete wearing-out and replacement are of exceedingly rare occurrence. The typical corporation does not accumulate a large cash fund over a stretch of years which is finally employed to replace the plant in its entirety at the end of its useful life. Factories do not actually wear out; they become obsolete. In nine cases out of ten, plants are given up because of changes in the character of the industry, or in the status of the corporation, or in the locality where the plant is situated, or for other reasons not related to actual depreciation.

These developments represent *business hazards*, the extent of which is not susceptible to any engineering or accounting measurement. Stated differently, the *long-term depreciation* factor is in reality overshadowed and absorbed by the *obsolescence* hazard. This risk is essentially an investment problem and not an accounting problem. It should not operate to reduce the *earnings* (as does a depreciation charge), but rather to reduce the *price* to be paid for an earning power subject to such a business risk.

Application of Foregoing in Determining Earning Power.—Let us endeavor to relate these conclusions to the Eureka Pipe Line example. The Expended Depreciation Charge has been found to average about $100,000 per annum. There are no indications that the entire plant will have to be replaced at any predictable date. On the contrary, the line appears to have an indefinite life, due to continuous expenditures on maintenance, repairs, and renewals. In this respect the enterprise resembles a railroad far more than it does a trucking company. According to our reasoning only the expended depreciation charge should be deducted from earnings. The remainder of the depreciation factor is actually the *obsolescence hazard*, which is related to the possible exhaustion of the tributary fields. This should be considered *after* the earnings are arrived at and not before. A proper statement of the case would appear as follows:

EUREKA PIPE LINE (1924–1926 BASIS)

Item	Total	Per share
Earnings before depreciation...............	$442,000	$8.84
Expended depreciation charge, estimated.....	100,000	2.00
Balance: Earning Power, subject to business hazards, including obsolescence..........	$342,000	$6.84

Problem of Valuing the Earning Power.—The company's figures showed no earning power for the period. Our figures show an earning power of about $7 per share, which clearly indicates substantial value for the enterprise. The *price* which may properly be paid for this earning power is subject to whatever considerations enter into buying a going business. This includes on the one hand the possibilities of increased profit, and on the other hand all the multitudinous risks of loss, of which obsolescence of the fixed assets is only one. If, for example, it seemed conservative to require earnings of 20% on the investment to cover these hazards adequately, then the indicated value of Eureka Pipe Line stock on the above showing would be about $35 per share. A detailed discussion of this point must be postponed, however, until we reach the topic of *valuation* of common stocks. For the purpose of this chapter it should suffice to point out that in the actual case of Eureka Pipe Line, as in the hypothetical case of Trucking Company *A*, it was both necessary and feasible for the investor to establish a depreciation allowance significantly different from that employed by the company itself.[1]

Inadequate Allowances for Depreciation.—Let us now consider examples involving the opposite type of situation, *viz.*, the use of accounting methods by corporations which give rise to inadequate allowances for depreciation. Particular attention must be given to the vogue for drastic write-offs of fixed assets for the admitted purposes of *reducing* the depreciation charges and thereby *increasing* the reported earnings. This practice had its inception during the 1927–1929 boom, but its widest development took place in the ensuing depression. Two typical cases are selected for discussion.

Examples: Early in 1933, the U. S. Industrial Alcohol Company and the Safety Car Heating and Lighting Company announced plans under which the property account was written down to a net value of $1, by means of a corresponding reduction in stated capital and surplus. The transactions may be summarized in the condensed balance sheets shown in the table on page 416.

The U. S. Industrial Alcohol revision was accompanied by a statement to the effect that by reducing the book value of fixed assets to $1 the necessity for future charges for depreciation would be eliminated. It was proposed, however, to set up a Reserve for Replacements account, by charges against income of amounts deemed sufficient to provide for the replacement of productive facilities. It was believed that for 1933 an adequate amount of such charge would be $300,000 which might be compared with approximately $900,000 charged against income for depreciation in 1932.

The Safety Car announcement carried the idea even further. No provision for depreciation was made in 1932, so that a net profit was reported for that year against a loss for 1931, although income before depreciation was smaller in 1932. It was stated in the annual report of the company for 1932 that "By the elimination of depreciation on Fixed Assets as of December 31, 1932, all profits above Operating Expense, and Depreciation on subsequently acquired Capital Assets, could be considered by

[1] That the official depreciation charges could stand revision in this case is evident from the fact that the corporation itself made several quite arbitrary changes in its methods of computation from year to year. In 1929, for example, the depreciation allowance was suddenly cut to $176,000. (Data given in reports to the Interstate Commerce Commission.)

EFFECT OF WRITING DOWN FIXED ASSETS
(Unit $1,000)

Item	Safety Car Heating & Lighting Co.		U.S. Industrial Alcohol Co.	
	Before write-downs	After write-downs	Before write-downs	After write-downs
Plant account...................	$ 9,578	$9,578	$29,116	$29,116
Less depreciation................	6,862	9,577	9,815	29,115
Plant account (net).............	$ 2,716	$ 1	$19,301	$ 1
Intangible and misc. assets (net)......	5,016	167	1,185	1,185
Investments in affiliates, etc..........	2,330	2,330	1,416	1,416
Net current assets................	4,379	4,379	6,891	6,891
Total...........................	$14,441	$6,877	$28,793	$ 9,493
Capital.........................	$ 9,862*	$4,931†	$22,585‡	$ 3,739
Surplus.........................	4,362	1,729	4,458	4,004
Contingency reserve..............	217	217	1,750	1,750
Total...........................	$14,441	$6,877	$28,793	$ 9,493

* 98,620 shares par $100.
† 98,620 shares, no par.
‡ 373,846 shares, no par.

your Directors for distribution to the stockholders without any decrease in the Company's current assets."

Earnings Manufactured from Depreciation Account.—The procedure followed by Safety Car is identical with that of our imaginary Trucking Company *C*, which wrote down its truck to $1 and thereby avoided charging depreciation to earnings. We have already pointed out that if depreciation must be allowed for in fact, it cannot be eliminated by bookkeeping entries. The Safety Car stockholder does not earn a dollar more on his investment because his fixed assets have been written down to nothing. Nor can necessary expenditures for plant upkeep or replacement be in any wise reduced by making believe that there no longer is any plant. Let us examine the Safety Car Heating and Lighting exhibit in somewhat the same manner as that of Eureka Pipe Line. Over a ten-year period the expended depreciation charge averaged about $500,000 per annum. The earnings record for the decade is approximately as follows:

Item	Annual average 1922–1931	Year 1931	Year 1932
Earnings before depreciation	$1,721,000	$336,000	$233,000
Depreciation charged	669,000	442,000	
Earnings as reported	1,052,000	106,000(d)	233,000
"Depreciation expended" (approximate)	$ 500,000	$130,000	$190,000
Cash earnings available for the stock ...	1,221,000	206,000	43,000

If this company were analyzed amidst the uncertainties of 1933, it would be impossible to determine whether the long-term or the recent figures are a better guide to the future. But whatever assumption is made on this score, it is quite clear that a depreciation charge must be allowed for. If no better than the 1932 results can be expected, then a very small earning power at best would be indicated, since actual expenditures on plant will no doubt come close to, if they do not exceed, the reported "earnings" of 233,000. If by any chance the profits should return to their ten-year average, the complete elimination of the former depreciation charge would result in a serious overstatement of the true earning power.

The U. S. Industrial Alcohol Company write-off did not result in the complete elimination of depreciation charges against earnings, but in lieu thereof it was proposed to set up a "replacement reserve" to be determined arbitrarily by the directors. For 1933 the amount was fixed at $300,000. A study of the approximate figures for the preceding five years would warrant grave doubts as to the adequacy of such a charge for replacements under normal conditions.

Item	Average 1928–1932 as reported*	Average 1928–1932, based on proposed 1933 replacement reserve
Net before depreciation	$2,090,000	$2,090,000
Depreciation charged...	1,350,000	300,000
Balance for common ...	740,000	1,790,000
Earned per share.......	$2	$5

* After deducting from earnings certain items charged by the company to surplus.

In this case the Net Plant account (Gross Plant less Depreciation) *increased* $500,000 during the five-year period (*i.e.*, from $18,800,000 at the end of 1927 to $19,300,000 at the end of 1932). In other words the money spent for property extensions and replacements somewhat exceeded the total depreciation allowance of $6,750,000. This development is characteristic of most of our large corporations, which tend to add to their facilities as the years pass. In all such cases it must be assumed that the depreciation charges based upon accepted accounting rules are the minimum necessary for properly reflecting the conditions of the business. They cannot soundly be reduced either by the corporation through arbitrary write-downs or by the investor in his individual calculations. Hence if the U. S. Industrial Alcohol Company should regain its former profit-making ability, a drastic reduction of the former depreciation reserves would in all probability result in a misleading overstatement of the true earning power.

Stock Watering Reversed.—The new policy of writing off fixed assets bears an interesting relationship to the recent conceptions of stock values. It is a direct outgrowth of the ignoring of asset values and the monopolizing of attention by the reported per-share earnings. A generation ago, when investors consulted balance sheets to ascertain the net worth behind their shares, this net worth was artificially inflated by writing *up* the book value of the fixed assets far above their actual cost. This in turn permitted a corresponding overstatement of the capitalization at par. "Stock watering," as this practice was called, constituted at that time one of the most severely criticized abuses of Wall Street.

It is a striking commentary on the change in our financial viewpoint that the term "stock watering" has practically disappeared from the investor's vocabulary. By a strange paradox the same misleading results which were obtained before the war by *overstating* property values are now sought by the opposite stratagem of *understating* these assets. Erase the plant account; thereby eliminate the depreciation charge; thereby increase the reported earnings; thereby enhance the value of the stock. The idea that such sleight-of-hand could actually add to the value of a security is nothing short of preposterous. Yet Wall Street solemnly accepts this topsy-turvy reasoning; and corporate managements are naturally not disinclined to improve their showing by so simple a maneuver.

Rules Summarized.—Let us summarize the foregoing discussion by stating the following rules:

Rule 1: The company's depreciation charges are to be accepted in analysis whenever (*both*):

    a. They are based on regular accounting rules applied to fair valuations of the fixed assets; and

    b. The *net* plant account has either increased or remained stationary over a period of years.

Rule 2: The company's charges may be *reduced* in the analyst's calculations if they regularly exceed the cash expenditures on the property. In such a case the average cash expenditures may be deducted from earnings as a provisional depreciation charge and the balance of depreciation included as part of the *obsolescence hazard*, which tends to reduce the valuation of this average cash earning power. The obsolescence allowance will be based upon the *price paid for the enterprise by the investor* and not upon either the book value or the reproduction cost of the fixed assets.

Rule 3: The company's charges must be *increased* in the analyst's calculations if they are both less than the average cash expenditures on the property and less than the reserve required by ordinary accounting rules applied to the fair value of the fixed assets used in the business.

CHAPTER XXXVI

RESERVES FOR DEPLETION, OTHER AMORTIZATION CHARGES AND CONTINGENCIES

DEPLETION OF ORE RESERVES

The distinction between the company's and the investor's allowance for amortization appears most clearly in cases involving depletion of ore reserves. As stated in the previous chapter, the amounts charged off by a mining company for depletion are based upon certain technical considerations which are likely to be quite irrelevant to the stockholders' situation.

Example: A study of the showing of Homestake Mining Company for the year 1925 and again for 1933 will illustrate this point.

Item	1933	Per share	1925	Per share
Gross earnings	$13,285,000	$53.00	$ 6,080,000	$24.32
Net earnings before depreciation and depletion	7,429,000	29.70	1,894,000	7.58
Depreciation and depletion	2,421,000	9.70	1,330,000	5.32
Balance for dividends	5,008,000	20.00	564,000	2.25
Market price (in March of following year)	360	......	50	
Market value of enterprise (250,000 shares)	$90,000,000	......	$12,500,000	
% earned on market value	5.6%	......	4.5%	

Superficially the price of 360 early in 1934 would seem to be somewhat better justified by the past year's earnings than the price of 50 in early 1926. But the reported earnings were based upon the company's charges for depreciation and depletion, which bear no relation to the price which the purchaser of the shares is actually paying for the mine. It will again be helpful to view the picture from the standpoint of a business man considering the purchase of the entire enterprise at the valuations indicated by the market price of the stock.

In 1926 the valuation would be $12,500,000. For this sum he would obtain about $2,500,000 in current assets (equivalent to cash), so that the mine and plant would cost him only $10,000,000. It is this capital investment which he would have to amortize, *i.e.* recover out of earnings, together with a suitable profit before the mine is exhausted. In 1926 the developed ore reserves indicated a minimum life of 11 years for the property at the current rate of production. Since new ore had continuously been developed in amounts very nearly equal to the tonnage mined, there was good reason to expect a life considerably longer than the minimum figure. It would not be conservative, however, to count on more than 20 years. In a mining venture of this type the same amortization rate should ordinarily be applied to the machinery and other equipment as to the mine proper, on the theory that the plant will last as long as the mine and will then have to be scrapped.

The Purchaser's Amortization Calculation.—The purchaser's amortization rate would therefore have to be somewhere between 5 and 9% annually on his $10,000,000 cost price for the mine. How this would work out is shown in the following table, which

BUYER'S AMORTIZATION CALCULATION

Item	1925 earnings basis, price 50	1933 earnings basis, price 360	1933 earnings adjusted for gold at $34 per ounce
Paid for entire company............	$12,500,000	$90,000,000	
Less cash assets included...........	2,500,000	12,000,000	
Paid for mining property...........	$10,000,000	$78,000,000	
(Value of mining property on balance sheet)........................	(20,960,000)	(5,860,000)	
Earnings before amortization........	1,900,000	7,430,000	$10,000,000 (est.)
Earnings required on cash assets (5 %)	125,000	600,000	600,000
Balance earned on mining investment.	$ 1,775,000	$ 6,830,000	$ 9,400,000
% earned before amortization........	17.8 %	8.8 %	12.1 %
(Company's amortization charge).....	($1,330,000)	($2,420,000)	($2,430,000)
Investor's amortization:			
Maximum 9 %...................	900,000	7,000,000	7,000,000
Minimum 5 %...................	500,000	3,900,000	3,900,000
Earned on mining investment after amortization:			
Minimum earnings...............	875,000	Nil	2,400,000
Maximum earnings...............	1,275,000	2,930,000	5,500,000
% earned on mining investment			
Minimum......................	8.8 %	Nil	3.1 %
Maximum......................	12.8 %	3.8 %	7.1 %

includes a corresponding analysis of the March 1934 situation. The same maximum and minimum figures for expected life are used in both cases because the reported ore reserves continued to show a life of at least 10 years. ′ •

From the business standpoint, the showing for 1926 (assuming it could be expected to continue) would indicate a satisfactory return on the investment at $50 per share.   This is by no means true, so far as the available facts are concerned, when dealing with the 1933 earnings and the related price of about 360.   The *company's* amortization charges for 1925 were considerably higher than required by a purchase of the shares at 50; but on the other hand the buyer at 360 could not be at all sure that the company's charges for 1933, even though increased over 1925, would be adequate to amortize his investment.[1]

In the more frequent case where a mining company's charge for depletion is not shown in its report, the same general approach must be used in attempting an analysis. This means that where the life of a property is limited, the stated *depreciation charge* should also be ignored and the "investor's amortization" charged against the earnings before depreciation. The three factors to be considered are: (1) the price paid for the mining property (total price less cash assets); (2) the earnings before depreciation and depletion; and (3) the *minimum* life of the mine, and, alternatively, its *probable* life.

AMORTIZATION OF PATENTS

In theory a patent should be dealt with in exactly the same manner as a mining property, *i.e.*, *its cost to the investor* should be written off against earnings during the period of its remaining life.   In practice, the issue is usually too complicated for any worthwhile analysis; first, because the investor cannot tell how much he is paying for the patents, as against the other assets of the company; and secondly because he cannot judge accurately the effect of the expiration of the patents upon the company's business.

---

[1] The large increase of Homestake's earnings in 1933 over 1925 was due not only to the higher price of gold but also to the mining of much richer ore. A change of the latter kind suggests the question whether the more favorable showing can be expected to continue in the future, a point which is discussed in a later chapter.

Examples: In the case of Gillette Safety Razor Company the expiration of the basic patents was followed unexpectedly by a number of years of largely increased earnings and by an enormous advance in the market value of the shares. The opposite development occurred in the case of American Arch Company, which supplied patented arch brick for locomotives to nearly all the railroads of the United States. Because of the technical nature of its business and its strong trade position, those identified with this company were confident that it would hold its customers after its patents expired in 1926. But immediately thereafter competition compelled a drastic cut in prices, the earnings dwindled, and the price of the stock collapsed.

The Investor's Calculation.—In the ordinary case the patents controlled by a manufacturing enterprise do not occupy so prominent a place as in the examples just discussed. A business purchaser would find it advisable to take the patent situation *generally* into account in considering a company's record and prospects, rather than to work out a specific basis for amortizing these patents by deductions from earnings. In other words patents must usually be dealt with in a manner similar to the method of dealing with long-term depreciation and obsolescence, *i.e.*, more as a business hazard than as a matter of accounting. Logic would require therefore that the charges for amortization of patents as given in corporate reports be ignored in analyzing the earning power, and that the patent item be considered only when trying to determine what price for the enterprise these earnings will justify. This suggestion is strengthened by the growing practice on the part of corporations to write down their patents to $1 in order to avoid annual charge-offs against earnings. There is obviously little sense in taking seriously an annual deduction for patent amortization when the company not only *may* but probably *will* eliminate this charge at any moment by a mere bookkeeping transfer.

Examples: For many years prior to 1933 United States Hoffman Machinery Corporation made an annual charge against earnings for amortization of its patents. This charge amounted to well over $200,000 a year and was equivalent to about $1 per share on the common stock. In 1933 the company reduced its stated capital by over $3,500,000 and through various surplus adjustments wrote off the entire book value of the patents and *restored to earned surplus* something over $1,500,000 which had been charged against earnings in previous years as amortization-of-patents expense. The result will be to increase the reported earnings per share by about $1 per annum.

American Laundry Machinery Company charges its small annual allowance for amortization of patents against surplus instead of against income. While this practice is exceptional and runs counter to accounting rules, it conforms with our reasoning on this subject.

AMORTIZATION OF LEASEHOLDS AND LEASEHOLD IMPROVEMENTS

The ordinary lease involves no capital investment by the lessee, who merely undertakes to pay rent in return for the use of property. But if the rental payments are considerably less than the use of the property is worth, and if the arrangement has a considerable period to run, the leasehold—as it is called—may have a substantial value. Oil lands are leased on a standard basis for a royalty amounting usually to one-eighth of the production. Leaseholds on which a substantial output is developed or assured are worth a large bonus above the rental payments involved, and they are bought and sold in the same way as the fee ownership of the property. Similar bonuses are paid—in boom times usually—for long-term leases on urban real estate.

If a company has paid money for a leasehold, the cost is regarded as a capital investment which should be written off during the life of the lease. (In the case of an oil lease the write-off is made against each barrel produced, rather than on a time basis, since the output declines rapidly from the initial flush figure.) These charges are in reality part of the rent paid for the property and must obviously be included in current operating expense.

When structures are built on leased property, or alterations made, or fixtures installed, they are designated as "leasehold improvements." Hence their cost must be written down to nothing during the life of the lease, since they belong to the landlord when the lease expires. The annual charge-off for this purpose is called "amortization of leasehold improvements." It partakes to some extent of the nature of a depreciation charge. Chain-store enterprises frequently invest considerable sums in such leasehold improvements and consequently the annual write-offs thereof may be of appreciable importance in their income accounts.

The December 31, 1932, balance sheet of F. W. Woolworth Company carried "Buildings Owned and Improvements on Leased Premises to be amortized over periods of leases" at a net valuation of $41,500,000. The charge against 1932 earnings for amortization of these buildings and leasehold improvements amounted to $2,678,400.

Since these items belong to the amortization group they lend themselves to the same kind of arbitrary treatment as do the others. By making the annual charge against surplus instead of income, or by writing down the entire capital investment to $1 and thus eliminating the annual charge entirely, a corporation can exclude these items of operating cost from its reported per-share earnings and thus make the latter appear deceptively large.

CONTINGENCY AND SIMILAR RESERVES

Conservatively managed companies in former days were wont to charge certain arbitrary amounts against the earnings of good years to absorb any special losses which might later arise, usually in a bad year. The intent of this policy was to equalize the earnings in prosperity and depression. In this respect it resembled the use of accumulated earnings of subsidiary companies discussed in Chap. XXXIII. Experience has shown that such devices for artificially modifying the actual earnings are too readily open to abuse. Intelligent financial opinion—as represented by the New York Stock Exchange—insists, therefore, that the management disclose the true results of each year and leave all equalization and averaging to be done by the stockholders.

Objects of Creating Contingency and Similar Reserves and an Example of Their Use.—During the years 1931 and 1932, however, contingency and similar reserves were resorted to by many companies with the effect of greatly obscuring and confusing their annual statements. These reserves were created for a threefold purpose: (1) to permit losses to be charged against surplus instead of against income; (2) to gloss over the actual taking of the loss; and (3) in some cases to lay the groundwork for inflated earnings in subsequent years. A detailed analysis of the reports of American Commercial Alcohol Corporation for

1931 and 1932 will serve to make these points clearer to the reader.

The results for the two years as given by the company in its annual statements were as follows:

Item	Total	Per share
1931 net loss....................	$597,000	$3.18(d)*
1932 net profit.................	586,000	3.01
Two years' net loss..............	$ 11,000	0.17(d)

* Adjusted to $20 par-value basis.

From these figures it would appear that the company had about broken even during the two depression years taken together, and that it had realized substantial earnings during 1932. But the balance sheets covering this period, which are presented in condensed form below, point to an entirely different conclusion. (Note that no dividends were paid during this time.)

CONDENSED BALANCE SHEETS OF AMERICAN COMMERCIAL ALCOHOL
CORPORATION, 1930–1932
(Unit $1,000)

Item	Dec. 31, 1930	Dec. 31, 1931	Dec. 31, 1932
Current assets.................	$2,657	$2,329	$2,588
Less current liabilities..........	294	1,225	1,327
Net working capital...........	$2,363	$1,104	$1,261
Fixed and miscellaneous assets less depreciation.............	6,440	6,126	6,220
Total net resources...........	$8,803	$7,230	$7,481
Capital......................	$3,775*	$3,764	$3,895
Miscellaneous reserves.........	256	416	413
Surplus......................	4,772	3,050	3,173
Total......................	$8,803	$7,230	$7,481

* Adjusted to $20 par value (report showed capital of $8,500,698 and surplus of $46,484).

These balance sheets show that instead of a merely nominal loss of $11,000 for the two years together, there was an actual shrinkage of $1,600,000 in the company's surplus, the greater part of which was represented by an increase in current debt.

The extraordinary discrepancy between these two exhibits was brought about by the exclusion from the income account of numerous losses and deductions, which were charged against surplus instead. This simple device was made more complicated —and therefore not so readily intelligible to stockholders—by the use of three stages of accounting procedure, *viz.*:

1. The transfer of a large amount from Capital to Capital Surplus.
2. The transfer of various sums from Capital Surplus to Reserves.
3. The charging of various losses against these Reserves, and of other losses directly against Surplus.

At the end of 1931 American Commercial Alcohol transferred the sum of $4,875,000 from Capital to Capital Surplus. It then used $576,000 of this Capital Surplus to cancel the accumulated profit-and-loss deficit. The entries in the surplus account for 1931 and 1932 show the following remarkable assortment of extraordinary losses and adjustments.

Reduction of inventory value under previous year's contracts	$ 145,000
Losses due to trading in corn options	88,000
Reduction in the value of fixed assets	157,000
Losses due to revaluation of containers	213,000
Balance of organization expenses	73,000
Income tax for prior years	54,000
Excess cost of raw materials 1932	255,000
Payment under salary contract	40,000
Loss on sale of treasury stock, etc	46,000
Miscellaneous items (10 debits and 1 credit)	117,000
Reserve for contingencies	400,000
Charges to surplus, 1931–1932	$1,588,000
Loss for two years, per income account	11,000
Total reduction in surplus, 1931–1932	$1,599,000

It is evident that a substantial part of these charges against Surplus actually represented operating losses, which were responsible in turn for the large increase in current liabilities. It should be noted furthermore that the company carried forward into 1933 a new contingency reserve of $400,000, against which might be charged future losses which properly should reflect themselves in the income account.

Hence the accounting procedure of this company—as well as of many others—in 1931 and 1932 not only concealed the true extent of the losses suffered, but also was calculated to understate the losses or to overstate the profits of succeeding years.[1]

[1] A Senate Investigating Committee (on Banking and Currency, investigating "Stock Exchange Practices") in February 1934 elicited the fact that there had been continuous pool activities in American Commercial Alcohol stock between February 1932 and July 1933.

CHAPTER XXXVII

SIGNIFICANCE OF THE EARNINGS RECORD

In the last six chapters our attention was devoted to a critical examination of the income account for the purpose of arriving at a fair and informing statement of the results for the period covered. The second main question confronting the analyst is concerned with the utility of this past record as an indicator of future earnings. This is at once the most important and the least satisfactory aspect of security analysis. It is the most important because the sole practical value of our laborious study of the past lies in the clue it may offer to the future; it is the least satisfactory because this clue is never thoroughly reliable and it frequently turns out to be quite valueless. These shortcomings detract seriously from the value of the analyst's work, but they do not destroy it. The past exhibit remains a sufficiently dependable guide, in a sufficient proportion of cases, to warrant its continued use as the chief point of departure in the valuation and selection of securities.

The Concept of Earning Power.—The concept of *earning power* has a definite and important place in investment theory. It combines a statement of actual earnings, shown over a period of years, with a reasonable expectation that these will be approximated in the future, unless extraordinary conditions supervene. The record must cover a number of years, first because a continued or repeated performance is always more impressive than a single occurrence, and secondly because the average of a fairly long period will tend to absorb and equalize the distorting influences of the business cycle.

A distinction must be drawn, however, between an average which is the mere arithmetical resultant of an assortment of disconnected figures, and an average which is "normal" or "modal," in the sense that the annual results show a definite tendency to approximate the average. The contrast between one type of earning power and the other may be clearer from the following examples:

ADJUSTED EARNINGS PER SHARE 1923–1932

Year	S. H. Kress	Hudson Motors
1932	$2.80	$ 3.54(d)
1931	4.19	1.25(d)
1930	4.49	0.20
1929	5.92	7.26
1928	5.76	8.43
1927	5.26	9.04
1926	4.65	3.37
1925	4.12	13.39
1924	3.06	5.09
1923	3.39	5.56
10-year average..........	$4.36	$ 4.75

The average earnings of about $4.50 per share shown by
S. H. Kress Company can truly be called its "indicated earning
power," for the reason that the figures of each separate year
show only moderate variations from this norm. On the other
hand the Hudson Motors average of $4.75 per share is merely
an abstraction from ten widely varying figures and there is no
convincing reason for believing that the earnings from 1933
onward will bear a recognizable relationship to this average. A
similar conclusion was drawn from our discussion of the exhibit
of J. I. Case Company in Chap. I.

**Quantitative Analysis Should Be Supplemented by Qualitative
Considerations.**—In studying earnings records an important
principle of security analysis must be borne in mind:

*Quantitative data are useful only to the extent that they are
supported by a qualitative survey of the enterprise.*

In order for a company's business to be regarded as reasonably
stable, it does not suffice that the past record should show
stability. The nature of the undertaking, considered apart
from any figures, must be such as to indicate an inherent permanence of earning power. The importance of this additional
criterion was well illustrated by the case of the Studebaker
Corporation which was used as an example in our discussion of
qualitative factors in analysis in Chap. II. It is possible, on the
other hand, that there may be considerable variation in yearly
earnings, but a reasonable basis nevertheless for taking the
average as a rough index at least of future performance. United

States Steel Corporation may be cited as a leading case in point. The annual earnings for 1923–1932 are given below.

UNITED STATES STEEL CORPORATION, 1923–1932

Year	Earnings per share of common*	Output of finished steel, tons	% of total output of country
1932	$11.08(d)	3,591,000	34.4
1931	1.40(d)	7,196,000	37.5
1930	9.12	11,609,000	39.3
1929	21.19	15,303,000	37.3
1928	12.50	13,972,000	37.1
1927	8.81	12,979,000	39.5
1926	12.85	14,334,000	40.4
1925	9.19	13,271,000	39.7
1924	8.41	11,723,000	41.7
1923	11.73	14,721,000	44.2
10-year average..............	$ 8.13	11,870,000	39.1

* Adjusted for changes in capitalization.

If compared with those of Studebaker for 1920–1929, the above earnings show much greater instability. Yet the average of about $8 per share for the ten-year period has far more significance as a guide to the future than had Studebaker's indicated earning power of about $6.75 per share. This greater dependability arises from the entrenched position of United States Steel in its industry; and also from the relatively narrow fluctuations in the annual output over most of this period, which thus affords a basis for calculating "normal earnings" of United States Steel. The calculation may be made as follows:

Approximate Figures

Normal or usual annual production of finished goods.....................	13,000,000 tons
Gross receipts per ton of finished products	$100.00
Net earnings per ton before depreciation..	$12.50
Net earnings on 13,000,000 tons.........	$160,000,000.00
Depreciation, bond interest, and preferred dividends.........................	90,000,000.00
Balance for 8,700,000 shares of common.	70,000,000.00
Normal earnings per share.............	$8.00

The average earnings for the 1923–1932 decade are thus seen to approximate a theoretical figure based upon a fairly well defined "normal output." While a substantial margin of error must be allowed for in such a computation, it at least supplies a starting point for an intelligent estimate of future probabilities.

Current Earnings Should Not Be the Primary Basis of Appraisal.—The *market level* of common stocks is governed more by their current earnings than by their long-term average. This fact accounts in good part for the wide fluctuations in common-stock prices, which largely (though by no means invariably) parallel the changes in their earnings between good years and bad. Obviously the stock market is quite irrational in thus varying its valuation of a company proportionately with the temporary changes in its reported profits. A private business might easily earn twice as much in a boom year as in poor times, but its owner would never think of correspondingly marking up or down the value of his capital investment.

This is one of the most important lines of cleavage between Wall Street practice and the canons of ordinary business. Because the speculative public is clearly wrong in its attitude on this point, it would seem that its errors should afford profitable opportunities to the more logically minded to buy common stocks at the low prices occasioned by temporarily reduced earnings and to sell them at inflated levels created by abnormal prosperity.

The Classical Formula for "Beating the Stock Market."—We have here the long-accepted and classical formula for "beating the stock market." Obviously it requires strength of character in order to think and to act in opposite fashion from the crowd; and also patience to wait for opportunities which may be spaced years apart. But there are still other considerations which greatly complicate this apparently simple rule for successful operations in stocks. In actual practice the selection of suitable buying and selling levels becomes a difficult matter. Taking the long market cycle of 1921–1933, an investor might well have sold out at the end of 1925, and remained out of the market in 1926–1930, and bought again in the depression year 1931. The first of these moves would later have seemed a bad mistake of judgment and the last would have had most disturbing consequences. In other market cycles of lesser amplitude such serious miscalculations are not so likely to occur, but there is always a good deal of doubt with regard to the correct time for applying the simple principle of "buy low and sell high."

It is true also that underlying values may change substantially from one market cycle to another, more so, of course, in the case of individual issues than for the market as a whole. Hence if a common stock is sold at what seems to be a generous price in relation to the average of past earnings, it may later so improve its position as to justify a still higher quotation even in the next depression. The converse may occur in the purchase of securities at subnormal prices. If such permanent changes did not frequently develop, it is doubtful whether the market would respond so vigorously to current variations in the business picture. The mistake of the market lies in its assumption that *in every case* changes of this sort are likely to go farther, or at least to persist; whereas experience shows that such developments are exceptional and that the *probabilities* favor a swing of the pendulum in the opposite direction.

The analyst cannot follow the stock market in its indiscriminate tendency to value issues on the basis of current earnings. He may on occasion attach predominant weight to the recent figures rather than to the average, but only when persuasive evidence is at hand pointing to the continuance of these current results.

Average vs. Trend of Earnings.—In addition to emphasizing strongly the current showing of a company, the stock market attaches great weight to the indicated *trend of earnings*. In Chap. XXVII we pointed out the twofold danger inhering in this magnification of the trend—the first being that the supposed trend might prove deceptive, and the second being that valuations based upon trend obey no arithmetical rules and therefore may too easily be exaggerated. There is indeed a fundamental conflict between the concepts of the *average* and of the *trend*, as applied to an earnings record. This may be illustrated by the following simplified example:

Company	Earned per share in successive years						7th (current)	Average of 7 years	Trend
	1st	2nd	3d	4th	5th	6th			
A	$ 1	$ 2	$ 3	$ 4	$5	$6	$7	$ 4	Excellent
B	7	7	7	7	7	7	7	7	Neutral
C	13	12	11	10	9	8	7	10	Bad

On the basis of these figures the better the trend, when compared with the same current earnings (in this case $7 per share), the poorer the average; and the higher the average, the poorer the trend. They suggest an important question respecting the theoretical and practical interpretation of earnings records: Is not the trend at least as significant for the future as the average? Concretely, in judging the probable performance of Companies A and C over the next five years, would not there be more reason to think in terms of a sequence of $8, $9, $10, $11, and $12 for A, and a sequence of $7, $6, $5, $4, and $3 for C, rather than in terms of the past average of $4 for A and $10 for C?

The answer to this problem derives from common sense rather than from formal or *a priori* logic. The favorable trend of Company A's results must certainly be taken into account, but not by a mere automatic projection of the line of growth into the distant future. On the contrary, it must be remembered that the automatic or normal economic forces militate *against* the indefinite continuance of a given trend.[1] Competition, regulation, the law of diminishing returns, etc., are powerful foes to unlimited expansion; and in smaller degree opposite elements may operate to check a continued decline. Hence instead of taking the maintenance of a favorable trend for granted—as the stock market is wont to do—the analyst must approach the matter with caution, seeking to determine the causes of the superior showing, and to weigh the specific elements of strength in the company's position against the general obstacles in the way of continued growth.

Attitude of Analyst Where Trend Is Upward.—If such a *qualitative* study leads to a favorable verdict—as frequently it should —the analyst's philosophy must still impel him to base his valuation upon the past *average* of earnings; but he will apply, of course, a more liberal multiplier to this average because of the excellent prospects. In the case of Company A, therefore, he would consider the reasonable value in terms of the $4 per-share average earnings, multiplied by a coefficient which may be as high as 16. This would result in a value of about 65. The stock market is now accustomed to applying a more liberal multiplier (say 20) to the high *current* earnings, thus reaching a value of say 140 for Company A shares. The divergence in method between the stock market and the analyst—as we define this viewpoint—would mean in general that the price levels ruling for the so-called "good stocks" under normal market conditions

[1] See our discussion of the Schletter and Zander example in Chap. XXVI.

are likely to appear overgenerous to the conservative student. This does not mean that the analyst is convinced that the market valuation is wrong, but rather that he is not convinced that its valuation is right. He would call a substantial part of the price a "speculative component," in the sense that it is paid not for *demonstrated* but for expected results. (This subject is discussed further in Chap. XXXIX.)

Attitude of Analyst Where Trend Is Downward.—Where the trend has been definitely downward, as that of Company *C*, the analyst will assign great weight to this unfavorable factor. He will not assume that the downcurve *must* presently turn upward, nor can he accept the past average—which is much higher than the current figure—as a normal index of future earnings. But he will be equally chary about any hasty conclusions to the effect that the company's outlook is hopeless, that its earnings are certain to disappear entirely, and that the stock is therefore without merit or value. Here again a qualitative study of the company's situation and prospects is essential to forming an opinion as to whether *at some price*, relatively low, of course, the issue may not be a bargain, despite its declining earnings trend. Once more we identify the viewpoint of the analyst with that of a sensible business man looking into the pros and cons of some privately owned enterprise.

To illustrate this reasoning, we append the record of net earnings for 1925–1933 of Continental Baking Corporation and American Laundry Machinery Company.

Year	Continental Baking	American Laundry Machinery
1933	$2,788,000	$1,187,000(d)
1932	2,759,000	986,000(d)
1931	4,243,000	772,000
1930	6,114,000	1,849,000
1929	6,671,000	3,542,000
1928	5,273,000	4,128,000
1927	5,570,000	4,221,000
1926	6,547,000	4,807,000
1925	8,794,000	5,101,000

The profits of American Laundry Machinery reveal an uninterrupted decline, and the trend shown by Continental Baking is almost as bad. It will be noted that in 1929—the peak of prosperity for most companies—the profits of these concerns were substantially less than they were four years earlier.

Wall Street reasoning would be prone to conclude from this exhibit that both enterprises are definitely on the downward path. But such extreme pessimism would be far from logical. A study of these two businesses from the qualitative standpoint would indicate first that the respective industries are permanent and reasonably stable; and secondly that each company occupies a leading position in its industry and is well fortified financially. The inference would properly follow that the unfavorable tendency shown during 1925–1932 was probably due to accidental or nonpermanent conditions, and that in gauging the future earning power more enlightenment will be derived from the substantial *average* than from the seemingly disastrous trend.

Deficits a Qualitative, Not a Quantitative Factor.—When a company reports a deficit for the year, it is customary to calculate the amount in dollars per share or in relation to interest requirements. The statistical manuals will state, for example, that in 1932 United States Steel Corporation earned its bond interest "deficit 12.40 times," and that it showed a deficit of $11.08 per share on its common stock. It should be recognized that such figures, when taken by themselves, *have no quantitative significance;* and that their value in forming an *average* may often be open to serious question.

Let us assume that Company A lost $5 per share of common in the last year and Company B lost $7 per share. Both issues sell at 25. Is this an indication of any sort that Company A stock is preferable to Company B stock? Obviously not; for assuming it were so, it would mean that the more shares there were outstanding, the more valuable each share would be. If Company B issues two shares for one, the loss would be reduced to $3.50 per share, and on the assumption just made, each new share would then be worth more than an old one. The same reasoning applies to bond interest. Suppose that Company A and Company B each lost $1,000,000 in 1932. Company A has $4,000,000 of 5% bonds and Company B has $10,000,000 of 5% bonds. Company A would then show interest earned "deficit 5 times" and Company B would earn its interest "deficit 2 times." These figures should not be construed as an indication of any kind that Company A's bonds are less secure than Company B's bonds. For, if so, it would mean that the smaller the bond issue, the poorer its position—a manifest absurdity.

When an *average* is taken over a period which includes a number of deficits, some question must arise as to whether the resultant figure is really indicative of the *earning power*. For the wide variation in the individual figures must detract from the representative character of the average. This point is of considerable importance in view of the prevalence of deficits during the depression of the nineteen-thirties. As a practical solution, we suggest that averages be employed in such cases only when they cover a full ten-year period. If earnings are available only for a shorter period, it would be preferable to present one average for the predepression years (*e.g.*, through 1930) and a separate statement of the 1931–1933 results. (The latter may be considered mainly as a *qualitative* indication of how the company fared during the depression.) The same division of results may be found a useful supplement to the ten-year average figure.

Intuition Not a Part of the Analyst's Stock in Trade.—In the absence of indications to the contrary we accept the past record as a basis for judging the future. But the analyst must be on the lookout for any such indications to the contrary. Here we must distinguish between vision or intuition on the one hand, and ordinary sound reasoning on the other. The ability to see what is coming is of inestimable value; but it cannot be expected to be part of the analyst's stock in trade. (If he had it he could dispense with analysis.) He can be asked only to show that moderate degree of foresight which springs from logic and from experience intelligently pondered. It was not to be demanded of the securities statistician, for example, that he foretell the enormous increase in cigarette consumption since 1915, or the decline in the cigar business, or the astonishing stability of the snuff industry; nor could he have predicted—to use another example—that the two large can companies would be permitted to enjoy the full benefits from the increasing demand for their product, without the intrusion of that demoralizing competition which ruined the profits of even faster growing industries, *e.g.*, radio.

Analysis of the Future Should Be Penetrating Rather than Prophetic.—Analytical reasoning with regard to the future is of a somewhat different character, being penetrating rather than prophetic.

Example: Let us take the situation presented by Mack Trucks, Inc., in 1933 when the shares were selling at an extremely low price in relation both to their asset values and to their average earnings. At the time the annual report was released early in March 1933 the common stock was selling at $15 per share. The report exhibited *net cash assets* available for the common stock of $12 per share and *net current assets* of $40 per share. The earnings exhibit was as follows:

Year	Available for common	Per share	Dividends paid
1932	$1,480,000(d)	$ 2.19(d)	$1.00
1931	2,150,000(d)*	2.90(d)*	2.25
1930	2,008,000	2.67	5.50
1929	6,841,000	9.05	6.00
1928	5,915,000	7.83	6.00
1927	4,707,000	6.60	6.00
1926	7,716,000	10.81	6.00
1925	8,331,000	13.64	6.00 and 50% in stock
1924	5,083,000	11.97†	6.00
1923	5,866,000	13.81†	5.00
Average.....	4,284,000	7.13	

* Before extraordinary write-down of tools, etc., to $1.
† Adjusted for 50 % stock dividend paid Dec. 31, 1925.

It will be observed from the above that the stock was selling in March 1933 at slightly in excess of one-third of the net current assets per share and at little more than twice the average earnings per share.

This company was the largest unit in an important industry, so that there was every reason to expect that it would again be able to earn a reasonable profit on its invested capital. But the low price of Mack Trucks presented another anomaly. The decline in the investment status of the railroads had been due largely to the growth of motor-truck competition and to the pervading fear that such competition would continue to attract traffic from the railways. On this premise the long-term outlook for heavy truck manufacturers should have seemed unusually good. Hence to the analyst the exceedingly subnormal price of Mack Truck shares had an especially illogical appearance.

Large Profits Frequently Transitory.—More frequently we have the opposite situation, where the analyst finds reason to question the indefinite continuance of past prosperity.

Examples: Consider companies like Gabriel Snubber Manufacturing Company (now The Gabriel Company) and J. W.

Watson ("Stabilator") Company, each engaged chiefly in the manufacture of a single type of automotive accessory. The success of such a "gadget" is normally short-lived; competition and changes in the art are an ever-present threat to the stability of earning power. Hence in these cases the student could have pointed out that the market price, bearing the usual ratio to current and average earnings, reflected a quite unwarranted confidence in the permanence of profits which by their nature were likely to be transitory. Some of the pertinent data relative to this judgment are given below with respect to these two companies.[1]

THE GABRIEL COMPANY

Year	Net for common	Per share	Price range for A stock	Dividend
1932	$ 107,939(d)	$0.54(d)	3½-¼	None
1931	377,844(d)*	1.89(d)*	6⅜- 1	None
1930	98,249(d)	0.49(d)	11¾- 2½	None
1929	401,427(d)	2.00(d)	33⅞- 5	None
1928	327,976	1.64	28½-15	None
1927	960,331	4.80	59 -22	$3.50
1926	1,033,631	5.16	42 -25⅝	$4.625
1925	1,334,082	6.67	39⅞-28⅞	$1.25
1924	1,086,195†	5.43†	(Issue not quoted prior to 1925)	
1923	1,237,595†	6.19†		
1922	1,161,751†	5.81†		
1921	569,959†	2.85†		
1920	698,158†	3.49†		

* After extraordinary write-offs amounting to about 90 cents per share.
† Data are for the predecessor company.

[1] The Class "A" nonvoting common stock of Gabriel Snubber Manufacturing Company was offered to the public in April 1925 at $25 per share, a price 5¼ times the average earnings of the predecessor company during the preceding five years. The common stock of the J. W. Watson ("Stabilator") Company was originally offered in September 1927 at $24.50 per share, a price 17.3 times the average earnings of the predecessor companies during the preceding five years. The difference between the two bases of appraisal at the time of original flotation is in part accounted for by the apparently favorable "trend" of earnings of the latter company and in part by the generally higher appraisal of common shares in relation to their earnings which prevailed in 1927 as compared with 1925. Moreover, the discrepancy was not nearly so great on the basis of the then current and most recent earnings of the two companies, to which the market was prone to pay most attention.

THE J. W. WATSON COMPANY

Year	Net for common	Per share	Price range for common	Dividend
1932	$214,026(d)	$1.07(d)	⅜– ⅛	None
1931	240,149(d)	1.20(d)	2 – ⅛	None
1930	264,269(d)	1.32(d)	6 – 1	None
1929	323,137(d)	1.61(d)	14⅞– 1⅝	None
1928	348,930(d)	1.74(d)	20 – 5¼	50 cents
1927	503,725	2.16	25¾–18⅞	50 cents
1926	577,450*	2.88*	(Issue not quoted	
1925	502,593*	2.51*	prior to 1927)	
1924	29,285*	0.15*		
1923	173,907*	0.86*		
1922	142,701*	0.71*		

* Data are for predecessor companies.

A similar consideration would apply to the exhibit of Coty, Inc., in 1928. Here was a company with an excellent earnings record, but the earnings were derived from the popularity of a trade-marked line of cosmetics. This was a field in which the variable tastes of femininity could readily destroy profits as well as build them up. The inference that rapidly rising profits in previous years meant much larger profits in the future was thus especially fallacious in this case, because *by the nature of the business* a peak of popularity was likely to be reached at some not distant point, after which a substantial falling off would be, if not inevitable, at least highly probable. Some of the data appearing on the Coty exhibit are as follows:

Year	Net income	Earned per share (adjusted)
1923	$1,070,000	$0.86
1924	2,046,000	1.66
1925	2,505,000	2.02
1926	2,943,000	2.38
1927	3,341,000	2.70
1928	4,047,000	3.09
1929	4,058,000	2.73

At the high price of 82 in 1929, Coty, Inc., was selling in the market for about $120,000,000 or thirty times its maximum earnings. The actual investment in the business (capital and surplus) amounted to about $14,000,000.

Subsequent earnings were as follows:

Year	Net income	Earned per share
1930	$1,318,000	$0.86
1931	991,000	0.65
1932	521,000	0.34 (low price in 1932–1½)

A third variety of this kind of reasoning could be applied to the brewery-stock flotations in 1933. These issues showed substantial current or prospective earnings based upon capacity operations and the indicated profit per barrel. Without claiming the gift of second sight, an analyst could confidently predict that the flood of capital being poured into this new industry would ultimately result in overcapacity and keen competition. Hence a continued large return on the actual cash investment was scarcely probable; it was likely moreover that many of the individual companies would prove financial failures, while most of the others would be unable to earn enough to justify the optimistic price quotations engendered by their initial success.

CHAPTER XXXVIII

SPECIFIC REASONS FOR QUESTIONING OR REJECTING THE PAST RECORD

In analyzing an individual company, each of the governing elements in the operating results must be scrutinized for signs of possible unfavorable changes in the future. This procedure may be illustrated by various examples drawn from the mining field. The four governing elements in such situations would be: (1) life of the mine, (2) annual output, (3) production costs, and (4) selling price. The significance of the first factor has already been discussed in connection with charges against earnings for depletion. Both the output and the costs may be affected adversely if the ore to be mined in the future differs from that previously mined in location, character, or grade.

Examples: **Homestake Mining Company.**—In our discussion of Homestake Mining in Chap. XXXVI we referred to the fact that the great advance in profits in the 1930–1932 period had been caused by the mining of ore of higher grade than in previous years. The relevant figures are as follows:

Year	Tons milled (in thousands)	Grade of ore (yield per ton mined-gold at $20.67 per oz.)	Total revenue (in thousands)	Net earnings before depreciation and depletion (in thousands)
1932	1,402	$7.07	$10,255	$4,838
1931	1,404	6.36	9,206	4,194
1930	1,364	6.18	8,668	3,307
1929	1,438	4.53	6,700	2,473
1928	1,417	4.63	6,730	2,897
1927	1,372	4.87	6,827	2,822
1926	1,416	4.11	5,924	1,880
1925	1,590	3.77	6,079	1,894
1924	1,670	3.67	6,213	2,007
1923	1,652	3.87	6,467	2,275

It is to be noted that the tonnage milled in 1932 was no greater than in 1926 so that the increase of more than 150% in the net profit was due entirely to the higher yield per ton. The question must therefore arise whether the richer ore can be expected to continue indefinitely, or whether it represents a temporary advantage caused by mining an unusually rich but limited ore body. The annual reports are silent on this important point. In the absence of such information the speculator is likely to hope for the best, but the conservatively minded analyst must point out the apparently serious danger of a return to the lower grade and lower yield of previous years.

Calumet and Hecla Consolidated Copper Company.—The report of Calumet and Hecla Consolidated Copper Company for 1927 illustrates another aspect of the question of character of ore. The income account showed earnings of about $2,500,000 (after depreciation, but before depletion). This equalled $1.24 per share on 2,006,000 shares, which sold in that year between 14 and 25. Further analysis of the report would disclose that about 60% of the year's profits were derived from operation of a reclamation plant which extracted copper from the accumulated waste of previous, less efficient milling operations, so-called "tailings." These tailings produced only about one-quarter of the total copper output, but the cost of production was much lower than that of metal taken from the mine. The figures are approximately as follows:

CALUMET AND HECLA RESULTS FOR 1927

Source of output	Quantity, pounds	Per pound		Profit about, cents	Total profit
		Selling price, cents	Cost* about, cents		
Copper produced by mine..	80,000,000	13.25	12	1.25	$1,000,000
Copper produced by reclamation plant...........	28,700,000	13.25	8	5.25	1,500,000
					$2,500,000

Earned per share from mining operations............ $0.50
Earned per share from reclamation plant............ 0.74

$1.24

* After depreciation, but before depletion.

This analysis indicates that a crucial point in the Calumet and Hecla situation was the amount of the cheaper copper available in the tailings pile. Investigation would have shown that the quantity was limited to only a few years' operations at the 1927 rate. It was clear that when these tailings were used up, there would then be both a smaller output and a higher unit cost. Hence the figures of 1927 could not be taken as fairly indicative of the future earning power of Calumet and Hecla, assuming the same selling price for its copper.

Freeport Texas Company.—The exhibit of Freeport Texas (Sulphur) Company in 1933 supplies the same type of problem for the analyst, and it also raises the question of the propriety of the use, under such circumstances, of the past earnings record to support the sale of new securities. An issue of $2,500,000 of 6% cumulative convertible preferred stock was sold at $100 per share in January 1933 in order to raise funds to equip a new sulphur property leased from certain other companies.

The offering circular stated among other things:

1. That the sulphur reserves had an estimated life of at least 25 years based upon the average annual sales for 1928–1932;
2. That the earnings for the period 1928–1932 averaged $2,952,501 or 19.6 times the preferred-dividend requirement.

The implication of these statements would be that, assuming no change in the price received for sulphur, the company could confidently be expected to earn over the next 25 years approximately the amounts earned in the past.

The facts in the case, however, do not warrant any such deduction. The company's past earnings were derived from the operation of the two properties, at Bryanmound and at Hoskins Mound, respectively. The Bryanmound area was owned by the company and contributed the bulk of the profits. But by 1933 its life was "definitely limited" (in the words of the listing application), *in fact the reserves were not likely to last more than about three years*. The Hoskins Mound was leased from the Texas Company. After paying $1.06 per ton fixed royalty, no less than 70% of the remaining profits were payable to Texas Company as rental. One half of Freeport's sales were required to be made from sulphur produced at Hoskins. The new property at Grande Ecaille, La., now to be developed, would require royalty payments amounting to some 40% of the net earnings.

When these facts are studied it will be seen that the earnings of Freeport Texas for 1928–1932 have no direct bearing on the results to be expected from future operations. The sulphur reserves, stated to be good for 25 years, represented mineral located in an entirely different place and to be extracted under entirely different conditions from those obtaining in the past. A large profit-sharing royalty will be payable on the sulphur produced from the new project, while the old Bryanmound was owned outright by Freeport and hence its profits accrued 100% to the company.

In addition to this known element of higher cost, great stress must be laid also upon the fact that the major future profits of Freeport were now expected from a new project. The Grande Ecaille property was not yet equipped and in operation, and hence it was subject to the many hazards that attach to enterprises in the development stage. The cost of production at the new mine might conceivably be much higher, or much lower, than at Bryanmound. From the standpoint of security analysis the important point is that where two quite different properties are involved, you have two virtually separate enterprises. Hence the 1928–1932 record of Freeport Texas was hardly more relevant to its future history than were the figures of some entirely different sulphur company, e.g., Texas Gulf Sulphur.

Returning once more to the business man's viewpoint on security values, the Freeport Texas exhibit suggests the following interesting line of reasoning. In June 1933 this enterprise was selling in the market for about $32,000,000 (25,000 shares of preferred at 125, and 730,000 shares of common at 40). The major portion of its future profits were expected to be derived from an investment of $3,000,000 to equip a new property leased from three large oil companies. Presumably these oil companies drove as good a bargain for themselves as possible in the terms of the lease. The market was in effect placing a valuation of some $20,000,000, or more, upon a new enterprise in which only $3,000,000 was to be invested. It was possible, of course, that this enterprise would prove to be worth much more than six times the money put into it. But from the standpoint of ordinary business procedure the payment of such an enormous premium for anticipated future results would appear imprudent in the extreme.[1]

---

[1] Since the Freeport Texas preferred issue was relatively small, representing less than one-tenth of the total market value of the company, this

Evidently the stock market—like the heart, in the French proverb—has reasons all its own. In the writers' view, where these reasons depart violently from sound sense and business experience, common-stock buyers must inevitably lose money in the end, even though large speculative gains may temporarily accrue, and even though certain fortunate purchases may turn out to be permanently profitable.

The Future Price of the Product.—The three preceding examples related to the future continuance of the rate of output and the operating costs upon which the past record of earnings was predicted. We must also consider such indications as may be available in regard to the *future selling price* of the product. Here we must ordinarily enter into the field of surmise or of prophecy. The analyst can truthfully say very little about future prices, except that they fall outside the realm of sound prediction. Now and then a more illuminating statement may be justified by the facts. Adhering to the mining field for our examples, we may mention the enormous profits made by zinc producers during the Great War, because of the high price of spelter. Butte and Superior Mining Company earned no less than $64 per share before depreciation and depletion in the two years 1915–1916, as the result of obtaining about 13 cents per pound for its output of zinc, against a prewar average of about $5\frac{1}{4}$ cents. Obviously the future earning power of this company was almost certain to shrink far below the war-time figures; nor could these properly be taken together with the results of any other years in order to arrive at the average or supposedly "normal" earnings.

Change in Status of Low-cost Producers.—The copper-mining industry offers an example of wider significance. An analysis of companies in this field must take into account the fact that since 1914 a substantial number of new low-cost producers have been developed and that other companies have succeeded in reducing extraction costs through metallurgical improvements. This means that there has been a definite lowering of the "center of gravity" of production costs for the entire industry. Other things being equal, this would make for a lower selling price in the future than obtained in the past. (Such a development is

---

analysis would not call into question the safety of the senior issue, but reflects only upon the soundness of the valuation accorded the common stock—judged by investment standards.

more strikingly illustrated by the crude-rubber industry.) Differently stated, mines which formerly rated as low-cost producers, *i.e.*, as having costs well below the average, may have lost this advantage, unless they have also greatly improved their technique of production. The analyst would have to allow for these developments in his calculations, by taking a cautious view of future copper prices.

Anomalous Prices and Price Relationships in the History of the I.R.T. System.—The checkered history of the Interborough Rapid Transit System in New York City has presented a great variety of divergences between market prices and the real or relative values ascertainable by analysis. Two of these discrepancies turn upon the fact that for specific reasons the then current and past earnings should not have been accepted as indicative of future earning power. In abbreviated form the details of these two situations are as follows:

For a number of years prior to 1918 the Interborough Rapid Transit Company was very prosperous. In the twelve months ended June 30, 1917 it earned $26 per share on its capital stock and paid dividends of $20 per share. Nearly all of this stock was owned by Interborough Consolidated Corporation, a holding concern (previously called Interborough-Metropolitan Corporation) which in turn had outstanding collateral trust bonds, 6% preferred stock and common stock. Including its share of the undistributed earnings of the operating company it earned about $11.50 per share on its preferred stock and about $2.50 on the common. The preferred sold in the market at 60 and the common at 10. These issues were actively traded in, and they were highly recommended to the public by various financial agencies which stressed the phenomenal growth of the subway traffic. In fact, a New York Stock Exchange house published an elaborate (and expensive) brochure dilating on the investment and speculative merits of Interborough Consolidated preferred and common stock.

A modicum of analysis would have shown that the real picture was entirely different than appeared on the surface. New rapid transit facilities were being constructed under contract between the City of New York and the Interborough (as well as others under contract between the City and the Brooklyn Rapid Transit Company). As soon as the new lines were placed in operation, which was to be the following year, the earnings available for

Interborough were to be limited under this contract to the figure prevailing in 1911–1913, *which was far less than the current profits.* The City would then be entitled to receive a high return on its enormous investment in the new lines. If and after all such payments were made in full, including back accruals, the City and the Interborough would then share equally in surplus profits. However, the preferential payments due the City would be so heavy that experts had testified that under the most favorable conditions it would be *more than* 30 *years* before there could be any surplus income to divide with the company.

The subjoined brief table shows the significance of these facts.

INTERBOROUGH RAPID TRANSIT SYSTEM

Item	Actual earnings 1917	*Maximum* earnings when contract with City became operative
Balance for I.R.T. stock................	$9,100,000	$5,200,000
Share applicable to Interborough Consolidated Corp.........................	8,800,000	5,000,000
Interest on Inter. Consol. bonds..........	3,520,000	3,520,000
Balance for Inter. Consol. pfd...........	5,280,000	1,480,000
Preferred dividend requirements..........	2,740,000	2,740,000
Balance for Inter. Consol. common........	2,540,000	*1,260,000(d)*
Earned per share, Inter. Consol. pfd.......	$11.50	$3.25
Earned per share, Inter. Consol. common..	2.50	nil

The underlying facts proved beyond question, therefore, that instead of a brilliant future being in store for Interborough, it was destined to suffer a severe loss of earning power within a year's time. It would then be quite impossible to maintain the $6 dividend on the holding company's preferred stock, and no earnings at all would be available for the common for a generation or more. On this showing it was mathematically certain that both Interborough Consolidated stock issues were worth far less than their current selling prices.[1]

[1] Indications pointed strongly to manipulative efforts by insiders in 1916–1917 to foist these shares upon the public at high prices before the period of lower earnings began. The payment of full dividends on the preferred stock, during an interlude of large earnings known to be temporary, was inexcusable from the standpoint of corporate policy, but understandable

The sequel not only bore out this criticism, which it was bound to do, but demonstrated also that where an *upper limit* of earnings or value is fixed, there is usually danger that the actual figure will be less than the maximum.   The opening of the new subway lines coincided with a large increase in operating costs, due to war-time inflation; and also, as was to be expected, it diminished the profits of the older routes.   Interborough Rapid Transit Company was promptly compelled to reduce its dividend, and it was omitted entirely in 1919.   In consequence the holding company, Interborough Consolidated, suspended its preferred dividends in 1918.   The next year it defaulted the interest on its bonds, became bankrupt, and disappeared from the scene, *with the complete extinction of both its preferred and common stock.* Two years later Interborough Rapid Transit Company, recently so prosperous, barely escaped an imminent receivership by means of a "voluntary" reorganization which extended a maturing note issue.   When this extended issue matured in 1932 the company was again unable to pay and this time receivers took over the property.

During the ten-year period between the two receivership applications another earnings situation developed, somewhat similar to that of 1917.[1]   In 1928 the Interborough reported earnings of $3,000,000 or $8.50 per share for its common stock, and the shares sold as high as 62.   But these earnings included $4,000,000 of "back preferentials" from the subway division. The latter represented a limited amount due the I.R.T. out of subway earnings to make good a deficiency in the profits of the early years of operating the new lines.   On June 30, 1928 the amount of back preferentials remaining to be paid the company

---

as a device to aid in unloading stock.   These dividend distributions were not only unfair to the 4½% bondholders, but, because of certain prior developments, they were probably illegal as well.   (Reference to this aspect of the case was made in Chap. XX).

[1] See Note 42 in the Appendix for a concise discussion of the numerous anomalies in price between various Interborough System securities, *viz.*,

   a. Between Interborough Metropolitan 4½s and Interborough Consolidated Preferred in 1919.

   b. Between I.R.T. 5s and I.R.T. 7s in 1920.

   c. Between I.R.T. stock and Manhattan "Modified" stock in 1929.

   d. Between I.R.T. 5s and I.R.T. 7s in 1933.

   e. Between Manhattan "Modified" and Manhattan "Unmodified" stock in 1933.

was only $1,413,000. *Hence all the profits available for Interborough stock were due to a special source of revenue which could continue for only a few months longer.* Heedless speculators, however, were capitalizing as permanent an earning power of Interborough stock which analysis would show was of entirely nonrecurrent and temporary character.

CHAPTER XXXIX

PRICE-EARNINGS RATIOS FOR COMMON STOCKS. ADJUSTMENTS FOR CHANGES IN CAPITALIZATION

In previous chapters various references have been made to Wall Street's ideas on the relation of earnings to values. A given common stock is generally considered to be worth a certain number of times its current earnings. This number of times, or multiplier, depends partly on the prevailing psychology and partly on the nature and record of the enterprise. Prior to the 1927–1929 bull market, ten times earnings was the accepted standard of measurement. More accurately speaking, it was the common point of departure for valuing common stocks, so that an issue would have to be considered exceptionally desirable to justify a higher ratio, and conversely. Beginning about 1927, and continuing through 1929, the ten-times-earnings standard was superseded by a rather confusing set of new yardsticks. On the one hand, there was a tendency to value common stocks in general more liberally than before. This was summarized in a famous dictum of a financial leader implying that good stocks were worth fifteen times rather than ten times their earnings.[1] There was also the tendency to make more sweeping distinctions in the valuations of different kinds of common stocks. Companies in especially favored groups, e.g., public utilities and chain stores, sold at a very high multiple of current earnings, say twenty-five to forty times. This was true also of the "blue chip" issues, which comprised leading units in miscellaneous fields. As pointed out before, these generous valuations were based upon the assumed continuance of the upward trend shown over a longer or shorter period in the past.

Exact Appraisal Impossible.—Security analysis cannot presume to lay down general rules as to the "proper value" of any given

[1] The wording of this statement, as quoted in the *Wall Street Journal* of March 26, 1928, was as follows: " 'General Motors shares, according to the Dow, Jones & Co. averages,' Mr. Raskob remarked, 'should sell at fifteen times earning power, or in the neighborhood of $225 per share, whereas at the present level of $180 they sell at approximately only twelve times current earnings.'"

451

common stock. Practically speaking, there is no such thing. The bases of value are too shifting to admit of any formulation which could claim to be even reasonably accurate. The whole idea of basing the value upon current earnings seems inherently absurd, since we know that the current earnings are constantly changing. And whether the multiplier should be ten or fifteen or thirty would seem at bottom a matter of purely arbitrary choice.

But the stock market itself has no time for such scientific scruples. It must make its values first and find its reasons afterwards. Its position is much like that of a jury in a breach-of-promise suit; there is no sound way of measuring the values involved, and yet they must be measured somehow and a verdict rendered. Hence the prices of common stocks are not carefully thought out computations, but the resultants of a welter of human reactions. The stock market is a voting machine rather than a weighing machine. It responds to factual data not directly, but only as they affect the decisions of buyers and sellers.

Limited Functions of the Analyst in Field of Appraisal of Stock Prices.—Confronted by this mixture of changing facts and fluctuating human fancies, the securities analyst is clearly incapable of passing judgment on common-stock prices generally. There are, however, some concrete, if limited, functions which he may carry on in this field, of which the following are representative:

1. He may set up a basis for *conservative* or *investment* valuation of common stocks, as distinguished from speculative valuations.
2. He may point out the significance of: (*a*) the capitalization structure; and (*b*) the source of income, as bearing upon the valuation of a given stock issue.
3. He may find unusual elements in the balance sheet which affect the implications of the earnings picture.

A Suggested Basis of Maximum Appraisal for Investment.—A conservative valuation of a stock issue must bear a reasonable relation to the *average* earnings. In addition it must be justified by whatever indications are available as to the future. This approach shifts the original point of departure, or basis of computation, from the current earnings to the average earnings, which should cover a period of not less than five years, and preferably seven to ten years. This does not mean that all common stocks with the same average earnings should have the same value. The common-stock investor (*i.e.*, the *conservative buyer*) will properly accord a more liberal valuation to those issues which have current earnings above the average, or which may reasonably be considered to possess better than average prospects. But it is the essence of our viewpoint that some moderate upper limit must *in every case* be placed on the multiplier in order to stay within the bounds of conservative valuation. We would suggest that *about sixteen times average earnings* is as high a price as can be paid in an *investment* purchase of a common stock.

While this rule is of necessity arbitrary in its nature, it is not entirely so. Investment presupposes demonstrable value, and the typical common stock's value can be demonstrated only by means of an established, *i.e.*, an average, earning power. But it is difficult to see how average earnings of *less than 6%* upon the market price could ever be considered as vindicating that price. No true investor could regard such a price-earnings ratio as satisfactory in itself. It would be acceptable only in the expectation that future earnings will be larger than in the past. In the original and most useful sense of the term such a basis of valuation is *speculative*. It falls outside the purview of common-stock investment.

Higher Prices May Prevail for Speculative Commitments.— The intent of this distinction must be clearly understood. We do not imply that it is a mistake to pay more than 16 times average earnings for any common stock. We do suggest that such a price would be speculative. The purchase may easily turn out to be highly profitable, but in that case it will have proved a wise or fortunate speculation. It is proper to remark, moreover, that very few people are consistently wise or fortunate in their speculative operations. Hence we may submit, as a corollary of no small practical importance, *that people who habitually purchase common stocks at more than about sixteen times their average earnings are likely to lose considerable money in the long run.* This is the more probable because, in the absence of such a mechanical check, they are prone to succumb recurrently to the lure of bull markets, which always find some specious argument to justify paying extravagant prices for common stocks.

Other Requisites for Common Stocks of Investment Grade and a Corollary Therefrom.—It should be pointed out that if 16 times average earnings is taken as the *upper limit* of price for an investment purchase, then ordinarily the price paid should be substantially less than this maximum. This brings us back to the formerly accepted *ten-times-earnings* ratio as suitable for the typical case. We must emphasize also that a reasonable ratio of average earnings to market price is not the only requisite for a common-stock investment. It is a necessary, but not a sufficient condition. The company must be satisfactory also in its financial set-up, its management, and its prospects.

From this principle there follows another important corollary, *viz.: An attractive common-stock investment is an attractive speculation.* This is true because if a common stock can meet the demand of a conservative investor that he get full value for his money *plus* satisfactory future prospects, then such an issue must also have a fair chance of appreciating in market value.

Examples of Speculative and Investment Common Stocks.— Our definition of an investment basis for common-stock purchases is at variance with the Wall Street practice in respect to common stocks of high rating. For such issues a price of considerably more than sixteen times average earnings is held to be warranted, and furthermore these stocks are designated as "investment issues" regardless of the price at which they sell. According to our view, the high prices paid for "the best common stocks" make these purchases essentially speculative, because they require future growth to justify them. Hence common-stock investment operations, as we define them, will occupy a middle ground in the market, lying between low-price issues which are speculative because of doubtful quality and well-entrenched issues which are speculative, none the less, because of their high price.

These distinctions are illustrated on the following pages by nine examples, taken as of July 31, 1933.

Comments on the Various Groups.—The companies listed in Group *A* are representative of the so-called first-grade or "blue chip" industrials, which were particularly favored in the great speculation of 1928–1929. They are characterized by relatively stable or growing earnings, a strong financial position, and presumably by excellent prospects. *The market price* of the shares, however, was higher than would be justified by their average earnings. In fact the profits of the *best* year in the 1923–1932 decade were less than 10% of the July 1933 market price.

GROUP A: COMMON STOCKS SPECULATIVE IN JULY 1933 BECAUSE OF THEIR
HIGH PRICE
(Figures adjusted to reflect changes in capitalization)

Item	National Biscuit	Air Reduction	Commercial Solvents
Amount earned per share of common:			
1932	$2.44	$2.73	$0.51
1931	2.86	4.54	0.84
1930	3.41	6.32	1.07
1929	3.28	7.75	1.45
1928	2.92	4.61	1.22
1927	2.84	3.58	0.84
1926	2.53	3.63	0.69
1925	2.32	3.33	0.37
1924	2.18	2.81	0.45
1923	2.02	4.14	*0.02(d)*
10-yr. averages......	$2.68	$4.34	$0.74
Pfd. stock..............	(248,000 sh. @ 140) $ 35,000,000		
Common stock.........	(6,289,000 sh. @ 53) 333,000,000	(841,000 sh. @ 90) $76,000,000	(2,495,000 sh. @ 30) $75,000,000
Total capitalization......	$368,000,000	$76,000,000	$75,000,000
Net tangible assets, 12/31/32.............	$129,000,000	$29,200,000	$ 8,700,000*
Net current assets, 12/31/32.............	36,000,000	9,800,000	6,000,000
Average earnings on common-stock price........	5.1%	4.8%	2.5%
Maximum earnings on common-stock price...	6.4%	8.6%	4.8%

* To this should be added an allowance for the plant and equipment written down on the
books to $1. In 1929 these fixed assets were valued at about $3,000,000, net.

It is also characteristic of such issues that they sell for enormous
premiums above the actual capital invested.

The companies analyzed in Group *B* are obviously speculative,
because of the great instability of their earnings records. They
show varying relationships of market price to average earnings
and to asset values.

The common stocks shown in Group *C* are examples of those
which meet specific and quantitative tests of investment quality.
These tests include the following:

1. The earnings have been reasonably stable, allowing for the tremendous
fluctuations in business conditions during the ten-year period.
2. The average earnings bear a satisfactory ratio to market price.
3. The financial set-up is conservative and the working-capital position
is strong.

GROUP B: COMMON STOCKS SPECULATIVE IN JULY 1933 BECAUSE OF THEIR
IRREGULAR RECORD

Item	B. F. Goodrich (Rubber)	Gulf States Steel	Standard Oil of Kansas
Earned per share of common*:			
1932	$ *6.73(d)*	$ *3.94(d)*	$0.23†
1931	*8.01(d)*	*5.89(d)*	*1.93(d)*
1930	*8.55(d)*	*4.84(d)*	1.19
1929	4.53	5.93	4.73
1928	1.50	6.28	0.91
1927	17.11	4.93	*2.59(d)*
1926	*4.15(d)*	5.28	0.51
1925	23.99	7.17	1.54
1924	11.10	7.48	*1.50(d)*
1923	*0.88(d)*	12.79	*0.88(d)*
10-yr. average.........	$ 2.99	$ 3.52	$0.22
Bonds (at par)............	$ 43,000,000	$ 5,200,000	
Pfd. stock................	(294,000 sh. @ 38) 11,200,000	(20,000 sh. @ 50) 1,000,000	
Common stock............	(1,156,000 sh. @ 15) 17,300,000	(198,000 sh. @ 28) 5,600,000	(269,000 sh. @ 20) $5,380,000
Total capitalization........	$ 71,500,000	$11,800,000	$5,380,000
Net tangible assets 12/31/32	105,300,000	27,000,000	5,290,000
Net current assets 12/31/32	43,700,000	2,230,000	3,980,000
Average earnings on common-stock price.........	19.9%	12.6%	1.1%
Maximum earnings on common-stock price.........	160%	45.7%	23.7%

* Adjusted in column 1 to reflect actual changes in inventory values.
† 9 months ended Dec. 31, 1932.

While we do not suggest that a common stock bought for
investment be *required* to show asset values equal to the price
paid, it is none the less characteristic of issues in Group *C* that,
as a whole, they will not sell for a huge premium above the
companies' actual resources.

Common-stock *investment*, as we envisage it, will confine itself
to issues making exhibits of the kind illustrated by Group *C*.
But the actual purchase of any such issues must require *also*
that the purchaser be satisfied in his own mind that the prospects
of the enterprise are at least reasonably favorable.

Allowances for Changes in Capitalization.—In dealing with
the past record of earnings, when given on a per-share basis,
it is elementary that the figures must be adjusted to reflect any
important changes in the capitalization which have taken place
during the period. In the simplest case these will involve a
change only in the number of shares of common stock due to
stock dividends, split-ups, etc. All that is necessary then is

GROUP C: COMMON STOCKS MEETING INVESTMENT TESTS IN JULY 1933
FROM THE QUANTITATIVE STANDPOINT*

Item	S. H. Kress	Island Creek Coal	Nash Motors
Earned per share of common:			
1932	$2.80	$1.30	$0.39
1931	4.19	2.28	1.78
1930	4.49	3.74	2.78
1929	5.92	5.05	6.60
1928	5.76	4.46	7.63
1927	5.26	5.64	8.30
1926	4.65	4.42	8.50
1925	4.12	3.22	5.57
1924	3.06	3.58	3.00
1923	3.39	4.08	2.96
10-yr. average.......	$4.36	$3.78	$4.75
Preferred stock..........	(372,000 sh. @ 10)	(27,000 sh. @ 90)	
	$ 3,700,000	$ 2,400,000	
Common stock..........	(1,162,000 sh. @ 33)	(594,000 sh. @ 24)	(2,646,000 sh. @ 19)
	38,300,000	14,300,000	$50,300,000
Total capitalization......	$42,000,000	$16,700,000	$50,300,000
Net tangible assets, 12/31/32.............	58,300,000	18,900,000	41,000,000
Net current assets, 12/31/32.............	15,200,000	7,500,000	33,000,000
Average earnings on common-stock price........	13.2%	15.8%	25.0%
Maximum earnings on common-stock price....	17.9%	23.5%	44.7%

* Island Creek Coal and Nash Motors figures adjusted for stock dividends.

to restate the capitalization throughout the period on the basis of the current number of shares. (Such recalculations are made by some of the statistical services, but not by others.)

When the change in capitalization has been due to the sale of additional stock at a comparatively low price (usually through the exercise of subscription rights or warrants), or to the conversion of senior securities, the adjustment is more difficult. In such cases the earnings available for the common during the earlier period must be increased by whatever gain would have followed from the issuance of the additional shares. When bonds or preferred stocks have been converted into common, the charges formerly paid thereon are to be added back to the earnings and the new figure then applied to the larger number of shares. If stock has been sold at a relatively low price, a proper adjustment would allow earnings of say 6 to 10% on the proceeds of the sale. (Such recalculations need not be made unless the changes indicated thereby are substantial.)

A corresponding adjustment of the per-share earnings must be made at times to reflect the possible *future* increase in the number of shares outstanding as a result of conversions or exercise of option warrants. When other security holders have a choice of any kind, sound analysis must allow for the possible adverse effect upon the per-share earnings of the common stock that would follow from the exercise of the option.

Examples: This type of adjustment is illustrated in an analysis of the showing of Barnsdall Corporation for the year 1926.

Earnings as reported: $6,077,000 = $5.34 per share on 1,140,000 shares outstanding.

But there were also outstanding warrants to purchase 1,000,000 shares at $25 per share, proceeds to be used to redeem $25,000,000 of 6% bonds. The analyst must *assume* exercise of these warrants, giving the following adjusted result: Earnings, $7,577,000 = $3.54 per share on 2,140,000 shares to be outstanding.

American Water Works and Electric Company can be used to illustrate both types of adjustment.

Year	Earnings* for common as reported		Adjustment A.			Adjustment B.		
	Amount	Number of shares	Per share	Number of shares	Earned per share	Amount	Number of shares	Earned per share
1933	$2,392	1,751	$1.37	1,751	$1.37	$3,140	2,501	$1.26
1932	2,491	1,751	1.42	1,751	1.42	3,240	2,501	1.30
1931	4,904	1,751	2.80	1,751	2.80	5,650	2,501	2.26
1930	5,424	1,741	3.10	1,751	3.10	6,170	2,501	2.47
1929	6,621	1,655	4.00	1,741	3.80	7,370	2,491	2.95
1928	5,009	1,432	3.49	1,739	2.88	5,760	2,491	2.30
1927	3,660	1,361	2.69	1,737	2.11	4,410	2,491	1.76
7-year average	......	.....	$2.70	.....	$2.50	......	.....	$2.04

* Number of shares and earnings in thousands.

Adjustment *A* reflects the payment of stock dividends in 1928, 1929, and 1930.

Adjustment *B* assumes conversion of the $15,000,000 of convertible 5s, issued in 1934, thus increasing the earnings by the amount of the interest charges, but also increasing the common-stock issue by 750,000 shares. (The above adjustments are independent of any possible modifications in the reported earnings arising from the questioning of the depreciation charges, etc., as previously discussed.)

Corresponding adjustments in book values or current-asset values per share of common stock should be made in analyzing the balance sheet. This technique is followed in our discussion of the Baldwin Locomotive Works exhibit in the Appendix, Note 48.

Allowances for Participating Interests.—In calculating the earnings available for the common, full recognition must be given to the rights of holders of participating issues, whether or not the amounts involved are actually being paid thereon. Similar allowances must be made for the effect of management contracts providing for a substantial percentage of the profits as compensation, as in the case of investment trusts. Unusual cases sometimes arise involving "restricted shares," dividends on which are contingent upon earnings or other considerations.

ADJUSTED EARNINGS: TRICO PRODUCTS CORPORATION

Year	Earnings for common	Earned per share on unrestricted stock		
		A. Ignoring restricted shares	*B.* Allowing for dividend rights of restricted shares	*C.* Allowing for release of restricted shares (*i.e.*, on total capitalization)
1924	$ 171,000	$0.76	$0.76	$0.25
1925	485,000	2.15	2.15	0.72
1926	807,000	3.59	2.86	1.19
1927	1,372,000	5.00	3.52	2.03
1928	1,778,000	5.27	3.88	2.64
1929	2,250,000	6.67	4.58	3.33
1930	1,908,000	5.09	3.94	2.83
1931	1,763,000	4.70	3.73	2.61
1932	965,000	2.57	2.54	1.44
1933	1,418,000	3.78	3.21	2.10
10-year average......	$1,292,000	$3.96	$3.12	$1.91

Example: Trico Products Corporation, a large manufacturer of automobile accessories, is capitalized at 675,000 shares of common stock, of which 450,000 shares (owned by the president) were originally "restricted" as to dividends. The unrestricted stock is first entitled to dividends of $2.50 per share, after which both classes share equally in further dividends. However, successive blocks of the unrestricted stock were to be released from the restriction, according as the earnings for 1925 and successive years reached certain stipulated figures. (To the end of 1933, a total of 150,000 shares had been released from the restriction.)

The computation shown on page 459 indicates the proper method of allowing for the facts as above stated.

A situation similar to that in Trico Products Corporation obtained in the case of Montana Power Company stock prior to June 1921.

CHAPTER XL

CAPITALIZATION STRUCTURE

The division of a company's total capitalization between senior securities and common stock has an important bearing upon the significance of the earning power per share. A set of hypothetical examples will help make this point clear. For this purpose we shall postulate three industrial companies, *A*, *B*, and *C*, each with an earning power (*i.e.*, with average and recent earnings) of $1,000,000. They are identical in all respects save capitalization structure. Company *A* is capitalized solely at 100,000 shares of common stock. Company *B* has outstanding $5,000,000 of 5% bonds and 100,000 shares of common stock. Company *C* has outstanding $10,000,000 of 5% bonds and 100,000 shares of common stock.

We shall assume that the bonds are worth par and that the common stocks are worth ten times their per-share earnings. Then the value of the three companies will work out as follows:

Company	Earnings for common stock	Value of common stock	Value of bonds	Total value of company
A	$1,000,000	$10,000,000	. . . .	$10,000,000
B	750,000	7,500,000	$ 5,000,000	12,500,000
C	500,000	5,000,000	10,000,000	15,000,000

These results challenge attention. Companies with identical earning power appear to have widely differing values, due solely to the arrangement of their capitalization. But the capitalization structure is itself a matter of voluntary determination by those in control. Does this mean that the fair value of an enterprise can be arbitrarily increased or decreased by changing around the relative proportions of senior securities and common stock?

Can the Value of an Enterprise Be Altered through Arbitrary Variations in Capital Structure?—To answer this question

461

properly we must scrutinize our examples with greater care. In working out the value of the three companies we assumed that the bonds would be worth par and that the stocks would be worth ten times their earnings. Are these assumptions tenable? Let us consider first the case of Company *B*. If there are no unfavorable elements in the picture the bonds might well sell at about 100, since the interest is earned four times. Nor would the presence of this funded debt ordinarily prevent the common stock from selling at ten times its established earning power.

It will be urged however, that if Company *B* shares are worth ten times their earnings, Company *A* shares should be worth more than this multiple because they have no debt ahead of them. The risk is therefore smaller and they are less vulnerable to the effect of a shrinkage in earnings than is the stock of Company *B*. This is obviously true, and yet it is equally true that Company *B* shares will be more responsive to an *increase* in earnings. The following figures bring this point out clearly:

Assumed earnings	Earned per share		Change in earnings per share from base	
	Co. *A*	Co. *B*	Co. *A*	Co. *B*
$1,000,000	$10.00	$ 7.50	(Base)	(Base)
750,000	7.50	5.00	−25%	−33⅓%
1,250,000	12.50	10.00	+25%	+33⅓%

Would it not be fair to assume that the greater sensitivity of Company *B* to a possible decline in profits is offset by its greater sensitivity to a possible increase? Furthermore, if the investor expects higher earnings in the future—and presumably he selects his common stocks with this in mind—would he not be justified in selecting the issue which will benefit more from a given degree of improvement? We are thus led back to the original conclusions that Company *B* may be worth $2,500,000, or 25%, more than Company *A* due solely to its distribution of capitalization between bonds and stock.

Principle of Optimum Capitalization Structure.—Paradoxical as this conclusion may seem, it is supported by the actual behavior of common stocks in the market. If we subject this contradiction to closer analysis we shall find that it arises from what may be called an *oversimplification* of Company *A*'s capital structure. Company *A*'s common stock evidently contains the two elements represented by the bonds and stock of Company *B*. Part of Company *A*'s stock is at bottom equivalent to Company *B*'s bonds and should *in theory* be valued on the same basis, *i.e.*, 5%. The remainder of Company *A*'s stock should then be valued on a 10% basis. This theoretical reasoning would give us a combined value of $12,500,000, *i.e.*, an average 8% basis, for the two components of Company *A* stock, which of course, is the same as that of Company *B* bonds and stock taken together.

But this $12,500,000 value for Company *A* stock would not ordinarily be realized *in practice*. The obvious reason is that the common-stock buyer will rarely recognize the existence of a "bond component" in a common-stock issue; and in any event, not wanting such a bond component, he is unwilling to pay extra for it. This fact leads us to an important principle, both for the security buyer and for corporate management, *viz.*:

The optimum capitalization structure for any enterprise includes senior securities to the extent that they may safely be issued and bought for investment.

Concretely this means that the capitalization arrangement of Company *B* is preferable to that of Company *A* from the stockholder's standpoint, assuming that in both cases the $5,000,000 bond issue would constitute a sound investment. (This might require, among other things, that the companies show a net working capital of not less than $5,000,000, in accordance with the stringent tests for sound industrial issues recommended in Chap. XIII.) Under such conditions the contribution of the entire capital by the common stockholders may be called an overconservative set-up, as it tends generally to make the stockholder's dollar less productive to him than if a reasonable part of the capital were borrowed. An analogous situation holds true in most private businesses, where it is recognized as profitable and proper policy to use a conservative amount of banking accommodation for seasonal needs rather than to finance operations entirely by owners' capital.

Corporate Practices Resulting in Shortage of Sound Industrial Bonds.—Furthermore, just as it is desirable from the bank's standpoint that sound businesses borrow seasonally, it is also desirable from the standpoint of investors generally that strong industrial corporations raise an appropriate part of their capital through the sale of bonds. Such a policy would increase the number of high-grade bond issues on the market, giving the bond investor a wider range of choice and making it deservedly difficult to sell unsound bonds. Unfortunately the practice of industrial corporations in recent years has tended to produce a shortage of good industrial bond issues. Strong enterprises have in general refrained from floating new bonds and in many cases have retired old ones. But this avoidance of bonded debt by the strongest industrial companies has in fact produced results demoralizing to investors and investment policies in a number of ways, *viz.:*

1. It has tended to restrict new industrial-bond financing to companies of weaker standing. The relative scarcity of good bonds impelled investment houses to sell and investors to buy inferior issues, with inevitably disastrous results.

2. The shortage of good bonds also tended to drive investors into the preferred-stock field. For reasons previously detailed (in Chap. XIV) preferred stocks are unsound in theory, and they are therefore certain to prove unsatisfactory investment media in the long run.

3. The elimination (or virtual elimination) of senior securities in the set-up of many large corporations has, of course, added somewhat to the investment quality of their common stocks, but it has added even more to the investor's demand for these common stocks. This in turn has resulted in a good deal of common-stock buying by people whose circumstances required that they purchase sound bonds. Furthermore it has supplied a superficial justification for the creation of excessive prices for these common stocks; and finally it contributed powerfully to that confusion between investment motives and speculative motives which during 1927–1929 served to debauch so large a proportion of the country's erstwhile careful investors.

Appraisal of Earnings Where Capital Structure Is Top-heavy.— In order to carry this theory of capitalization structure a step further, let us examine the case of Company *C*. We arrived at a valuation of $15,000,000 for this enterprise by assuming that its $10,000,000 bond issue would sell at par and the stock would sell for ten times its earnings of $5 per share. But this assumption as to the price of the bond is clearly fallacious. Earnings of twice interest charges are not sufficient protection for an industrial bond, and hence investors would be unwise to purchase such an issue at par. In fact this very example supplies a useful demonstration of our contention that a coverage of two times interest is inadequate. If it were ample—as some investors seem to believe—the owners of any reasonably prosperous business, earning 10% on the money invested, could get back their entire capital by selling a 5% bond issue, and they would still have control of the business together with one-half of its earnings. Such an arrangement would be exceedingly attractive for the proprietors but idiotic from the standpoint of those who buy the bonds.

Our Company *C* example also sheds some light on the effect of the rate of interest on the apparent safety of the senior security. If the $10,000,000 bond issue had carried a 7% coupon, the interest charges of $700,000 would then be earned less than $1\frac{1}{2}$ times. Let us assume that Company *D* had such a bond issue. An unwary investor, looking at the two exhibits, might reject Company *D*'s 7% bonds as unsafe because their interest coverage was only 1.43; but yet accept the Company *C* bonds at par because he was satisfied with earnings of twice fixed charges. Such discrimination would be scarcely intelligent. Our investor would be rejecting a bond *merely because* it pays him a generous coupon rate and he would be accepting another bond *merely because* it pays him a low interest rate. The real point, however, is that the minimum margin of safety behind bond issues must be set high enough to avoid the possibility that safety may even *appear* to be achieved by a mere lowering of the interest rate. The same reasoning would apply of course to the dividend rate on preferred stocks.

Since Company *C* bonds are not safe, because of the excessive size of the issue, they are likely to sell at a considerable discount from par. We cannot suggest the proper price level for such an issue, but we have indicated in Chap. XXVI that a bond speculative because of inadequate safety should not ordinarily be purchased above 70. It is also quite possible that the presence of this excessive bond issue might prevent the stock from selling at ten times its earnings, because conservative stock buyers would avoid Company *C* as subject to too great hazard of financial difficulties in the event of untoward developments. The result may well be that, instead of being worth $15,000,000 in the market as originally assumed, the combined bond and stock issues of Company *C* will sell for less than $12,500,000 (the Company *B* valuation), or even for less than $10,000,000 (the value of Company *A*).

As a matter of cold fact, it should be recognized that this unfavorable result may not necessarily follow. If investors are insufficiently careless and if speculators are sufficiently enthusiastic, the securities of Company C may conceivably sell in the market for $15,000,000 or even more. But such a situation would be unwarranted and unsound. Our theory of capitalization structure could not admit a Company C arrangement as in any sense standard or suitable. This indicates that there are definite limits upon the advantages to be gained by the use of senior securities. We have already expressed this fact in our principle of the optimum capitalization structure, for senior securities cease to be an advantage at the point where their amount becomes larger than can safely be issued or bought for investment.

We have characterized the Company A type of capitalization arrangement as "overconservative"; the Company C type may be termed "speculative," while that of Company B may well be called "suitable" or "appropriate."

The Factor of Leverage in Speculative Capitalization Structure.
While a speculative capitalization structure throws all the company's securities outside the pale of investment, it may give the common stock a definite speculative advantage. A 25% increase in the earnings of Company C (from $1,000,000 to $1,250,000) will mean a 50% increase in the earnings per share of common (from $5 to $7.50). Because of this fact there is a tendency for speculatively capitalized enterprises to sell at relatively high values in the aggregate during good times or good markets. Conversely, of course, they may be subject to a greater degree of undervaluation in depression. There is, however, a real advantage in the fact that such issues, when selling on a deflated basis, can advance much further than they can decline.

The record of American Water Works and Electric Company common stock between 1921 and 1929 presents an almost fabulous picture of enhancement in value, a great part of which was due to the influence of a highly speculative capitalization structure. Four annual exhibits during this period are summarized in the table on page 467.

The purchaser of one share of American Water Works common stock at the high price of 6½ in 1921, if he retained the distributions made in stock, would have owned about 12½ shares when the common sold at its high price of 199 in 1929. His $6.50 would have grown to about $2,500. While the market value of the common shares was thus increasing some 400-fold, the gross earnings had expanded to only 2.6 times the earlier figure. The tremendously disproportionate rise in the common-stock value was due to the following elements, in order of importance.

AMERICAN WATER WORKS AND ELECTRIC COMPANY

Item	1921	1923	1924	1929	Ratio of 1929 figures to 1921 figures
Gross earnings*	$20,574	$ 36,380	$ 38,356	$ 54,119	2.63 :1
Net for charges*	6,692	12,684	13,770	22,776	3.40 :1
Fixed charges and preferred dividends*	6,353	11,315	12,780	16,154	2.54 :1
Balance for common*	339	1,369	990	6,622	19.5 :1
1921 basis:†					
Number of shares of common	92,000	100,000	100,000	130,000	1.35 :1
Earned per share	$3.68	$13.69	$9.90	$51.00	13.8 :1
High price of common	6½	44¾	209	about 2500	385 :1
% earned on high price of common	56.6%	30.6%	4.7%	2.1%	0.037:1
As reported:					
Number of shares of common	92,000	100,000	500,000	1,657,000	
Earned per share	$3.68	$13.69	$1.98	$4.00	
High price of common	6½	44¾	41⅞	199	

* In thousands.
† Number of shares and price adjusted to eliminate effect of stock dividends and split-ups.

1. A much higher valuation placed upon the per-share earnings of this issue. In 1921 the company's capitalization was recognized as top-heavy; its bonds sold at a low price, and the earnings per share of common were not taken seriously, especially since no dividends were being paid on the second preferred. In 1929 the general enthusiasm for public-utility shares resulted in a price for the common issue of nearly 50 times its highest recorded earnings.

2. The speculative capitalization structure allowed the common stock to gain an enormous advantage from the expansion of the company's properties and earnings. Nearly all the additional funds needed were raised by the sale of senior securi-

ties. It will be observed that whereas the gross revenues increased about 160% from 1921 to 1929, the balance per share of old common stock grew fourteen-fold during the same period.

3. The margin of profit improved during these years, as shown by the higher ratio of net to gross. The speculative capital structure greatly accentuated the benefit to the common stock from the additional net profits so derived.

Other Examples: The behavior of speculatively capitalized enterprises *under varying business conditions* is well illustrated by the appended analysis of A. E. Staley Manufacturing Company, manufacturers of corn products. For comparison there is given also a corresponding analysis of American Maize Products Company, a conservatively capitalized enterprise in the same field.

A. E. STALEY

Year	Net before depreciation*	Depreciation*	Fixed charges and pfd. dividends*	Balance for common*	Earned per share
1933	$2,563	$743	$652	$1,168	$55.63
1932	1,546	753	678	114	5.43
1931	892	696	692	*496(d)*	*23.60(d)*
1930	1,540	753	708	79	3.74
1929	3,266	743	757	1,766	84.09
1928	1,491	641	696	154	7.35
1927	1,303	531	541	231	11.01
1926	2,433	495	430	1,507	71.77
1925	792	452	358	*18(d)*	*0.87(d)*
1924	1,339	419	439	481	22.89

* 000 omitted.

AMERICAN MAIZE PRODUCTS

Year	Net before depreciation*	Depreciation*	Fixed charges and pfd. dividends*	Balance for common*	Earned per share
1933	$1,022	$301	.....	$ 721	$2.40
1932	687	299	.....	388	1.29
1931	460	299	.....	161	0.54
1930	1,246	306	22	918	3.06
1929	1,835	312	80	1,443	4.81
1928	906	317	105	484	1.61
1927	400	318	105	*23(d)*	*0.08(d)*

* 000 omitted.

CAPITALIZATION (AS OF JANUARY 1933)

Item	A. E. Staley	American Maize Products
6% bonds.....................	($4,000,000* @ 75) $3,000,000	
$7 pfd. stock.................	(50,000 sh. @ 44) 2,200,000	
Common stock................	(21,000 sh. @ 45) 950,000	(300,000 sh. @ 20) $6,000,000
Total capitalization...........	$ 6,150,000	$6,000,000
Average earnings, 1927–1932, about......................	900,000	615,000
% of these earnings on 1933 capitalization................	14.6%	10.3%†
Average earnings per sh. of common......................	$14.76	$1.87
% earned on price of common...	32.8%	9.4%†
Working capital, Dec. 31, 1932..	$ 3,664,000	$2,843,000
Net assets, Dec. 31, 1932......	$15,000,000	$4,827,000

* Deducting estimated amount of bonds in treasury.

† The difference between these two figures is due to the varying treatment of the preferred stock outstanding during 1927–1930.

The most striking aspect of the Staley exhibit is the extraordinary fluctuation in the yearly earnings per share of common stock. The business itself is evidently subject to wide variations in net profit, and the effect of these variations on the common stock is immensely magnified by reason of the small amount of common stock in comparison with the senior securities.[1] The large depreciation allowance acts also as the equivalent of a heavy fixed charge. Hence a decline in net before depreciation from $3,266,000 in 1929 to $1,540,000 the next year, somewhat over 50%, resulted in a drop in earnings *per share of common* from $84 to only $3.74. The net profits of American Maize Products were fully as variable, but the small amount of prior charges made the fluctuations in common-stock earnings far less spectacular.

Speculative Capitalization May Cause Valuation of Total Enterprise at an Unduly Low Figure.—The market situation of the Staley securities in January 1933 presents a practical confirmation of our theoretical analysis of Company C above.

[1] In 1934 the company declared a 100% stock dividend, thus doubling the number of shares of common.

The top-heavy capitalization structure resulted in a low price for the bonds and the preferred stock, the latter being affected particularly by the suspension of its dividend. The result was that, instead of showing an increased total value by reason of the presence of senior securities, the company sold in the market at a much lower relative price than the conservatively capitalized American Maize Products. (The latter company showed a normal relationship between average earnings and market value. It should not properly be termed *overconservatively* capitalized because the variations in its annual earnings would constitute a good reason for avoiding any substantial amount of senior securities. A bond or preferred stock issue of very small size, on the other hand, would be of no particular advantage or disadvantage.)

The indication that the A. E. Staley Company was undervalued in January 1933 in comparison with American Maize Products is strengthened by reference to the relative current-asset positions and total resources. The fixed assets of Staley probably include some unstated intangibles, but in all probability the actual investment in the business was not less than twice the valuation of all its securities in the market.

The overdeflation of a speculative issue like Staley common in unfavorable markets creates the possibility of an amazing price advance when conditions improve, because the earnings per share then show so violent an increase. Note that at the beginning of 1927 Staley common was quoted at about 75; and a year later it sold close to 300.

A Corresponding Example.—A more spectacular instance of tremendous price changes for the same reason is supplied by Mohawk Rubber. In 1927 the common sold at 15, representing a valuation of only $300,000 for the junior issue, which followed $1,960,000 of preferred. The company had lost $610,000 in 1926 on $6,400,000 of sales. In 1927 sales dropped to $5,700,000, but there was a net profit of $630,000. This amounted to over $23 per share on the small amount of common stock. The price consequently advanced from its low of 15 in 1927 to a high of 251 in 1928. In 1930 the company again lost $669,000, and the next year the price declined to the equivalent of only $4.

In a speculatively capitalized enterprise, the common stockholders benefit—or have the possibility of benefiting—at the expense of the senior security holders. The common stockholder is operating with a little of his own money and with a great deal of the senior security-holder's money; as between him and them it is a case of "heads I win, tails you lose." This strategic position of the common stockholder with relatively small commitment is an extreme form of what is called "trading on the equity." Using another expression, he may be said to have a "cheap call" on the future profits of the enterprise.

Speculative Attractiveness of "Shoe-string" Common Stocks Considered.—Our discussion of fixed-value investment has emphasized as strongly as possible the disadvantage (amounting to unfairness) which attaches to the senior security holder's position where the junior capital is proportionately slight. The question would logically arise whether there are not corresponding *advantages* to the common stock in such an arrangement, from which it gains a very high degree of speculative attractiveness. This inquiry would obviously take us entirely outside the field of common-stock *investment*, but would represent an expedition into the realm of intelligent or even scientific speculation.

We have already seen from our A. E. Staley example that in bad times a speculative capitalization structure may react adversely on the market price of both the senior securities and the common stock. During such a period, then, the common stockholders do not derive a present benefit at the expense of the bondholder. This fact clearly detracts from the speculative advantage inherent in such common stocks. It is easy to suggest that these issues be purchased only when they are selling at abnormally low levels due to temporarily unfavorable conditions. But this is really begging the question, because it assumes that the intelligent speculator can consistently detect and wait for these abnormal and temporary conditions. If this were so, he could make a great deal of money regardless of what type of common stock he buys; and under such conditions he might be better advised to select high-grade common stocks at bargain prices, rather than these more speculative issues.

Practical Aspects of the Foregoing.—To view the matter in a practical light, the purchase of speculatively capitalized common stocks must be considered under general or market conditions which are supposedly normal, *i.e.*, under those which are not obviously inflated or deflated. Assuming (1) diversification, and (2) reasonably good judgment in selecting companies with satisfactory prospects, it would seem that the speculator should be able to profit rather substantially in the long run from commitments of this kind. In making such purchases, partiality should evidently be shown to those companies in which most of the senior capital is in the form of preferred stock rather than bonds. Such an arrangement removes or minimizes the danger of extinction of the junior equity through default in bad times, and thus permits the shoe-string common stockholder to maintain his position until prosperity returns. (But just because the preferred-stock contract benefits the common shareholder in this way, it is clearly disadvantageous to the preferred stockholder himself.)

We must not forget, however, the peculiar practical difficulty in the way of realizing the full amount of prospective gain in any one of the purchases. As we pointed out in the analogous case of convertible bonds, as soon as a substantial profit appears the holder is in a dilemma, because he can hold for a further gain only by risking that already accrued. Just as a convertible bond losses its distinctive advantages when the price rises to a point which carries it clearly outside of the straight investment class, so a shoe-string common-stock commitment is transformed into a more and more substantial commitment as the price continues to rise. In our Mohawk Rubber example the intelligent purchaser at 15 could not have expected to hold it beyond 100—even though its quotation did reach 250—because at 100, or before, the shares had lost the distinctive characteristics of a speculatively capitalized junior issue.

CHAPTER XLI

LOW-PRICED COMMON STOCKS. ANALYSIS OF THE SOURCE OF INCOME

LOW-PRICED STOCKS

The characteristics discussed in the preceding chapter are generally thought of by the public in connection with *low-priced stocks*. The majority of issues of the speculatively capitalized type do sell within the low-priced range. The definition of "low-priced" must, of course, be somewhat arbitrary. Prices below $10 per share belong to this category beyond question; those above $20 are ordinarily excluded; so that the dividing line would be set somewhere between $10 and $20.

Arithmetical Advantage of Low-priced Issues.—Low-priced common stocks appear to possess an inherent arithmetical advantage arising from the fact they can advance so much more than they can decline. It is a commonplace of the securities market that an issue will rise more readily from 10 to 40 than from 100 to 400. This fact is due in part to the preferences of the speculative public, which generally is much more partial to issues in the 10-to-40 range than to those selling above 100. But it is also true that in many cases low-price common stocks give the owner the advantage of an interest in, or "call" upon, a relatively large enterprise at relatively small expense.

In January 1931 J. H. Holmes and Co., members of the New York Stock Exchange, published a statistical study entitled "Low-priced Common Stocks for a Diversified Investment," which pointed to the foregoing advantages of low-priced issues as contrasted with the high-priced or "blue-chip" issues, when purchased at or near the bottom of a depression in security prices. The following quotation from the study summarizes the results and indicated conclusions:

"An investigation of the trend of stock prices in various other depression years [tabulated in the study and covering the price-behavior of the low- and high-priced groups in the following years and three years subsequent thereto in each case: 1897, 1907, 1914, 1921] shows that the lower priced stocks

473

(for which we took $12 per share as an optional high limit) tend to increase in price relatively more than the so-called leading issues. Three years after the periods taken, the low priced stocks had advanced, on the average, 3.45 times, against 1.71 for the leading issues. We have also tried to show that an arbitrary amount invested in a group of the low priced stocks will show better results over a period than the same amount of money invested in the leading issues."[1]

Some Reasons Why Most Buyers of Low-priced Issues Lose Money.—The pronounced liking of the public for "cheap stocks" would therefore seem to have a sound basis in logic. Yet it is undoubtedly true that most people who buy low-priced stocks lose money on their purchases. Why is this so? The underlying reason is that the public buys issues which are *sold* to it, and the sales effort is put forward to benefit the seller and not the buyer. In consequence the bulk of the low-priced purchases made by the public are of the wrong kind, *i.e.*, they do not provide the real advantages of this security type. The reason may be either because the companies are in bad financial condition, or because the common stock is low-priced in appearance only and actually represents a full or excessive commitment in relation to the size of the enterprise. The latter is preponderantly true of *new* security offerings in the low-priced range. In such cases, a *pseudo*-low price is accomplished by the simple artifice of creating so large a number of shares that even at a few dollars per share the total value of the common issue is excessive. This has been true of mining-stock flotations from of old and was encountered again in the liquor stock offerings of 1933.

A *genuinely* low-priced common stock will show an aggregate value for the issue which is small in relation to the company's assets, sales, and past or prospective profits. The examples shown on page 475 will illustrate the difference between a "genuine" and "pseudo-low" price.

The Wright-Hargreaves issue was low-priced in appearance only, for in fact the price registered a very high valuation for the company as compared with all parts of its financial exhibit. The opposite was true of Barker Bros. because here the $743,000 valuation represented by the common stock was exceedingly small in relation to the size of the enterprise. (Note also that

---

[1] In the interest of strict accuracy it should be observed that the study did not show an "advance" of 345%, but an advance from a level of 100% to a level of 345%. This fact was clearly indicated in the tabular presentation, but not in the quoted text.

the same statement could be applied to Barker Bros. Preferred, which at its quotation of 18 partook of the qualities of a low-priced common stock.)

Item	Wright-Hargreaves Mines, Ltd. (gold mining)	Barker Bros. Corp. (retail store)
July 1933:		
Price of common stock........	7	5
Number of shares outstanding	5,500,000	148,500
Total value of common......	$38,500,000	$ 743,000
Preferred stock at par........	....	2,815,000
Preferred stock at market.....	....	500,000
Year 1932:		
Sales.....................	$ 3,983,000	8,154,000
Net earnings...............	2,001,000*	703,000(d)
Period 1924–1932:		
Maximum sales.............	$ 3,983,000	$16,261,000
Maximum net earnings.......	2,001,000*	1,100,000
Maximum earnings per share of common...............	0.36*	$7.59
Working capital, Dec. 1932...	$ 1,930,000	$ 5,010,000
Net tangible assets, Dec. 1932	4,544,000	7,200,000

* Before depletion.

Observation of the stock market will show that the stocks of companies facing receivership are likely to be more active than those which are very low in price merely because of poor current earnings. This phenomenon is caused by the desire of insiders to dispose of their holdings before the receivership wipes them out, thus accounting for a large supply of these shares at a low level, and also sometimes for unscrupulous efforts to persuade the unwary public to buy them. But where a low-priced stock fulfills our conditions of speculative attractiveness, there is apt to be no pressure to sell and no effort to create buying. Hence the issue is inactive and attracts little public attention. This analysis may explain why the public almost always buys the wrong low-priced issues and ignores the really promising opportunities in this field.

Low Price Coupled with Speculative Capitalization.—Speculatively capitalized enterprises, according to our definition, are marked by a relatively large amount of senior securities and a comparatively small issue of common stock. While in most cases the common stock will sell at a low price per share, it need not necessarily do so if the number of shares is small. In the Staley case, for example (referred to on page 468) even at $50 per share for the common the capitalization structure would still be speculative, since the bonds and preferred at par would represent over 90% of the total. It is also true that even where there are no senior securities the common stock may have possibilities equivalent to those in a speculatively capitalized enterprise. These possibilities will occur wherever the market value of the common issue represents a small amount of money in relation to the size of the business, regardless of how it is capitalized.

To illustrate this point we append a condensed analysis of Mandel Bros., Inc., and Gimbel Bros., Inc., two department-store enterprises. The common stock of each was selling at 5 in July 1933.

Item	Gimbel Bros., Inc.	Mandel Bros., Inc.
July, 1933:		
Bonds at par............	$ 29,100,000	
Preferred stock at par....	16,100,000	
Common stock.........	(960,000 sh. @ 5)	(307,000 sh. @ 5)
	4,800,000	$ 1,535,000
Total capitalization......	$ 50,000,000	$ 1,535,000
1932 results:		
Sales..................	$ 72,200,000	$14,800,000
Net before interest.......	*2,710,000*(d)	*579,000*(d)
Balance for common......	*5,537,000*(d)	*579,000*(d)
Period 1924–1932:		
Maximum sales (1930)....	124,600,000	(1924) 29,100,000
Maximum net earnings for common (1924)........	6,118,000	(1925) 2,003,000
Maximum earnings per share of common (1924).	$10.19	(1925) $6.40
Jan. 31, 1933:		
Net current assets........	$ 19,100,000	$ 3,669,000
Net tangible assets.......	76,000,000	5,650,000

Gimbel Bros. presents a typical picture of a speculatively capitalized enterprise. On the other hand Mandel Bros. has no senior securities ahead of the common; but despite this fact the relatively small market value of the entire issue imparts to the shares the same sort of speculative possibilities (though in somewhat lesser degree) as are found in the Gimbel Bros. set-up.[1]

Large Volume and High Production Cost Equivalent to Speculative Capital Structure.—This example should lead us to widen our conception of a speculatively situated common stock. The speculative or *marginal* position may arise from any cause which reduces the percentage of gross available for the common to a subnormal figure, and which therefore serves to create a subnormal value for the common stock in relation to the volume of business. Unusually high operating or production costs have the identical effect as excessive senior charges in cutting down the percentage of gross available for common. The following hypothetical examples of three copper producers will make this point more intelligible and also lead to some conclusions on the subject of large output versus low operating costs.

Item	Company *A*	Company *B*	Company *C*
Capitalization:			
6% Bonds............	....	$50,000,000	
Common stock........	1,000,000 sh.	1,000,000 sh.	1,000,000 sh.
Output................	100,000,000 lb.	150,000,000 lb.	150,000,000 lb.
Cost of production (before interest).............	7¢	7¢	9¢
Interest charge per pound	....	2¢	
Total cost per pound.....	7¢	9¢	9¢
A			
Assumed price of copper.	10¢	10¢	
Profit per pound........	3¢	1¢	
Output per share........	100 lb.	150 lb.	
Profit per share.........	$3	$1.50	
Value of stock at 10 times earnings.............	$30	$15	
Output per $1 of market value of stock.........	3⅓ lb.	10 lb.	
B			
Assumed price of copper.	13¢	13¢	
Profit per pound........	6¢	4¢	
Profit per share........	$ 6	$ 6	
Value per share at ten times earnings........	$60	$60	
Output per $1 of market price of stock.........	1⅔ lb.	2½ lb.	

---

[1] While the Mandel Bros. income account and balance sheet show no prior securities, it should be recognized that the company has fixed rental obligations corresponding in part to the Gimbel Bros. bonds. This point was discussed in Chap. XVII.

It is scarcely necessary to point out that the higher production cost of Company C will have exactly the same effect as the bond-interest requirement of Company B (so long as the production remains at 150,000,000 pounds for each company and so long as the relative production costs remain constant.)

General Principle Derived.—The above table is perhaps more useful in showing concretely the inverse relationship which usually exists between profit per unit and "output per dollar of stock value."

The general principle may be stated that the lower the unit cost the lower the production per dollar of market value of stock and *vice versa*. Since Company A has a 7-cent cost, its stock naturally sells at a higher price *per pound of output* than Company C with its 9-cent cost. Conversely, Company C produces more pounds per dollar of stock value than Company A. This fact is not without significance from the standpoint of speculative technique. When a rise in the price of the commodity occurs, there will ordinarily be a larger advance, percentagewise, in the shares of high-cost producers than in the shares of low-cost producers. The above table indicates that a rise in the price of copper from 10 to 13 cents would increase the value of Company A shares by 100% and the value of Company B and C shares by 300%. Contrary to the general impression in Wall Street, the stocks of high-cost producers are more logical commitments than those of the low-cost producers when the buyer is convinced that a rise in the price of the product is imminent and he wishes to exploit this conviction to the utmost.[1] Exactly the same advantage attaches to the purchase of speculatively capitalized common stocks when a pronounced improvement in sales and profits is confidently anticipated.

THE SOURCES OF INCOME

The "source of income" will ordinarily be thought of as meaning the same thing as the "type of business." This consideration enters very largely into the basis on which the public will value the earnings per share shown by a given common stock.

---

[1] The action of the market in advancing Company B shares from 15 to 60 because copper rises from 10 to 13 cents is in itself extremely illogical, for there is ordinarily no warrant for supposing that the higher metal price will be *permanent*. However, since the market does in fact behave in this irrational fashion, the speculator must recognize this behavior in his calculation.

Different "multipliers" are used for different sorts of enterprise; but we must point out that these distinctions are themselves subject to change with the changing times. Prior to the war the railroad stocks were valued most generously of all, because of their supposed stability. In 1927–1929 the public-utility group sold at the highest ratio to earnings, because of their record of steady growth. In 1933 rate regulation and the fear of higher operating costs reduced the popularity of the utility stocks, and the most liberal valuations were then accorded to the large and strong industrials which had been able to maintain substantial earnings even during the depression. Because of these repeated variations in relative behavior and popularity, security analysis must hesitate to prescribe any definitive rules for valuing one type of business as against another. It is a truism to say that the more impressive the record and the more promising the prospects of stability and growth, the more liberally the per-share earnings should be valued, subject always to our principle that a multiplier higher than about 16 (*i.e.*, an "earnings basis" of less than 6%) will carry the issue out of the *investment* price range.

A Special Phase: Three Examples.—A more fruitful field for the technique of analysis is found in those cases where the source of income must be studied in relation to specific assets owned by the company, instead of in relation merely to the general nature of the business. This point may be quite important when a substantial portion of the income accrues from investment holdings or from some other fixed and dependable source. Three examples will be used to illuminate this rather subtle aspect of common-stock analysis.

1. *Northern Pipe Line Company.*—For the years 1923–1925 the Northern Pipe Line Company reported earnings and dividends as follows:

Year	Net earnings	Earned per share*	Dividend paid
1923	$308,000	$7.70	$10, plus $15 extra
1924	214,000	5.35	8
1925	311,000	7.77	6

* Capitalization, 40,000 shares of common stock.

In 1924 the shares sold as low as 72; in 1925 as low as 67½; and in 1926 as low as 64. These prices were on the whole somewhat less than ten times the reported earnings and reflected a lack of enthusiasm for the shares, due to a pronounced decline in profits from the figures of preceding years, and also to the reductions in the dividend.

Analysis of the income account however, would have revealed the following division of the *sources of income:*[1]

Income	1923		1924		1925	
	Total	Per share	Total	Per share	Total	Per share
Earned from:						
Pipe-line operations	$179,000	$4.48	$ 69,000	$1.71	$103,000	$2.57
Interest and rents..	164,000	4.10	159,000	3.99	170,000	4.25
Nonrecurrent items	dr. 35,000	dr. 0.88	dr. 14,000	0.35	cr. 38,000	cr. 0.95
	$308,000	$7.70	$214,000	$5.35	$311,000	$7.77

This income account is exceptional in that the greater part of the profits were derived from sources other than the pipe-line business itself. About $4 per share were regularly received in interest on investments and rentals. The balance sheet showed holdings of nearly $3,200,000 (or $80 per share) in Liberty Bonds and other gilt-edged marketable securities on which the interest income was about 4%.

This fact meant that a special basis of valuation must be applied to the per-share earnings, inasmuch as the usual "ten-times-earnings" basis would result in a nonsensical conclusion. Gilt-edged investments of $80 per share would yield an income of $3.20 per share; and at ten times earnings this $80 would be "worth" only $32 per share, a *reductio ad absurdum.* Obviously, that part of the Northern Pipe Line income which was derived from its bond holdings should logically be valued at a higher basis than the portion derived from the fluctuating pipe-line business. A sound valuation of Northern Pipe Line stock would therefore have to proceed along the lines shown at the top of page 481. The pipe-line earnings would have to be valued at a low basis because of their unsatisfactory trend. The interest and rental income must presumably be valued on a basis corresponding with

---

[1] Although the company's reports to its stockholders contained very little information, complete financial and operating data were on file with the Interstate Commerce Commission and open to public inspection.

Average 1923–1925*	Valuation basis	Value per share
Earned per share from pipe line. $2.92	15% (6⅔ times earnings)	$ 20
Earned per share from interest and rentals............... 4.10	5% (20 times earnings)	80
Total................. $7.02		$100

* The nonrecurrent profits and losses are not taken into account

the actual value of the assets producing the income. This analysis indicated clearly that at the price of 64 in 1926, Northern Pipe Line stock was selling considerably below its intrinsic value.

2. *Lackawanna Securities Company.*—This company was organized to hold a large block of Glen Alden Coal Company 4% bonds formerly owned by the Delaware, Lackawanna and Western Railroad Company, and its shares were distributed pro rata to the Delaware, Lackawanna and Western stockholders. The Securities Company had outstanding 844,000 shares of common stock. On December 31, 1931 its sole asset—other than about $1 per share in cash—consisted of $51,000,000 face value of Glen Alden 4% first mortgage bonds. For the year 1931, the income account was a follows:

Interest received on Glen Alden bonds......... $2,084,000
Less:
    Expenses............................... 17,000
    Federal taxes........................... 250,000
    Balance for stock....................... 1,817,000
    Earned per share....................... $2.15

Superficially, the price of 23 in 1932 for a stock earning $2.15 did not appear out of line. But these earnings were derived, not from ordinary commercial or manufacturing operations, but from the holding of a bond issue which presumably constituted a high-grade investment. (In 1931 the Glen Alden Coal Company earned $9,550,000 available for interest charges of $2,151,-000, thus covering the bond requirements 4½ times.) By valuing this interest income on about a 10% basis the market was in fact valuing the Glen Alden bonds at only 37 cents on the dollar. (The price of 23 for a share of Lackawanna Securities was equivalent to $60 face value of Glen Alden bonds at 37, plus $1 in cash.

Here again, as in the Northern Pipe Line example, analysis would show convincingly that the customary ten-times-earnings basis resulted in a glaring undervaluation of this specially situated issue.

3. *Tobacco Products Corporation of Virginia.*

Item	Price: December 1931	Market value
Capitalization		
2,240,000 shares of 7% Class A (par $20).................	$6	$13,440,000
3,300,000 shares common.....	2¼	7,425,000
Total......................	....	$20,825,000
Net income for the year 1931..	....	about $ 2,200,000
Earned per share of Class A...	....	about $1
Earned for common after Class A dividends...............	....	nil
Dividend paid on Class A.....	....	$0.80

In this example, as in the other two, the company was selling in the market for about ten times the latest reported earnings. But the 1931 earnings of Tobacco Products were derived entirely from a lease of certain of its assets to American Tobacco Company, which provided for an annual rental of $2,500,000 for 99 years from 1923. Since the American Tobacco Company was able to meet its obligation without question, this annual rental income was equivalent to interest on a high-grade investment. Its value was therefore much more than ten times the income therefrom. This meant that the market valuation of the Tobacco Products stock issues in December 1931 was far less than was justified by the actual position of the company. (The value of the lease was in fact calculated to be about $35,600,000 on an amortized basis. The company also owned a large amount of United Cigar Stores' stock, which later proved to be practically worthless, but these additional holdings did not, of course, detract from the value of its American Tobacco lease.)

Relative Importance of Situations of This Kind.—The field of study represented by the above examples is not important quantitatively, because, after all, only a very small percentage of the companies examined will fall within this group. Situations of this kind arise with sufficient frequency, however, to give this discussion practical value. It should be useful also in illustrating again the wide technical difference between the critical approach of security analysis and the highly superficial reactions and valuations of the stock market.

Two Lines of Conduct Suggested.—When it can be shown that certain conditions, such as those last discussed, tend to give rise to undervaluations in the market, two different lines of conduct are thereby suggested. We have first an opportunity for the securities analyst to detect these undervaluations and eventually to profit from them. But there is also the indication that the financial set-up which causes this undervaluation is erroneous and that the stockholders' interests require the correction of this error. The very fact that a company constituted like Northern Pipe Line or Lackawanna Securities tends to sell in the market far below its true value, proves as strongly as possible that the whole arrangement is wrong from the standpoint of the owners of the business.

At the bottom of these cases there is a basic principle of consistency involved. It is inconsistent for most of the capital of a pipe-line enterprise actually to be employed in the ownership of gilt-edged bonds. The whole set-up of Lackawanna Securities was also inconsistent, because it replaced a presumably high-grade bond issue, which investors might be willing to buy at a fair price, by a nondescript stock issue which no one would purchase except at an especially low price. (In addition a heavy and needless burden of corporate income tax was involved, as was true also in the Tobacco Products case.)

Illogical arrangements of this kind should be recognized by the real parties in interest, *i.e.*, the stockholders, and they should insist that the anomaly be rectified. This was finally done in the three examples just given. In the case of Northern Pipe Line the capital not needed in the pipe-line business was returned to the stockholders by means of special distributions aggregating $70 per share. The Lackawanna Securities Company was entirely dissolved and the Glen Alden bonds in its treasury distributed pro rata to the stockholders in lieu of their stock. Finally, the Tobacco Products Corporation was recapitalized on a basis by which $6\frac{1}{2}\%$ bonds were issued against the American Tobacco lease, so that this asset of fixed value was represented by a fixed-value security (which later sold at par) instead of by shares of stock in a corporation subject to highly speculative influences. By means of these corporate rearrangements the real values were speedily established in the market price.

The situations which we have just analyzed required a transfer of attention from the income account figures to certain related features revealed in the balance sheet. Hence the foregoing topic—Sources of Income—carries us over into our next field of inquiry: The Balance Sheet.

PART VI

BALANCE-SHEET ANALYSIS. IMPLICATIONS OF ASSET VALUES

CHAPTER XLII

BALANCE-SHEET ANALYSIS: SIGNIFICANCE OF BOOK VALUE

Before discussing the role that the balance sheet plays in security analysis we shall first present certain definitions. The *book value* of a stock is the value of the assets applicable thereto as shown in the balance sheet. It is customary to restrict this value to the tangible assets, *i.e.*, to eliminate from the calculation such items as good-will, trade names, patents, franchises, leaseholds, etc. The book value is also referred to as the "asset value," and sometimes as the "tangible-asset value," to make clear that intangibles are not included. In the case of common stocks, it is also frequently termed the "equity."

Computation of Book Value.—The *book value per share* of a common stock is found by adding up all the tangible assets, subtracting all liabilities and stock issues ahead of the common, and then dividing by the number of shares.

In many cases the following formula will be found to furnish a short cut to the answer:

Book value per share of common

$$= \frac{\text{Common Stock} + \text{Surplus Items} - \text{Intangibles}}{\text{Number of shares outstanding}}$$

By Surplus Items are meant not only items clearly marked as surplus but also premiums on capital stock and such reserves as are really part of the surplus. This would include, for example, reserves for preferred-stock retirement, for plant improvement, for contingencies (unless known to be actually needed), etc. Reserves of this character may be termed "Voluntary Reserves."

CALCULATION OF BOOK VALUE OF UNITED STATES STEEL COMMON ON
DECEMBER 31, 1932
CONDENSED BALANCE SHEET DECEMBER 31, 1932 (*in millions*)

Assets		Liabilities	
1. Property Investment Account (less depreciation)..............	$1,651	7. Common Stock........ $	870
		8. Preferred Stock........	360
2. Mining Royalties.......	69	9. Premium on Common Stock..............	81
3. Deferred Charges[1]......	2	10. Bonded Debt..........	96
4. Miscellaneous Investments..............	19	11. Mining Royalty Notes .	19
		12. Installment Deposits ...	2
5. General Reserve Fund Assets..............	20	13. Current Liabilities......	47
6. Current Assets.........	398	14. Contingency and Other Reserves...........	39
		15. Insurance Reserves.....	46
		16. Appropriated Surplus...	270
		17. Undivided Surplus......	329
	$2,159		$2,159

Tangible assets.......................... $2,159,000,000
Less: All liabilities ahead of common
(Sum of items 8, 10, 11, 12, and 13)... 524,000,000
Accumulated dividends on preferred
stock........................... 4,504,000

Net assets for common stock.............. 1,630,496,000
Book value per share (on 8,700,000 shares). $187.40

[1] Considerable argument could be staged over the question whether Deferred Charges are intangible or tangible assets, but as the amount involved is almost always small, the matter has no practical importance. It is more convenient, of course, to include the Deferred Charges with the other assets. Standard Statistics Company, Inc., however, rules them out.

The alternative method of computation, which is usually shorter than the foregoing, is as follows:

Common stock......................... $ 870,000,000
Surplus and voluntary Reserves
(Sum of items 9, 14, 15, 16, and 17)...... 765,000,000

$1,635,000,000
Less accumulated dividends on preferred
stock............................... 4,504,000

Net assets for common stock.............. $1,630,496,000

Treatment of Preferred Stock When Calculating Book Value of Common.

—In calculating the assets available for the common stock, care must be taken to subtract preferred stock at its proper valuation. Ordinarily, this will be the par or stated value of the preferred stock as it appears in the balance sheet. But there is a growing number of cases in which preferred stock is carried in the balance sheet at arbitrary values far lower than the real liability attaching thereto.

Island Creek Coal Company has a preferred stock of $1 par, which is entitled to annual dividends of $6 and to $120 per share in the event of dissolution. In 1933 the price of this issue ruled about 90. In the calculation of the asset-value of Island Creek Coal Common the preferred stock should be deducted not at $1 per share but at $100 per share, its "true" or "effective" par. Coca-Cola Company issued in 1929, as a stock dividend, one million shares of Class A (really preferred) stock, entitled to cumulative dividends of $3 and redeemable at $52.50. The shares were of no par, but were given a stated value of only $5 per share. The balance sheet on December 31, 1929, had the following appearance:

Fixed Assets, less Depreciation................	$ 6,306,000
Miscellaneous Assets........................	427,000
Current Assets..............................	16,964,000
Formulas, Trade-mark, Good-will.............	21,931,000
Class A stock (194,000 shares at cost).........	9,434,000
Total................................	$55,062,000
Current Liabilities..........................	2,729,000
Reserve for Contingencies and Miscellaneous Operations..............................	6,687,000
Class A Stock (1,000,000 shares).............	5,000,000
Common Stock (1,000,000 shares).............	25,000,000
Surplus....................................	15,646,000
Total................................	$55,062,000

It is clear that the book valuation of $5,000,000 given to the Class A stock is entirely unrelated to its real position. There is indeed a striking contradiction in this balance sheet, since the assets include 194,000 shares of Class A stock at a value of $9,434,000, while the entire 1,000,000-share issue is carried as a liability at only $5,000,000. This is nonsensical accounting. The true par value of Coca-Cola Class A stock is obviously $50 per share, corresponding to the ordinary $3 preferred stock. The reason for not listing the Class A issue at the proper figure in the balance sheet is seen at once in the fact that this would exceed the net amount of all the company's assets, including intangibles, and thus leave a negative book value for the common stock.

In all instances such as the above an "effective par value" must be set up for the preferred stock which will correspond properly to its dividend rate. A strong argument may be advanced in favor of valuing all preferred stocks on a uniform dividend basis, say 6%. This would mean that a $1,000,000 six per cent issue would be valued at $1,000,000, a $1,000,000 four per cent issue would be given an effective value of $667,000 and a $1,000,000 seven per cent issue would be given an effective value of $1,167,000. But it is more convenient, of course, to use the par value, and in most cases the result will be sufficiently accurate.[1]

Calculation of Book Value of Preferred Stocks.—In calculating the book value of a preferred stock issue it is treated as a common stock and the issues junior to it are left out of consideration. The following computations from the December 31, 1932 balance sheet of Tubize Chatillon Corporation will illustrate the principles involved.

TUBIZE CHATILLON CORPORATION
Balance Sheet December 31, 1932

Assets		Liabilities	
Property and Equipment.............	$19,009,000	7% First Preferred Stock (par $100)....	$ 2,500,000
Patents, Processes, etc.	802,000	$7 Second Preferred Stock (par $1)......	136,000
Miscellaneous Assets..	478,000	Common Stock (par $1)................	294,000
Current Assets........	4,258,000	Bonded Debt.........	2,000,000
		Current Liabilities. ...	613,000
		Reserve for Depreciation, etc...........	11,456,000
		Surplus..............	7,548,000
Total assets......	$24,547,000	Total liabilities...	$24,547,000

[1] Standard Statistics Company, Inc., follows the practice of deducting preferred stock at its value *in case of dissolution*, when computing the book value of the common. This is scarcely logical, because dissolution or liquidation is almost always a remote contingency, and would take place under conditions quite different from those obtaining at the time of analysis. The Standard Statistics Company method results in placing a "value" of $115 per share on Procter and Gamble Company $5 Second Preferred and a value of only $100 per share on the same company's $8 First Preferred. The real or practical value of the preferred stockholder's claims in this case would be much nearer in the proportion of 135 for the First Preferred against 85 for the Second Preferred, a 6% dividend basis for both.

The book value of the First Preferred is computed as follows:

Total Assets.		$24,547,000
Less: Intangible Assets.	802,000	
Reserve for Depreciation, etc.	11,456,000	
Bonds.	2,000,000	
Current Liabilities.	613,000	14,871,000

Net assets for First Preferred.		$ 9,676,000
Book value per share.		$387

Alternative method:

Capital Stock at par.	$ 2,930,000
Surplus.	7,548,000
	$10,478,000
Less Intangible Assets.	802,000
Net assets for First Preferred.	$ 9,676,000

The Reserve for Depreciation and Miscellaneous Purposes is very large and probably includes arbitrary allowances belonging in Surplus. But in the absence of details the entire reserve must be deducted from the assets.

The book value of the Second Preferred stock is readily computed from the above as follows:

Net assets for First Preferred.	$9,676,000
Less: First Preferred at par.	2,500,000
Net assets for Second Preferred.	$7,176,000
Book value per share.	$52.75

In computing the book value of the common it would be an obvious error to deduct the Second Preferred at its nonrepresentative par value of $1. The "effective par" should be taken at $100 per share, in view of the $7 dividend. Hence there are no assets available for the common stock and its book value is nil.

Current-asset Value and Cash-asset Value.—In addition to the well-known concept of book value, we wish to suggest two others of similar character, *viz.*, current-asset value and cash-asset value.

The current-asset value of a stock consists of the quick assets alone, minus all liabilities and claims ahead of the issue. It excludes not only the intangible assets but the fixed and miscellaneous assets as well.

The cash-asset value of a stock consists of the cash assets alone, minus all liabilities and claims ahead of the issue. Cash assets, other than cash itself, are defined as those directly equivalent to and held in place of cash. They include certificates of deposit, call loans, marketable securities at market value, and cash-surrender-value of insurance policies.

The following is an example of the computation of the three categories of asset value:

OTIS COMPANY (COTTON GOODS)

MARKET PRICE OF COMMON STOCK IN JUNE 1929, 35

BALANCE SHEET JUNE 29, 1929

Assets		Liabilities	
1. Cash................	$ 532,000	8. Accounts Payable.. $	79,000
2. Call Loans.........	1,200,000	9. Accrued Items, etc..	291,000
3. Accounts Receivable		10. Reserve for Equip-	
(less reserve)......	1,090,000	ment, etc........	210,000
4. Inventory (less re-		11. Preferred Stock....	400,000
serve of $425,000)*	1,648,000	12. Common Stock....	4,079,000
5. Prepaid Items.....	108,000	13. Earned Surplus....	1,944,000
6. Investments........	15,000	14. Paid-in Surplus....	1,154,000
7. Plant (less Deprecia-			
tion)............	3,564,000		
	$8,157,000		$8,157,000

* Inventories before reserves are valued at cost or market, whichever is lower.

A. Calculation of book value of common stock:

Total assets.....................................		$8,157,000
Less: Payables........................... $ 79,000		
Accrued items..................... 291,000		
Preferred stock.................... 400,000		770,000
		$7,387,000
Add voluntary reserve of $425,000 subtracted from inventory.................................		425,000
Net assets for common stock.........................		$7,812,000
Book value per share (on 40,790 shares)..............		$191

B. Calculation of current-asset value of the common stock:

Total current assets (items 1, 2, 3, and 4)..............	$4,470,000
Add voluntary reserve against inventory..............	425,000
	$4,895,000
Less liabilities ahead of common (items 8, 9, and 11)....	770,000
Current assets available for common.................	$4,125,000
Current-asset value per share.................,......	$101

C. Calculation of cash-asset value of the common stock:

Total cash assets (items 1 and 2)....................	$1,732,000
Less liabilities ahead of common (items 8, 9, and 11)....	770,000
Cash assets available for common....................	$ 962,000
Cash-asset value per share..........................	$23.50

In these calculations it will be noted, first, that the inventory is increased by restoring the Reserve of $425,000 subtracted therefrom in the balance sheet. This is done because the deduction taken by the company is clearly a reserve for contingent decline in value which has not yet taken place. As such it is entirely arbitrary or voluntary, and consistency of method would require the analyst to regard it as a surplus item. The same is true of the $210,000 "Reserve for Equipment and Other Expenses," which, as far as can be seen, represents neither an actual liability nor a necessary deduction from the value of any specific asset. The reader will note an extraordinary divergence between the current-asset value and the market price of Otis Company in June 1929. Its significance will engage our attention later.

Practical Significance of Book Value.—The book value of a common stock was originally the most important element in its financial exhibit. It was supposed to show "the value" of the shares in the same way as a merchant's balance sheet shows him the value of his business. This idea has almost completely disappeared from the financial horizon. The value of a company's assets as carried in its balance sheet has lost practically all its significance. This change arose from the fact, first, that the value of the fixed assets, as stated, frequently bore no relationship to the actual cost; and secondly, that in an even larger proportion of cases these values bore no relationship to the figure at which they would be sold or the figure which would be justified by the earnings. The practice of inflating the book value of the fixed property has been succeeded by the opposite artifice of cutting it down to nothing in order to avoid depreciation charges; but both have the same consequence of depriving the book-value figures of any real significance. It is a bit strange, like a quaint survival from the past, that as late as 1933 the Standard Statistics Company, Inc., still maintained the old procedure of calculating the book value per share of common stock from each balance sheet that it published.

Before we discard completely this time-honored conception of book value, let us ask whether it may ever have practical significance for the analyst. In the ordinary case, probably not. But what of the extraordinary or extreme case? Let us consider the following four exhibits as representative of extreme relationships between book value and market price:

Item	General Electric	Pepperell Manufacturing
Price.........................	(1930) 95	(1932) 18
Number of shares................	28,850,000	97,600
Market value of common.........	$2,740,000,000	$ 1,760,000
Balance sheet...................	(Dec. 1929)	(June 1932)
Fixed assets (less depreciation)...	$ 52,000,000	$ 7,830,000
Miscellaneous assets............	183,000,000	230,000
Net quick assets...............	206,000,000	9,120,000
Total net assets...............	$ 441,000,000	$17,180,000
Less bonds and preferred........	45,000,000	
Book value of common..........	$ 396,000,000	$17,180,000
Book value per share...........	$13.75	$176

Item	Commercial Solvents	Pennsylvania Coal and Coke
Price.........................	(July 1933) 57	(July 1933) 3
Number of shares................	2,493,000	165,000
Market value of common.........	$142,000,000	$ 495,000
Balance sheet...................	(Dec. 1932)	(Dec. 1932)
Fixed assets (less depreciation)...	....	6,500,000
Miscellaneous assets............	2,600,000	990,000
Net quick assets...............	6,000,000	740,000
Total assets for common.........	$ 8,600,000	$8,230,000
Book value per share...........	$3.50	$50

No thoughtful observer could fail to be impressed by the disparities revealed in the above examples. In the case of General Electric and Commercial Solvents the figures proclaim more than the bare fact that the market was valuing the shares at many times their book value. The stock ticker seems here to register an aggregate valuation for these enterprises which is totally unrelated to their standing as ordinary business enterprises. In other words, these are in no sense *business valuations;* they are products of Wall Street's legerdemain, or possibly of its clairvoyance.

Financial Reasoning vs. Business Reasoning.—We have here the point that brings home more strikingly perhaps than any other the widened rift between financial thought and ordinary business thought. It is an almost unbelievable fact that Wall Street never asks, "How much is the business selling for?" Yet this should be the first question in considering a stock purchase. If a business man were offered a 5% interest in some concern for $10,000, his first mental process would be to multiply the asked price by 20 and thus establish a proposed value of $200,000 for the entire undertaking. The rest of his calculation would turn about the question whether the business was a "good buy" at $200,000.

This elementary and indispensable approach has been practically abandoned by those who purchase stocks. Of the thousands who "invested" in General Electric in 1929–1930 probably only an infinitesimal number had any idea that they were paying on the basis of two and three-quarter billions of dollars for the company, of which over two billions represented a premium above the money actually invested in the business. The price of 57 established for Commercial Solvents in July 1933 was more of a gambling phenomenon, induced by the expected repeal of prohibition. But the gamblers in this instance were acting no differently from those who call themselves investors, in their blithe disregard of the fact that they were paying 140 millions for an enterprise with about 10 millions of resources. (The fixed assets of Commercial Solvents, written down to nothing in the balance sheet, had real value, of course, but not in excess of a few millions.)

The contrast in the other direction shown by our examples is almost as impressive. A going but unsuccessful concern like Pennsylvania Coal and Coke can be valued in the market at about one-sixteenth of its stated resources almost on the same day as a speculatively attractive issue is bid for at sixteen times its net worth. The Pepperell example is perhaps more striking still, because of the unquestioned reality of the figures of book value, and also because of the high reputation, large earnings, and liberal dividends of the enterprise covering a long stretch of years. Yet part owners of this business—under the stress of depression, it is true—were willing to sell out their interest at one-tenth of the value which a single private owner would have unhesitatingly placed upon it.

Recommendation.—These examples, extreme as they are, suggest rather forcibly that the book value deserves at least a fleeting glance by the public before it buys or sells shares in a business undertaking. In any particular case the message which the book value conveys may well prove to be inconsequential and unworthy of attention. But this testimony should be examined before it is rejected. Let the stock buyer, if he lays any claim to intelligence, at least be able to tell himself, first, how much he is actually paying for the business, and secondly, what he is actually getting for his money in terms of tangible resources.

There are indeed certain presumptions in favor of purchases made far below asset value and against those made at a high premium above it. (It is assumed that in the ordinary case the book figures may be accepted as roughly indicative of the actual cash invested in the enterprise.) A business which sells at a premium does so because it earns a large return upon its capital; this large return attracts competition; and, generally speaking, it is not likely to continue indefinitely. Conversely in the case of a business selling at a large discount because of abnormally low earnings. The absence of new competition, the withdrawal of old competition from the field, and other natural economic forces, should tend eventually to improve the situation and restore a normal rate of profit on the investment.

While this is orthodox economic theory, and undoubtedly valid in a broad sense, we doubt if it applies with sufficient certainty and celerity to make it useful as a governing factor in common-stock selection. It may be pointed out that under modern conditions the so-called "intangibles," *e.g.*, good-will, or even a highly efficient organization, are every whit as real from a dollars-and-cents standpoint as are buildings and machinery. Earnings based on these intangibles may be even less vulnerable to competition than those which require only a cash investment in productive facilities. Furthermore, when conditions are favorable the enterprise with the relatively small capital investment is likely to show a more rapid rate of growth. Ordinarily it can expand its sales and profits at slight expense and therefore more rapidly and profitably for its stockholders than a business requiring a large plant investment per dollar of sales.

We do not think, therefore, that any rules may reasonably be laid down on the subject of book value in relation to market price, except the strong recommendation already made that the purchaser know what he is doing on this score and be satisfied in his own mind that he is acting sensibly.

CHAPTER XLIII

SIGNIFICANCE OF THE CURRENT-ASSET VALUE

The current-asset value of a common stock is more likely to be an important figure than the book value, which includes the fixed assets. Our discussion of this point will develop the following theses:

1. The current-asset value is generally a rough index of the *liquidating value*.

2. A large number of common stocks sell for less than their current-asset value, and therefore sell below the amount realizable in liquidation.

3. The phenomenon of many stocks selling persistently below their liquidating value is fundamentally illogical. It means that a serious error is being committed, either: (a) in the judgment of the stock market; (b) in the policies of the company's management; or (c) in the attitude of the stockholders toward their property.

Liquidating Value.—By the liquidating value of an enterprise we mean the money which the owners could get out of it if they wanted to give it up. They might sell all or part of it to some one else, on a going-concern basis. Or else they might turn the various kinds of assets into cash, in piecemeal fashion, taking whatever time is needed to obtain the best realization of each. Such liquidations are of everyday occurrence in the field of private business. By contrast, however, they are very rare indeed in the field of publicly owned corporations. It is true that one company often sells out to another, usually at a handsome price; also that insolvency will at times result in the piecemeal sale of the assets; but the voluntary withdrawal from an unprofitable business, accompanied by the careful liquidation of the assets is an infinitely more frequent happening among private than among publicly owned concerns. This divergence is not without its cause and meaning, as we shall show later.

Realizable Value of Assets Varies with Their Character.—A company's balance sheet does not convey exact information as to its value in liquidation, but it does supply clues or hints which may prove useful. The first rule in calculating liquidating value is that the liabilities are real but the assets are of question-

495

able value. This means that all true liabilities shown on the books must be deducted at their face amount. The value to be ascribed to the assets however, will vary according to their character. The following schedule indicates fairly well the relative dependability of various types of assets in liquidation.

Type of asset	% of liquidating value to book value	
	Normal range	Rough average
Current assets:		
Cash assets (including securities at market)........................	100	100
Receivables (less usual reserves)*....	75–90	80
Inventories (at lower of cost or market)......................	50–75	66⅔
Fixed and miscellaneous assets:		
(Real estate, buildings, machinery, equipment, nonmarketable investments, intangibles, etc.)...........	1–50	15 (approx.)

* Note: Retail installment accounts must be valued for liquidation at a lower rate. Range about 30 to 60 %. Average about 50 %.

Calculation Illustrated.—The calculation of approximate liquidating value in a specific case is illustrated as follows:

Example: (See page 497.)

Object of This Calculation.—In studying this computation it must be borne in mind that our object is not to determine the exact liquidating value of White Motor, but merely to form a rough idea of this liquidating value *in order to ascertain whether or not the shares are selling for less than the stockholders could actually take out of the business.* The latter question is answered very definitely in the affirmative. With full allowance for possible error, there was no doubt at all that White Motor would liquidate for a great deal more than $8 per share or $5,200,000 for the company. The striking fact that the cash assets alone considerably exceed this figure, *after deducting all liabilities,* completely clinched the argument on this score.

Current-asset Value a Rough Measure of Liquidating Value.— The estimated values in liquidation as given for White Motor are somewhat lower in respect of inventories and somewhat higher as regards the fixed and miscellaneous assets than one

WHITE MOTOR COMPANY

Capitalization: 650,000 shares of common stock.
Price in December 1931: $8 per share.
Total market value of the company: $5,200,000.

BALANCE SHEET, DECEMBER 31, 1931 (000 omitted)

Item	Book value	Estimated liquidating value	
		% of book value	Amount
Cash......................................	$ 4,057 ⎱	100	$ 8,600
U.S. Govt. and New York City bonds ...	4,573 ⎰		
Receivables (less reserves)..............	5,611	80	4,500
Inventory (lower of cost or market)	9,219	50	4,600
Total current assets...................	$23,460	...	$17,700
Less current liabilities.................	1,353	...	1,400
Net current assets....................	$22,107	...	$16,300
Plant account........................	16,036 ⎫		
Less depreciation.....................	7,491 ⎬		
Plant account, net....................	$ 8,545 ⎱	20	4,000
Investments in subsidiaries, etc..........	4,996		
Deferred charges......................	388		
Good-will............................	5,389 ⎰		
Total net assets for common stock.......	$41,425	...	$20,300

Estimated liquidating value per share.............. $31
Book value per share........................... 55
Current-asset value per share.................... 34
Cash-asset value per share...................... $11
Market price per share......................... 8

might be inclined to adopt in other examples. We are allowing for the fact that motor-truck inventories are likely to be less salable than the average. On the other hand some of the assets listed as noncurrent, in particular the investment in White Motor Securities Corporation, would be likely to yield a larger proportion of their book values than the ordinary property account. It will be seen that White Motor's estimated liquidating value (about $31 per share) is not far from the current-asset value ($34 per share). In the typical case it may be said that the noncurrent assets are likely to realize enough to make up most of the shrinkage suffered in the liquidation of the quick assets. Hence our first thesis, *viz.*, that the current-asset value affords a rough measure of the liquidating value.

Prevalence of Stocks Selling below Liquidating Value.—Our second point is that for some years past a considerable number of common stocks have been selling in the market well below their liquidating value. Naturally the percentage was largest during the depression. But even in the bull market of 1926–1929 instances of this kind were by no means rare. It will be noted that the striking case of Otis Company, presented in the last chapter, occurred during June 1929, at the very height of the boom. The Northern Pipe Line example, given in Chap. XLI, dates from 1926. On the other hand, our Pepperell and White Motor illustrations were phenomena of the 1931–1933 collapse.

It seems to us that the most distinctive feature of the stock market of those two years was the large proportion of issues which sold below their liquidating value. Our computations indicate that over 40% of all the industrial companies listed on the New York Stock Exchange were quoted at some time in 1932 at less than their net quick assets. A considerable number actually sold for less than their cash-asset value, as in the case of White Motor.[1] On reflection this must appear to be an extraordinary state of affairs. The typical American corporation was apparently worth more dead than alive. The owners of these great businesses could get more for their interest by shutting up shop than by selling out on a going-concern basis.

It is important to observe that these widespread discrepancies between price and current-asset value are a comparatively recent development. In the severe market depression of 1921 the proportion of industrial stocks in this class was quite small. Evidently the phenomenon of 1932 was the direct outgrowth of the new-era doctrine which transferred *all* the tests of value to the income account and completely ignored the balance-sheet picture. In consequence, a company without current earnings was regarded as having very little real value, and it was likely to sell in the market for the merest fraction of its realizable resources. Most of the sellers were not aware that they were disposing of their interest at far less than its scrap value. Many,

---

[1] See Appendix, Note 43, for a representative list of issues selling for less than liquidating value in 1932.

however, who might have known the fact would have justified
the low price on the ground that the liquidating value was of no
practical importance, since the company had no intention of
liquidating.

Logical Significance of This Phenomenon.—This brings us to
the third point, *viz.*, the logical significance of this "subliquidat-
ing-value" phenomenon from the standpoint of the market, of
the managements, and of the stockholders. The whole issue
may be summarized in the form of a basic principle, *viz.*:

When a common stock sells persistently below its liquidating
value, then either the price is too low or the company should be
liquidated. Two corollaries may be deduced from this principle:

Corollary I. Such a price should impel the stockholders to raise the
question whether it is in their interest to continue the business.

Corollary II. Such a price should impel the management to take all
proper steps to correct the obvious disparity between market quotation and
intrinsic value, including a reconsideration of its own policies and a frank
justification to the stockholders of its decision to continue the business.

The truth of the principle above stated should be self-evident.
There can be no *sound* reason for a stock's selling continuously
below its liquidation value. If the company is not worth more
as a going concern than in liquidation, it should be liquidated.
If it is worth more as a going concern, then the stock should sell
for more than its liquidating value. Hence, on either premise, a
price below liquidating value is unjustifiable.

Twofold Application of Foregoing Principle.—Stated in the
form of a logical alternative, our principle invites a twofold
application. Stocks selling below liquidation value are in many
cases too cheap, and so offer an attractive medium for purchase.
We have thus a profitable field here for the technique of security
analysis. But in many cases also the fact that an issue sells
below liquidating value is a signal that a mistaken policy is
being followed, and that therefore the management should take
corrective action, if not voluntarily, then under pressure from
the stockholders. Let us consider these two lines of inquiry
in order.

ATTRACTIVENESS OF SUCH ISSUES AS COMMITMENTS

Common stocks in this category practically always have an
unsatisfactory trend of earnings. If the profits had been
increasing steadily it is obvious that the shares would not sell

at so low a price. The objection to buying these issues lies in the probability, or at least the possibility, that earnings will decline or losses continue, and that the resources will be dissipated and the intrinsic value ultimately become less than the price paid. It may not be denied that this does actually happen in individual cases. On the other hand, there is a much wider range of potential developments which may result in establishing a higher market price. These include the following:

1. The creation of an earning power commensurate with the company's assets. This may result from:
   a. General improvement in the industry.
   b. Favorable change in the company's operating policies, with or without a change in management. These changes include more efficient methods, new products, abandonment of unprofitable lines, etc.
2. A sale or merger, because some other concern is able to utilize the resources to better advantage and hence can pay at least liquidating value for the assets.
3. Complete or partial liquidation.

Examples of Effect of Favorable Developments on Such Issues. *General Improvement in the Industry.*—Examples already given, and certain others, will illustrate the operation of these various kinds of favorable developments. In the case of Pepperell the low price of 17½ coincided with a large loss for the year ended June 30, 1932. In the following year conditions in the textile industry improved; Pepperell earned over $9 per share and resumed dividends; consequently the price of the stock advanced to 100 in January 1934.

Changes in Operating Policies.—Hamilton Woolen Company, another example in the textile field, is a case of individual rather than of general improvement. For several years prior to 1928 the company had operated at substantial losses which amounted to nearly $20 and $12 a share in 1926 and 1927, respectively. Late in 1927 the common stock sold at $13 a share, although the company had net current assets of $38.50 per share at that time. In 1928 and 1929 changes in management and in managerial policies were made, new lines of product and direct sales methods were introduced, and certain phases of production were reorganized. This resulted in greatly improved earnings which averaged about $5.50 per share during the succeeding four years, and within a single year the stock had risen to a price of about $40.

Sale or Merger.—The White Motor instance is typical of the genesis and immediate effect of a sale or merger, as applied to a subliquid-value issue. (The later developments, however, were quite unusual.) The heavy losses of White Motor in 1930–1932 impelled the management to seek a new alignment. Studebaker Corporation believed it could combine its own operations with those of White to mutual advantage, and it was greatly attracted by White's large holdings of cash. Hence, in September 1932 Studebaker offered to purchase all White Motor's stock, paying for each share as follows:

$5 in cash.
$25 in 10-year 6% notes of Studebaker Corporation.
1 share of Studebaker common, selling for about $10.

It will be seen that these terms of purchase were based not on the recent market price of White—below $7 per share—but primarily upon the current-asset value. White Motor shares promptly advanced to 27 and later sold at the equivalent of 31½.[1]

Complete Liquidation.—Mohawk Mining Company supplies an excellent example of a large advance in market value caused by the actual liquidation of the enterprise.

In December 1931 the stock sold at $11 per share, representing a total valuation of $1,230,000 for the 112,000 shares outstanding. The balance sheet at the end of 1931 showed the following:

Cash and marketable securities at market.........	$1,381,000
Receivables..............................	9,000
Copper at market value, about...............	1,800,000
Supplies.................................	71,000
	$3,261,000
Less current liabilities......................	68,000
Net current assets.........................	$3,193,000
Fixed assets, less depreciation and depletion.....	2,460,000
Miscellaneous assets........................	168,000
Total assets for common stock................	$5,821,000

[1] An extraordinary sequel of this transaction was the receivership of Studebaker Corporation in April 1933, ostensibly caused by the opposition of minority stockholders of White Motor to a merger of the two companies. But this development is quite unrelated to our point of discussion, which turns upon the fact that in a sale or merger full recognition should always be, and is ordinarily, given to liquidating value, even though the current market price may be much lower.

Book value per share[1]................... $52
Current-asset value per share[1]............ 28.50
Cash-asset value per share[1]............... 11.75
Market price per share..................... 11

[1] After reducing securities and copper inventory to market value.

Shortly thereafter the management decided to liquidate the property. Within the next two years several liquidating dividends were paid, aggregating $26.50 per share. It will be noted that the amount actually received in liquidation closely approximated the current-asset value just before the liquidation began, and it was 2½ times the then ruling market price.

Partial Liquidation.—Northern Pipe Line Company and Otis Company, already discussed, are examples of the establishment of a higher market value through partial liquidation. The two companies made the following exhibits:

Item	Northern Pipe Line	Otis Company
Date............................	1926	June 1929
Market price.....................	$ 64	$ 35
Cash-asset value per share.........	79	23½
Current-asset value per share........	82	101
Book value per share..............	116	191

In September 1929 Otis Company paid a special dividend of $4 per share and in 1930 it made a distribution of $20 in partial liquidation, reducing the par value from $100 to $80. In April 1931 the shares sold at 45 and in April 1932 at 41. These prices were higher than the quotation in June 1929, despite the distributions of $24 per share made in the interim, and despite the fact also that the *general* market level had changed from fantastic inflation to equally fantastic deflation.

Northern Pipe Line Company distributed $50 per share to its stockholders in 1928, as a return of capital, *i.e.*, partial liquidation. This development resulted in an approximate doubling of the market price between 1926 and 1928. Later a second distribution of $20 per share was made, so that the stockholders received more in cash than the low market price of 1925 and 1926, and they also retained their full interest in the pipe-line business.

Discrimination Required in Selecting Such Issues.—There is scarcely any doubt that common stocks selling well below liquidating value represent on the whole a class of undervalued securities. They have declined in price more severely than the actual conditions justify. This must mean that *on the whole* these stocks afford profitable opportunities for purchase. Nevertheless, the securities analyst should exercise as much discrimination as possible in the choice of issues falling within this category. He will lean toward those for which he sees a fairly imminent prospect of some one of the favorable developments listed above. Or else he will be partial to such as reveal other attractive statistical features besides their liquid-asset position, *e.g.*, satisfactory current earnings and dividends, or a high average earning power in the past. The analyst will avoid issues which have been losing their quick assets at a rapid rate and show no definite signs of ceasing to do so.

Examples: This point will be illustrated by the following comparison of two companies, the shares of which sold well below liquidating value early in 1933.

Item	Manhattan Shirt Company		Hupp Motor Car Corporation	
Price, January 1933	6		2½	
Total market value of Company.	$1,476,000		$3,323,000	
Balance sheet:	Nov. 30, 1932	Nov. 30, 1929	Dec. 31, 1932	Dec. 31, 1929
Preferred stock at par.........	....	$ 300,000		
Number of shares of common..	246,000	281,000	1,329,000	1,475,000
Cash assets.................	$1,961,000	$ 885,000	$ 4,615,000	$10,156,000
Receivables.................	771,000	2,621,000	226,000	1,246,000
Inventories.................	1,289,000	4,330,000	2,115,000	8,481,000
Total current assets.........	$4,021,000	$7,836,000	$ 6,956,000	$19,883,000
Current liabilities...........	100,000	2,574,000	1,181,000	2,541,000
Net quick assets............	$3,921,000	$5,262,000	$ 5,775,000	$17,342,000
Other tangible assets.........	1,124,000	2,066,000	9,757,000	17,870,000
Total assets for common (and preferred)................	$5,045,000	$7,328,000	$15,532,000	$35,212,000
Cash-asset value per share.....	$ 7.50	Nil	$2.625	$ 5.125
Current-asset value per share..	16.00	$17.50	4.375	11.75

Both of these companies disclose an interesting relationship of quick assets to market price at the close of 1932. But a comparison with the balance-sheet situation of three years previously will yield much more satisfactory indications for Manhattan Shirt than for Hupp Motors. The latter concern had lost more than half of its cash assets and more than 60% of its net quick assets during the depression period. On the other hand the quick-asset value of Manhattan Shirt common was reduced by only 10% during these difficult times; and furthermore its cash-asset position was greatly improved. The latter result was obtained through the liquidation of receivables and inventories, the proceeds of which paid off the 1929 bank loans and largely increased the cash resources.

From the viewpoint of past indications, therefore, the two companies must be placed in different categories. In the Hupp Motors case, we should have to take into account the possibility that the remaining excess of current assets over market price might soon be dissipated. This is not true so far as Manhattan Shirt is concerned; and in fact the achievement of the company in strengthening its cash position during the depression must be given favorable consideration. We shall recur later to this phase of security analysis, *viz.*, the comparison of balance sheets over a period in order to determine the true progress of an enterprise.

Bargains of This Type.—Common stock which: (*a*) are selling below liquid-asset values; (*b*) are apparently in no danger of dissipating these assets; and (*c*) have formerly shown a large earning power on the market price, may be said truthfully to constitute a class of *investment bargains*. They are indubitably *worth* considerably more than they are selling for, and there is a reasonably good chance that this greater worth will sooner or later reflect itself in the market price. At their low price these bargain stocks actually enjoy a high degree of safety, meaning by safety a relatively small risk of loss of principal.

A Common Stock Representing the Entire Business Cannot Be Less Safe Than a Bond Having a Claim to Only a Part Thereof.—In considering these issues it will be helpful to apply the converse of the proposition developed earlier in this book with reference to senior securities. We pointed out (Chap. XXVI) that a bond or preferred stock could not be worth more than its value would be if it represented full ownership of the company, *i.e.*, if it were a common stock without senior claims ahead of it. The converse is also true. A common stock cannot be *less safe* than it would be if it were a bond, *i.e.*, if instead of representing full ownership of the company it were given a fixed and limited claim, with some new common stock created to own what was left. This idea, which may appear somewhat abstract at first, can be clarified by the use of a concrete example, taken in January 1933:

Item	The American Laundry Machinery Company		Western Electric Company
	A Actual exhibit	B "Recapitalized"	
Bonds......................	None	Assume: $4,500,000 5 % bonds @ 94 = $4,300,000	$35,000,000 5 % bonds @ 94 = $33,000,000 Other funded debt $39,000,000
Common stock..............	614,000 sh. @ 7 = $4,300,000	614,000 sh. @ ? = ?	Total funded debt $72,000,000 6,000,000 sh. @ 20 = $120,000,000
Total capitalization..........	$4,300,000	More than $4,300,000	$192,000,000
Balance sheet, Dec. 31, 1932:			
Cash assets...............		$ 4,134,000	$ 4,920,000
Other current assets........		17,386,000	97,678,000
Total current assets........		$21,520,000	$102,598,000
Current liabilities..........		220,000	10,693,000
Net quick assets...........		$21,300,000	$ 91,905,000
Other tangible assets (net)...		5,947,000	155,000,000
Total assets for capitalization		$27,247,000	$246,905,000
Net earnings before interest:			
1932......................		985,000(d)	9,032,000(d)
1931......................		772,000	15,558,000
1930......................		1,849,000	20,298,000
1929......................		3,542,000	31,556,000
1928......................		4,128,000	22,023,000
1927......................		4,221,000	19,339,000
1926......................		4,806,000	16,432,000
1925......................		5,101,000	16,074,000
1924......................		3,977,000	14,506,000
1923......................		4,055,000	10,079,000
10-yr. average..............		$ 3,150,000	$ 15,700,000
Interest charges, 1932 basis....	....	$225,000	3,600,000
10-yr. average:			
Interest charges earned......	....	14 times	4.4 times
Earned per share of stock....	$5.13	$4.76	$2.02

The purpose of this analysis is to show that at $7 per share for American Laundry Machinery stock in early 1933—equivalent to only $4,300,000 for the entire business—the purchaser was getting as much *safety of principal* as would be required of a good bond, and in addition he was obtaining all the profit opportunities attaching to common stock ownership. Our contention is that if American Laundry Machinery had happened to have outstanding a $4,500,000 bond issue, this issue would have been considered adequately secured by the standards of fixed-value investment. It is true that the recent earnings were bad—due to the depression—but this disadvantage would be fully offset by the enormous coverage shown in the 10-year average. There would have been no question about the continuance of interest payments, in view of the powerful cash position revealed by the balance sheet. Nor could the investor fail to be impressed by the fact that net current assets alone were nearly five times the amount of the (putative) bond issue. The relative safety of this American Laundry Machinery "bond" is well brought out by the appended comparison with the actual showing of Western Electric 5s due 1944. Clearly the former bonds would be entitled to sell as high as the latter.

If a $4,500,000 bond issue of American Laundry Machinery would have been safe, then *the purchase of the entire company* for $4,300,000 would also have been safe. For a bondholder can enjoy no right or protection which the full owner of the business, without bonds ahead of him, does not also enjoy. Stated somewhat fancifully, the owner (stockholder) can write out his own bonds, if he pleases, and give them to himself. We pointed out an actual illustration of this reasoning in the United States Express Company case, discussed in an early chapter (see Chap. V and Note 2 of the Appendix). Now, near the end of this work, we are brought back to the same basic point—that safety does not reside in titles, or forms, or legal rights, but in the values behind the security issue.

Wall Street would have considered American Laundry Machinery stock "unsafe" at 7, but it would unquestionably have accepted a $4,500,000 bond issue of the same company. Its "reasoning" would have run that the interest on the bond was sure to be continued but that the 40-cent dividend then being paid on the stock was very insecure. In one case the directors had no choice but to pay interest, and therefore would surely do so; in the other case the directors could pay or not as they saw fit, and therefore would very likely suspend the dividend. But Wall Street is here confusing the temporary continuance of income with the more fundamental question of safety of principal. Dividends paid to common stockholders do not in themselves make the stock any safer. The directors are merely turning over to the stockholders part of their own property; if the money were left in the treasury it would still be the stockholder's property. There must therefore be an underlying fallacy in assuming that if the stockholders were given the power to compel payment of income—*i.e.*, if they were made bondholders in whole or in part—their position would thus be made intrinsically sounder. It is little short of idiocy to assume that the stockholders would be better off if they surrendered their complete ownership of the company in exchange for a limited claim against the same property at the rate of 5% or 6% on the investment. This is exactly what the public would do if they were willing to buy a $4,500,000 bond issue of American Laundry Machinery, but would reject as "unsafe" the present common stock at $7 per share.

Nevertheless, Wall Street persists in thinking in these irrational terms, and it does so in part with practical justification. Somehow or other, common-stock ownership does not seem to give the public the same powers and possibilities—the same *values*, in short—as are vested in the private owners of a business. This brings us to the second line of reasoning on the subject of stocks selling below liquidating value.

CHAPTER XLIV

IMPLICATIONS OF LIQUIDATING VALUE. STOCKHOLDER-MANAGEMENT RELATIONSHIPS

Wall Street holds that liquidating value is of slight importance because the typical company has no intention of liquidating. This view is logical, as far as it goes. When applied to a stock selling below break-up value, the Wall Street view may be amplified into the following: "Although this stock would liquidate for more than its market price, it is not worth buying because: (1) the company cannot earn a satisfactory profit; and (2) it is not going to liquidate. In the previous chapter we suggested that the first assumption is likely to be wrong in a number of instances; for although past earnings may have been disappointing, there is always a chance that through external or internal changes the concern may again earn a reasonable amount on its capital. But in a considerable proportion of cases the pessimism of the market will at least *appear* to be justified. We are led, therefore, to ask the question: "Why is it that no matter how poor a corporation's prospects may seem, its owners permit it to remain in business until its resources are exhausted?"

The answer to this question takes us into the heart of one of the strangest phenomena of American finance—the relations of stockholders to the businesses which they own. The subject transcends in its scope the narrow field of security analysis, but we shall discuss it here briefly because there is a distinct relationship between the value of securities and the intelligence and alertness of those who own them. The choice of a common stock is a single act; its ownership is a continuing process. Certainly there is just as much reason to exercise care and judgment in *being* as in *becoming* a stockholder.

Typical Stockholder Apathetic and Docile.—It is a notorious fact, however, that the typical American stockholder is the most docile and apathetic animal in captivity. He does what the board of directors tell him to do, and never thinks of asserting his individual rights as owner of the business and employer of

508

its paid officers. The result is that the effective control of many, perhaps most, large American corporations is exercised not by those who together own a majority of the stock but by a small group known as "the management." This situation has been effectively described by Berle and Means in their significant work *The Modern Corporation and Private Property*. In Chap. I of Book IV the authors say:

It is traditional that a corporation should be run for the benefit of its owners, the stockholders, and that to them should go any profits which are distributed. We now know, however, that a controlling group may hold the power to divert profits into their own pockets. There is no longer any certainty that a corporation will in fact be run primarily in the interests of the stockholders. The extensive separation of owner-ship and control, and the strengthening of the powers of control, raise a new situation calling for a decision whether social and legal pressure should be applied in an effort to insure corporate operation primarily in the interests of the owners or whether such pressure shall be applied in the interests of some other or wider group.

Again, page 355, the authors restate this view in their con-cluding chapter as follows:

. . . A third possibility exists, however. On the one hand, the owners of passive property, by surrendering control and responsibility over the active property, have surrendered the right that the corporation should be operated in their sole interest—they have released the com-munity from the obligation to protect them to the full extent implied in the doctrine of strict property rights. At the same time, the con-trolling groups, by means of the extension of corporate powers, have in their own interest broken the bars of tradition which require that the corporation be operated solely for the benefit of the owners of passive property. Eliminating the sole interest of the passive owner, however, does not necessarily lay a basis for the alternative claim that the new powers should be used in the interest of the controlling groups. The latter have not presented, in acts or words, any acceptable defense of the proposition that these powers should be so used. No tradition supports that proposition. The control groups have, rather, cleared the way for the claims of a group far wider than either the owners or the control. They have placed the community in a position to demand that the modern corporation serve not alone the owners or the control but all society.

Plausible but Partly Fallacious Assumptions by Stock-holders.—Alert stockholders—if there are any such—are not likely to agree fully with the conclusion of Messrs. Berle and

Means that they definitely have "surrendered the right that the corporation should be operated in their sole interest." After all, the American stockholder has abdicated not intentionally, but by default. He can reassert the rights of control which inhere in ownership. Quite probably he would do so if he were properly informed and guided. In good part his docility and seeming apathy are results of certain traditional but unsound viewpoints which he seems to absorb by inheritance or by contagion. These cherished notions include the following:

1. The management knows more about the business than the stockholders do, and therefore its judgment on all matters of policy is to be accepted.
2. The management has no interest in or responsibility for the prices at which the company's securities sell.
3. If a stockholder disapproves of any major policy of the management, his proper move is to sell his stock.

Assumed Wisdom and Efficiency of Management Not Always Justified.—These statements sound plausible, but they are in fact only half truths—the more dangerous because they are not wholly false. It is nearly always true that the management is in the best position to judge which policies are most expedient. But it does not follow that it will always either recognize or adopt the course most beneficial to the shareholders. It may err grievously through incompetence. All stockholders seem to take it for granted that their management is capable. Yet in selecting stocks, great emphasis is laid on the question whether the company enjoys efficient management. This must imply that many companies are poorly directed. Should not this mean also that the stockholders of any company should be open-minded on the question whether its management is efficient or the reverse?

Interests of Stockholders and Officers Conflict at Certain Points.—But a second reason for not always accepting implicitly the decisions of the management is that *on certain points* the interests of the officers and the stockholders may be in conflict. This field includes the following:

1. Compensation to officers—Comprising salaries, bonuses, options to buy stock.
2. Expansion of the business—Involving the right to larger salaries, and the acquisition of more power and prestige by the officers.
3. Payment of dividends—Should the money earned remain under the control of the management, or pass into the hands of the stockholders?

4. Continuance of the stockholders' investment in the Company—Should the business continue as before, although unprofitable, or should part of the capital be withdrawn, or should it be wound up completely?

5. Information to stockholders—Should those in control be able to benefit through having information not given to stockholders generally?

On all of these questions the decisions of the management are *interested* decisions, and for that reason they require scrutiny by the stockholders. We do not imply that corporate managements are not to be trusted. On the contrary, the officers of our large corporations constitute a group of men above the average in probity as well as in ability. But this does not mean that they should be given *carte blanche* in all matters affecting their own interests. A private employer hires only men he can trust, but he does not let these men fix their own salaries or decide how much capital he should place or leave in the business.

Directors Not Always Free from Self-interest in Connection with These Matters.—In publicly owned corporations these matters are passed on by the board of directors, whom the stockholders elect and to whom the officials are responsible. Theoretically, the directors will represent the stockholders' interests, when need be, as against the opposing interests of the officers. But this cannot be counted upon in practice. The individual directors are frequently joined by many close ties to the chief executives. It may be said in fact that the officers choose the directors more often than the directors choose the officers. Hence the necessity remains for the stockholders to exercise critical and independent judgments on all matters where the personal advantage of the officers may conceivably be opposed to their own. In other words, in this field the usual presumption of superior knowledge and judgment on the part of the management should not obtain, and any criticism offered in good faith deserves careful consideration by the stockholders.

Abuse of Managerial Compensation.—Numerous cases have come to light in which the actions of the management in the matter of its own compensation have been open to serious question. In the case of Bethlehem Steel Corporation, cash bonuses clearly excessive in amount were paid. In the case of American Tobacco Company, rights to buy stock below the market price, of an enormous aggregate value, were allotted to the officers. These privileges to buy stock are readily subject to abuse. In the case of Electric Bond and Share Company, the management permitted itself to buy many shares of stock at far below market price. When later the price of the stock collapsed to a figure less than the subscription price, the obligation to pay for the shares was cancelled and the sums already paid were returned to the officers. A similar procedure was followed in the case of White Motor Company, which will be more fully discussed later in this chapter.

Some of these transactions are explained, and partly justified, by the extraordinary conditions of 1928–1932. Others are inexcusable from any point of view. Nevertheless, human nature being what it is, such developments are not in the least surprising. They do not really reflect upon the character of corporate managements, but rather on the patent unwisdom of leaving such matters within the virtually uncontrolled discretion of those who are to benefit by their own decisions.

Fuller Disclosure of Managerial Compensation Desirable.—It would be a salutary step if the law required corporations to disclose annually the salaries and other compensation paid to their officers. (Data of this kind must now be filed in connection with new security offerings under the Securities Act of 1933.) It would also be helpful if the stock holdings of the officers and directors were published regularly. Such information may now be obtained by inspection of the stockholders' list, and at times directors' holdings are so obtained and reported in the financial press. It might be well if the statistical services made an effort to secure this information for as many corporations as possible and included it in the detailed descriptions appearing in their manuals.

In recent years the question of excessive compensation to management has excited considerable attention, and the public understands fairly well that here is a field where the officers' views do not necessarily represent the highest wisdom. It is not so clearly realized that to a considerable extent the same limitations apply in matters affecting the use of the stockholders' capital and surplus. We have alluded to certain aspects of this subject in our discussion of dividend policies (Chap. XXIX). It should be evident also that the matter of raising *new capital* for expansion is affected by the same reasoning as applies to the withholding of dividends for this purpose.

Wisdom of Continuing the Business Should Be Considered.—
A third question, *viz.*, that of retaining the stockholder's capital
in the business, involves considerations which are basically
identical. Managements are naturally loath to return any
part of the capital to its owners, even though this capital may
be far more useful—and therefore valuable—outside of the
business than in it. Returning *a portion* of the capital (*e.g.*,
excess cash holdings) means curtailing the resources of the
enterprise, perhaps creating financial problems later on, and
certainly reducing somewhat the prestige of the officers. Complete liquidation means the loss of the job itself. It is scarcely
to be expected, therefore, that the paid officers will consider
the question of continuing or winding up the business from the
standpoint solely of what is in the best interests of the owners.
We must emphasize again that the directors are often so closely
allied with the officers—who are themselves members of the
board—that they too cannot be counted upon to consider such
problems purely from the stockholders' point of view.

Thus it appears that the question whether a business should
be continued is one which at times may deserve independent
thought by its proprietors, the stockholders. (It should be
pointed out also that this is, by its formal or legal nature, an
ownership problem and not a *management problem*.) And a logical
reason for devoting thought to this question would arise precisely from the fact that the stock has long been selling considerably below its liquidating value. After all, this situation
must mean that either the market is wrong in its valuation, or
the management is wrong in keeping the enterprise alive. It is
altogether proper that the stockholders should seek to determine
which of these is wrong. In this determination the views and
explanations of the management deserve the most appreciative
attention; but the whole proceeding would be stultified if the
management's opinion on this subject were to be accepted as
final *per se*.

It is an unhappy fact that in many cases where a management's
policies are attacked the critic has some personal axe to grind.
This too is perhaps inevitable. There is very little altruism
in finance. Wars against corporate managements take time,
energy, and money. It is hardly to be expected that individuals
will expend all these merely to see the right thing done. In
such matters the most impressive and creditable moves are those

made by a group of substantial stockholders, having an important stake of their own to protect and impelled thereby to act in the interests of the shareholders generally. Representations from such a source, *in any matter where the interest of the officers and the owners may conceivably be opposed*, should gain a more respectful hearing from the rank and file of stockholders than has hitherto been accorded them in most cases.

Broadcast criticisms initiated by stockholders, proxy battles, and various kinds of legal proceedings are exceedingly vexatious to managements; and in many cases they are unwisely or improperly motivated. Yet these should be regarded as one of the drawbacks of being a corporate official, and as part of the price of a vigilant stock ownership. The public must learn to judge such controversies on their merits, as developed by statements of fact and by reasoned argument. It must not allow itself to be swayed by mere accusation or by irrelevant personalities.

The subject of liquidation must not be left without some reference to the employees' vital interest therein. It seems heartless in the extreme to discuss such a decision solely from the standpoint of what will be best for the stockholder's pocketbook. Yet nothing is to be gained by confusing the issue. If the reason for continuing the business is primarily to keep the workers employed, and if this means a real sacrifice by the owners, they are entitled to know and to face the fact. They should not be told that it would be unwise for them to liquidate, when in truth it would be profitable but inhumane. It is fair to point out that under our present economic system the owners of a business are not expected to dissipate their capital for the sake of continuing employment. In privately owned enterprises such philanthropy is rare. Whether a sacrifice of capital for this purpose is conducive to the economic welfare of the country as a whole, is a moot point also; but it is not within our province to discuss it here. Our object has been to clarify the issue, and to stress the fact that a market price below liquidating value has special significance to the stockholders and should lead them to ask their management some searching questions.

Management May Properly Take Some Interest in Market Price for Shares.—Managements have succeeded very well in avoiding these questions with the aid of the time-honored principle that market prices are no concern or responsibility of theirs. It is true, of course, that a company's officers are not responsible for fluctuations in the price of its securities. But this is very far from saying that market prices should *never* be a matter of concern to the management. This idea is not only basically wrong, but it has the added vice of being thoroughly hypocritical. It is wrong because the marketability of securities is one of the chief qualities considered in their purchase. But marketability must presuppose not only a place where they can be sold but also an opportunity to sell them at a *fair price*. It is at least as important to the stockholders that they be able to obtain a fair price for their shares as it is that the dividends, earnings, and assets be conserved and increased. It follows that the responsibility of managements to act in the interest of their shareholders includes the obligation to prevent—in so far as they are able—the establishment of either absurdly high or unduly low prices for their securities.

It is difficult not to lose patience with the sanctimonious attitude of many corporate executives who profess not even to know the market price of their securities. In many cases they have a vital personal interest in these very market prices, and at times they use their inside knowledge to take advantage in the market of the outside public and of their own stockholders. Not as a startling innovation but as a common-sense recognition of things as they are, we recommend that directors be held to the duty of observing the market price of their securities and of using all proper efforts to correct patent discrepancies, in the same way as they would endeavor to remedy any other corporate condition inimical to the stockholders' interest.

Various Possible Moves for Correcting Market Prices for Shares.—The forms which these proper efforts might take are various. In the first place the stockholders' attention may be called officially to the fact that the liquidating, and therefore the minimum, value of the shares is substantially higher than the market price. If, as will usually be the case, the directors are convinced that continuance is preferable to liquidation, the reasons leading to this conclusion should at the same time be supplied. A second line of action is in the direction of dividends. A special endeavor should be made to establish a dividend rate proportionate at least to the liquidating value, in order that the stockholders should not suffer a loss of income through keeping the business alive. This may be done even if current earnings are insufficient, provided there are accumulated profits and provided also the cash position is strong enough to permit such payments.

A third procedure consists of returning to the stockholders such cash capital as is not needed for the conduct of the business. This may be done through a pro rata distribution, accompanied usually by a reduction in par value; or through an offer to purchase a certain number of shares pro rata at a fair price. Finally, a careful consideration by the directors of the discrepancy between earning power and liquidating value may lead them to conclude that a sale or winding up of the enterprise is the most sensible corrective step—in which case they should act accordingly.

Examples: Otis Company, 1929–1930.—The course of action followed by the Otis Company management in 1929–1930 combined a number of these remedial moves. In July 1929 the president circularized the shareholders, presenting an intermediate balance sheet as of June 30, and emphasizing the disparity between the current bid price and the liquidating value. In September of that year—although earnings were no larger than before—dividend payments were resumed, a step permitted by the company's large cash holdings and substantial surplus. In 1930 a good part of the cash, apparently not needed in the business, was returned to the stockholders through the redemption of the small preferred issue and the repayment of $20 per share of common stock on account of capital. As we pointed out in our last chapter, these steps were highly effective in improving the status of the Otis stockholders during a period when most other issues were suffering a shrinkage in value.

Hamilton Woolen Company.—The action of the Hamilton Woolen Company directors was even more interesting, because of its frank and intelligent handling of a difficult problem. Continued operating losses had resulted in a market price well below liquidating value. There was danger that the losses might continue and wipe out the capital. On the other hand, there was a possibility of much better results in the future, especially if new policies were adopted. A statement of the arguments for and against liquidation was forwarded to the stockholders and they were asked to vote on the question. They voted to continue the business, with a new operating head, and the decision proved a wise one. This is an admirable example of the proper procedure when such conditions arise. The ultimate decision—to continue or to quit—is put up to the stockholders, in whose province it lies; the management supplies information, expresses its own opinion, and permits an adequate statement of the other side of the case. There are, of course, a number of instances during the same period of the stockholders' voting in favor of liquidation, although in practically every case the move was recommended by the directors.

Lyman Mills and Other Examples.—Lyman Mills affords one of the most interesting examples. Liquidation was recommended by the directors in 1927, a time of prosperity for most industries but of unsatisfactory results for textile manufacturers. Previous to the decision to liquidate, the stock sold at 110. A total of $220.25 was received by the stockholders in liquidating dividends. In the meantime, conditions had been growing steadily worse in the textile field and the shares of most other cotton-mill concerns had suffered a disastrous decline. Other examples of voluntary liquidation, resulting in a recovery by stockholders far exceeding the previous market price, are:

Company	Year liquidation voted	Price before vote to liquidate	Amount paid to stockholders in liquidation
American Glue.....................	1930	$53	about $140
Mohawk Mining...................	1933	11	about $ 27
Signature Hosiery (Preferred).........	1931	3⅛	$17 +

Repurchase of Shares Pro Rata from Shareholders.—The Hamilton Woolen management is also to be commended for its action during 1932 and 1933 in employing excess cash capital to repurchase pro rata a substantial number of shares at a reasonable price. This reversed the procedure followed in 1929 when additional shares were offered for subscription to the stockholders. The contraction in business which accompanied the depression made this additional capital no longer necessary; and it was therefore a logical move to give most of it back to the stockholders, to whom it was of greater benefit when in their own pockets than in the treasury of the corporation.[1]

[1] Hamilton Woolen sold 13,000 shares pro rata to stockholders at $50 per share in 1929. It repurchased, pro rata, 6,500 shares at $65 in 1932 and

Abuse of Shareholders through Open-market Purchase of Shares.—During the depression period, repurchases of their own shares were made by many industrial companies out of their surplus cash assets,[1] but the procedure generally followed was open to grave objection. The stock was bought in the open market without notice to the shareholders. This method introduced a number of unwholesome elements into the situation. It was thought to be "in the interest of the corporation" to acquire the stock at the lowest possible price. The consequence of this idea is that those stockholders who sell their shares back to the company are made to suffer as large a loss as possible, for the presumable benefit of those who hold on. While this is a proper viewpoint to follow in purchasing other kinds of assets for the business, there is no warrant in logic or in ethics for applying it to the acquisition of shares of stock from the company's own stockholders. The management is the more obligated to act fairly toward the sellers because the company is itself on the buying side.

But in fact the desire to buy back shares cheaply may lead to a determination to reduce or pass the dividend, especially in times of general uncertainty. Such conduct would be injurious to nearly all the stockholders, whether they sell or not, and it is for that reason that we spoke of the repurchase of shares at an unconscionably low price as only *presumably* to the advantage of those who retained their interest.

Example: White Motor Company.—In the previous chapter attention was called to the extraordinary discrepancy between the market level of White Motor's stock in 1931–1932 and the minimum liquidating value of the shares. It will be instructive to see how the policies followed by the management contributed mightily to the creation of a state of affairs so unfortunate for the stockholders.

White Motor Company paid dividends of $4 per share (8%) practically from its incorporation in 1916 through 1926. This period included the depression year 1921, in which the company

---

1,200 shares at $50 in 1933. Faultless Rubber Company followed a similar procedure in 1934. Simms Petroleum Company reacquired stock both directly from the shareholders on a pro rata basis and in the open market. Its repurchases by both means between 1930 and 1933 aggregated nearly 45% of the shares outstanding at the end of 1929.

[1] Figures published by the New York Stock Exchange in February 1934 revealed that 259 corporations with shares listed thereon had reacquired portions of their own stock.

reported a loss of nearly $5,000,000. It drew, however, upon its accumulated surplus to maintain the full dividend, a policy which prevented the price of the shares from declining below 29. With the return of prosperity the quotation advanced to 72½ in 1924 and 104½ in 1925. In 1926, the stockholders were offered 200,000 shares at par ($50), increasing the company's capital by $10,000,000. A stock dividend of 20% was paid at the same time.

Hardly had the owners of the business paid in this additional cash, when the earnings began to shrink and the dividend was reduced. In 1928 about $3 were earned (consolidated basis) but only $1 was disbursed. In the 12 months ending June 30, 1931 the company lost about $2,500,000. The next dividend payment was omitted entirely, and the price of the stock collapsed to 7½.

The contrast between 1931 and 1921 is striking. In the earlier year the losses were larger, the profit-and-loss surplus was smaller, and the cash holdings far lower than in 1931. But in 1921 the dividend was maintained and the price thereby supported. A decade later, despite redundant holdings of cash and the presence of substantial undistributed profits, a single year's operating losses sufficied to persuade the management to suspend the dividend and permit the establishment of a grotesquely low market price for the shares.

During the period before and after the omission of the dividend the company was active in buying its own shares in the open market. These purchases began in 1929 under a plan adopted for the benefit of "those filling certain managerial positions." By June 1931 about 100,000 shares had been bought in at a cost of $2,800,000. With the passing of the dividend, the officers and employees were relieved of whatever obligations they had assumed to pay for these shares and the plan was dropped. In the next six months, aided by the collapse in the market price, the company acquired 50,000 additional shares in the market at an average cost of about $11 per share. The total holdings of 150,000 shares were then retired and cancelled.

These facts, thus briefly stated, illustrate the vicious possibilities inherent in permitting managements to exercise discretionary powers to purchase shares with the company's funds. We note first the painful contrast between the treatment accorded to the White Motor managerial employees and to its stockholders. An extraordinarily large amount of stock was bought for the benefit of these employees at what seemed to be an attractive price. All the money to carry these shares was supplied by the stockholders. If the business had improved, the value of the stock would have advanced greatly and all the benefits would have gone to the employees. When things became worse, "those in managerial positions" were relieved of any loss and the entire burden fell upon the stockholders.

In its transactions *directly with its stockholders,* we see White Motor soliciting $10,000,000 in new capital in 1926. We see some of this additional capital (not needed to finance sales) employed to buy back many of these very shares at one-fifth of the subscription price. The passing of the dividend was a major factor in making possible these repurchases at such low quotations. The facts just related without further evidence might well raise a suspicion in the mind of a stockholder that the omission of the dividend was in some way related to a desire to depress the price of the shares. If the reason for the passing of the dividend was a desire to preserve cash then it is not easy to see why, since there was money available to buy in stock, there was not money available to continue a dividend paid without interruption for 15 years.

The spectacle of a company overrich in cash passing its dividend, in order to impel desperate stockholders to sell out at a ruinous price, is not pleasant to contemplate.

Nash Motors—a Companion Case.—In April 1933 Nash Motors "deferred" (which means omitted) its dividend, also paid continuously up to then for 15 years. Despite the remarkable achievements of this company's management, the circumstances surrounding this dividend omission, in our opinion, made it an astonishing and inexcusable act. The following figures are instructive on this point:

Amount required for the regular 25-cent dividend	$ 660,000
November 30, 1932:	
Company's holdings of cash and Liberty Bonds	32,500,000
Inventories and receivables.................	1,600,000
Total liabilities............................	1,150,000
Profit and loss surplus.....................	26,300,000
Net sales, 1932...........................	15,331,000
Earned for the stock......................	1,030,000
Earned per share.........................	0.39
Results for quarter ended February 1933:	
Before depreciation, profit about............ $	100,000
After depreciation, loss....................	134,000

The cash holdings of Nash were exceedingly large in relation to the needs of the business. (This is manifest from the very small amount required for inventories and receivables.) The company had an enormous surplus, built up for the most part out of undistributed profits; relatively speaking, it had done quite well in the depression; the previous quarter's loss had been the first in its history and negligible in amount. It is difficult to escape the conclusion that the omission of dividends on such an exhibit arises from a complete misconception of what is in the stockholders' interest.[1]

Summary and Conclusion.—The behavior of corporate managements in general during the hectic years from 1928 to 1933 did not reach dazzling heights of wisdom and punctilious rectitude. In fact this behavior was itself responsible in part for the excesses of both optimism and pessimism that marked the period. In our view much of the harm may be traced to the forgetting on all sides of certain elementary facts, *viz.*, that corporations are the mere creatures and property of the stockholders who own them; that the officers are only the paid employees of the stockholders; and that the directors, however chosen, are virtually trustees, whose legal duty it is to act solely on behalf of the owners of the business. In addition to realizing these general truths, and in order to make them practically effective, it is necessary that the public be educated to a clear idea of what are the true interests of the stockholders, in such matters as dividend policies, expansion policies, the use of corporate cash to repurchase shares, the various methods of compensating management, and the fundamental question of whether the owners' capital shall remain in the business or be taken out by them in whole or in part.

[1] The Nash dividend was resumed in August 1933 after an interval of only three months, but was omitted again in April 1934.

CHAPTER XLV

BALANCE-SHEET ANALYSIS (*Concluded*)

Our discussion in the preceding chapters has related chiefly to situations in which the balance-sheet exhibit apparently justified a higher price than prevailed in the market. But the more usual purpose of balance-sheet analysis is to detect the opposite state of affairs, *viz.*, the presence of financial weaknesses which may detract from the investment or speculative merits of an issue. Careful buyers of securities scrutinize the balance sheet to see whether the cash is adequate, whether the current assets bear a suitable ratio to the current liabilities, and whether there is any indebtedness of near maturity which may threaten to develop into a refinancing problem.

WORKING-CAPITAL POSITION AND DEBT MATURITIES

Nothing useful may be said here on the subject of how much cash a corporation should hold. The investor must form his own opinion as to what is needed in any particular case, and also as to how seriously an apparent deficiency of cash should be regarded. On the subject of the *working-capital ratio*, a minimum of $2 of quick assets for $1 of current liabilities was formerly regarded as a standard for industrial companies. An additional rule-of-thumb test for industrials would demand that the current assets *exclusive of inventories* be at least equal to the current liabilities. Failure to meet either of these tests would in most cases reflect strongly upon the investment standing of a common-stock issue—as it would in the case of a bond or preferred stock—and it would supply an argument against the security from the speculative standpoint as well.

As in all arbitrary rules of this kind, exceptions must be allowed if justified by special circumstances. Consider, for example, the current position of Archer-Daniels-Midland Company on June 30, 1933, as compared with the previous year's figures.

The position of this company on June 30, 1933 was evidently much less comfortable than a year before; and judged by the usual standards, it might appear somewhat overextended. But in this case the increase in payables represented a return to the normal practice in the vegetable-oil industry, under which fairly large seasonal borrowings are regularly incurred to carry grain and flaxseed supplies. Upon investigation, therefore, the analyst would not consider the financial condition shown in the 1933 balance sheet as in any sense disturbing.

ARCHER-DANIELS-MIDLAND COMPANY

Item	June 30, 1933	June 30, 1932
Cash assets.........................	$ 1,392,000	$3,230,000
Receivables.	4,391,000	2,279,000
Inventories.........................	12,184,000	4,081,000
Total current assets.................	$17,967,000	$9,590,000
Current liabilities...................	8,387,000	778,000
Working capital.....................	$ 9,580,000	$8,812,000
Working capital excluding inventories..	−2,604,000	+4,731,000

As we pointed out in our discussion of bond selection (Chap. XIII), no such standard requirements are recognized as applicable to railroads and public utilities. It must not be inferred therefrom that the working-capital exhibit of these companies is entirely unimportant—the contrary will soon be shown to be true—but only that it is not to be tested by any cut-and-dried formulas.

Large Bank Debt Frequently a Sign of Weakness.—Financial difficulties are almost always heralded by the presence of bank loans or of other debt due in a short time. In other words, it is rare for a weak financial position to be created solely by ordinary trade accounts payable. This does not mean that bank debt is a bad sign in itself; the use of a reasonable amount of bank credit —particularly for seasonal needs—is not only legitimate but even desirable. But whenever the statement shows Notes or Bills Payable the analyst will subject the financial picture to a somewhat closer scrutiny than in cases where there is a "clean" balance sheet.

The postwar boom in 1919 was marked by an enormous expansion of industrial inventories carried at high prices and financed largely by bank loans. The 1920–1921 collapse of commodity prices made these industrial bank loans a major

problem. But the depression of the 1930's had different characteristics. Industrial borrowings in 1929 had been remarkably
small, due first to the absence of commodity or inventory speculation and secondly to the huge sales of stock to provide additional
working capital. (Naturally there were exceptions, such as,
notably, Anaconda Copper Mining Company which owed
$35,000,000 to banks at the end of 1929, increased to $70,500,000
three years later.) The large bank borrowings were shown more
frequently by the railroads and public utilities. These were
contracted to pay for property additions, or to meet maturing
debt, or—in the case of some railways—to carry unearned fixed
charges. The expectation in all these cases was that the bank
loans would be refunded by permanent financing, but in many
instances such refinancing proved impossible and receivership
resulted. The collapse of the Insull system of public-utility
holding companies was precipitated in this way.

Examples: It is difficult to say exactly how apprehensively the
investor or speculator should have viewed the presence of $68,-
000,000 of bank loans in the New York Central balance sheet at
the end of 1932, or the Bills Payable of $69,000,000 owed by
Cities Service Company on December 31, 1931. But certainly
this adverse sign should not have been ignored. The more
conservatively minded would have taken it as a strong argument
against any and all securities of companies in such a position,
except possibly issues selling at so low a price as to constitute an
admitted but attractive gamble. An improvement in conditions
will, of course, permit the bank loans to be refunded; but logic
requires us to recognize that the improvement is prospective
while the bank loans themselves are very real and very menacing.

When a company's earnings are substantial it rarely becomes
insolvent because of bank loans. But if refinancing is impracticable—as frequently it was in the 1931–1933 period—the lenders
may require suspension of dividends in order to make all the
profits available to reduce the debt. It is for this reason that
the dividend on Brooklyn-Manhattan Transit Corporation
common was passed in 1932 and the preferred dividend of New
York Water Service Corporation was passed in 1931, although
both companies were reporting earnings about as large as in
previous years.

Intercorporate Indebtedness May Cause Trouble.—Current
indebtedness to a parent or to affiliated companies is ordinarily

not so serious a matter as loans from banks, but such balance-sheet items are important nevertheless.

Examples: On December 31, 1932, United Gas Corporation owed $26,000,000 "on open account" to Electric Bond and Share Company, which indirectly held a good deal of its stock; and it also owed $21,000,000 to banks. This was a highly uncomfortable position, which necessitated the passing of the preferred dividend, although earned, and which was likely to interfere with the resumption of dividends even if profits should improve.

A rather peculiar situation was shown in the case of Metro-Goldwyn Pictures Corporation Preferred stock. (This 7% issue has a unique par value of $27 and thus is entitled to annual dividends of $1.89.) The company is a subsidiary of Loew's, Inc., and earned the preferred dividend many times over, even in 1931 and 1932. But its balance sheet as of August 31, 1932, showed $22,000,000 currently due to the parent company, a sum about equal to its current assets less other current liabilities. Although it was unlikely that Loew's would seek to use this claim to freeze out the preferred stockholders—and while an attempt of this sort could no doubt be successfully resisted—the presence of so large an unfunded debt must be regarded as disquieting. If Loew's, Inc., were to find itself hard pressed for cash, it might levy on Metro-Goldwyn Pictures Corporation to repay the current debt, to the extent at least of imperiling the subsidiary's preferred dividend, even should it still be amply earned.

The Danger of Early Maturing Funded Debt.—A large bond issue coming due in a short time constitutes a critical financial problem when operating results are unfavorable. Investors and speculators should both give serious thought to such a situation when revealed by a balance sheet. Maturing funded debt is a frequent cause of insolvency.

Examples: Fisk Rubber Company was thrown into receivership by its inability to pay off an $8,000,000 note issue at the end of 1930. The insolvency of Colorado Fuel and Iron Company and of the Chicago, Rock Island and Pacific Railway Company in 1933 were both closely related to the fact that large bond issues fell due in 1934. The heedlessness of speculators is well shown by the price of $54 established for Colorado Fuel and Iron Preferred, in June 1933 when its short-term bond issue (Colorado Industrial Company 5s, due 1934, guaranteed by the parent company) were selling at 45, *an indicated yield of well over* 100% *per annum.* This price for the bonds was an almost certain sign of trouble ahead. Failure to meet the maturity would in all likelihood mean insolvency (for a voluntary extension could by no means be counted upon) and the danger of complete extinction of the stock issues. It was typical of the speculator to ignore so obvious a hazard, and typical also that he suffered a large loss for his carelessness. (Two months later, on announcement of the receivership, the price of the preferred stock dropped to 17¼.)

Even when the maturing debt can probably be taken care of in some way, the possible cost of the refinancing must be taken into account.

Examples: This point is well illustrated by the $14,000,000 issue of American Rolling Mill Company 4½% Notes, due November 1, 1933. In June 1933 the notes were selling at 80, which meant an annual yield basis of about 75%. At the same time the common stock had advanced from 3 to 24 and then represented a total valuation for the common stock of over $40,000,000. Speculators buying the stock because of improvement in the steel industry failed to consider the fact that in order to refund the notes in the poor market then existing for new capital issues, a very attractive conversion privilege would have to be offered. This would necessarily react against the profit possibilities of the common stock. As it happened, a new 5% note issue, convertible into stock at 25, was offered in exchange for the 4½% notes. The result was the establishment of a price of 101 for the notes, in August 1933, against a coincident price of 21 for the common stock; and a price of 15 for the stock on November 1, 1933 while the notes were taken care of at par.

The impending maturity of a bond issue is of importance to the holders of all the company's securities, including mortgage debt ranking ahead of the maturing issue. For even the prior bonds will in all likelihood be seriously affected if the company is unable to take care of the junior issue. This point is illustrated in striking fashion by the Fisk Rubber Company First Mortgage 8s, due 1941. Although they were deemed to be superior in their position to the 5½% unsecured notes, their holders suffered grievously from the receivership occasioned by the maturity of the 5½s. The price of the 8s declined from 115 in 1929 to 16 in 1932.[1]

[1] See other references to the two Fisk bond issues in Chaps. VI, XVIII, and L.

It should not be necessary to dilate further upon the prime necessity of examining the balance sheet for any possible adverse features in the nature of bank loans or other short term debt.

COMPARISON OF BALANCE SHEETS OVER A PERIOD OF TIME

This part of security analysis may be considered under three aspects, *viz.*:

1. As a check-up on the reported earnings per share.
2. To determine the effect of losses (or profits) on the financial position of the company.
3. To trace the relationship between the company's resources and its earning power over a long period.

Check-up on Reported Earnings Per Share, Via the Balance Sheet.—Some of this technique has already been used in connection with related phases of security analysis. In Chap. XXXVI, for instance, we gave an example of the first aspect, in checking the reported earnings of American Commercial Alcohol Corporation for 1931 and 1932. As an example covering a larger stretch of years we submit the following contrast between the average earnings of Stewart Warner Corporation for the eight years 1925–1932 as shown by the reported per-share figures, and as indicated by the changes in its net worth in the balance sheet.

STEWART-WARNER CORPORATION, 1925–1932
1. NET EARNINGS AS REPORTED

1925	$ 7,544,000	Per share: $ 5.80*
1926	5,109,000	3.89
1927	5,210,000	3.99
1928	7,753,000	5.97
1929	6,839,000	5.26
1930	1,262,000	0.98
1931	*1,830,000(d)*	*1.44(d)*
1932	*2,445,000(d)*	*1.96(d)*
Total for 8 years...	$29,442,000	$22.49

* 1925–1929 figures adjusted to allow for stock dividends and split-up.

2. DISCREPANCY BETWEEN EARNINGS AS ABOVE AND CHANGES IN THE SURPLUS ACCOUNT

Net earnings 1925–1932, as reported.......... $29,442,000
Less cash dividends paid..................... 21,463,000
Indicated increase in capital and surplus....... 7,979,000

COMPARATIVE BALANCE SHEET

Item	Dec. 31, 1924 (including Bassick-Alemite Corporation)	Dec. 31, 1932
Fixed and miscellaneous assets ...	$13,100,000	$11,900,000
Current assets..................	12,500,000	6,200,000
Patents and good-will...........	10,800,000	
Total assets....................	$36,400,000	$18,100,000
Current liabilities..............	$ 2,000,000	$ 800,000
Bonds and preferred stock.......	2,500,000	200,000
Common stock.................	19,200,000	12,400,000
Surplus.......................	12,700,000	4,700,000
Total liabilities.................	$36,400,000	$18,100,000

Capital and surplus, Dec. 31, 1924............ $31,900,000
Less good-will and patents included therein....   10,800,000

Capital and surplus, Dec. 31, 1924, adjusted... $21,100,000
Capital and surplus, Dec. 31, 1932............   17,100,000

Decrease for 8 years per balance sheet........ $ 4,000,000
Increase for eight years per income accounts.... $ 8,000,000
Discrepancy between earnings shown by income
accounts and those indicated by balance sheets: 12,000,000

3. EXPLANATION OF DISCREPANCY

Charges made to surplus and not deducted in income account:
Patents and development expense written off
(applicable to 1925–1932).................. $ 4,400,000
Royalty litigation expense, plant reappraisal,
prior years' taxes.......................   6,200,000
Premium on securities retired, and other charges,
(net)...................................   1,400,000

$12,000,000

4. RESTATEMENT OF STEWART-WARNER CORPORATION'S EARNINGS FOR 1925–1932

Earnings, per income account................ $29,400,000
Less charges made to surplus................   12,000,000

Earnings for the period as corrected........... $17,400,000
Reported earnings were overstated by nearly 70%.

The above analysis does not require extended discussion, since most of the points involved were covered in Chap. XXXI to XXXVI. By far the larger part of the charges made direct to surplus undoubtedly represented a real diminution of the earning power of Stewart-Warner during the eight-year period. Most of the expenditures for patents and development work would have had to be written off on any proper basis of amortization. The back taxes and royalty payments were certainly chargeable against earnings of the years to which they applied; and the "plant reappraisal" charges were, in part at least, equivalent to necessary depreciation charges. The fact that the company's working capital decreased by $5,000,000 during this period (and by $2,800,000 after deducting bonds and preferred stock) is conclusive evidence that the "surplus earnings" of $8,000,000 above dividends, as shown by the income accounts, simply did not exist.

Checking the Effect of Losses or Profits on the Financial Position of the Company.—An example of the second aspect was given in Chap. XLIII, in the comparison of the 1929–1932 balance sheets of Manhattan Shirt Company and Hupp Motor Car Corporation respectively. A similar comparison is appended herewith, covering the exhibit of Plymouth Cordage Company and H. R. Mallinson and Company during the same period.

Examples:

Item	Plymouth Cordage Co.	H. R. Mallinson & Co.
Earnings reported:		
1930.................................	$ 288,000	*$1,457,000(d)*
1931.................................	25,000	*561,000(d)*
1932.................................	*233,000(d)*	*200,000(d)*
Total (3 years) profit.............	$ 80,000	*$2,218,000(d)*
Dividends......................	1,348,000	66,000
Charges to surplus and reserves....	2,733,000	116,000
Decrease in surplus and reserve for 3 years......................	$4,001,000	$2,400,000

COMPARATIVE BALANCE SHEETS

(000 omitted)

Item	Plymouth Cordage		H. R. Mallinson	
	Sept. 30, 1929	Sept. 30, 1932	Dec. 31, 1929	Dec. 31, 1932
Fixed and miscellaneous assets (net)...................	$ 7,211	$ 5,157	$2,539	$2,224
Cash assets...................	1,721	3,784	526	20
Receivables..................	1,156	668	1,177	170
Inventories..................	8,059	3,150	3,060	621
Total assets..................	$18,147	$12,759	$7,302	$3,035
Current liabilities..............	$ 982	$ 309	$2,292	$ 486*
Preferred stock................	....	....	1,342	1,281
Common stock.................	8,108	7,394	500	500
Surplus and miscellaneous reserves	9,057	5,056	3,168	768
Total liabilities................	$18,147	$12,759	$7,302	$3,035
Net current assets..............	$ 9,954	$ 7,298	$2,471	$ 357
Net current assets excluding inventory..................	1,895	4,143	*589(d)*	*264(d)*

* Including $32,000 of "deferred liabilities."

Despite the large reduction in the surplus of Plymouth Cordage during these years, its financial position was even stronger at the end of the period than at the beginning, and the liquidating value per share (as distinct from book value) was probably somewhat higher.   On the other hand, the losses of Mallinson almost denuded it of working capital and thereby created an extremely serious obstacle to a restoration of its former earnings power.

Taking Losses on Inventories May Strengthen Financial Position. It is obvious that losses which are represented solely by a decline in the inventory account are not so serious as those which must be financed by an increase in current liabilities.   If the shrinkage in the inventory exceeds the losses, so that there is an actual increase in cash or reduction in payables, it may then be proper to say—somewhat paradoxically—that the company's financial position has been strengthened even though it has been suffering losses.   This reasoning has a concrete application in analyzing issues selling at less than liquidating value.   It will be recalled that in estimating break-up value inventories are ordinarily taken at about 50 to 75% of the balance sheet figure, even though the latter is based on the lower of cost or market. The result is that what appears as an operating loss in the company's statement may have the actual effect of a profit from the standpoint of the investor who has valued the inventory in his own mind at considerably less than the book figure. This idea is concretely illustrated in the Manhattan Shirt Company example.

MANHATTAN SHIRT COMPANY
(000 omitted)

Item	Balance sheet Nov. 30, 1929		Balance sheet Nov. 30, 1932	
	Book value	Estimated liquidating value	Book value	Estimated liquidating value
Cash and bonds at market......	$ 885	$ 885	$ 1,961	$ 1,961
Receivables.................	2,621	2,100	771	620
Inventories.................	4,330	2,900	1,289	850
Fixed and other assets.........	2,065*	500	1,124	300
Total assets.................	$ 9,901	$ 6,385	$ 5,145	$ 3,731
Current liabilities..............	2,574	2,574	100	100
Preferred stock...............	299	299		
Balance for common............	$ 7,028	$ 3,513	$ 5,045	$ 3,631
Number of shares..............	281,000	281,000	246,000	246,000
Value per share...............	$25.00	$12.50	$20.50	$14.75

* Excluding good-will.

INCOME ACCOUNT 1930–1932
Balance after preferred dividends:
1930................................	*318,000*(d)
1931 Profit..........................	93,000
1932................................	*139,000*(d)
3 years.............................	*364,000*(d)
Charges to surplus......................	505,000*
Common dividends paid..................	723,000
	$1,592,000
Less discount on common stock bought......	481,000
Decrease in surplus for period..............	$1,111,000*

* Eliminating transfer of $100,000 to Contingency Reserve.

If we consider only the company's figures, there was evidently a loss for the period, with a consequent shrinkage in the value of the common stock. But if an investor had bought the stock, say at $8 per share in 1930 (the low price in that year was 6⅛), he would more logically have appraised the stock in his own mind on the basis of its liquidating value, rather than its book value. From his point of view therefore, the intrinsic value of his holdings would have *increased* during the depression period, from $12.50 to $14.75 per share, even after deducting the substantial dividends paid. What really happened was that Manhattan Shirt turned the larger portion of its assets into cash during these three years and sustained a much smaller loss in so doing than a conservative buyer of the stock would have anticipated. This accomplishment can be summarized as follows:

Assets turned into cash and application of proceeds	Amount	"Expected loss" thereon and application of difference	
Reduction in inventory.......	$3,000,000	$1,000,000	
Reduction in receivables......	1,800,000	350,000	
Reduction in plant, etc.......	1,000,000	750,000	
	$5,800,000	$2,100,000	
Actual loss sustained.........	800,000	800,000	
Net amount realized..........	$5,000,000	"Gain" on basis of liquidation values	$1,300,000
Applied as follows:		Applied as follows:	
To common dividends....	$ 700,000	To common dividends...	$ 700,000
To payment of liabilities..	2,500,000	To increase liquidating value......	$ 600,000
To redemption of preferred	300,000		
To retirement of common.	500,000		
To increase in cash assets.	1,000,000		
	$5,000,000		

We have here a direct contrast between the superficial indications of the income account and the truer story told by the successive balance sheets. Situations of this kind justify our repeated assertion that income-account analysis must be supplemented and confirmed by balance-sheet analysis.

Is Shrinkage in Value of Normal Inventory an Operating Loss?—
A further question may be raised with respect to changes in the
inventory account, and that is whether a mere reduction in the
carrying price should be regarded as creating an operating loss.
In the case of Plymouth Cordage we note the following comparative figures:

Inventory, Sept. 30, 1929	$8,059,000
Inventory, Sept. 30, 1932	3,150,000
Decrease	60%

In the meantime the price of fibers had declined more than
50%, and there was good reason to believe that the actual
number of pounds of fiber, rope, and twine contained in the
company's inventory was not very much smaller in 1932 than in
1929. At least half of the decline in the inventory account was
therefore due solely to the fall in unit prices. Did this portion
of the shrinkage in inventory values constitute an operating loss?
Could it not be argued that its fixed assets had suffered a similar
reduction in their appraisal value, and that there was as much
reason to charge this shrinkage against earnings as to charge
the shrinkage in the carrying price of a certain physical amount
of inventory?

The analogy between inventory value and fixed property is a
fairly close one. Either account lends itself to arbitrary treatment which may have an important effect upon the reported
per-share earnings. We have discussed at length the anomaly
that writing down fixed assets to nothing can apparently increase
the value of a common stock by relieving the earnings of the
annual depreciation charge. Somewhat similarly, by the artifice
of carrying the inventory at zero—or at exceedingly low base
prices—a good part of the losses suffered in a period like that of
1929–1932 could be avoided, at least to the extent of eliminating
them from the income account or balance-sheet figures.

Examples: As long ago as 1913 the National Lead Company
adopted a policy designed to eliminate from its earnings statement the effect of variations in metal prices, with respect to
certain "normal stocks" of the three principal constituents of its
inventory, namely, lead, tin, and antimony. These were considered in the light of permanent holdings, necessary for the
conduct of the business; and it was contended, quite logically,

that there was no more reason to vary each year the value of this fixed inventory than to vary the value of the manufacturing plants. Some of the New England cotton mills had followed a like policy, prior to the collapse in the cotton market in 1930, by carrying their raw cotton and work in process at very low base prices.

It is not certain that any uniform accounting methods of this kind could be developed for industrial companies generally, in order to eliminate from the income account the somewhat unreal and distorting effect of price changes. But the competent analyst will note this factor whenever it bulks large in the picture, and will endeavor to give it suitable weight in arriving at the true meaning of the reported profits.

Profits from Inventory Inflation.—That the importance of inventory price changes is not confined to a depression period is emphatically shown by the events of 1919 and 1920. In 1919 the profits of industrial companies were very large; in 1920 the reported earnings were irregular but in the aggregate quite substantial. Yet the gains shown in these two years were in many cases the result of an *inventory inflation, i.e.,* a huge and speculative advance in commodity prices. Not only was the authenticity of these profits thereby made open to question, but the situation was replete with danger because of the large bank loans contracted to finance these overvalued inventories.

Examples: The following tabulation, which covers a number of the leading industrial companies, will bring out the significant contrast between the apparently satisfactory earnings developments and the undoubtedly disquieting balance-sheet developments between the end of 1918 and the end of 1920.

TWELVE INDUSTRIAL COMPANIES

	Year 1919	Year 1920	Years 1919–1920
Earned for common stock........	$100,000,000	$ 48,000,000	$148,000,000
Dividends paid.................	35,000,000	68,000,000	103,000,000
Charges to surplus.............	5,000,000	10,000,000	15,000,000
Added to surplus...............	60,000,000	*30,000,000* (decr.)	30,000,000
Inventories increased...........	57,000,000	84,000,000	141,000,000
Change in other net current assets.	+30,000,000	*131,000,000* (decr.)	*101,000,000* (decr.)
Plant, etc. increased...........	33,000,000	169,000,000	202,000,000
Capitalization increased........	69,000,000	141,000,000	210,000,000
Reserves increased.............	....	12,000,000	12,000,000

The companies included in the above computation were: American Can, American Smelting and Refining, American Woolen, Baldwin Locomotive Works, Central Leather, Corn Products Refining, General Electric, B. F. Goodrich, Lackawanna Steel, Republic Iron and Steel, Studebaker, U.S. Rubber. We append also the individual figures for U.S. Rubber, in order to add concreteness to our illustration:

U.S. RUBBER (1919–1920)

Earned for common stock:

1919..............................	$12,670,000	Per share: $17.60
1920..............................	16,002,000	19.76
Total..........................	$28,672,000	$37.36
Cash dividends paid..................	8,580,000	
Stock dividend paid..................	9,000,000	
Transferred to contingency reserve.....	6,000,000	
Adjustments of surplus and reserves....	cr. 2,210,000	
Net increase in surplus and miscellaneous reserves......................	$7,300,000	

BALANCE SHEET
(000 omitted)

Item	Dec. 31, 1918	Dec. 31, 1920	Increase
Plant and miscellaneous assets (net)...	$131,000	$185,500	$ 54,500
Inventories.......................	70,700	123,500	52,800
Cash and receivables................	49,500	63,600	14,100
Total assets....................	$251,200	$372,600	$121,400
Current liabilities...................	$ 26,500	$ 74,300	$ 47,800
Bonds............................	68,600	87,000	18,400
Preferred and common stock.........	98,400	146,300	49,900
Surplus and miscellaneous reserves.....	57,700	65,000	7,300
Total liabilities.................	$251,200	$372,600	$121,400
Working capital....................	93,700	112,800	19,100
Working capital excluding inventory...	23,000	10,700(d)	33,700(d)

The U. S. Rubber figures for 1919–1920 present the complete reverse of Manhattan Shirt's exhibit for 1930–1932. In the Rubber example we have large earnings but a coincident deterioration of the financial position due to heavy expenditures on plant and a dangerous expansion of inventory. The stock buyer would have been led astray completely had he confined his attention solely to U. S. Rubber's reported earnings of nearly $20 per share in 1920; and, conversely, the securities markets were equally mistaken in considering only the losses reported during 1930–1932, without reference to the favorable changes occurring at the same time in the balance-sheet position of many companies.

Long-range Study of Earning Power and Resources.—The third aspect of the comparison of successive balance sheets is of restricted interest because it comes into play only in an exhaustive study of a company's record and inherent characteristics. The purpose of this kind of analysis may best be conveyed by means of the followings applications to the long-term exhibits of U. S. Steel Corporation and Corn Products Refining Company.

I. United States Steel Corporation: Analysis of Operating Results and Financial Changes by Decades, 1903–1932

The balance sheets are adjusted to exclude an intangible item ("water"), amounting to $508,000,000, originally added to the Fixed Property Account. This was subsequently written off between 1902 and 1929 by means of an annual sinking-fund charge (aggregating $182,000,000) and by special appropriations from surplus. The sinking-fund charges in question are also eliminated from the income account.

A. Operating Results
(In millions)

Item	First decade 1903–1912	Second decade 1913–1922	Third decade 1923–1932	Total for 30 years
Finished goods produced.......	93.4 tons	123.3 tons	118.7 tons	335.4 tons
Gross sales (excluding intercompany items).............	$4,583	$9,200	$9,185	$22,968
Net earnings*................	979	1,674	1,096	3,749
Bond interest................	303	301	184	788
Preferred dividends...........	257	252	252	761
Common dividends............	140	356	609†	1,105†
Balance to surplus and "voluntary reserves"..............	279	765	51	1,095

* After depreciation, but eliminating parent company sinking-fund charges.
† Including $204,000,000 paid in stock.

B. BALANCE-SHEET CHANGES
(All figures in millions)

Item	Dec. 31, 1902	Dec. 31, 1912	Changes in first decade	Dec. 31, 1922	Changes in second decade	Dec. 31, 1932	Changes in third decade	Changes in 30 years
Assets:								
Fixed (less deprec.) and misc.*	$820	$1,160	+$340	$1,466	+$306	$1,741	+$275	+$ 921
Net current assets	167	256	+ 89	606	+ 350	371	- 235	+ 204
Total	$987	$1,416	+$429	$2,072	+$656	$2,112	+$ 40	+$1,125
Liabilities:								
Bonds	$380	$ 680	+$300	$ 571	-$109	$ 116	-$455	-$ 264
Preferred stock	510	360	- 150	360	......	360	......	- 150
Preferred dividends accrued	......	......	......	......	......	5	+ 5	+ 5
Common stock	508	508	......	508	......	952†	+ 444	+ 444
Surplus and "voluntary" reserves*	411(d)	132(d)	+ 279	633	+ 765	679	+ 46	+ 1,090
Total	$987	$1,416	+$429	$2,072	+$656	$2,112	+$ 40	+$1,125

* Eliminating initial mark-up of $508,000,000, later written off.
† Including premiums of $81,000,000 and stock dividend of $204,000,000.

C. Relations of Earnings to Average Capital
(All dollar figures in millions)

Item	First decade	Second decade	Third decade	Total for 30 years
Capital at beginning................	$ 987	$1,416	$2,072	$ 987
Capital at end.....................	1,416	2,072	2,112	2,112
Average capital about.............	1,200	1,750	2,100	1,700
% earned on average capital, per year..........................	8.1%	9.6%	5.2%	7.4%
% paid per year in interest and dividends on average capital.........	5.8%	5.2%	4.0%*	5.2%*
Average common stock equity (common stock, surplus, and reserves).	$ 237	$ 620	$1,389	$ 816
% earned on common stock equity..	17.7%	18.3%	4.8%	9.0%
% paid on common stock equity....	5.9%	5.7%	2.9%*	3.7%*
Depreciation per year..............	$24	$34	$46	$35
Average fixed property account.....	1,000	1,320	1,600	1,300
Ratio of depreciation to fixed property...........................	2.4%	2.6%	2.9%	2.7%

* Excluding stock dividend.

D. The Significance of the Above Figures.—The three decades had, superficially at least, a somewhat equal distribution of good years and bad. In the first decade, 1904 and 1908 were depression years, while 1911 and 1912 were subnormal. The second period had three bad years, namely 1914, 1921, and 1922—the last due to high costs rather than to small volume. The third decade was made up of eight years of prosperity followed by two of unprecedented depression.

The figures show that the war period, which occurred in the middle decade, was a windfall for United States Steel, and added more than three hundred millions to profits, as compared with the rate established in the first ten years. On the other hand, the last ten years were marked by a drastic falling off in the rate of earnings on the invested capital. The difference between the 5.2% actually earned and the 8% which might be regarded as a satisfactory annual average, amounted to close to six hundred million dollars for the ten-year period.

Viewing the picture from another angle, we note that in the thirty years the actual investment in United States Steel Corporation was more than doubled and its productive capacity was increased threefold. Yet the average annual production was only 27% higher, and the average annual earnings before interest charges were only 12% higher, in 1923–1932 than in 1903–1912. This analysis would serve to raise the question: (a) whether, since the end of the war, steel production has been transformed from a reasonably prosperous into a relatively unprofitable industry; and (b) whether this transformation is due in good part to excessive reinvestment of earnings in additional plant, thus creating a condition of overcapacity with resultant reduction in the margin of profit.

II. Corn Products Refining Company
Feb. 28, 1906 to Dec. 31, 1932
(000 omitted)

A. Average Annual Income Account

	1906–1912	1913–1922	1923–1932
Earned before depreciation.............	$3,750	$10,188	$14,490
Depreciation.........................	770	1,937	2,862
Balance for interest and dividends......	2,980	8,251	11,628
Bond interest........................	500	400	115
Preferred dividends (paid or accrued)...	2,070	1,983	1,749
Balance for common..................	410	5,868	9,764
Common dividends...................	....	1,294	7,461
Balance to surplus...................	410	4,574	2,303
Balance to surplus for period..........	2,808	45,740	23,030
Adjustment of common stock, surplus and reserves......................	cr. 898	cr. 8,091	dr. 16,881
Increase in common stock, surplus and reserves.........................	3,706	53,831	6,149

B. Balance Sheets

	Feb. 28, 1906	Dec. 31, 1912	Dec. 31, 1922	Dec. 31, 1932
Plant (less depreciation) and miscellaneous assets......................	$49,000	$56,478	$52,222	$ 37,431
Investment in affiliates...............	2,000	3,620	8,080	32,045
Net current assets...................	1,000	5,010	38,905	33,386
Total......................	$52,000	$65,108	$99,207	$102,862
Bonds.............................	$ 9,571	$14,039	$ 2,807	$ 1,766
Preferred stock.....................	28,293	29,827	24,827	24,374
Common stock, surplus and miscellaneous reserves......................	14,136	17,742	71,573	76,722
Preferred dividend accrued...........	....	3,500		
Total......................	$52,000	$65,108	$99,207	$102,862

C. PERCENTAGE EARNED* AND PAID ON TOTAL CAPITALIZATION AND ON
COMMON STOCK EQUITY

Item	1906–1912	1915–1922	1923–1932	26⅚ years
Average capitalization	$58,600,000	$82,200,000	$101,000,000	$77,400,000
Earned thereon	5.1%	10.1%	11.5%	10.5%
Paid thereon	4.4%	4.5%	9.3%	7.2%
Average common equity	$15,800,000	$44,500,000	$74,200,000	$45,400,000
Earned thereon	2.6%	13.2%	13.2%	13.1%
Paid thereon	nil	2.9%	10.1%	7.2%

* Adjustments to Surplus and Reserves are excluded from earnings.

NOTES ON ABOVE COMPUTATION

1. The plant account and common stock equity are corrected throughout to reflect a write-down of $36,000,000 made in 1922.

2. Bonds outstanding are increased in 1906 and 1912 to reflect liability for issues of subsidiaries.

3. Estimates considered to be sufficiently accurate are used in the initial balance sheet.

4. Deductions for bond interest are partly estimated for the first two periods.

5. The Adjustment of Common Stock, Surplus and Reserves for 1913–1922 represent chiefly an increase in Miscellaneous Reserves. The decrease for 1923–1932 is due mainly to losses on marketable securities sold and a write-down of the balance held to the market. (The 1932 balance sheet has been adjusted to reflect this write-down.)

Comment on the Corn Products Refining Company Exhibit.—The early period was one of subnormal earnings, which would have been still poorer if more nearly adequate depreciation charges had been made. As in the case of United States Steel, the war period brought enormous earnings to Corn Products. The decade 1913–1922 was marked as a whole by a great increase in working capital and a substantial reduction in funded debt and preferred stock. Depreciation charges exceeded expenditures on new plant.

In the 1923–1932 period we note a striking divergence from the exhibit of United States Steel. Corn Products was able to increase its earning power proportionately with its enlarged capital investment. Its annual profits (both before and after depreciation) were about four times as large in this decade as in the period ending in 1912. (In the case of United States Steel the increase was only some 10%.) The balance-sheet changes were marked by a further substantial shrinkage in the property account (due to the liberal depreciation charged), but by a much larger increase in the investment in affiliated companies—indicating a broad expansion of the company's activities. The decline in working capital was occasioned by an extraneous factor, *viz.*, a severe fall in the value of marketable securities owned.

It is clear that the record of Corn Products Refining Company does not suggest the same questions or doubts as arise from an examination of the United States Steel Corporation's exhibit.

PART VII

ADDITIONAL ASPECTS OF SECURITY-ANALYSIS. DISCREPANCIES BETWEEN PRICE AND VALUE

CHAPTER XLVI

STOCK-OPTION WARRANTS

During the last two decades the use of stock-option warrants has passed through an extraordinary development. They were devised originally as a form of participating privilege for bonds and preferred stocks to which they were attached. In this form they were commonly regarded only as a feature of the senior security, similar to a conversion right, and the warrants themselves had little significance in relation to the company's capitalization structure. Later the idea was hit upon of creating stock-option warrants separately from other securities, and delivering them as compensation to underwriters, promoters, and executives. From this point the inevitable next development was the issuance, through sale or exchange, of separate option warrants to the general public in the same manner as common stocks. They thus attained full stature as an independent form of "security," as an important part of the financial set-up of many corporations, and as a popular and prominent medium of speculative activity.

In a previous chapter we considered the technical aspects of option warrants as an adjunct of senior securities. In this chapter we shall discuss the more important role of option warrants as a separate financial instrument. Our treatment falls into three sections: (1) description; (2) technical characteristics of warrants as a vehicle of speculation; (3) their significance as a part of the financial structure.

DESCRIPTIVE SUMMARY

A (detachable) option warrant is a transferable right to buy stock, originally running for a considerable period of time.

542

(Warrants attached to a debenture bond issue are sometimes called "Debenture Rights." A third name for the same thing is "Stock-purchase Warrant.") Its terms include: (1) the kind of stock; (2) the amount; (3) the price; (4) the method of payment; (5) the duration of the privilege; and (6) antidilution provisions. (The last were described in Chap. XXV.)

Kind of Stock Covered by the Privilege.—Nearly all option warrants call for common stock of the issuing company. In rare instances they apply to preferred stock (*e.g.*, American Solvents and Chemical), or for stock of some other concern (*e.g.*, warrants attached to Central States Electric Corporation Preferred called for North American Company stock; and warrants attached to Solvay American Investment Corporation preferred stock called for Allied Chemical and Dye Corporation stock). Warrants have no right to receive interest, dividends, or payments on account of principal, nor have they the right to cast any vote.

Resemblance to Subscription "Rights."—Option warrants bear some resemblance to the "subscription rights" which are issued by corporations to their stockholders in connection with the sale of additional stock. There are two significant differences, however, between warrants and rights. Warrants run for a long period and the stock-purchase price is almost always set higher than the quotation at the time of their issuance. Moreover, the price is frequently varied in accordance with the terms of the warrant. Subscription rights run for a short time and call for a fixed price, usually under the market at the time of their authorization. Subscription rights are devised, therefore, with the intent of assuring their exercise and the prompt receipt of funds by the company. Option warrants generally have no relation to the financial needs of the company, and they are not expected to be exercised in short order. Stated in a different way, a subscription right *will* be exercised unless the market declines substantially before they expire; option warrants *will not* be exercised unless the market price advances substantially in the near or distant future. Subscription rights generally run for about sixty days; the original duration of option warrants is rarely, if ever, less than a year, and many of them are perpetual.

Method of Payment.—Most option warrants require payment of the subscription price in cash. Those originally attached to bonds or preferred shares may permit payment either in cash or by tender of the senior security which is accepted at its face value. This alternative may be of considerable practical importance.

Example: American and Foreign Power Company Warrants are a perpetual call on common stock at $25 per share. Payment may be made either in cash or by tendering second preferred stock at $100 per share. In October 1933 the common sold at 10 and the second preferred at 12. Because of the very low price of the senior issue, the warrants had an "exercisable value," even though the common was selling 15 points below the option price. The calculation is as follows:

One warrant $+ \frac{1}{4}$ share of second preferred = one share of common.

Value of option warrant = $10 - \frac{1}{4}(12) = 7$

Basis of Trading in Warrants.—Option warrants are bought and sold in the market in the same way as common stocks. Up to the end of 1933 only one issue of warrants had been listed on the New York Stock Exchange (Commercial Investment Trust Corporation Warrants), but many were actively dealt in on the New York Curb Exchange and other exchanges. The basis of trading in these instruments is somewhat eccentric, and at times conducive to serious error. Under the standard rule, "one warrant" means the right to buy one share of stock, and *not* the right originally attached to one share of stock.

Examples: Walgreen (Drug) Company preferred stock was sold with warrants entitling the holder to buy two shares of common for each preferred share. Under the regular rule of trading, "one Walgreen Warrant" meant the right to buy one share of common, *i.e.*, each share of preferred was said to carry "two warrants."

Similarly, Consolidated Cigar Corporation $6\frac{1}{2}\%$ Preferred Stock was issued with a warrant attached to each share calling for the purchase of one-half share of common. These warrants were also traded in on the basis that one warrant was the right to buy one share of common, *i.e.*, each share of $6\frac{1}{2}\%$ preferred was said to carry "half a warrant."

But the exceptions to this standard rule are numerous.

Examples: Commercial Investment Trust Corporation $6\frac{1}{2}\%$ Preferred carried warrants to buy one-half share of common for each share of preferred (the same ratio as in the case of Consolidated Cigar Preferred). But these were dealt in on the New York Stock Exchange on the basis that "one warrant" meant the warrant originally attached to one share of preferred, *i.e.*, it called for half a share of common. Similar departures were made in the rules of trading for Niagara Hudson Power Corporation *B* Warrants; Loew's, Inc., Preferred Warrants; Safeway Stores, Inc., "Old Series" Warrants, etc.

When a change is made in the number of shares called for by the warrant, the customary procedure is to continue to trade in "one old warrant" as "one warrant."

Example: "One Loew's Bond Warrant" originally called for one share of common at $55. It represented the warrant attached to $200 of Loew's 6% Debentures, due 1941. When a 25% stock dividend was paid in 1928, the antidilution provision required that an additional quarter share be given free with each share subscribed for under the warrant. "One Loew's Bond Warrant" remained physically unchanged and thereafter represented the right to purchase 1¼ shares for $55. Similarly in the case of Commercial Investment Trust Warrants when the common stock was split 2½ for one. One warrant thereafter represented the right to buy 1¼ new shares instead of half an old share.

But the opposite practice is sometimes followed.

Example: Niagara Hudson Power *A* Warrants. These called for one share of common at $35. The company recapitalized in 1932 and issued one new share for three old. Hence what was formerly "one warrant" now called for one-third of a new share for $35, *i.e.*, at $105 per share. The N. Y. Curb Exchange thereupon redefined "one *A* warrant" as representing the right to buy one *new* share. Hence three old warrants became one new warrant.

These technical details are given here because they are not available in standard descriptive textbooks. Those buying or selling a particular option warrant are cautioned to make careful inquiry into the basis of trading therein.[1]

Examples of Warrants Issued for Various Purposes. *A. Attached to Senior Securities.*—Perhaps the earliest instance is

---

[1] Subscription rights are invariably dealt in in New York on the basis of "one right" meaning the right *received* by the owner of one share of stock. This is the opposite idea from that ordinarily followed in option warrants. See Appendix, Note 44, for a rapid method of calculating the value of subscription rights.

the issue of Sinclair Oil and Refining Corporation 7s in 1917. By far the most prominent is the sale by American and Foreign Power Company of $270,000,000 of Second Preferred stock carrying warrants for no less than 7,100,000 shares of common.

B. *As Compensation to Underwriters.*—The first important case seems to have been the $25,000,000 Barnsdall Corporation 6% bond issue of 1926. Here the bankers received, as *part* of their compensation, warrants for 500,000 shares of common. At the subsequent high price these warrants would have been worth $13,000,000.

C. *As Compensation to Promoters and Management.*—A striking case was the formation of Petroleum Corporation of America in January 1929. The public was offered 3,250,000 shares of stock at $34 per share. Five-year warrants to buy 1,625,000 shares at 34 were issued to the promoters and management.

D. *Issued in a Merger or Reorganization Plan, in Exchange for Other Securities.*—Commonwealth and Southern Corporation issued about 17,500,000 warrants, together with 34,000,000 shares of common and 1,500,000 shares of preferred, mainly in exchange for securities of six constituent companies. It is interesting to note that it issued *common stock* and warrants in exchange for Penn-Ohio Edison Company and Southeastern Power and Light Company option warrants.

In 1933 Armour and Company (Ill.) proposed a recapitalization plan under which its Class *A* and Class *B* common stock would be exchanged chiefly for option warrants. There would have been issued about 5,000,000 warrants in all. The plan encountered opposition, however, and was abandoned.

E. *Attached to an Original Issue of Common Stock.*—Public Utility Holding Corporation of America sold 2,500,000 shares of common stock, carrying warrants to buy an equal number of shares of additional common. In addition, the organizing interests purchased 500,000 shares of Class *A* stock (with voting control) together with warrants to buy 1,000,000 shares of either Class *A* or common stock.

F. *Sold Separately for Cash.*—In 1929, Fourth National Investors Corporation sold to its parent company 750,000 option warrants for $3,000,000.

WARRANTS AS A VEHICLE OF SPECULATION

In a broad sense, option warrants possess the same general characteristics as low-priced common stocks, the theory of which was discussed in Chap. XLI.   Warrants are in name and in form, as low-priced stocks frequently are in essence, a long-term call upon the future of a business.   It is true also that the relationship between a warrant and its common stock is roughly similar to that between a common stock and a speculative senior security of the same company.   It may be contended that warrants have a generic advantage over low-priced stocks, because the latter are identified for the most part with unsuccessful businesses while the former are usually created by companies which are successful and expanding.   While such an advantage may be admitted, it is certainly not dependable enough to justify any indiscriminate partiality to warrants or rejection of low-priced shares.

The Qualitative Element.—As with all other speculative commitments, the attractiveness of a given warrant depends upon two entirely dissimilar factors: the qualitative element, being the nature of the enterprise, in relation particularly to its supposed chance of great improvement; and the quantitative element, being the terms on which the warrant is offered, including its price and the price of the common stock it calls for. Security analysis cannot be counted upon to reveal those businesses which are most likely to forge ahead in the years to come. There is not much we can say, therefore, about the qualitative element in selecting warrants for speculation.   Since ordinarily a warrant can attain tangible value only through an *increase* in earnings, emphasis must be laid upon the prospects of *change* rather than upon stability.   Public-utility warrants, for example, became extremely popular in 1928–1929 not because of the superior stability of utility enterprises but because the market was convinced that their earnings would continue to expand indefinitely.

As far as the arithmetical chance of a large price advance is concerned, we have already shown that this is most likely to be found in the common stock of speculatively capitalized enterprises (*e.g.*, A. E. Staley Company and American Water Works and Electric, discussed in Chap. XL.)   Hence warrants to buy common stocks of this kind may also be said to have a special speculative advantage.   But this is at bottom a quantitative rather than a qualitative matter.   In our view, it is rarely possible to say *with assurance* that the long-term prospects of a particular line of business are so much better than the average as to make warrants connected with that field more attractive than any others. But if the individual speculator has definite opinions and preferences on this score, it is perfectly logical for him to follow them.

Quantitative Considerations: Importance of Low Price.—It is an easier matter to point out the elements which govern the relative attractiveness of warrants from a quantitative standpoint. The desirable qualities are: first, a low price; second, a long duration; and thirdly, an option (or purchase) price close to the market. From the standpoint of speculative theory, the most important of the three no doubt is a low price for the warrant. This may be brought out by a comparison of the situation existing in the Sinclair Oil and Refining Corporation Warrants in 1917 and Niagara Hudson Power Corporation B Warrants in 1929.

Examples: The warrant attached to each $1,000 Sinclair Oil and Refining Corporation note, issued in 1917, entitled the holder to buy 25 shares of stock at $45 per share until August 1, 1918; at $47\frac{1}{2}$ until August 1, 1919; and at 50 until February 1, 1920. In December 1917 the stock had declined to $25\frac{1}{4}$ and a warrant for 25 shares could be bought at $20, *i.e.*, at a cost of only 80 cents per share. Here the market price of the stock was far below the option price, but the option could be acquired at a very low cost per share. The sequel was quite characteristic of speculative markets. In less than eighteen months Sinclair Oil stock rose to $69\frac{3}{4}$ giving a warrant for 25 shares a realizable value of over $550. An increase of 175% in the price of the stock produced an increase of 2,680% in the price of the warrant.

The Niagara Hudson Power Corporation B Warrants entitled the holder to buy $3\frac{1}{2}$ shares of common for $50, *i.e.*, at $14.285 per share. When the warrants were admitted to trading on the New York Curb in 1929 they sold at 60—equivalent to 17 for a one-share warrant—while the stock was selling at $22\frac{1}{2}$. In this case the speculator was paying nearly as much per share for the warrants as for the stock. When the latter advanced to its high of 31 later in the year, the warrants rose by a much smaller percentage, to 21. Still later in the same year, the price of the stock broke to $11\frac{1}{4}$, and then the warrants collapsed to a low of 2. These comparative figures show that at the equivalent of 17 the Niagara Hudson B Warrants were selling at an extraordinarily unattractive price.

Low Relative Price Important.—It is technically desirable that the price of a warrant be low not only in itself but also in relation to the price of the common stock. This point may be shown by a comparison of Commercial Investment Trust Corporation warrants in 1928 with American and Foreign Power Company warrants in 1933.

Examples: Commercial Investment Trust Corporation Warrants sold at $6 each in August 1928. They entitled the holder to buy one-half share of common at $90 per share until the end of 1929, and at 100 thereafter until January 1, 1931. The common was then selling at about 70. The warrant for one share thus represented a commitment of $12, or about ⅙ the current value of the stock. Despite the relatively high purchase price specified in the warrant, the latter might be considered as having a speculative advantage over the stock because of the much smaller money cost involved. (As it happened, the price of the warrants advanced elevenfold in 1928–1929 as against a threefold rise in the common.) In the American and Foreign Power example, given earlier in this chapter, we note that the warrants for one share could be bought at 7, representing exact parity with the common. But the fact that the common was itself selling at only 10 removed any special speculative advantage from the warrants at 7. As we shall see later, it throws the stock and the warrants together into the category of "pseudo" low-priced speculations, of the kind discussed at the beginning of Chap. XLI.

The foregoing discussion leads to the conclusion that a given option warrant has speculative attractiveness, in a technical sense, only if it constitutes a low-cost, long-term right to purchase a stock at a price not unimaginably remote from the current market.

Examples: The Sinclair Oil and Commercial Investment Trust warrants referred to above, are examples that met these requirements. An unusual example is furnished by the Barnsdall Oil warrants in 1927. These were a call on the stock at 25. When the shares were selling at 31, the warrants sold at 6, exactly at parity. In this case, any rise in the value of the stock would have meant—and later did mean—a much larger proportionate rise in the price of the warrants.

Technical Advantages Often Absent.—During 1928–1929 when trading in warrants was most active, there was a tendency for these instruments to sell at high levels, both relatively and absolutely, so that they could not be said to possess any technical advantage over the typical common stock.   During the ensuing depression, many warrant issues were obtainable at very low prices; but here again the related common shares were also quoted so low as to call into question the comparative attractiveness of the warrant.   This is illustrated by the following representative list of warrants, as of the close of 1933.

Name of corporation issuing warrant	Duration	Purchase price of stock named in warrant	Market price of stock	Market price of warrant
United Aircraft and Transport............	To Nov. 1, 1938	30	31½	12¾
Atlas Corp................	Perpetual	25	12	4¼
United Corp...............	Perpetual	27.50	4⅞	2⅛
Tri-Continental Corp.......	Perpetual	18.46	4¾	1¾
Niagara Hudson Power *A* Warrants...............	To Oct. 1, 1944	105	5½	½
Petroleum Corp...........	To Feb. 1, 1934	34	9½	¹⁄₆₄

None of these warrants appears especially attractive from the technical angle.   In the case of the Petroleum Corporation issue, the short duration spoils an otherwise interesting price relationship.   (In other words, if the warrants were to run five years instead of one month they would appear an excellent speculation at 1½ cents each.)

WARRANTS AS PART OF THE CAPITALIZATION STRUCTURE

Option warrants are essentially a device to give separate embodiment to the element of future prospects.   But the right to benefit from future improvement or enhancement belongs inherently to the common stockholder.   It is one of the important considerations which he receives in return for putting up his money and taking the "first risk" of loss.   The basic fact about an option warrant, therefore, is that it represents *something which has been taken away* from the common stock.   The equation is a simple one:

Value of common stock + value of warrants = value of common stock alone (*i.e.*, if there were no warrants).

Warrants Represent a Subtraction from the Related Stock.— To illustrate this point concretely, let us assume that Company *A*

has an invested capital of $2,000,000 and annual earnings of
$200,000. It has outstanding 100,000 shares of common stock,
which may be assigned a fair value of $20 per share. The
management issues to itself, or to the stockholders without
consideration, option warrants to buy 100,000 additional shares
at $20. What is the effect of this development on the value of
the stock?

The warrants will have no "exercisable value" at the time of
issuance; but they would have a real value nevertheless, and they
would command a market price. For the right to benefit from
any increase in the price of the stock is well worth owning and is
therefore worth paying for. It is possible that the warrants
would be fairly worth $5 each. If that is so, then it follows from
our principle that the value of the common stock would be
reduced by $5 per share by reason of the creation of the warrants
—because whatever value attaches to the warrants must have
been subtracted from the common stock.

The specific reason for the reduction in the value of the stock
is that the shares now have only a 50% interest in any future
enhancement instead of a 100% interest as previously. The
effect of this dilution may be shown by assuming that the value
of the enterprise is doubled. The results would be as follows:

A. 100,000 shares of stock; no warrants;
  Former value $2,000,000; value per share $20,
  Earnings $ 200,000; Earnings per share $ 2
  New value $4,000,000; Value per share $40
  Earnings $ 400,000; Earnings per share $ 4
B. 100,000 shares of stock, and warrants to buy 100,000 shares additional
  at $20 per share
  Former value $2,000,000; value per share $20, *less deduction for*
  warrants.
  New value $4,000,000; value per share $40, *less deduction for*
  warrants.
  This increase in value causes the warrants to be exercised. Company
  receives $2,000,000 cash and issues 100,000 new shares. Assume addi-
  tional earnings of 10% on the new capital, or $200,000.
  New value of enterprise $6,000,000; value per share $30; earnings
  $600,000; earnings per share $3.

Under *A* the value of the stock (assuming a 10% basis of
capitalization) would advance to $40 per share; under *B* it rises
only to $30. The latter figure would confirm our hypothetical,
initial values of 5 for the warrants and 15 for the stock; for at

30 there would be a 100% increase in the value of both the stock and warrants.

This illustration shows clearly that the effect of the creation of warrants is to diminish the benefits realized by the common from a large increase in the earnings and in the value of the business. Warrants to buy stock, even at a price above the market, therefore detract from the present value of the common stock, because part of this present value is based upon the right to benefit from future improvement.

A Dangerous Device for Diluting Stock Values.—The option warrant is a fundamentally dangerous and objectionable device because it effects an indirect and usually unrecognized dilution of common-stock values. The stockholders view the issuance of warrants with indifference, failing to realize that part of their equity in the future is being taken from them. The stock market, with its usual heedlessness, applies the same basis of valuation to common shares whether warrants are outstanding or not. Hence warrants may be availed of to pay unreasonable bonuses to promoters or other insiders without fear of comprehension and criticism by the rank and file of stockholders. Furthermore, the warrant device facilitates the establishment of an artificially high aggregate market valuation for a company's securities, because (with a little manipulation) large values can be established for a huge issue of warrants without reducing the quotation of the common shares.

A Reductio ad Absurdum.—The public's failure to comprehend that all the value of option warrants is derived at the expense of the common stock has led to a practice which would be ridiculous if it were not so mischievous. We refer to the original sale of common stock carrying warrants to buy additional common stock. This arrangement gives nothing to the stockholders which they would not have without the warrant, and it violates an obvious rule of sound corporate financing. A properly managed business sells additional stock *only when new capital is needed*, and in that event the stockholders are usually entitled to subscribe pro rata to the offering.[1] To give subscription rights to stockholders when the money is not needed is non-

---

[1] It has become fashionable to insert charter provisions which deprive stockholders of this so-called "preemptive right." It is claimed that the surrender of this right is necessary in order to give the directors more flexible powers in making corporate deals involving issuance of stock. We are very sceptical of the soundness of this argument.

sensical from all viewpoints except that of deceiving people into believing that something attractive is being offered them. It resembles the practice, sometimes indulged in, of declaring dividends in "scrip" which is redeemable at the pleasure of the directors. This "scrip" is an unnecessary expression in separate form of a right which the common stock possesses inherently, *viz.*, to receive future dividends when the directors see fit to pay them.[1] Similarly these option warrants attached to original issues of common stock are a superfluous expression of the stockholders' inherent right to participate in future stock offerings.

A further study of the unwholesome implications of the warrant device is integrated with two broader lines of inquiry into financial practices—the first relating to the price paid by the public for the financing and management of business; the second relating to that group of manipulative and dangerous corporate practices referred to as "pyramiding." These aspects of security analysis will be considered in the ensuing chapters.

[1] Cities Service Company paid dividends in scrip of this kind between 1921 and 1925, redeeming it in the latter year. Since its value depended almost entirely on the whim of the directors, it was the sort of speculative medium which gives an enormous advantage to insiders. Gas Securities Company, a subsidiary of Cities Service, paid dividends in scrip of this kind during 1933.

CHAPTER XLVII

COST OF FINANCING AND MANAGEMENT

Let us consider in more detail the organization and financing of Petroleum Corporation of America, mentioned in the last chapter. This was a large investment company formed for the purpose of specializing in securities of enterprises in the oil industry. The public was offered 3,250,000 shares of capital stock at $34 per share. The company received therefor a net amount of $31 per share, or $100,750,000 in cash. It issued to unnamed recipients—presumably promoters, investment bankers, and the management—warrants, good for five years, to buy 1,625,000 shares of additional stock, also at $34 per share.

This example is representative of the investment trust financing of the period. Moreover, as we shall see, the technique on this score which developed in boom years was carried over through the ensuing depression, and it threatens to be accepted as the standard practice for stock financing of all kinds of enterprises. But there is good reason to ask the real meaning of a set-up of this kind, first, with respect to what the buyer of the stock gets for his money, and second, with respect to the position occupied by the investment banking houses floating these issues.

Cost of Management; Three Items.—A new investment trust—such as Petroleum Corporation in January 1929—starts with two assets: cash and management. Buyers of the stock at $34 per share were asked to pay for the management in three ways, viz.:

1. By the difference between what the stock cost them and the amount received by the corporation.

It is true that this difference of $3 per share was paid not to the management but to those underwriting and selling the shares. But from the standpoint of the stock buyer the only justification for paying more for the stock than the initial cash behind it would lie in his belief that the management was worth the difference.

2. By the value of the option warrants issued to the organizing interests.

554

These warrants in essence entitled the owners to receive one-third of whatever appreciation might take place in the value of the enterprise over the next five years. (From the 1929 viewpoint a five-year period gave ample opportunity to participate in the future success of the business.) This block of warrants had a real value, and that value in turn was taken out of the initial value of the common stock.

A calculation such as we used for our hypothetical example in the previous chapter would indicate that the 1,625,000 warrants would take about one-sixth of the value away from the common stock. This estimate is borne out fairly well by the actual market prices of warrants in relation to their common stocks. On this basis, one-sixth of the $100,750,000 cash originally received by the company would be applicable to the warrants, and five-sixths to the stock.

3. By the salaries which the officers were to receive, and also by the extra taxes incurred through the use of the corporate form.

Summarizing the above analysis, we find that buyers of Petroleum Corporation shares were paying the following price for the managerial skill to be applied to the investment of their money:

1. Cost of financing ($3 per share)................... $ 9,750,000
2. Value of warrants (⅙th of remaining cash)...... about 16,790,000
3. Future deductions for managerial salaries, etc........   ?
                                                         _____
   Total.. $26,500,000+

The three items together may be said to absorb between 25 and 30% of the amount contributed by the public to the enterprise. By this we mean not merely a deduction of that percentage of future profits, but an actual sacrifice of invested *principal* in return for management.

What Was Received for the Price Paid?—Carrying the study a step farther, let us ask what kind of managerial skill was this enterprise to enjoy? The board of directors consisted of many men prominent in finance, and their judgment on investments was considered well worth having. But two serious limitations on the value of this judgment must here be noted. The first is that the directors were not obligated to devote themselves exclusively or even preponderantly to this enterprise. They were permitted, and seemingly intended, to multiply these activities indefinitely. Common sense would suggest that the value of their expert judgment to Petroleum Corporation would be greatly diminished by the fact that so many other claims were being made upon it at the same time.

A more obvious limitation appears from the Corporation's projected activities. It proposed to devote itself to investments in a single field—petroleum. The scope for judgment and analysis was thereby greatly circumscribed. As it turned out, the funds were largely concentrated, first in two related companies—Prairie Pipe Line Company and Prairie Oil and Gas Company—and then in a single successor enterprise (Consolidated Oil Corporation). Thus Petroleum Corporation took on the complexion of a holding company, in which the exercise of managerial skill appears to be reduced to a minimum once the original acquisitions are made.[1]

We are forced to conclude that financial schemes of the kind illustrated by Petroleum Corporation of America are unsatisfactory from the standpoint of the stock buyer. This is true not only because the total cost to him for management is excessive in relation to the value of the services rendered, but also because the cost is not clearly disclosed, being concealed in good measure by the use of the warrant artifice. (The above reasoning does not rest in any way upon the fact that Petroleum Corporation's investments proved unprofitable.)

Position of Investment Banking Firms in This Connection.— The second line of inquiry suggested by this example is also of major importance. What is the position occupied by the investment banking firms floating an issue such as Petroleum Corporation of America, and how does this compare with the practice of former years? Prior to the late 1920s, the sale of stock to the public by reputable houses of issue was governed by the following three important principles:

1. The enterprise must be well established, and offer a record and financial exhibit adequate to justify the purchase of the shares at the issue price.
2. The investment banker must act primarily as the representative of the buyers of the stock, and he must deal at arm's-length with the company's management. His duty includes protecting his clients against the payment of excessive compensation to the officers, or any other policies inimical to the stockholders' interest.

---

[1] The same logical objection to the payment of a large "managerial bonus," in the form of option warrants to those organizing a holding company, may be urged against the set-up of Alleghany Corporation and United Corporation.

3. The compensation taken by the investment banker must be reasonable. It represents a fee paid by the corporation for the service of raising capital.

These rules of conduct afforded a clear line of demarcation between responsible and disreputable stock financing. It was an established Wall Street maxim that capital for a new enterprise must be raised from private sources.[1] These private interests would be in a position to make their own investigation, work out their own deal, and keep in close touch with the enterprise, all of which safeguards (in addition to the chance to make a large profit) were considered necessary to justify a commitment in any new venture. Hence the public sale of securities in a *new enterprise* was confined almost exclusively to "blue sky" promoters and small houses of questionable standing. The great majority of such flotations were either downright swindles or closely equivalent thereto by reason of the unconscionable financing charges taken out of the price paid by the public.

Investment-trust financing, by its very nature, was compelled to contravene these three established criteria of reputable stock flotations. The investment trusts were *new* enterprises; their management and their bankers were generally *identical;* the compensation for financing and management had to be determined solely by the recipients, without accepted standards of reasonableness to control them. In the absence of such standards, and in the absence also of the invaluable arm's-length bargaining between corporation and banker, it was scarcely to be hoped that the interests of the security buyer would be adequately protected. Allowance must be made besides for the generally distorted and egotistical views prevalent in the financial world during 1928 and 1929.

Demoralizing Results.—The result of these various factors was demoralizing in the extreme, and its unfortunate effects were not confined to the financial set-ups of investment trusts organized during the boom period. For indeed these objectionable features in common-stock financing appear to be establishing and extending themselves. The character of such flotations in 1933 is

---

[1] An apparent exception might be made sometimes in a case such as Chile Copper Company where the demonstrated presence of huge bodies of ore was regarded as justifying public financing to bring the mine into production. The sale of stock of the Lincoln Motor Company in 1920 was one of the few real exceptions to the rule as here stated. In this instance an unusually high personal reputation was behind the enterprise, but it resulted in disastrous failure.

anything but reassuring from the viewpoint of the conservative analyst. We find that reputable houses are now willing to sell shares of new or virtually new commercial enterprises, without past records and on the basis entirely of their expected future earnings. This means inevitably that the investment banker is not acting primarily on behalf of the buyers of the stock. For on the one hand the new corporation is not an independent entity, which can negotiate at arm's-length with various bankers representing clients with money to invest; and on the other hand, the banker is himself in part a promoter, in part a proprietor of the new business. In an important sense, he is raising funds from the public *for himself*.

New Role of Investment Banker.—More exactly stated, the investment banker now appears to be operating in a double guise. He makes a deal on his own behalf with the originators of the enterprise, and then he makes a separate deal with the public to raise from them the funds he has promised the business. He demands—and no doubt is entitled to—a liberal reward for his pains. But the very size of his compensation introduces a significant change in his relationship to the public. For it makes a very real difference whether a stock buyer can consider the investment banker as essentially his agent and representative, or whether he must view the issuing house as a promoter-proprietor-manager of a business, endeavoring to raise funds to carry it on.

If investment banking becomes identified with the latter approach, the interests of the general public are certain to suffer. The Securities Act of 1933 aims to safeguard the security buyer by requiring full disclosure of the pertinent facts and by extending the previously existing liability for concealment or misrepresentation. While full disclosure is undoubtedly desirable, it may not be of much practical help except to the skilled and shrewd investor or to the trained analyst. It is to be feared that the typical stock buyer will neither read the long prospectus carefully nor understand the implications of all it contains. Modern financing methods are not far different from a magician's bag of tricks; they can be executed in full view of the public without it being very much the wiser. The use of stock options as part of the underwriter-promoter's compensation is one of the newer and more deceptive tricks of the trade. An example of new-enterprise stock financing during

1933 will be discussed in some detail, with the object of illustrating both the character of these flotations and the technique of analysis required to appraise them.

Example: Mouquin, Inc.—This company was organized in July 1933, for the chief purpose of reestablishing a liquor importing business which had been carried on by the Mouquin family for many years before the advent of Prohibition. The public was offered 55,000 shares of common stock at $6.75 per share, of which sum $5 was paid into the company's treasury and $1.75 was retained by the underwriters to cover selling expense and profit. This financing charge of about 26% was evidently far higher than that paid by established corporations for new capital. But the Mouquin financing included other elements of more serious import to the stock buyer. These may be summarized as follows:

1. An option was given to the underwriters to buy 14,000 shares additional at $1 per share.
2. A second option was given to the underwriters to buy 30,000 shares at $7 per share, good for nearly two years.
3. The company had issued 176,000 shares to the owners of the predecessor company for all of the assets, which included tangible assets of only $108,000, and trade names, *etc.*, which were "appraised" at $876,000 and entered on the balance sheet at $438,000.

The option to buy 14,000 shares at $1 was an artifice with no understandable object other than to mislead the public as to the amount of the underwriting fee being paid. Obviously these shares were expected to be taken up promptly and sold with the others. This arrangement meant simply that instead of receiving $5 per share for 55,000 shares, the company was to receive $4.20 per share for 69,000 shares. Hence the real selling commission amounted to 38% of the offering price, instead of 26% as it was made to appear. Furthermore the additional option to buy 30,000 shares at $7 had a cash value of some sort which must be included in the cost of financing.

The terms under which the major part of the stock was issued to the owners of the predecessor business are of still greater importance in the analysis. If the tangible assets only are considered, these proprietors may be said to have received 176,000 shares upon payment of only 62 cents per share. Undoubtedly the Trade Names, etc., had a value because of their prospective earning power, but whether this value was fairly equal to the difference between 62 cents and $6.75 must be considered as doubtful.

Summarizing the flotation in its entirety, and omitting reference to a minor detail, we may say that in paying $6.75 per share for the stock the public was asked to place a valuation of about $1,670,000 for an enterprise with tangible assets of $424,000 and no earnings record. This means that *nearly 75%* of the offering price really represented either selling commission or a premium paid to the organizing interests for their trade connections and their expected ability to produce profits in excess of a normal return on the invested capital. This analysis should bring into sharp relief the twofold disadvantage to the public in such financing methods—first, through the use of artifices to cloak the true terms of the deal; and secondly, through the absence of that arm's-length, mutually critical, relationship between house of issue and corporate management upon which we have previously laid such stress.

We must reluctantly conclude also that no amount of disclosure made compulsory by law will afford the security buyer that high degree of protection which he was formerly wont to obtain through the experienced judgment and scrupulously fair dealing of the old established investment houses.[1]

Blue-sky Promotions.—In the "good old days" fraudulent stock promoters relied so largely upon high pressure salesmanship that they rarely bothered to give their proposition any semblance of serious merit. They could sell shares in a mine which was not even a "hole in the ground," or in an invention the chief recommendation for which was the enormous profit made by Henry Ford's early partners. The victim was in fact buying "blue sky" and nothing else. Any one with the slightest business sense could have detected the complete worthlessness of these ventures almost at a glance; in fact, the glossy paper used for the prospectus was in itself sufficient to identify the proposition as fraudulent.

[1] Two other examples will further illustrate 1933 stock financing methods. The offering of Speculative Profit Shares, Inc., disclosed a threefold profit to the organizers: (1) a 15% selling commission; (2) additional compensation in the form of option warrants; (3) a management fee equal to 20% of the net cash income.

In the case of Barium Steel Corporation Class *A* stock, the underwriting-selling commission was 11%; but the underwriters received also about 40% of the Class *B* stock as additional compensation.

The tightening of federal and state regulations against these swindles has led to a different type of security promotion. Instead of offering something entirely worthless, the promoter selects a real enterprise which he can sell at many times its fair value. By this means the law can be obeyed and the public fleeced just the same. Oil and mining ventures lend themselves best to such stock flotations, because it is easy to instill in the uninitiated an exaggerated notion of their true worth. The typical oil-stock flotation of this kind capitalizes the current income from flush production as if it were permanent, which decidedly it is not. The prospects of maintaining or increasing the production are glowingly depicted, though in fact this would require an unusual stroke of good fortune.

Most of the numerous gold-mine flotations of 1933 involved properties which had formerly produced but had been abandoned when the good ore was exhausted. These situations permitted legally correct but entirely misleading statements to be made about the large amounts of gold produced "to date" by the mine, and about the substantial sums actually expended upon its development and equipment. The higher price of gold could properly be emphasized; but it was certain to be over-exploited by the promoters and their salesmen. Hence in the typical case the unwary buyer was persuaded to pay perhaps $5 for each $1 of value fairly existing behind the shares. Needless to say, the public cannot possibly get its money's worth (except by a miracle—and these are rare even in the oil and mining business) when much less than half of the price it pays for these flotations finds its way into the treasury of the corporation—the rest going for promotion and sales expense.

Promotional activities are attracted also to any new industry which is in the public eye. Profits made by those first in the field, or even currently by the enterprise floated, can be given a fictitious guise of permanence and of future enhancement. Hence gross overvaluations can be made plausible enough to sell. In the liquor flotations of 1933 the degree of overvaluation depended entirely upon the conscience of the sponsors. Accordingly, the list of stock offerings showed all gradations from the thoroughly legitimate down to the almost completely fraudulent. The participation of reputable houses in new flotations of this sort, even though there was nothing unfair in their own propositions, must still be criticized because it made the undiscriminating public much more receptive to the apparently similar offerings of unscrupulous promoters.

Repercussions of Unsound Investment Banking.—The relaxation of investment bankers' standards in the late 1920s, and their use of ingenious means to enlarge their compensation, had unwholesome repercussions in the field of corporate management. Operating officials felt themselves entitled not only to handsome salaries but also to a substantial participation in the profits of the enterprise. In this respect the investment-trust arrangements, devised by the banking houses for their own benefit, set a stimulating example to the world of "big business."

Whether or not it is proper for executives of a large and prosperous concern to receive annual compensation running into hundreds of thousands or even millions of dollars, is perhaps an open question. Its answer will depend upon the extent to which the corporation's success is due to their unique or surpassing ability, and this must be very difficult to determine with assurance. But it may not be denied that devious and questionable means were frequently employed to secure these large bonuses to the management without full disclosure of their extent to the stockholders. Stock-option warrants (or long-term subscription rights) to buy shares at low prices, proved an excellent instrument for this purpose—as we have already pointed out in our discussion of stockholder-management relationships. In this field complete and continued publicity is not only theoretically desirable but of practical utility as well. If managerial compensation is fully disclosed,[1] and adequately analyzed, public opinion may be relied upon fairly well to prevent it from passing all reasonable limits.

[1] A report of the Federal Trade Commission to the United States Senate in February 1934 sets forth data regarding the salaries and bonuses paid officers of many large corporations in 1929–1932.

CHAPTER XLVIII

SOME ASPECTS OF CORPORATE PYRAMIDING

Pyramiding in corporate finance is the creation of a speculative capital structure by means of a holding company or a series of holding companies. Usually the predominating purpose of such an arrangement is to enable the organizers to control a large business with the investment of little or no capital, and also to secure to themselves the major part of its surplus profits and increased going-concern value. The device is most often utilized by dominant interests to "cash in" speculative profits on their holdings and at the same time to retain control. With the funds so provided, these successful captains of finance generally endeavor to extend their control over additional operating enterprises. The technique of pyramiding is well illustrated by the successive maneuvers of O. P. and M. J. Van Sweringen, which started with purchase of control of the then relatively unimportant New York, Chicago, and St. Louis Railroad and rapidly developed into a far-flung railroad "empire."[1]

Example: The Van Sweringen Pyramid.—The original transaction of the Van Sweringens in the railroad field took place in 1916. It consisted of the purchase from the New York Central Railroad Company, for the sum of $8,500,000, of common and preferred stock constituting control of the New York, Chicago, and St. Louis Railroad Company (known as the "Nickel Plate").

[1] The complete story of how this pyramiding was effected is told in the *Hearings before the Committee on Banking and Currency, United States Senate*, 73d Congress, 1st Session, on Senate Resolution 84 of the 72d Congress and Senate Resolution 56 of the 73d Congress, Part 2, pp. 563-777, June 5 to 8, 1933—on "Stock Exchange Practices." The story is also set forth in greater detail and with graphic portrayal in *Regulation of Stock Ownership in Railroads*, Part 2, pp. 820-1173 (House Report No. 2789, 71st Congress, 3d Session), especially the inserts at p. 878 thereof.

The most notorious pyramided structure of recent years was the Insull set-up. An interesting example of a different type is presented by the U.S. and Foreign Securities Corporation-U.S. and International Securities Corporation relationship. These two situations are briefly described in Note 45 of the Appendix.

This purchase was financed by giving a note to the seller for $6,500,000, and by a cash payment of $2,000,000, which in turn was borrowed from a Cleveland bank. Subsequent acquisitions of control of many other companies were effected by various means, including the following:

1. The formation of a private corporation for the purpose (*e.g.*, Western Corporation to acquire control of Lake Erie and Western Railroad Company, and Clover Leaf Corporation to acquire control of Toledo, St. Louis and Western Railroad Company—both in 1922).

2. The use of the resources of one controlled railroad to acquire control of others (*e.g.*, the New York, Chicago and St. Louis Railroad Company purchased large amounts of stock of Chesapeake and Ohio Railway and Pere Marquette Railway Company during 1923–1925).

3. The formation of a holding company to control an individual road, with sale of the holding company's securities to the public (*e.g.*, Chesapeake Corporation, which took over control of Chesapeake and Ohio Railway Company and sold its own bonds and stock to the public, in 1927).

4. Formation of a general holding company (*e.g.*, Alleghany Corporation, chartered in 1929. This ambitious project took over control of many railroad, coal, and miscellaneous enterprises).

The report on the "Van Sweringen Holding Companies" made to the House of Representatives in 1930[1] includes an interesting chart showing the contrast between the control exercised by the Van Sweringens and their relatively small equity or financial interest in the capital of the enterprises controlled. On page 565 we append a summary of these data. The figures in Column *A* show the percentage of voting securities held or controlled by the Van Sweringens; the figures in Column *B* show the proportion of the "contributed capital" (bonds, stock, and surplus) actually owned directly or indirectly by them.

It is worth recalling that similar use of the holding company for pyramiding control of railroad properties had been made before the war—notably in the case of the Rock Island Company. This enterprise was organized in 1902. Through an intermediate subsidiary it acquired nearly all the common stock of the Chicago, Rock Island and Pacific Railway Company and about 60% of the capital stock of the St. Louis and San Francisco Railway Com-

[1] House Report 2789, 71st Congress, 3d Session, Part 2, pp. 820–1173.

pany. Against these shares the two holding companies issued large amounts of collateral trust bonds, preferred stock, and common stock. In 1909 the stock of the St. Louis and San Francisco was sold. In 1915 the Rock Island Company and its intermediate subsidiary both went into bankruptcy; the stock of the operating company was taken over by the collateral trust bondholders; and the holding company stock issues were wiped out completely.

Companies	A. Control, %	B. Equity, %
Holding companies:		
The Vaness Co............................	80.0	27.7
General Securities Corp..................	90.0	51.8
Geneva Corp.............................	100.0	27.7
Alleghany Corp..........................	41.8	8.6
The Chesapeake Corp.....................	71.0	4.1
The Pere Marquette Corp.................	100.0	0.7
Virginia Transportation Corp............	100.0	0.8
The Pittston Co.........................	81.8	4.3
Railroad Companies:		
The New York, Chicago and St. Louis R.R. Co.............................	49.6	0.7
The Chesapeake and Ohio Railway Co....	54.4	1.0
Pere Marquette Railway Co..............	48.3	0.6
Erie Railroad Co........................	30.8	0.6
Missouri Pacific Railroad Co............	50.5	1.7
The Hocking Valley Railway Co.........	81.0	0.2
The Wheeling and Lake Erie Railway Co.	53.3	0.3
Kansas City Southern Railway Co.......	20.8	0.9

The ignominious collapse of this venture was accepted at the time as marking the end of "high finance" in the railroad field. Yet some ten years later the same unsound practices were introduced once again, but on a larger scale and with correspondingly severer losses to investors. It remains to add that the Congressional investigation of railroad holding companies instituted in 1930 had its counterpart in a similar inquiry into the finances of the Rock Island Company made by the Interstate Commerce Commission in 1914. The memory of the financial community is proverbially and distressingly short.

Evils of Corporate Pyramiding.—The pyramiding device is harmful to the security-buying public from several standpoints.

It results in the creation and sale to investors of large amounts of unsound senior securities. It produces common stocks of holding companies which are subject to deceptively rapid increases in earning power in favorable years and which are invariably made the vehicle of wild and disastrous public speculation. The possession of control by those who have no real capital investment is inequitable and makes for irresponsible and unsound managerial policies. Finally the holding company device permits of financial practices which exaggerate the indicated earnings, dividend return, or "book value," during boom times, and thus intensify speculative fervor and facilitate market manipulation. Of these four objections to corporate pyramiding, the first three are plainly evident, but the last one requires a certain amount of analytical treatment in order to present its various implications.

Overstatement of Earnings.—Holding companies can overstate their apparent earning power by valuing at an unduly high price the stock dividends they receive from subsidiaries, or by including in their income profits made from the sale of stock of subsidiary companies.

Examples: The chief asset of Central States Electric Corporation was a large block of North American Company common on which regular stock dividends were paid. Prior to the end of 1929, these stock dividends were reported as income by Central States at the current market value. As explained in our chapter on stock dividends, such market prices averaged far in excess of the value at which North American charged the stock dividends against its surplus, and also far in excess of the distributable earnings on North American common. Hence the income account of Central States Electric gave a misleading impression of the earnings accruing to the company.

A transaction of somewhat different character but of similar effect to the foregoing was disclosed by the report of American Founders Trust for 1927. In November 1927 American Founders offered its shareholders the privilege of buying about 88,400 shares of International Securities Corporation of America Class *B* Common at $16 a share. International Securities Corporation was a subsidiary of American Founders, and the latter had acquired the Class *B* stock of the former at a cash cost of $3.70 per share in 1926. American Founders reported net earnings for common stock in 1927 amounting to $1,316,488, most of which was created by its own stockholders through their purchase of shares of the subsidiary as indicated above.

Distortion of Dividend Return.—The dividend return may be distorted in the public's mind by paying periodic stock dividends with a market value exceeding the current earnings. This device has been resorted to most frequently by holding companies and was discussed in detail in an earlier chapter. People are readily persuaded also to regard the value of frequent subscription rights as equivalent to an income return on the common stock. Pyramided enterprises are prodigal with subscription rights, for they flow naturally from the succession of new acquisitions and new financing which both promote the ambitions of those in control and maintain speculative interest at fever heat—until the inevitable collapse.

The issuance of subscription rights sometimes gives the stock market an opportunity to indulge in that peculiar circular reasoning which is the joy of the manipulator and the despair of the analyst. Company *A*'s stock is apparently worth no more than 25. Speculation or pool activity has advanced it to 75. Rights are offered to buy additional shares at 25, and the rights have a market value of, say, $10 each. To the speculative fraternity these rights are practically equivalent to a special dividend of $10. It is a bonus which not only justifies the rise to 75 but warrants more optimism and a still higher price. To the analyst the whole proceeding is a delusion and a snare. Whatever value the rights command is manufactured solely out of speculator's misguided enthusiasm; yet this chimerical value is accepted as tangible income and as vindication of the enthusiasm which gave it birth. Thus, with the encouragement of the manipulator, the speculative public pulls itself up by its bootstraps to dizzier heights of irrationality.

Example: Between August 1928 and February 1929 American and Foreign Power Company common stock advanced from 33 to 138⅞, although paying no dividend. Rights were offered to the common stockholders (and other security holders) to buy second preferred stock with detached stock-purchase warrants. The offering of these rights, which had an initial market value of about $3 each, was construed by many as the equivalent of a dividend on the common stock.

Exaggeration of Book Value.—The exaggeration of book value may be effected in cases where a holding company owns most of the shares of a subsidiary, and where consequently an artificially high quotation may readily be established for the subsidiary issue by manipulating the small amount of stock remaining in the market. This high quotation is then taken as the basis of figuring the book value (sometimes called the "break-up value") of the share of the holding company. For an early example of these practices we may point to Tobacco Products Corporation (Va.) which owned about 80% of the common stock of United Cigar Stores Company of America. An unduly high market price seems to have been established in 1927 for the small amount of Cigar Stores stock available in the market and this high price was used to make Tobacco Products shares appear attractive to the unwary buyer. The thoroughly objectionable accounting and stock dividend policies of United Cigar Stores, which we have previously discussed, were adjuncts to this manipulative campaign.

The most extraordinary example of such exaggeration of the book value is found perhaps in the case of Electric Bond and Share Company and was founded on its ownership of most of the American and Foreign Power Company warrants. The whole set-up seems to have been contrived to induce the public to pay absolutely fantastic prices without their complete absurdity being too apparent. A brief review of the various steps in this phantasmagoria of inflated values should be illuminating to the student of security analysis.

First, American and Foreign Power Company issued in all 1,600,000 shares of common and warrants to buy 7,100,000 more shares at $25. This permitted a price to be established for the common stock which generously capitalized its earnings and prospects, but paid no attention to the existence of the warrants. The quotation of the common was aided by the issuance of rights, as explained above.

Second, the high price registered for the relatively small common-stock issue automatically created a correspondingly high value for the millions of warrants.

Third, Electric Bond and Share could apply these high values to its large holdings of American and Foreign Power common and its enormous block of warrants, thus setting up a correspondingly inflated value for its own common stock.

Exploitation of the Stock-purchase-warrant Device.—The result of this process, at its farthest point in 1929, was almost incredible.

The earnings available for American and Foreign Power common stock had shown the following rising trend (due in good part, however, to continuous new acquisitions):

Year	Earnings for common	Number shares	Earned per share
1926	$ 216,000	1,243,988	0.17
1927	856,000	1,244,388	0.69
1928	1,528,000	1,248,930	1.22
1929	6,510,000	1,624,357	4.01

On the theory that a "good public-utility stock is worth up to 50 times its current earnings," a price of 199¼ per share was recorded for American and Foreign Power common. This produced in turn a price of 174 for the warrants. Hence, by the insane magic of Wall Street, earnings of $6,500,000 were transmuted into a market value of $320,000,000 for the common shares and $1,240,000,000 for the warrants, a staggering total of $1,560,000,000.

Since over 80% of the warrants were owned by Electric Bond and Share Company, the effect of these absurd prices for American and Foreign Power junior securities was to establish a correspondingly absurd "break-up" value for Electric Bond and Share common. This "break-up" value was industriously exploited to justify higher and higher quotations for the latter issue. In March 1929 attention was called to the fact that the market value of this company's portfolio was equivalent to about $108 per share (of new stock), against a range of 91 to 97 for its own market quotation. The implication was that Electric Bond and Share stock was "undervalued." In September 1929 the price had advanced to 184½. It was then computed that the "break-up value" amounted to about 150, "allowing no value for the company's supervisory and construction business." The public did not stop to reflect that a considerable part of this "book value" was based upon an essentially fictitious market quotation for an asset which the company had received *for nothing* only a few years before (as a bonus with American and Foreign Power Second Preferred stock).

This exploitation of the warrants had a peculiar vitality which made itself felt even in the depth of the depression in 1932–1933.

Time having brought its usual revenge, the once dazzling American and Foreign Power Company had trembled on the brink of receivership, as shown by a price of only 15¼ for its 5% bonds. Nevertheless, in November 1933 the highly unsubstantial warrants still commanded an aggregate market quotation of nearly $50,000,000, a figure which bore a ridiculous relationship to the exceedingly low values placed upon the senior securities. The following table shows how absurd this situation was, the more so since it existed in a time of deflated stock prices, when relative values are presumably subjected to more critical appraisal.

Issue	Amount out- standing	Price Nov. 1933	Total market value
5% Debentures......................	$50,000,000	40	$20,000,000
$7 First Preferred...............shares	480,000	21	10,100,000
$6 First Preferred...............shares	387,000	15	5,800,000
$7 Second Preferred.............shares	2,655,000	12	31,900,000
Common......................shares	1,850,000	10	18,500,000
Warrants.....................shares	6,874,000	7	48,100,000

Some Holding Companies Not Guilty of Excessive Pyramiding.—To avoid creating a false impression, we must point out that while pyramiding is usually effected by means of holding companies, it does not follow that all holding companies are created for this purpose and are therefore reprehensible. The holding company is often utilized for entirely legitimate purposes, *e.g.*, to permit unified and economical operations of separate units, to diversify investment and risk, and to gain certain technical advantages of flexibility and convenience. Many sound and important enterprises are in holding company form.

Examples: United States Steel Corporation is entirely a holding company; while originally there was some element of pyramiding in its capital set-up, this defect disappeared in later years. American Telephone and Telegraph Company is preponderantly a holding company, but its financial structure has never been subject to serious criticism. General Motors Corporation is largely a holding company. A holding-company exhibit must therefore be considered on its merits. American Light and

Traction Company is a typical example of the holding company organized entirely for legitimate purposes. On the other hand the acquisition of control of this enterprise by United Light and Railways Company (Del.) must be regarded as a pyramiding move on the part of the United Light and Power interests.

Speculative Capital Structure May Be Created in Other Ways.—It may be pointed out also that a speculative capital structure can be created without the use of a holding company.

Examples: The Maytag Company recapitalization, discussed in an earlier chapter, yielded results usually attained by the formation of a holding company and the sale of its senior securities. In the case of Continental Baking Corporation—to cite another example—the holding company form was not an essential part of the pyramided result there attained. The speculative structure was due entirely to the creation of large preferred issues by the parent company, and it would still have existed if Continental Baking had acquired all its properties directly, eliminating its subsidiaries.

CHAPTER XLIX

COMPARATIVE ANALYSIS OF COMPANIES IN THE SAME FIELD

Statistical comparisons of groups of concerns operating in a given industry are more or less a routine part of the analyst's work. Such tabulations permit each company's showing to be studied against a background of the industry as a whole. They frequently bring to light instances of undervaluation or overvaluation, or lead to the conclusion that the securities of one enterprise should be replaced by those of another in the same field.

In this chapter we shall suggest standard forms for such comparative analyses, and we shall also discuss the significance of the various items included therein. Needless to say, these forms are called "standard" only in the sense that they can be used generally to good advantage; no claim of perfection is made for them, and the student is free to make any changes which he thinks will serve his particular purpose.

FORM I. RAILROAD COMPARISON

A. Capitalization:
  1. Net deductions.
  2. Effective debt (net deductions multiplied by 20).
  3. Preferred stock at market (number of shares × market price).
  4. Common stock at market (number of shares × market price).
  5. Total capitalization.
  6. Ratio of effective debt to total capitalization.
  7. Ratio of preferred stock to total capitalization.
  8. Ratio of common stock to total capitalization.

B. Income Account:
  9. Gross revenues.
  10. Ratio of maintenance to gross.
  11. Ratio of railway operating income (net after taxes) to gross.
  12. Ratio of net deductions to gross.
  13. Ratio of preferred dividends to gross.
  14. Ratio of balance for common to gross.

572

C. Calculations:

    15. Number of times net deductions earned

    15. I.P. Number of times net deductions plus preferred dividends earned.

    16. Earned on common stock, per share.

    17. Earned on common stock, % of market price.

    18. Ratio of gross to aggregate market value of common stock (9 ÷ 4).

    16. S.P. Earned on preferred stock, per share.

    17. S.P. Earned on preferred stock, % of market price.

    18. S.P. Ratio of gross to aggregate market value of preferred stock (9 ÷ 3).

    19. Credit or debit to earnings for undistributed profit or loss of subsidiaries.

D. Seven-year average figures:

    20. Earned on common stock, per share.

    21. Earned on common stock, % of current market price of common.

    20. S.P. Earned on preferred stock, per share.

    21. S.P. Earned on preferred stock, % of current market price of preferred.

    22. Number of times net deductions earned.

    23. Number of times fixed charges earned.

    22. I.P. Number of times net deductions plus preferred dividends earned.

    23. I.P. Number of times fixed charges plus preferred dividends earned.

E. Trend figure:

    24 to 30. Earned per share on common stock each year for past seven years. (Where necessary, earnings should be adjusted to present capitalization.)

    24. S.P. to 30. S.P. Same data for speculative preferred stock, if wanted.

F. Dividends:

    31. Dividend rate on common.

    32. Dividend yield on common.

    31. P. Dividend rate on preferred.

    32. P. Dividend yield on preferred.

Observations on the Railroad Comparison.[1]—The earnings figures used would ordinarily be those of the previous calendar year. Computations may be available in the statistical manuals which will simplify the analyst's work. If more up-to-date figures are desired it is feasible, by using the monthly reports, to calculate twelve-month totals ending just prior to the date of the comparison. If the net deductions are

---

[1] Reference is made to earlier chapters for explanation of the terminology and the critical tests referred to in this discussion.

not reported, they may be closely estimated from the figures of the last calendar year.

Our table includes a few significant calculations based on the seven-year average. In an intensive study average results should be scrutinized in more detail. To save time, it is suggested that additional average figures be computed only for those roads which the analyst selects for further investigation after he has studied the exhibits in the "standard form."

Figures relating to preferred stocks fall into two different classes, depending on whether the issue is considered for fixed-value investment or as a speculative commitment. (Usually the market price will indicate clearly enough in which category a particular issue belongs.) The items marked "I.P." are to be used in studying an investment preferred stock, and those marked "S.P." in studying a speculative preferred. Where there are junior income bonds, the simplest and most satisfactory procedure will be to treat them in all respects as a preferred stock issue, with a footnote referring to their actual title. Such contingent bond interest will therefore be excluded from the net deductions or the fixed charges.

For convenience we use the "net-deductions basis" for stating the true debt. The seven-year coverage of fixed charges is also listed, so that both guides may be available for determining whether the bond issues (or preferred shares) are adequately protected. In using the table as an aid to the selection of senior issues for investment, chief attention will be paid to items 22 and 23 (or 22 "I.P." and 23 "I.P."), showing the average margin above interest (and preferred dividend) requirements. Consideration should be given also to items 6, 7, and 8, showing the division of total capitalization between senior securities and junior equity. (In dealing with bonds, the preferred stock is part of the junior equity; in considering a preferred stock for investment, it must be included with the effective debt.) Items 10 and 19 should also be examined to see if the earnings have been overstated by reason of inadequate maintenance or by the inclusion of unearned dividends from subsidiaries.

Speculative preferred stocks will ordinarily be analyzed in much the same way as common stocks, and the similarity becomes greater as the price of the preferred stock is lower. It should be remembered, however, that a preferred stock is always less attractive, logically considered, than a common stock making the same showing. For example, a 6% preferred earning $5 per share is intrinsically less desirable than a common stock earning $5 per share, since the latter is entitled to all the present and future equity, while the preferred stock is strictly limited in its claim upon the future.

In comparing railroad common stocks (and preferred shares equivalent thereto), the point of departure is the percentage earned on the market price. This may be qualified, to an extent more or less important, by consideration of items 10 and 19. Items 12 and 18 will indicate at once whether the company is speculatively or conservatively capitalized, relatively speaking. A speculatively capitalized road will show a large ratio of net deductions to gross, and (ordinarily) a small ratio of common stock at market value to gross. The converse will be true for a conservatively capitalized road.

Limitation upon Comparison of Speculatively and Conservatively Capitalized Companies in the Same Field.—The analyst must beware of trying to draw conclusions as to the relative attractiveness of two railroad common stocks when one is speculatively and the other is conservatively capitalized. Two such issues will respond quite differently to changes for the better or the worse, so that an advantage possessed by one of them under current conditions may readily be lost if conditions should change.

Example: The example shown on page 576 illustrates in a twofold fashion the fallacy of comparing a conservatively capitalized with a speculatively capitalized common stock. In 1922 the earnings of Union Pacific common were nearly four times as high in relation to market price as were those of Rock Island common. A conclusion that Union Pacific was "cheaper," based on these figures, would have been fallacious, because the relative capitalization structures were so different as to make the two companies noncomparable. This fact is shown graphically by the much larger expansion of the earnings and the market price of Rock Island common which accompanied the moderate rise in gross business during the five years following.

The situation in 1927 was substantially the opposite. At that time Rock Island common was earning proportionately more than Union Pacific common. But it would have been equally fallacious to conclude that Rock Island common was "intrinsically cheaper." The speculative capitalization structure of the latter road made it highly vulnerable to unfavorable developments, so that it was unable to withstand the post-1929 depression.

COMPARISON OF UNION PACIFIC AND ROCK ISLAND COMMON STOCKS

Item	Union Pacific R.R.	Chicago, Rock Island, & Pacific Ry.
A. Showing the effect of general improvement:		
Average price of common, 1922.............	140	40
Earned per share, 1922....................	$12.76	$0.96
% earned on market price, 1922...........	9.1%	2.4%
Fixed charges and preferred dividends earned, 1922.................................	2.39 times	1.05 times
Ratio of gross to market value of common, 1922.................................	62%	419%
Increase in gross, 1927 over 1922...........	5.7%	12.9%
Earned per share of common, 1927..........	$16.05	$12.08
Increase in earnings on common, 1927 over 1922.................................	26%	1,158%
Average price of common, 1927.............	179	92
Increase in average price, 1927 over 1922....	28%	130%
B. Showing the effect of a general decline in business:		
Earned on average price, 1927.............	9.0%	13.1%
Fixed charges and preferred dividends earned, 1927.................................	2.64 times	1.58 times
Ratio of gross to market value of common, 1927.................................	51%	204%
Decrease in gross, 1933 below 1927.........	46%	54%
Earned on common, 1933..................	$7.88	$20.40(d)
Decrease in earnings for common, 1933 below 1927.................................	51%	269%
Average price of common, 1933.............	97	6
Decrease in average price, 1933 below 1927...	46%	93%

NOTE: In June 1933 trustees in bankruptcy were appointed for the Rock Island.

Other Illustrations in Appendix.—The practical approach to comparative analysis of railroad stocks (and bonds) may best be illustrated by the reproduction of several such comparisons made by one of the authors a number of years ago, and published as part of the service rendered to clients by a New York Stock Exchange firm. These will be found in the Appendix, Note 46. It will be observed that the comparisons were made between roads in approximately the same class as regards capitalization structure, with the exception of the comparison between Atchison and New York Central, in which instance special reference was made to the greater sensitivity of New York Central to changes in either direction.

FORM II. PUBLIC-UTILITY COMPARISON

The public-utility comparison form is practically the same as that for railroads. The only changes are the following: "fixed charges" should be substituted for "net deductions," in items 1, 2, 12 and 15 and items 22 and 22 I.P. should be deleted. The effective debt is found by multiplying the fixed charges by 18. Item 10 becomes "ratio of depreciation to gross." An item, 10M, may be included to show "ratio of maintenance to gross" for the companies which publish this information. Item 19 may be omitted, since information on this point is seldom available.

Our observations regarding the use of the railroad comparison apply as well to the public-utility comparison. Variations in the depreciation rate are fully as important as variations in the railroad maintenance ratios. When a wide difference appears, it should not be taken for granted that one property is unduly conservative or the other not conservative enough; but a *presumption* to this effect does arise, and the question should be investigated as thoroughly as possible. A statistical indication that one utility stock is more attractive than another should not be acted upon until (among other qualitative matters) some study has been made of the rate situation and the relative prospects for favorable or unfavorable changes therein.

FORM III. INDUSTRIAL COMPARISON (FOR COMPANIES IN THE SAME FIELD)

Since this form differs in numerous respects from the two preceding, it is given in full herewith:

A. Capitalization:
   1. Bonds at par.
   2. Preferred stock at market value (number of shares × market price).
   3. Common stock at market value (number of shares × market price).
   4. Total capitalization.
   5. Ratio of bonds to capitalization.
   6. Ratio of aggregate market value of preferred to capitalization.
   7. Ratio of aggregate market value of common to capitalization.

B. Income Account (most recent year).
  8. Gross sales.
  9. Depreciation.
  10. Net available for bond interest.
  11. Bond interest.
  12. Preferred dividend requirements.
  13. Balance for common.
  14. Margin of profit (ratio of 10 to 8).
  15. % earned on total capitalization (ratio of 10 to 4).

C. Calculations.
  16. Number of times interest charges earned.
  16. I.P. Number of times interest charges plus preferred dividends earned.
  17. Earned on common, per share.
  18. Earned on common, % of market price.
  17. S.P. Earned on preferred, per share.
  18. S.P. Earned on preferred, % of market price.
  19. Ratio of gross to aggregate market value of common.
  19. S.P. Ratio of gross to aggregate market value of preferred.

D. Seven-year average:
  20. Number of times interest charges earned.
  21. Earned on common stock per share.
  22. Earned on common stock, % of current market price.
  (20 I.P., 21 S.P. and 22 S.P.—Same calculation for preferred stock if wanted).

E. Trend figure:
  23. Earned per share of common stock each year for past seven years (adjustments in number of shares outstanding to be made where necessary).
  23. S.P. Same data for speculative preferred issues, if wanted.

F. Dividends:
  24. Dividend rate on common.
  25. Dividend yield on common.
  24. P. Dividend rate on preferred.
  25. P. Dividend yield on preferred.

G. Balance sheet:
  26. Cash assets.
  27. Receivables (less reserves).
  28. Inventories (less proper reserves).
  29. Total current assets.
  30. Total current liabilities.
  30. N. Notes Payable (Including "Bank Loans" and "Bills Payable").
  31. Net current assets.
  32. Ratio of current assets to current liabilities.
  33. Net tangible assets available for total capitalization.

34. Cash-asset-value of common per share (deducting all prior obligations).
35. Net-current-asset-value of common per share (deducting all prior obligations).
36. Net-tangible-asset-value of common per share (deducting all prior obligations).
(34 S.P., 35 S.P., 36 S.P.—Same data for speculative preferred issues, if wanted).

H. Supplementary data (when available):
1. Physical output:
   Number of units; receipts per unit; cost per unit; profit per unit; total capitalization per unit; common stock valuation per unit.
2. Miscellaneous:
   For example: number of stores operated; sales per store; profit per store; ore reserves; life of mine at current (or average) rate of production.

Observations on the Industrial Comparison.—Some remarks regarding the use of this suggested form may be helpful. The net earnings figure must be corrected for any known distortions or omissions. If it appears to be misleading and cannot be adequately corrected, it should not be used as a basis of comparisons. (Inferences drawn from unreliable figures must themselves be unreliable.) No attempt should be made to subject the depreciation figures to exact comparisons; they are useful only in disclosing wide and obvious disparities in the rates used. The calculation of bond-interest-coverage is subject to the qualification, discussed in Chap. XVII, with respect to companies which may have important rental obligations equivalent to interest charges.

While the percentage earned on the market price of the common (item 18) is a leading figure in all comparisons, almost equal attention must be given to item 15, showing the percentage earned on total capitalization. These figures, together with items 7 and 19 (ratio of aggregate market value of common stock to sales and to capitalization), will indicate the part played by conservative or speculative capitalization structures among the companies compared. (The theory of capitalization structure was considered in Chap. XL.)

As a matter of practical procedure it is not safe to rely upon the fact that the earnings ratio for the common stock (item 18) is higher than the average for the industry, unless the percentage earned on the total capitalization (item 15) is also higher. Furthermore, if the company with the poorer earnings exhibit shows much larger sales-per-dollar-of-common-stock (item 19), it may have better speculative possibilities in the event of general business improvement.

The balance-sheet computations do not have primary significance unless they indicate either definite financial weakness or a substantial excess of quick-asset-value over the market price. The division of importance as between the current results, the seven-year average, and the trend, is something entirely for the analyst's judgment to decide. Naturally, he will have the more confidence in any suggested conclusion if it is confirmed on each of these counts.

Example of the Use of Standard Forms.—An example of the use of the standard form to reach a conclusion concerning comparative values should be of interest. A survey of the common stocks of the leading steel producers in July 1927 indicated that Youngstown Sheet and Tube Company common made a better showing and Colorado Fuel and Iron Company common made a poorer showing than the average. In the tabulation shown on page 581 we supply the comparative figures for these two companies. (A few of the items in the standard form are omitted as immaterial to this analysis.)

Comments on This Comparison.—This comparison revealed only one point of superiority for Colorado Fuel and Iron, *viz.*, the larger earnings for the current six-months period. The other figures favored Youngstown Sheet and Tube so strongly as to leave no doubt as to the relative merits of the two common issues at their respective prices. The exhibit for 1926, and the average figures for four years, showed a statistical advantage for Youngstown on each of the following important points:

Earnings on common stock.
Earnings on total capitalization.
Interest coverage.
Ratio of gross to common.
Margin of profit.
Adequacy of depreciation charges.
Working capital position.
Dividend return.

It was obvious that Youngstown common had greater investment merit than Colorado Fuel and Iron common. But from the speculative standpoint as well, Youngstown appeared the more attractive issue, since each dollar paid for the common stock

(000 omitted)

Item	Youngs-town	Colorado Fuel & Iron
Market price of common, July, 1927..........	84	92
1. Bonds at par........................	$ 67,200	$33,945
2. Preferred stock at market.............	15,665	2,600
3. Common stock at market..............	82,908	31,326
4. Total capitalization..................	165,773	67,871
7. Ratio of common to total capitalization..	50.1%	46.2%
Calendar year 1926:		
8. Gross sales........................	$150,023	$35,758
9. Depreciation........................	9,167	1,860
10. Net available for bond interest..........	19,439	4,556
11. Bond interest.......................	4,391	1,808
12. Preferred dividends...................	997	160
13. Balance for common..................	14,151	2,588
14. Margin of profit.....................	13.0%	12.7%
15. % earned on total capitalization.........	11.8%	6.7%
16. Interest charges earned...............	4.54 times	2.50 times
17. Earned on common, per share..........	$14.33	$7.60
18. Earned on common, % of market price...	17.06%	8.26%
19. Ratio of gross to market value of common	161%	114%
Four-year average:		
20. Interest charges earned................	4.1 times	1.8 times
21. Earned on common, per share..........	$12.08	$3.74
22. Earned on common, % of market price...	14.4%	4.1%
Trend figures:		
23. Earned per share by years:		
1927 (first six months)................	$ 3.87	$6.98
1926.............................	14.33	7.60
1925.............................	12.38	4.65
1924.............................	6.68	1.05
1923.............................	14.94	1.67
Dividends:		
24. Dividend rate on common..............	$ 4.00	None
25. Dividend yield on common.............	4.8%	None
Financial position:		
29. Total current assets...................	$80,381	$16,020
30. Total current liabilities...............	16,640	8,004
31. Net current assets....................	63,741	8,016
33. Net tangible assets for total capitalization	199,079	71,888

commanded an interest in a substantially higher volume of sales than in the case of the other company.

Study of Qualitative Factors Also Necessary.—Conclusions suggested by comparative tabulations of this sort should not be accepted until careful thought has been given to the qualitative factors. When one issue seems to be selling much too low on the basis of the exhibit in relation to that of another in the same field, there may be adequate reasons for this disparity which the statistics do not disclose. Among such valid reasons may be a definitely poorer outlook or a questionable management. A lower dividend return for a common stock should not ordinarily be considered as a strong offsetting factor, since the dividend is usually adjusted to the earning power within a reasonable time. While overconservative dividend policies are sometimes followed for a considerable period (a subject referred to in Chap. XXIX), there is a well-defined tendency even in these cases for the market price to reflect the earning power sooner or later.

Relative popularity and relative market activity are two elements not connected with intrinsic value which nevertheless exert a powerful and often a continuing effect upon the market quotation. The analyst must give these factors respectful heed, but his work would be stultified if he always favored the more active and the more popular issue.

The recommendation of an exchange of one security for another seems to involve a greater personal accountability on the part of the analyst than the selection of an issue for original purchase. The reason is that holders of securities for investment are loath to make changes, and thus they are particularly irritated if the subsequent market action makes the move appear to have been unwise. Speculative holders will naturally gauge all advice by the test of market results—usually immediate results. Bearing these human-nature factors in mind, the analyst must avoid suggesting common-stock exchanges to speculators (except possibly if accompanied by an emphatic disclaimer of responsibility for subsequent market action); and he must hesitate to suggest such exchanges to holders for investment unless the statistical superiority of the issue recommended is quite impressive. As an arbitrary rule, we might say that there should be good reason to believe that by making the exchange the investor would be getting at least 50% more for his money.

Variations in Homogeneity Affect the Values of Comparative Analysis.—The dependability of industrial comparisons will vary with the nature of the industry considered. The basic question, of course, is whether future developments are likely to affect all the companies in the group similarly or dissimilarly. If similarly, then substantial weight may be accorded to the relative performance in the past, as shown by the statistical exhibit. An industrial group of this type may be called "homogeneous." But if the individual companies in the field are likely to respond quite variously to new conditions, then the relative showing must be regarded as a much less reliable guide. A group of this kind may be termed "heterogeneous."

The railroads are a highly "homogeneous" group, despite their geographical variations. The same is true of the larger light, heat and power utilities. In the industrial field the best examples of homogeneous groups are afforded by the producers of raw materials and of other standardized products in which the trade name is a minor factor. These would include producers of sugar, coal, metals, steel products, cement, cotton print cloths, etc. The larger oil companies may be considered as fairly homogeneous; the smaller concerns are not well suited to comparison because they are subject to sudden important changes in production, reserves, and relative price received. The larger baking, dairy, and packing companies fall into fairly homogeneous groups. The same is true of the larger chain-store enterprises when compared with other units in the same subgroups, *e.g.*, grocery, five-and-ten-cent, restaurant, etc. Department stores are less homogeneous, but comparisons in this field are by no means far-fetched.

Makers of manufactured goods sold under advertised trademarks must generally be regarded as belonging to heterogeneous groups. In these fields one concern frequently prospers at the expense of its competitors, so that the units in the industry do not improve or decline together. Among automobile manufacturers, for example, there have been continuous and pronounced variations in relative standing. Producers of all the various classes of machinery and equipment are subject to somewhat the same conditions. This is true also of the proprietary drug manufacturers. Intermediate positions from this point of view are occupied by such groups as the larger makers of tires, of tobacco products, of shoes, etc., wherein changes of relative position are not so frequent.

The analyst must be most cautious about drawing comparative conclusions from the statistical data when dealing with companies in a heterogeneous group. No doubt preference may properly be accorded in these fields to the companies making the best quantitative showing (if not offset by known qualitative factors)—for this basis of selection would seem sounder than any other—but the analyst and the investor should be fully aware that such superiority may prove evanescent. As a general rule, the less homogeneous the group, the more attention must be paid to the qualitative factors in making comparisons.

CHAPTER L

DISCREPANCIES BETWEEN PRICE AND VALUE

Our exposition of the technique of security analysis has included many different examples of overvaluation and undervaluation. Evidently the processes by which the securities market arrives at its appraisals are frequently illogical and erroneous. These processes, as we pointed out in our first chapter, are not automatic or mechanical, but psychological, for they go on in the minds of people who buy or sell. The mistakes of the market are thus the mistakes of groups or masses of individuals. Most of them can be traced to one or more of three basic causes: exaggeration, oversimplification, or neglect.

In this chapter and the next we shall attempt a concise review of the various aberrations of the securities market. We shall approach the subject from the standpoint of the practical activities of the analyst, seeking in each case to determine the extent to which it offers an opportunity for profitable activity on his part. This inquiry will thus constitute an amplification of our early chapter on the scope and limitations of security analysis, drawing upon the material developed in the succeeding discussions, to which a number of references will be made.

The best understood disparities between price and value are those which accompany the recurrent broad swings of the market through boom and depression. It is a mere truism that stocks sell too high in a bull market and too low in a bear market. For at bottom this is simply equivalent to saying that any upward or downward movement of prices must finally reach a limit, and since prices do not remain at such limits (or at any other level) permanently, it must turn out retrospectively that prices will have advanced or declined too far.

Can Cyclical Swings of Prices Be Exploited?—Can the analyst exploit successfully the repeated exaggerations of the general market? Experience suggests that a procedure somewhat like the following should turn out to be reasonably satisfactory:

585

1. Select a diversified list of leading common stocks.

2. Make composite purchases of this list when the shares can be bought at substantially less than ten times average earnings of the past seven years—say at seven times such earnings.

3. Sell out such purchases when a price is reached substantially higher than ten times the seven-year average, say at fifteen times such earnings.

This was the general scheme of operations developed by Roger Babson many years ago. It yielded quite satisfactory results prior to 1925. But—as we pointed out in Chap. XXXVII— during the 1921–1933 cycle (measuring from low point to low point), it would have called for purchasing during 1921, selling out probably in 1925, thus requiring complete abstinence from the market during the great boom of 1927–1929; and repurchasing in 1931, to be followed by a severe shrinkage in market values. A program of this character would have made far too heavy demands upon human fortitude. But assuming that future market cycles will be neither so wide nor so long as was this extraordinary one (apparently a reasonable expectation), the general plan of "buying cheap and selling dear" may prove quite workable.

"Catching the Swings" on a Marginal Basis Impracticable.— From the ordinary speculative standpoint, involving purchases on margin and short sales, this method of operation must be set down as impracticable. The outright owner can afford to buy too soon and to sell too soon. In fact he must expect to do both, and to see the market decline farther after he buys and advance farther after he sells out. But the margin trader is necessarily concerned with immediate results; he swims with the tide, hoping to gauge the exact moment when the tide will turn and to reverse his stroke the moment before. In this he rarely succeeds, so that his typical experience is temporary success ending in complete disaster. It is the essential character of the speculator that he buys because he thinks stocks are going up, not because they are cheap; and conversely when he sells. Hence there is a fundamental cleavage of viewpoint between the speculator and the securities analyst, which militates strongly against any enduringly satisfactory association between them.

Bond prices tend undoubtedly to swing through cycles in somewhat the same way as stocks; and it is frequently suggested that bond investors follow the policy of selling their holdings near the top of these cycles and repurchasing them near the bottom. We are doubtful if this can be done with satisfactory results in the typical case. There are no well-defined standards as to when high-grade bond prices are cheap or dear corresponding to the earnings-ratio test for common stocks, and the operations have to be guided chiefly by a technique of gauging market moves which seems rather far removed from "investment." The loss of interest on funds between the time of sale and repurchase is a strong debit factor; and in our opinion the net advantage is not sufficient to warrant incurring the psychological dangers that inhere in any placing of emphasis by the investor upon market movements.

Opportunities in "Secondary" or Little-known Issues.—Returning to common stocks, while overvaluation or undervaluation of leading issues occurs only at certain points in the stock-market cycle, the large field of "nonrepresentative" or "secondary" issues is likely to yield instances of undervaluation at all times. When the market leaders are cheap, some of the less prominent common stocks are likely to be a good deal cheaper. During 1932–1933, for example, stocks such as Plymouth Cordage, Pepperell Manufacturing, American Laundry Machinery, and many others, sold at unbelievably low prices in relation to their past records and current financial exhibits. It is probably a matter for individual preference whether the investor should purchase an outstanding issue like General Motors at about 50% of its conservative valuation or a less prominent stock like Pepperell at about 25% of such value.

During the intermediate periods when average prices show no definite signs of being either too low or too high, shrewd stock investors may favor common stocks which seem clearly undervalued, even if these lack prominence or great market activity.

Examples: The example of Wright Aeronautical in 1923, referred to in Chap. I, falls in this class. We append a few facts bearing upon two other examples of this kind:

BANGOR & AROOSTOOK R.R. COMMON		FIRESTONE TIRE & RUBBER COMMON	
Average price in 1926.......	39½	Price in Nov. 1925........	120
Dividend................	$3	Dividend...............	$6
Earned per share:		Earned per share year ended Oct.:	
1926....................	$8.69	1925..................	$32.57*
1925....................	6.22	1924..................	16.92
1924....................	6.21	1923..................	14.06
1923....................	4.55	1922..................	17.08

* Earnings before contingency reserves were $40.95 per share.

In these cases the market price had failed to reflect adequately the indicated earning power.

Market Behavior of Standard and Nonstandard Issues.—A close study of the market action of common stocks suggest the following general observations:

1. Standard or leading issues almost always respond rapidly to changes in their reported profits—so much so that they tend regularly to exaggerate marketwise the significance of year-to-year fluctuations in earnings.

2. The action of the less familiar issues depends largely upon what attitude is taken towards them by professional market operators. If interest is lacking, the price may lag far behind the statistical showing. If interest is attracted to the issue, either manipulatively or more legitimately, the opposite result can readily be attained and the price will respond in extreme fashion to changes in the company's exhibit.

Examples of Behavior of Nonstandard Issues.—The following two examples will illustrate this diversity of behavior of non-representative common stocks.

BUTTE AND SUPERIOR COPPER (ACTUALLY ZINC) COMPANY COMMON

Period	Earnings per share	Dividend per share	Price range
Year, 1914......................	$ 5.21	$ 2.25	44–24
1st quarter, 1915..............	4.27	0.75	50–36
2d quarter, 1915..............	7.73	3.25	80–45
3d quarter, 1915..............	10.13	5.75	73–57
4th quarter, 1915..............	11.34	8.25	75–59
Year 1915...................	$33.47	$18.00	80–36
Year 1916...................	30.58	34.00	105–42

These were extraordinarily large earnings and dividends. Even allowing for the fact that they were due to wartime prices for zinc, the market price showed none the less a striking disregard of the company's spectacular exhibit. The reason was lack of general interest or of individual market sponsorship.

Contrast the foregoing with the appended showing of the common stock of Mullins Body (later Mullins Manufacturing) Corporation.

Year	Earned per share	Dividend	Price range
1924	$1.91	None	18–9
1925	2.47	None	22–13
1926	1.97	None	20–8
1927	5.13	None	79–10
1928	6.53	None	95–69
1929	2.67	None	82–10

Between 1924 and 1926 we note the characteristic market swings of a low-priced "secondary" common-stock issue. At the beginning of 1927 the shares were undoubtedly attractive, speculatively, at about 10; for the price was low in relation to the earnings of the three years previously. A substantial, but by no means spectacular, rise in profits during 1927–1928 resulted in a typical stock-market exploitation. The price advanced from 10 in 1927 to 95 in 1928, and fell back again to 10 in 1929.

Relationship of the Analyst to Such Situations.—The analyst can deal intelligently and fairly successfully with situations such as Wright Aeronautical, Bangor and Aroostook, Firestone, and Butte and Superior at the periods referred to. He could even have formed a worthwhile opinion about Mullins early in 1927. But once this issue fell into market operators' hands it passed beyond the pale of analytical judgment. As far as Wall Street was concerned, Mullins had ceased to be a business and had become a symbol on the ticker tape. To buy it or to sell it was equally hazardous; the analyst could warn of the hazard but he could have no idea of the limits of its rise or fall. (As it happened, however, the company issued a convertible preferred stock in 1928 which made possible a profitable hedging operation, consisting of the purchase of the preferred and the sale of the common.)

When the general market appears dangerously high to the analyst, he must be hesitant about recommending unfamiliar common stocks, even though they may seem to be of the bargain type. A severe decline in the general market will affect all stock prices adversely, and the less active issues may prove especially vulnerable to the effects of necessitous selling.

Market Exaggerations Due to Factors Other than Changes in Earnings: *Dividend Changes.*—The inveterate tendency of the stock market to exaggerate extends to factors other than changes in earnings. Overemphasis is laid upon such matters as dividend changes, stock split-ups, mergers, segregations, and what not. An increase in the cash dividend is a favorable development, but it is absurd to add $20 to the price of a stock just because the dividend rate is advanced from $5 to $6 annually. The buyer at the higher price is paying out in advance all the additional dividends which he will receive at the new rate *over the next 20 years*. The excited responses often made to stock dividends are even more illogical, since they are in essence nothing more than pieces of paper. The same is true of split-ups, which create more shares but give the stockholder nothing he did not have before.[1]

Mergers and Segregations.—Wall Street becomes easily enthusiastic over mergers and just as ebullient over segregations, which are the exact opposite. Putting two and two together frequently produces five in the stock market; and this five may later be split up into three and three. Such inductive studies as have been made of the results following mergers seem to cast considerable doubt upon the efficacy of consolidation as an aid to earning power.[2] There is also reason to believe that the personal element in corporate management often stands in the way of really advantageous consolidations, and that those which are consummated are due sometimes to knowledge by those in control of unfavorable conditions ahead.

The exaggerated response made by the stock market to developments which seem relatively unimportant in themselves is readily explained in terms of the psychology of the speculator. He wants "action," first of all; and he is willing to contribute to this action

---

[1] In the Atlas Tack manipulation of 1933 an effort was made to attract public buying by promising a split-up of the stock, three shares for one. Obviously, such a move could make no real difference of any kind in the case of an issue selling in the 30s. The circumstances surrounding the rise of Atlas Tack from 1½ to 34¾ in 1933 and its precipitous fall to 10 are worth studying as a perfect example of the manipulative pattern. It is illuminating to compare the price-earnings and the price-assets relationships of the same stock prior to 1929.

[2] See, for example, Arthur S. Dewing, "A Statistical Test of the Success of Consolidations," published in *Quarterly Journal of Economics*, November 1921 and reprinted in his *Financial Policy of Corporations*, pp. 885–898, New York, 1926. But see Henry R. Seager and Charles A. Gulick, *Trust and Corporation Problems*, pp. 659–661, New York, 1929, and *Report of the Committee on Recent Economic Changes*, Vol. I, pp. 194ff., New York, 1929.

if he can be given any pretext for bullish excitement. (Whether through hypocrisy or self-deception, brokerage-house customers generally refuse to admit they are merely gambling with ticker quotations, and insist upon some ostensible "reason" for their purchases.) Stock dividends and other "favorable developments" of this character supply the desired pretexts, and they are fully exploited by the professional market operators, sometimes with the connivance of the corporate officials. The whole thing would be childish if it were not so vicious. The securities analyst should understand how these absurdities of Wall Street come into being, but he would do well to avoid any form of contact with them.

Litigation.—The tendency of Wall Street to go to extremes is illustrated in the opposite direction by its tremendous dislike of litigation. A lawsuit of any significance casts a damper on the securities affected, and the extent of the decline may be out of all proportion to the merits of the case. Developments of this kind may offer real opportunities to the analyst, though of course they are of a specialized nature. The aspect of broadest importance is that of receivership. Since the undervaluations resulting therefrom are almost always confined to bond issues, we shall discuss this subject later in the chapter in connection with senior securities.

Example: A rather striking example of the effect of litigation on common-stock values is afforded by the Reading Company case. In 1913, the United States government brought suit to compel separation of the company's railroad and coal properties. The stock market, having its own ideas of consistency, considered this move as a dangerous attack on Reading, despite the fact that the segregation would in itself ordinarily be considered as "bullish." A plan was later agreed upon (in 1921) under which the coal subsidiary's stock was in effect to be distributed pro rata among the Reading Company's common and preferred shareholders. This was hailed in turn as a favorable development, although in fact it constituted a victory for the government against the company.

Some common stockholders, however, objected to the participation of the preferred stock in the coal company "rights." Suit was brought to restrict these rights to the common stock. Amusingly, but not surprisingly, the effect of this move was to depress the price of Reading common. In logic, the common should have advanced, since if the suit were successful there would be more value for the junior shares, and if it failed (as it did) there would be no less value than before. But the stock market reasoned merely that here was some new litigation and hence Reading common should be "let alone."

Situations involving litigation frequently permit the analyst to pursue to advantage his quantitative approach in contrast with the qualitative attitude of security holders in general. Assume the assets of a bankrupt concern have been turned into cash and there is available for distribution to its bondholders the sum of say 50% net. But there is a suit pending, brought by others, to collect a good part of this money. It may be that the action is so far-fetched as to be almost absurd; it may be that it has been defeated in the lower courts, and even on appeal, and that it has now but a microscopic chance to be heard by the United States Supreme Court. Nevertheless, the mere pendency of this litigation will severely reduce the market value of the bonds. Under the conditions named, they are likely to sell as low as 35 instead of 50 cents on the dollar. The anomaly here is that a remote claim, which the plaintiff can regard as having scarcely any real value to him, is made the equivalent in the market to a heavy liability on the part of the defendant. We thus have a mathematically demonstrable case of undervaluations; and taking these as a class, they lend themselves exceedingly well to exploitation by the securities analyst.

Example: Island Oil and Transport 8% Notes.—In June 1933 these notes were selling at 18. The receiver held a cash fund equivalent to about 45% on the issue, from which were deductible certain fees and allowances, indicating a net distributable balance of about 30 for the notes. The distribution was being delayed by a suit for damages which had been repeatedly unsuccessful in its various legal stages and was now approaching final determination. This suit was exerting an adverse effect upon the market value of the notes out of all proportion to its merits, a statement which is demonstrable from the fact that the litigation could have been settled by payment of a relatively small amount. After the earlier decisions were finally sustained by the higher courts, the noteholders received a distribution of $290 per $1,000 in April 1934. A small additional distribution was indicated.

Undervalued Investment Issues.—In Chap. XXII we discussed the subject of undervalued bonds and preferred stocks of investment caliber, and in Chap. XXVI we considered this element in relation to speculative senior issues. Undervalued investment issues may be discovered in any period by means of assiduous search. In many cases the low price of a bond or preferred stock is due to a poor market, which in turn results from the small size of the issue; but this very small size may make for greater inherent security. The Electric Refrigeration Building Corporation 6s, due 1936, described in Chap. XXVI, are a good example of this paradox.

At times some specific development greatly strengthens the position of a senior issue; but the price is slow to reflect this improvement, and thus a bargain situation is created. These developments relate usually to the capitalization structure or to corporate relationships. Several examples will illustrate our point.

Examples: In 1923 Youngstown Sheet and Tube Company purchased the properties of Steel and Tube Company of America and assumed liability for the latter's General Mortgage 7s, due 1951. Youngstown sold a 6% debenture issue at 99 to supply funds for this purchase. The following price relationship obtained at the time:

Company	Price	Yield, %
Youngstown Sheet and Tube Debenture 6s.....	99	6.02
Steel and Tube General 7s.................	102	6.85

The market failed to realize the altered status of the Steel and Tube bonds and thus they sold illogically at a higher yield than the unsecured issue of the same obligor company. This presented a clear-cut opportunity to the analyst to recommend a purchase or an exchange.

In 1922 the City of Detroit purchased the urban lines of Detroit United Railway Company and agreed to pay therefor sums sufficient to retire the Detroit United Railway First 4½s, due 1932. Unusually strong protective provisions were inserted in the purchase contract which practically, if not technically, made the City of Detroit liable for the bonds. But after the deal was consummated, the bonds sold at 82, yielding more than 7%. The bond market failed to recognize their true status as virtual obligations of the City of Detroit.

In 1924, Congoleum Company had outstanding $1,800,000 of 7% preferred stock junior to $2,890,000 of bonds and followed by 960,000 shares of common stock having an average market value of some $48,000,000. In October of that year the company issued 681,000 additional shares of common for the business of the Nairn Linoleum Company, a large unit in the same field, with $15,000,000 of tangible assets. The enormous equity thus created for the small senior issues made them safe beyond question, but the price of the preferred stock remained under par.

In 1927 Electric Refrigeration Corporation (now Kelvinator Corporation) sold 373,000 shares of common stock for $6,600,000, making a total of 1,000,000 shares of common stock, with average market value of about $21,000,000, coming behind only $2,880,000 of 6% notes, due in 1936. The notes sold at 74, however, to yield 11%. The low price was due to a large operating deficit incurred in 1927, but the market failed to take into account the fact that the receipt of a much greater amount of new cash from the sale of additional stock had established a very strong backing for the small note issue.

These four senior issues have all been paid off at par or higher. (The Congoleum-Nairn Preferred was called for payment at 107 in 1934.) Examples of this kind are convenient for the authors since they do not involve the risk of some later mischance casting doubt upon their judgment. To avoid loading the dice too heavily in our favor, we add another illustration which is current as this chapter is written.

A Current Example.—Fox Film Corporation, following large losses in 1931–1932, recapitalized as of April 1933 by persuading the holders of about 95% of its debt to take common stock in exchange therefor. As a result its bank loans were eliminated and its note issue, due April 1936, was reduced from $30,000,000 to less than $1,800,000. In December 1933 the 6% notes sold at 75, yielding over 20% to maturity. The market value of the common stock was about $35,000,000 and the net current assets were about $10,000,000. The quantitative signs certainly pointed to the conclusion that the note issue was amply protected, and cheap in consequence at 75.

How dependable was this conclusion? It is certainly safe to say that either the stock was not worth anywhere near $35,000,000 or else the $1,800,000 note issue must be entirely safe. But a statement of this kind is less conclusive than it sounds, because ordinarily there is no way of taking advantage of a discrepancy between the relative prices of a highly speculative stock and a senior issue of investment grade.[1] The analyst must decide whether the issue is an attractive purchase, considered by itself. If the business is highly unstable even an enormous junior equity might disappear entirely and the note issue fail to be paid off despite its small size.[2] In the case of Fox Film we have on the one hand a large factor in an important industry, which should argue for sufficient stability at least to assure discharge of this small obligation. On the other hand, the moving-picture business has been highly speculative and the record of Fox Film since 1930 has not been confidence-inspiring.

Our conclusion must be, however, that the extraordinarily large quantitative backing for these notes in December 1933 reduced the risk of nonpayment to very minor proportions. Emphasizing once again the element of diversification as a safeguard in all such operations, we express the view that a number of purchases of this type will in all probability turn out quite satisfactorily in the aggregate. That some losses will occur goes without saying, but the proportion of such losses should undoubtedly be much lower in a reasonably normal period such as 1923–1927 than in cataclysmic years like 1930–1933.

Price-value Discrepancies in Receiverships.—In Chap. XVIII, dealing with reorganization procedure, we gave two diverse examples of disparities arising under a receivership: The Fisk Rubber case, in which the obligations sold at a ridiculously low price compared with the current assets available for them; and the Studebaker case, in which the price of the 6% notes was clearly out of line with that of the stock. A general statement may fairly be made that in cases where substantial values are ultimately realized out of a receivership, the senior securities will be found to have sold at much too low a price. This characteristic has a twofold consequence. It has previously led us to advise strongly against buying at investment levels *any* securities

---

[1] In the Fox Film case, the 6% notes were still exchangeable for stock on the basis of the recapitalization plan, *i.e.*, at $18.90 per share. If this were a contractual instead of merely a voluntary conversion privilege, the Fox notes would have been demonstrably superior at 75 to the Fox stock at 14, *from all standpoints.*

[2] The Willys Overland Company First 6½s, due Sept. 1, 1933, discussed in the Appendix, Note 32, are an example of such a calamitous outcome.

of a company which is likely to fall into financial difficulties; it now leads us to suggest that *after* these difficulties have arisen they may produce attractive analytical opportunities.

This will be true not only of issues so strongly entrenched as to come through reorganization unscathed (*e.g.*, Brooklyn Union Elevated 5s, as described in Chap. II), but also of senior securities which are "scaled down" or otherwise affected in a readjustment plan. It seems to hold most consistently in cases where liquidation or a sale to outside interests results ultimately in a cash distribution or its equivalent.

Examples: Three typical examples of such a consummation are given herewith.

1. *Ontario Power Service Corporation First 5½s, Due 1950.*— This issue defaulted interest payment on July 1, 1932. About this time the bonds sold as low as 21. The Hydro-Electric Commission of Ontario purchased the property soon afterwards, on a basis which gave $900 of new debentures, fully guaranteed by the Province of Ontario, for each $1,000 Ontario Power Service bond. The new debentures were quoted at 90 in December 1933, equivalent to 81 for the old bonds. The small number of bondholders not making the exchange received 70% in cash.

2. *Amalgamated Laundries, Inc., 6½s, Due 1936.*—Receivers were appointed in February 1932. The bonds were quoted at 4 in April 1932. In June 1932 the properties were sold to outside interests, and liquidating dividends of 12½% and 2% were paid in August 1932 and March 1933. In December 1933 the bonds were still quoted at 4, indicating expectation of at least that amount in further distributions.

3. *Fisk Rubber Company First 8s and Debenture 5½s, Due 1941 and 1931.*—Information regarding these issues was given in Chap. XVIII. Receivership was announced in January 1931. In 1932 the 8s and 5½s sold as low as 16 and 10½ respectively. In 1933 a reorganization was effected, which distributed 40% in cash on the 8s and 37% on the 5½s, together with securities of two successor companies. The aggregate values of the cash and the new securities at the close of 1933 came close to 100% for the 8% bonds and 70% for the debenture 5½s.

Price Patterns Produced by Receivership.—Certain price patterns are likely to be followed in receiverships, especially if they are protracted. In the first place, there is often a tendency for the stock issues to sell too high, not only in relation to the price of the bond issues, but also absolutely, *i.e.*, in relation to their probable ultimate value. This is due to the incidence of speculative interest, which is attracted by a seemingly low price range. In the case of senior issues, popular interest steadily decreases, and the price tends to decline accordingly, as the receivership wears on. Consequently, the lowest levels are likely to be reached a short time before a reorganization plan is ready to be announced.

A profitable field of analytical activity should be found therefore in keeping in close touch with receivership situations, endeavoring to discover securities which appear to be selling far under their intrinsic value, and to determine approximately the best time for making a commitment in them. But in these, as in all analytical situations, we must warn against an endeavor to gauge too nicely the proper time to buy. An essential characteristic of security analysis, as we understand it, is that the time factor is a subordinate consideration. Hence our use of the qualifying word "approximately," which is intended to allow a leeway of several months, and sometimes even longer, in judging the "right time" to enter upon the operation.

CHAPTER LI

DISCREPANCIES BETWEEN PRICE AND VALUE

(Continued)

The practical distinctions drawn in our last chapter between leading and secondary common stocks have their counterpart in the field of senior securities as between seasoned and unseasoned issues. A seasoned issue may be defined as an issue of a company long and favorably known to the investment public. (The security itself may be of recent creation so long as the company has a high reputation among investors.) Seasoned and unseasoned issues tend at times to follow divergent patterns of conduct in the market, viz.:

1. The price of seasoned issues is often maintained despite a considerable weakening of their investment position.

2. Unseasoned issues are very sensitive to adverse developments of any nature. Hence they often fall to prices far lower than seem to be warranted by their statistical exhibit.

Price Inertia of Seasoned Issues.—These opposite characteristics are due, in part at least, to the inertia and lack of penetration of the typical investor. He buys by reputation rather than by analysis and he holds tenaciously to what he has bought. Hence holders of long-established issues do not sell them readily, and even a small decline in price attracts buyers long familiar with the security.

Example: This trait of seasoned issues is well illustrated by the market history of the U. S. Rubber Company 8% Noncumulative Preferred. The issue received full dividends between 1905 and 1927. In each year of this period except 1924 there were investors who paid higher than par for this stock. Its popularity was based entirely upon its reputation and its dividend record, for the statistical exhibit of the company during most of the period was anything but impressive, even for an industrial bond, and hence ridiculously inadequate to justify the purchase of a noncumulative industrial preferred stock. Between the years

598

1922 and 1927, the following coverage was shown for interest charges and preferred dividends combined:

1922	1.20 times
1923	1.18 times
1924	1.32 times
1925	1.79 times
1926	1.00 times
1927	1.01 times

In 1928 the stock sold as high as 109. During that year the company sustained an enormous loss and the preferred dividend was discontinued. Despite the miserable showing and the absence of any dividend, the issue actually sold at 92½ in 1929. (In 1932 it sold at 3⅛.)

Vulnerability of Unseasoned Issues.—Turning to unseasoned issues, we may point out that these belong almost entirely to the industrial field. The element of seasoning plays a very small part as between the various senior issues of the railroads; and in the public-utility group proper (*i.e.*, electric, manufactured gas, telephone, and water companies) price variations will be found to follow the statistical showing fairly closely, without being strongly influenced by the factor of popularity or familiarity— except in the case of very small concerns.

Industrial financing has brought into the market a continuous stream of bond and preferred stock issues of companies new to the investment list. Investors have been persuaded to buy these offerings largely through the appeal of a yield moderately higher than the standard rate for seasoned securities of comparable grade. If the earning power is maintained uninterruptedly after issuance, the new security naturally proves a satisfactorily commitment. But any adverse development will ordinarily induce a severe decline in the market price. This vulnerability of unseasoned issues gives rise to the practical conclusion that it is unwise to buy a *new* industrial bond or preferred stock for straight investment.

Since such issues are unduly sensitive to unfavorable developments it would seem that the price would often fall too low and in that case they would afford attractive opportunities to purchase. This is undoubtedly true, but there is great need of caution in endeavoring to take advantage of these disparities. In the first place, the disfavor accorded to unseasoned securities in the market is not merely a subjective matter, due to lack of knowledge. Seasoning is usually defined as an objective quality, arising from a demonstrated ability to weather business storms. While this definition is not entirely accurate, there is enough truth in it to justify in good part the investor's preference for seasoned issues.

More important, perhaps, is the broad distinction of size and prominence that can be drawn between seasoned and unseasoned securities. The larger companies are generally the older companies, having senior issues long familiar to the public. Hence unseasoned bonds and preferred stocks are for the most part issues of concerns of secondary importance. But we have pointed out, in our discussion of industrial investments (Chap. VII), that in this field dominant size may reasonably be considered a most desirable trait. It follows, therefore, that in this respect unseasoned issues must suffer as a class from a not inconsiderable disadvantage.

Unseasoned Industrial Issues Rarely Deserve an Investment Rating.—The logical and practical result is that unseasoned industrial issues can very rarely deserve an investment rating, and consequently they should only be bought on an admittedly speculative basis. This requires in turn that the market price be low enough to permit of a substantial rise, *e.g.*, the price must ordinarily be below 70.

It will be recalled that in our treatment of speculative senior issues (Chap. XXVI), we referred to the price sector of about 70 to 100 as the "range of subjective variation," in which an issue might properly sell because of a legitimate difference of opinion as to whether or not it was sound. It seems, however, that in the case of unseasoned industrial bonds or preferred stocks the analyst should not be attracted by a price level within this range, even though the quantitative showing be quite satisfactory. He should favor such issues only when they can be bought at a frankly speculative price.

Exception may be made to this rule when the statistical exhibit is extraordinarily strong, as perhaps in the case of the Fox Film 6% notes mentioned in the preceding chapter. We doubt whether such exceptions can prudently include any unseasoned industrial preferred stocks, because of the contractual weakness of such issues. (In the case of Congoleum preferred, described above, the company was of dominant size in its field, and the preferred stock was not so much "unseasoned" as it was inactive marketwise.)

Discrepancies in Comparative Prices.—Comparisons may or may not be odious, but they hold a somewhat deceptive fascination for the analyst. It seems a much simpler process to decide that issue *A* is preferable to issue *B*, than to determine that issue *A* is an attractive purchase in its own right. But in our chapter on comparative analysis we have alluded to the particular responsibility that attaches to the recommendation of security exchanges, and we have warned against an overready acceptance of a purely quantitative superiority. The future is often no respecter of statistical data. We may frame this caveat in another way by suggesting that the analyst should not urge a security exchange unless either: (1) the issue to be bought is attractive, regarded by itself; or (2) there is a definite contractual relationship between the two issues in question. Let us illustrate consideration (1) by two examples of comparisons taken from our records.

Examples: I. Comparison Made in March 1932.

Item	Ward Baking First 6s, due 1937. Price 85¼, yield 9.70%	Bethlehem Steel First & Ref. 5s, due 1942. Price 93, yield 5.90%
Total interest charges earned:		
1931.............................	8.1 times	1.0 times
1930.............................	8.2 times	4.3 times
1929.............................	11.0 times	4.8 times
1928.............................	11.2 times	2.7 times
1927.............................	14.0 times	2.3 times
1926.............................	14.5 times	2.6 times
1925.............................	12.6 times	2.1 times
Seven-year average................	11.4 times	2.8 times
Amount of bond issues.............	$ 4,546,000	$145,000,000*
Market value of stock issues (March '32 average)....................	12,200,000	116,000,000
Cash assets.......................	3,438,000	50,300,000
Net working capital...............	3,494,000	116,300,000

* Including guaranteed stock.

In this comparison the Ward Baking issue made a far stronger statistical showing than the Bethlehem Steel bonds. Furthermore, it appeared sufficiently well protected to justify an investment rating, despite the high return. The qualitative factors, while not impressive, did not suggest any danger of collapse of the business. Hence the bonds could be recommended either as an original purchase or as an advantageous substitute for the Bethlehem Steel 5s.

II. Comparison Made in March 1929.

Item	Spear & Co. (Furniture Stores) 7% First Preferred. Price 77, yielding 9.09 %	Republic Iron & Steel 7% Preferred. Price 112, yielding 6.25%
(Interest and) preferred dividends earned:		
1928...........................	2.4 times	1.9 times
1927...........................	4.0 times	1.5 times
1926...........................	3.0 times	2.1 times
1925...........................	2.5 times	1.7 times
1924...........................	4.7 times	1.1 times
1923...........................	6.5 times	2.5 times
1922...........................	4.3 times	0.5 times
Seven-year average.................	3.9 times	1.6 times
Amount of bond issues..............	None	$32,700,000
Amount of (1st) preferred issue.......	$ 3,900,000	25,000,000
Market value of junior issues.........	3,200,000*	62,000,000
Net working capital.................	10,460,000	21,500,000

* Includes Second Preferred estimated at 50.

In this comparison the Spear and Company issue undoubtedly made a better statistical showing than Republic Iron and Steel Preferred. Taken by itself, however, its exhibit was not sufficiently impressive to carry conviction of investment merit, considering the type of business and the fact that we were dealing with a preferred stock. The price of the issue was not low enough to warrant recommendation on a fully speculative basis, *i.e.,* with prime emphasis on the opportunity for enhancement of principal. This meant in turn that it could not consistently be recommended in exchange for another issue, such as Republic Iron and Steel Preferred.

Comparison of Definitely Related Issues.—When the issues examined are definitely related, a different situation obtains.

An exchange can then be considered solely from the standpoint of the respective merits within the given situation; the responsibility for entering into or remaining in the situation need not be assumed by the analyst. In our previous chapters we have considered a number of cases in which relative prices were clearly out of line, permitting authoritative recommendations of exchange. These disparities arise from the frequent failure of the general market to recognize the effect of contractual provisions, and often also from a tendency for speculative markets to concentrate attention on the common stocks and to neglect the senior securities. Examples of the first type were given in our discussion of price discrepancies involving guaranteed issues in Chap. XVII. The price discrepancies between various Interborough Rapid Transit Company issues, discussed in the Appendix, Note 42, and between Brooklyn Union Elevated Railroad 5s and Brooklyn-Manhattan Transit Corporation 6s, referred to in Chap. II, are other illustrations in this category.[1]

The illogical price relationships between a senior convertible issue and the common stock, discussed in Chap. XXV, are examples of opportunities arising from the concentration of speculative interest on the more active junior shares. A different manifestation of the same general tendency is shown by the spread of 7 points existing in August 1933 between the price of American Water Works and Electric Company "free" common and the less active voting trust certificates for the same issue. Such phenomena invite not only direct exchanges but also hedging operations.

Other and Less Certain Discrepancies.—In these examples the aberrations are mathematically demonstrable. There is a larger class of disparities between senior and junior securities which may not be proved quite so conclusively but are sufficiently certain for practical purposes. As an example of these, consider Colorado Industrial Company 5s, due August 1, 1934, guaranteed by Colorado Fuel and Iron Company, which in May 1933 sold at 43, while the Colorado Fuel and Iron 8% Preferred, paying

---

[1] The student is invited to consider the price relationships between Pierce Petroleum and Pierce Oil Preferred and Common in 1929; between Central States Electric Corporation 5½% bonds and North American Company Common in January 1934; and between the common issues of Advance-Rumely Corporation and Allis-Chalmers Manufacturing Company in 1933; as examples of disparities arising from ownership by one company of securities in another.

no dividend, sold at 45. The bond issue had to be paid off in full within fourteen months' time, or else the preferred stock was faced with the possibility of complete extinction through receivership. In order that the preferred stock might prove more valuable than the bonds bought at the same price, it would be necessary not only that the bonds be paid off at par in little over a year, but that preferred dividends be resumed and back dividends discharged within that short time. This was almost, if not quite inconceivable.

A similar comparison could be made between Southern Railway Company 5% Noncumulative Preferred, paying no dividend and selling at 32⅞ in May 1933 and the Atlantic and Danville Railway Company Second 4s, due 1948, guaranteed as to interest under a lease to Southern Railway, selling at the same time at 30½.[1] Another comparison is that between New York, New Haven and Hartford Railroad Company 7% Preferred, paying no dividend and selling at 47 in May 1933 and the New York, Westchester, and Boston Railway Company 4½s, due 1946, completely guaranteed by the New Haven, paying their interest and selling at 48.

In comparing nonconvertible preferred stocks with common stocks of the same company, we find the same tendency for the latter to sell too high, relatively, when both issues are on a speculative basis. Comparisons of this kind can be safely drawn, however, only when the preferred stock bears cumulative dividends. (The reason for this restriction should be clear from our detailed discussion of the disabilities of noncumulative issues in Chap. XV.) A price of 10 for American and Foreign Power Company Common when the $7 Cumulative Second Preferred was selling at 11 in April 1933 was clearly unwarranted. A similar remark may be made of the price of 21½ for Chicago Great Western Railroad Company Common in February 1927, against 32½ for the 4% preferred stock, on which dividends of $44 per share had accumulated.

It is true that if extraordinary prosperity should develop in situations of this kind, the common shares might eventually be worth substantially more than the preferred. But even if this

---

[1] The Southern Railway-Mobile and Ohio Stock Trust Certificates, bearing a perpetual guarantee of 4% dividends, sold at 39¾ in July 1933, when Southern Railway Preferred sold at 49. The former issue was thus demonstrably cheaper than the latter.

should occur, the company is bound to pass through an intermediate period during which the improved situation permits it to resume preferred dividends and then to discharge the accumulations. Since such developments benefit the preferred stock directly, they are likely to establish (for a while at least) a market value for the senior issues far higher than that of the common stock. Hence, assuming any appreciable degree of improvement, a purchase of the preferred shares at the low levels should fare better than one made in the common stock.

Discrepancies Due to Special Supply and Demand Factors.— The illogical relationships which we have been considering grow out of supply and demand conditions which are in turn the product of unthinking speculative purchases. Sometimes discrepancies are occasioned by special and temporary causes affecting either demand or supply.

Examples: In the illogical relationship between the prices of Interboro Rapid Transit Company 5s and 7s in 1933, the operations of a substantial sinking fund, which purchased the 5s and not the 7s, were undoubtedly instrumental in raising the price of the former disproportionately. An outstanding example of this kind is found in the market action of United States Liberty 4¼s during the postwar readjustment of 1921–1922. Large amounts of these bonds had been bought during the war for patriotic reasons and financed by bank loans. A general desire to liquidate these loans later on induced a heavy volume of sales which drove the price down. This special selling pressure actually resulted in establishing a lower price basis for Liberty bonds than for high-grade railroad issues, which were, of course, inferior in security, and at a greater disadvantage also in the matter of taxation. Compare the following simultaneous prices in September 1920.

Issue	Price	Yield
United States Liberty Fourth 4¼s, due 1938.....	84½	5.64%*
Union Pacific First 4s, due 1947...............	80	5.42%

* Not allowing for tax exemption.

This situation supplied an excellent opportunity for the securities analyst to advise exchanges from the old-line railroad issues into Liberty Bonds.

A less striking disparity appeared a little later between the price of these Liberty bonds and of U. S. Victory 4¾s, due 1923. This state of affairs is discussed in a circular, prepared by one of the authors and issued at that time, a copy of which is given in the Appendix, Note 47, as an additional example of "practical security analysis."

CHAPTER LII

MARKET ANALYSIS AND SECURITY ANALYSIS

Forecasting security prices is not properly a part of security analysis. However, the two activities are generally thought to be closely allied, and they are frequently carried on by the same individuals and organizations. Endeavors to predict the course of prices have a variety of objectives and a still greater variety of techniques. Most emphasis is laid in Wall Street upon the science, or art, or pastime, of prophesying the immediate action of the "general market," which is fairly represented by the various averages used in the financial press. Some of the services or experts confine their aim to predicting the longer term trend of the market, purporting to ignore day-to-day fluctuations and to consider the broader "swings" covering a period of, say, several months. A great deal of attention is given also to prophesying the market action of individual issues, as distinct from the market as a whole.

Market Analysis as a Substitute For or Adjunct to Security Analysis.—Assuming that these activities are carried on with sufficient seriousness to represent more than mere guesses, we may refer to all or any of them by the designation of "market analysis." In this chapter we wish to consider the extent to which market analysis may seriously be considered as a substitute for or a supplement to security analysis. The question is important. If, as many believe, one can dependably foretell the movements of stock prices without any reference to the underlying values, then it would be sensible to confine security analysis to the selection of fixed-value investments only. For when it comes to the common-stock type of issue, it would manifestly be more profitable to master the technique of determining when to buy or sell, or of selecting the issues which are going to have the greatest or quickest advance, than to devote painstaking efforts to forming conclusions about intrinsic value. Many other people believe that the best results can be obtained by an analysis of the market position of a stock in conjunction with an analysis of its intrinsic value. If this is so, the securities

607

analyst who ventures outside the fixed-value field must qualify as a market analyst as well, and be prepared to view each situation from both standpoints at the same time.

It is not within our province to attempt a detailed criticism of the theories and the technique underlying all the different methods of market analysis. We shall confine ourselves to considering the broader lines of reasoning which are involved in the major premises of price forecasting. Even with this sketchy treatment it should be possible to reach some useful conclusions on the perplexing question of the relationship between market analysis and security analysis.

Two Kinds of Market Analysis.—A distinction may be made between two kinds of market analysis. The first finds the material for its predictions exclusively in the past action of the stock market. The second considers all sorts of economic factors, *e.g.*, business conditions, general and specific; money rates; the political outlook, etc. The underlying theory of the first approach may be summed up in the declaration that "the market is its own best forecaster." The behavior of the market is generally studied by means of charts on which are plotted the movements of individual stocks or of "averages."

Implication of the First Type of Market Analysis.—It must be recognized that the vogue of such "technical study" has increased immensely during the past ten years. Whereas security analysis suffered a distinct and continued loss of prestige beginning about 1927, chart reading apparently increased the number of its followers even during the long depression. Many sceptics, it is true, are inclined to dismiss the whole procedure as akin to astrology or necromancy;[1] but the sheer weight of its importance in Wall Street requires that its pretensions be examined with some degree of care. In order to confine our discussion within the framework of logical reasoning, we shall purposely omit

---

[1] Apropos of this attitude, we refer to a statement made by Frederick R. Macaulay at a meeting of the American Statistical Association in 1925, to the effect that he had plotted the results of tossing a coin several thousand times (heads = "one point up"; tails = "one point down") and had thereby obtained a graph resembling in all respects the typical stock chart—with resistance points, trend lines, double tops, areas of accumulation, *etc.* Since this graph could not possibly hold any clue as to the future sequence of heads or tails, there was a rather strong inference that stock charts are equally valueless. Mr. Macaulay's remarks were summarized in *Journal of the American Statistical Association*, Vol. 20, p. 248, June 1925.

even a condensed summary of the main tenets of chart reading.[1] We wish to consider only the implications of the general idea that a study of past price movements can be availed of profitably to foretell the movements of the future.

Such consideration, we believe, should lead to the following conclusions:

1. Chart reading cannot possibly be a science.

2. It has not proved itself in the past to be a dependable method of making profits in the stock market.

3. Its theoretical basis rests upon faulty logic or else upon mere assertion.

4. Its growing vogue is due to certain advantages it possesses over haphazard speculation; but these advantages tend to diminish as the number of chart students increases.

1. *Chart Reading Not a Science and Its Practice Cannot Be Continuously Successful.*—That chart reading cannot be a science is clearly demonstrable. If it were a science, its conclusions would be as a rule dependable. In that case everybody could predict tomorrow's or next week's price changes, and hence every one could make money continuously by buying and selling at the right time. This is patently impossible. A moment's thought will show that there can be no such thing as a scientific prediction of economic events under human control. The very "dependability" of such a prediction will cause human actions which will invalidate it. Hence thoughtful chartists admit that continued success is dependent upon keeping the successful method known to only a few people.

2. Because of this fact it follows that there is no generally known method of chart reading which has been continuously successful for a long period of time. If it were known, it would be speedily adopted by numberless traders. This very following would bring its usefulness to an end.

3. *Theoretical Basis Open to Question.*—The theoretical basis of chart reading runs somewhat as follows:

a. The action of the market (or of a particular stock) reflects the activities and the attitude of those interested in it.

[1] For detailed statements concerning the theory and practice of chart reading the student is referred to: R. W. Shabacker, *Stock Market Theory and Practice*, Chaps. XXIV ff., B. C. Forbes, New York, 1930; Robert Rhea, *The Dow Theory, passim.* Barron's, New York, 1932. The writers are informed that an exhaustive study on the subject by Harold M. Gartley is shortly to appear under the title *Charting The Stock Market*, Harper & Brothers, New York, 1934.

b. Therefore, by studying the record of market action, we can tell what is going to happen next in the market.

The premise may well be true, but the conclusion does not necessarily follow. You may learn a great deal about the technical position of a stock by studying its chart, and yet you may not learn *enough* to permit you to operate profitably in the issue. A good analogy is provided by the "past performances" of race horses, which are so assiduously studied by the devotees of the race track. Undoubtedly these charts afford considerable information concerning the relative merits of the entries; they will often enable the student to pick the winner of a race; but the trouble is that they do not furnish that valuable information *often enough* to make betting on horse races a profitable diversion.

Coming nearer home, we have a similar situation in security analysis itself. The past earnings of a company supply a useful indication of its future earnings—useful, but not infallible. Security analysis and market analysis are alike, therefore, in the fact that they deal with data which are not conclusive as to the future. The difference, as we shall point out, is that the securities analyst can protect himself by a *margin of safety* which is denied to the market analyst.

4. *Other Theoretical and Practical Weaknesses.*—The appeal of chart reading to the stock-market trader is something like that of a patent medicine to an incurable invalid. The stock speculator does suffer, in fact, from a well-nigh incurable ailment. The cure he seeks, however, is not abstinence from speculation, but profits. Despite all experience, he persuades himself that these can be made and retained; he grasps greedily and uncritically at every plausible means to this end.

The plausibility of chart reading, in our opinion, derives largely from its insistence on the sound gambling maxim that losses should be cut short and profits allowed to run. This principle usually prevents sudden large losses and at times it permits a large profit to be taken. The results are likely to be better, therefore, than those produced by the haphazard following of "market tips." Traders, noticing this advantage, are certain that by developing the technique of chart reading farther they will so increase its reliability as to assure themselves continued profits.

But in this conclusion there lurks a double fallacy. Many players at roulette follow a similar system, which limits their losses at any one session and permits them at times to realize a substantial gain. But in the end they always find that the aggregate of small losses exceeds the few large profits. (This must be so, since the mathematical odds against them are inexorable over a period of time.) The same is true of the stock trader, who will find that the expense of trading weights the dice heavily against him. A second difficulty is that as the methods of chart reading gain in popularity, the amount of the loss taken in unprofitable trades tends to increase, and the profits also tend to diminish. For as more and more people, following the same system, receive the signal to buy at about the same time, the result of this competitive buying must be that a higher average price is paid by the group. Conversely, when this larger group decides to sell out at the same time, either to cut short a loss or to protect a profit, the effect must again be that a lower average price is received. (The growth in the use of "stop-loss orders," formerly a helpful technical device of the trader, had this very effect of detracting greatly from their value as a protective measure.)

The more intelligent chart students recognize these theoretical weaknesses, we believe, and take the view that market forecasting is an *art* which requires talent, judgment, intuition, and other personal qualities. They admit that no rules of procedure can be laid down, the automatic following of which will insure success.

The Second Type of Mechanical Forecasting.—Before considering the significance of this admission let us pass on to the other type of mechanical forecasting, which is based upon factors outside of the market itself. As far as the general market is concerned the usual procedure is to construct indices representing various economic factors, *e.g.*, money rates, carloadings, steel production, and to deduce impending changes in the market from an observation of a recent change in these indices. One of the earliest methods of the kind, and a very simple one, was based upon the percentage of blast furnaces in operation.

This theory was developed by Col. Leonard P. Ayers of the Cleveland Trust Company and ran to the effect that security prices usually reached a bottom when blast furnaces in operation declined through 60% of the total, and that conversely they usually reached a top when blast furnaces in operation passed through the 60% mark on the upswing in use thereof.[1] A companion theory of Colonel Ayres was that the high point in bond prices is reached about fourteen months subsequent to the low point in pig-iron production, and that the peak in stock prices is reached about two years following the low point for pig-iron production.[2]

This simple method is representative of all mechanical forecasting systems, in that: (1) it sound vaguely *plausible* on the basis of *a priori* reasoning; and (2) it relies for its *convincingness* on the fact that it has "worked" for a number of years past. The necessary weakness of all these systems lies in the time element. It is easy and safe to prophesy, for example, that a period of high interest rates will lead to a sharp decline in the market. The question is, "How soon?" There is no scientific way of answering this question. Many of the forecasting services are therefore driven to a sort of pseudo-science, in which they take it for granted that certain time lags or certain coincidences which happened to occur several times in the past (or have been worked out laboriously by a process of trial and error), can be counted upon to occur in much the same way in the future.

Broadly speaking, therefore, the endeavor to forecast security-price changes by reference to mechanical indices is open to the same objections as the methods of the chart readers. They are not truly scientific, because there is no convincing reasoning to support them; and because, furthermore, really scientific (*i.e.*, entirely dependable) forecasting in the economic field is a logical impossibility.

Disadvantages of Market Analysis as Compared with Security Analysis.—We return in consequence to our earlier conclusion that market analysis is an art, for which special talent is needed in order to pursue it successfully. Security analysis is also an art; and it, too, will not yield satisfactory results unless the analyst has ability as well as knowledge. We think, however, that security analysis has several advantages over market analysis, which are likely to make the former a more successful

[1] See *Bulletin of the Cleveland Trust Company*, July 15, 1924, cited by David F. Jordan, in *Practical Business Forecasting*, p. 203n., New York, 1927.

[2] See *Business Recovery Following Depression*, a pamphlet published by the Cleveland Trust Company in 1922. The conclusions of Colonel Ayres are summarized on p. 31 of the pamphlet.

field of activity for those with training and intelligence. In security analysis the prime stress is laid upon protection against untoward events. We obtain this protection by insisting upon margins of safety, or values well in excess of the price paid. The underlying idea is that even if the security turns out to be less attractive than it appeared, the commitment might still prove a satisfactory one. In market analysis there are no margins of safety; you are either right or wrong, and if you are wrong, you lose money.

The cardinal rule of the market analyst that losses should be cut short and profits safeguarded (by selling when a decline commences) leads in the direction of active trading. This means in turn that the cost of buying and selling becomes a heavily adverse factor in aggregate results. Operations based on security analysis are ordinarily of the investment type and do not involve active trading.

A third disadvantage of market analysis is that it involves essentially a battle of wits. Profits made by trading in the market are for most part realized at the expense of others who are trying to do the same thing. The trader necessarily favors the more active issues, and the price changes in these are the resultant of the activities of numerous operators of his own type. The market analyst can be hopeful of success only upon the assumption that he will be more clever or perhaps luckier than his competitors.

The work of the securities analyst, on the other hand, is in no similar sense competitive with that of his fellow analysts. In the typical case the issue which he elects to buy is not sold by some one who has made an equally painstaking analysis of its value. We must emphasize the point that the security analyst examines a far larger list of securities than does the market analyst. Out of this large list, he selects the exceptional cases in which the market price falls far short of reflecting intrinsic value, either through neglect or because of undue emphasis laid upon unfavorable factors which are probably temporary.

Market analysis seems easier than security analysis, and its rewards may be realized much more quickly. For these very reasons, it is likely to prove more disappointing in the long run. There are no dependable ways of making money easily and quickly, either in Wall Street or anywhere else.

Prophesies Based on Near-term Prospects.—A good part
of the analysis and advice supplied in the financial district rests
upon the near-term business prospects of the company considered.
It is assumed that if the outlook favors increased earnings, the
issue should be bought in the expectation of a higher price when
the larger profits are actually reported. In this reasoning,
security analysis and market analysis are made to coincide. The
market prospect is thought to be identical with the business
prospect.

But to our mind the theory of buying stocks chiefly upon the
basis of their immediate outlook makes the selection of speculative securities entirely too simple a matter. Its weakness lies
in the fact that the current market price already takes into
account the consensus of opinion as to future prospects. And
in many cases the prospects will have been given *more* than their
just need of recognition. When a stock is recommended for the
reason that next year's earnings are expected to show improvement, a twofold hazard is involved. First, the forecast of next
year's results may prove incorrect; second, even if correct, it
may have been discounted or even overdiscounted in the current
price.

If markets generally reflected only this year's earnings, then a
good estimate of next year's results would be of inestimable
value. But the premise is not correct. We append a table
showing on the one hand the annual earnings per share of United
States Steel Corporation common and on the other hand the
price range of that issue for the years 1902–1933. Excluding
the 1928–1933 period (in which business changes were so extreme
as necessarily to induce corresponding changes in stock prices),
it is difficult to establish any definite correlation between fluctuations in earnings and fluctuations in market quotations.

In the Appendix, Note 48, we reproduce significant parts of
the analysis and recommendation concerning two common
stocks made by an important statistical and advisory service in
the latter part of 1933. The recommendations are seen to be
based largely upon the apparent outlook for 1934. There is no
indication of any endeavor to ascertain the fair value of the
business and to compare this value with the current price. A
thoroughgoing statistical analysis would point to the conclusion
that the issue of which the sale is advised was selling below its
intrinsic value, just because of the unfavorable immediate

UNITED STATES STEEL COMMON, 1901–1933

Year	Earned per share	Range of market price		
		High	Low	Average
1901	$ 9.1	55	24	40
1902	10.7	47	30	39
1903	4.9	40	10	25
1904	1.0	34	8	21
1905	8.5	43	25	34
1906	14.3	50	33	42
1907	15.6	50	22	36
1908	4.1	59	26	48
1909	10.6	95	41	68
1910	12.2	91	61	76
1911	5.9	82	50	66
1912	5.7	81	58	70
1913	11.0	69	50	60
1914	*0.3(d)*	67	48	58
1915	10.0	90	38	64
1916	48.5	130	80	105
1917	39.2	137	80	109
1918	22.1	117	87	102
1919	10.1	116	88	102
1920	16.6	109	76	93
1921	2.2	87	70	79
1922	2.8	112	82	97
1923	16.4	110	86	98
1924	11.8	121	94	108
1925	12.9	139	112	126
1926	18.0	161	117	139
1927*	12.3	246	155	201
1927†	8.8	176	111	144
1928	12.5	173	132	153
1929	21.2	262	150	206
1930	9.1	199	134	167
1931	*1.4(d)*	152	36	99
1932	*11.1(d)*	53	21	37
1933	*5.4(d)‡*	68	23	46

* Before allowing for 40 % stock dividend.
† After allowing for 40 % stock dividend.
‡ Nine months to Sept. 30, 1933.

prospects; and that the opposite was true of the common stock recommended as worth holding because of its satisfactory outlook.

We are sceptical of the ability of the analyst to forecast with a fair degree of success the market behavior of individual issues over the near-term future—whether he base his predictions upon the technical position of the market, or upon the general outlook for business, or upon the specific outlook for the individual companies. More satisfactory results are to be obtained, in our opinion, by confining the positive conclusions of the analyst to the following fields of endeavor:

1. The selection of standard senior issues which meet exacting tests of safety.

2. The discovery of senior issues which merit an investment rating but which also have opportunities of an appreciable enhancement in value.

3. The discovery of common stocks, or speculative senior issues, which appear to be selling at far less than their intrinsic value.

4. The determination of definite price discrepancies existing between related securities, which situations may justify making exchanges or initiating hedging or arbitrage operations.

APPENDIX

NOTE 1

An outstanding example of funded indebtedness retired through the sale of stock is afforded by the action of United States Steel Corp. in 1929. In that year about $143,000,000 was raised through the sale of additional common shares, and the proceeds, together with about $195,000,000 of treasury cash, were employed for the retirement of the entire bonded indebtedness of the parent company and also the funded debt of two subsidiaries.

The rise and decline in the vogue of stock financing, particularly through the issuance of stockholders' "rights," are indicated by the following table which summarizes new security offerings in the United States, as compiled and reported by Standard Statistics Co. Federal, state, and municipal issues are not included.

NEW SECURITY OFFERINGS, 1924–1933
(Unit: $1,000,000)

Year	Common and preferred stock through "rights"	All bonds	Common stock by public offer	Preferred stock by public offer	Total stocks and bonds
1924	$ 444.7	$4,186.0	$ 111.8	$ 283.5	$ 5,026.0
1925	752.9	4,605.3	220.9	730.4	6,309.5
1926	901.2	4,857.0	234.4	687.5	6,580.1
1927	1,442.4	6,813.3	263.7	942.1	9,461.5
1928	2,399.5	5,132.2	918.1	1,620.4	10,070.2
1929	4,205.0	3,375.1	2,021.7	1,808.2	11,410.0
1930	1,135.2	4,686.2	249.0	496.0	6,566.4
1931	196.3	2,643.0	27.4	106.4	2,973.1
1932	40.0	933.0	14.4	5.7	993.1
1933	82.1	172.9	110.9	23.0	388.9

Standard Statistical Bulletin (Base Book). January 1932, p. 49, and monthly supplements thereto, both published by Standard Statistics Co., Inc.

NOTE 2

A part of the financial history of the United States Express Co. shows how the conversion of an interest in property from the stock form into the bond form obtained buyers for the new securities which were both less safe and less profitable than the stock issue.

United States Express Co. entered upon liquidation in 1914. By May 1918 it had paid out $52 per share in liquidating dividends on its 100,000 shares of capital stock. Its remaining assets consisted mainly of complete ownership, free and clear, of the United States Express Co. building at 2 Rector Street, New York City, together with other real-estate holdings valued on the books at about $680,000. Liabilities were nominal. The stock sold at about $15 per share, representing a valuation of some $1,500,000 for all the assets.

In December 1919 the Rector Street building and adjoining lots were sold for $3,725,000. The buyer financed the purchase in part through the sale at par of $3,000,000 Two Rector Street Corp. First Mortgage 6s, due April 1, 1935, secured by a first mortgage on the land and building owned in fee. The offering circular of the National City Company stated that on the basis of appraisals by four leading real-estate operators in the city the bonds were legal for the investment of trust funds under the laws of the State of New York. This latter statement implied an "appraised value" of at least $5,000,000 in view of the terms of the New York law.

There is a striking contrast between the essential merits of the United States Express stock at 15 and of these Two Rector Street Corp. first-mortgage bonds at par. Buyers of the former were paying the equivalent of $1,500,000 for *complete ownership* of the Rector Street property, plus other assets. Buyers of the latter were paying $3,000,000 for a *limited interest* in the Rector Street property alone. At one half the price, the stock buyer obtained *all the security* later given the bond buyer, *plus* the right to all the surplus profits and value.

Owing to the boom in real estate which developed in New York City shortly after the bonds were floated, the Rector Street building was resold and the bond issue called at 103 in April 1925. This netted the holders a 3% profit. Meanwhile, however, the United States Express Co. had paid, out of the proceeds of the sale of the building and other assets, additional liquidating dividends of $37.50 per share, followed by another dividend of $1.75 in 1929. Obviously the stock at 15 was both a safer and a more attractive commitment than the bonds were at par. Apparently the public regarded the stock as a speculation; whereas the bonds, representing only a part interest in the assets behind the stock, were regarded as an investment. There seems to be little doubt that at least a part of the explanation of this anomaly lies in the magic influence of the title, "Bond." The announcement that the issue was eligible for the investment of trust funds also undoubtedly played a role.

The history of this case can be traced through the financial manuals and through the following references: 109 *Chronicle* 2466; 110 *Chronicle* 86, 1421, 1533; 116 *Chronicle* 833; 120 *Chronicle* 1215; 128 *Chronicle* 267.

NOTE 3

A PARTIAL LIST OF SECURITIES WHICH DEVIATE FROM THE NORMAL PATTERNS AS OUTLINED IN CHAP. V

In assembling the material presented herewith it has not been our purpose to present a complete list of all types of securities which vary from the customary contractual arrangements between the issuing corporation and the holder. Such a list would extend the size of this volume beyond reasonable limits. We have, however, attempted to give a reasonably complete sample of deviations from the standard patterns as the latter have been outlined in the text. It is hoped that the ensuing list will at least serve as a useful starting point in the development of an intelligent scepticism concerning the sanctity of titles which are assigned to securities, and that the reader will hereby be assisted toward a realization of the necessity for careful inquiry concerning the terms of any particular issue which he contemplates purchasing.

BONDS WHICH DEVIATE FROM THE STANDARD PATTERN

A. Modifications of the unqualified right to a fixed-interest payment on fixed dates:

1. Income bonds: These are sometimes called "adjustment bonds." The amount of interest payable thereon is dependent upon the amount of earnings available therefor up to a stated maximum. Payment of interest is sometimes wholly or partly discretionary with the directors. Interest may be cumulative or noncumulative. *Typical examples:* Atchison, Topeka & Santa Fe Ry. Adjustment 4s, due 1995; Chicago, Milwaukee, St. Paul & Pacific R.R. Adjustment 5s, due 2000; Third Avenue Ry. Adjustment Income 5s, due 1960; Chicago Rapid Transit Co. Adjustment 6s, due 1963.

2. Hybrid issues on which interest is contingent for a time and a fixed charge thereafter: *Examples:* Denver & Rio Grande Western R.R. General 5s, due 1955. Prior to Feb. 1, 1929, interest payments on this issue were discretionary with the directors. No interest was paid during the preceding five-year period. 25% accumulated on the issue as a charge prior to dividends. Subsequent to Feb. 1, 1929, interest thereon became a fixed charge. Three issues of Sacramento Northern R.R. maturing in 1937 have similar provisions. Eitingon Schild Co. Income 5s, Series *A*, due 1938. These bonds carry contingent interest until Jan. 1, 1937, and a fixed charge thereafter until maturity.

3. Bonds on which a part of the interest burden is a fixed charge and a part contingent upon earnings and/or upon the discretion of the directors: *Example:* Cincinnati, Hamilton & Dayton Ry. General Mortgage Bonds, due 1939. See status prior to 1916. The bonds were also irregular in other respects.

4. Bonds on which interest payments may be deferred: *Example:* Boston, Worcester & New York Street Ry. Co. Reorganization 5s, due 1947. During the first six years from 1927 varying proportions of the interest were subordinated to payment of preferred dividends. Payments were also in part discretionary with the directors.

5. Issues on which the rate of interest is graduated up or down: *Examples:* Kansas City Public Service First Mortgage Bonds, Series *B*, due 1951. Rate varies from 3 to 7%. Public Service Corp. of New Jersey Perpetual 6% Certificates. Interest rate graduated

from 1904 to 1913. Boston & Maine R.R. General Mortgage Bonds. About a dozen series of this issue carried interest graduated upward. Two series carried interest graduated downward.

6. Bonds with option to the holder of receiving his interest payments in cash or in stock of the company: *Examples:* Central States Electric Corp. Optional 5½s, due 1954; Warner Brothers Pictures Optional 6s, Series due 1939.

7. Non-interest-bearing obligations: *Examples:* Mexican Light & Power Co. Notes, due 1937. Issued to fund accumulated interest on Income 6s of the company. All held by the parent company (Mexico Tramways Co.). Non-interest-bearing Convertible Debentures of Alabama Water Service Co. No maturity date. All held by the parent company (Federal Water Service Co.).

8. Rate of interest contingent upon rate paid on preferred stock of the obligor: *Example:* Associated Gas & Electric Co. 6½% Convertible Obligations. Issue was also convertible at company's option into $6.50 dividend preferred stock. Rate and form of payment depended upon rate and form of payment made to holders of this stock. Company has since exercised its option to call this issue for conversion into stock.

9. Interest payment subordinated to dividends on preferred stock of obligor: *Examples:* Chicago Rapid Transit Co. Adjustment Debenture 6s, due 1963. Interest subordinated to dividends on two preferred stock issues. Boston, Worcester & New York Street Ry. Co. Reorganization 5s, due 1947. Alabama Water Service Co. nonnegotiable, non-interest bearing notes issued in 1927 and held by parent company.

B. Modifications of the unqualified right to repayment of a fixed principal amount on a fixed date:

1. Noncallable, perpetual bonds: *Examples:* Lehigh Valley R.R. Consolidated 4½s and 6s; Public Service Corp. of New Jersey Perpetual 6% Certificates; Canadian Pacific Ry. 4% Irredeemable Consolidated Debenture Stock.

2. Callable bonds without specified maturity date: *Examples:* United Kingdom 2½% Consols of 1888; also Consolidated 4s of 1927. Nova Scotia Steel & Coal Co. 6% Perpetual Debenture Stock. Atlantic Coast Line Co. 4% Class B Certificates.

3. Bonds payable at option of the obligor after a specified date: *Example:* United Railway & Electric Co. of Baltimore Income 4s, dated Mar. 3, 1899. Payable at company's option after Mar. 1, 1949. Maturity accelerated by default.

4. Bonds payable at the option of the holder on or after a specified date: *Examples:* Bermuda Traction, Ltd., 7% Participating Debentures. No specified maturity, but payable at the option of the holder at 105 in 1968. Kreuger & Toll Co. 5% Participating Debentures. No specified maturity date, but holder may demand par and interest on or after July 1, 2003. "American Certificates" for these "bonds" were really a participating stock. Minimum call price of five times par value and maximum dependent upon average

market price for preceding three months. *Cf.* Siemens & Halske A. G. Participating Debentures, Series *A*, due 2930.

5. Bonds falling due on sale or reorganization of company: *Examples:* Green Bay & Western R.R. Class *A* and Class *B* Debentures. Provision is equivocal. Issues are really prior participating preferred and common stock respectively. Associated Gas & Electric Co.—numerous issues outstanding early in 1932, with no fixed maturity. Were also convertible into stock at company's option. See 1931 and 1932 Manuals.

6. Bonds optional as to character of principal payment: *Examples:* Railroad Securities Co.-Illinois Central 4% Stock Trust Certificates, due 1952. Company has option of paying par or delivering holder his pro rata share of deposited collateral. United States Securities Corp. Collateral Trust Income 7s, due 1962. Company had option during last five years before maturity of paying to the holders their pro rata share of the proceeds from the sale of the deposited collateral. See 135 *Chronicle* 147.

7. Bonds with repayment of principal subject to nullification through option of obligor to convert them into stock: *Examples:* Associated Gas & Electric Co.—numerous issues listed in 1931 Manuals; Notes of certain Tudor City Units; see *Moody's Manual (Banks and Finance)* (1932), pp. 2595–2597.

8. Bonds maturing at a premium above par (sum is fixed but nonstandard): *Examples:* Detroit International Bridge Co. Participating 7s, due 1952. Mature at 125% of their par value. Call price also increases as maturity approaches. Warner Brothers Pictures, Inc., 6½s, due 1928. Payable at maturity at 105 or at par plus 7½ shares of common stock at the option of the holder.

9. Bonds redeemable prior to maturity at a discount from par: *Examples:* Eitingon Schild Co. Income Debenture 5s, Series *A*, due 1938. These bonds are redeemable at discounts until Mar. 1, 1937; Brooklyn-Manhattan Transit Corp. 6s, due 1934. These bonds were redeemable at discounts prior to Feb. 1, 1934.

10. Bonds payable at an amount dependent upon general level of prices: *Example:* Rand Kardex 7% Thirty-year Stabilized Debentures, due 1955. This issue was unique and was retired shortly after original issuance. The amount of the current interest payments and the sum payable at maturity were made to depend upon the level of the United States Bureau of Labor Statistics index number of wholesale prices. See 26 *Annalist* 603 (Nov. 13, 1925).

C. Modifications of the rule that the bondholder has no further interest in assets or profits beyond the promise to pay a principal sum at maturity and a definite rate of interest meanwhile, and no voice in the management:

1. Participating bonds (note great variation in basis of participation): *Examples:* Kreuger & Toll Co. 5% Participating Debentures. Note also the special arrangement concerning redemption at a minimum of 500% of par value. Siemens & Halske A. G. Participating Debentures, Series *A*, due in 2930. Siemens & Halske

A. G.—Siemens-Schuckertwerke A. G. 6½s, due in 1951. A joint obligation carrying warrants for contingent additional interest. United Steel Works Corp. 6½s, due in 1947. Associated Gas & Electric Corp. Participating 8s, due in 1940. Participating distributions dependent upon rates of dividends paid on various classes of stock of the Associated Gas & Electric Co. Green Bay & Western R.R. Class *A* Income Debentures. United Oil Producers Corp. First Mortgage Participating 8s, due in 1931. No longer outstanding. Amount of participating distributions depended on sum realized from a fixed number of barrels of oil. White Sewing Machine Corp. Participating 6s, due 1940.

2. Convertible bonds: These constitute the largest single group of exceptions to the rule that the bondholder has a definitely limited interest in assets and earnings. The number of variations in the terms of convertible issues is practically unlimited. The variations play mainly upon the conversion ratio, the duration of the privilege, the security into which the issue is convertible, special limitations upon the right to convert, and combinations of these factors. The following classified list is only a partial one. It should be noted that the terms of the privilege are usually subject to alteration through operation of antidilution clauses.

(1) Duration of the privilege subject to termination through reserved right of the corporation to call the issue: *Example:* The Texas Corporation Convertible Debenture 5s, due 1944.

(2) Same as foregoing, but right to call issue is suspended for a stated period following issuance: *Example:* Baltimore & Ohio R.R. Co. Convertible 4½s, due 1960.

(3) Noncallable convertible issues: *Example:* New York, New Haven & Hartford R.R. Co. Convertible 6s, due 1948.

(4) Convertible issues not subject to redemption until after the privilege has expired: *Example:* Atchison, Topeka & Santa Fe Ry. Co. Convertible 4½s, due in 1948.

(5) Duration of privilege limited only by maturity or prior redemption: *Example:* California Packing Corp. Convertible 5s, due 1940.

(6) Privilege expires on definite date prior to maturity: *Example:* Baltimore & Ohio R.R. Co. Convertible 4½s, due 1960.

(7) Privilege is not operative from the date of issuance, but begins only after the lapse of a specified period of time: *Example:* Chesapeake Corp. Convertible 5s, due 1947.

(8) Issues convertible into stock or other bonds *at the option of the obligor: Examples:* See numerous issues of Associated Gas & Electric Co. listed in 1931 Manuals. Western New York Water Co. First 5s, due 1951, convertible into like amount of New York Water Service Corp. First 5s, Series *A*, due 1951.

(9) Issues convertible by either the holder or the obligor under special limitations upon the right to convert: *Example:* London Terrace Apartments 6s, due Apr. 15, 1940.

(10) Issues convertible only when certain conditions with reference to earnings have been met: *Examples:* North American Edison Co. Convertible 5s, Series *A*, due in 1957. Power Securities Corp. Secured Income 6s, due 1949.

(11) Only a part of the issue may be converted: *Example:* Dodge Brothers, Inc., Convertible 6s, due 1940. New England Gas & Electric Assoc. Convertible 5s, due 1947.

(12) Amount convertible in any one year limited to stated maximum: *Examples:* Boston & Maine R.R. General Mortgage Bonds, Series *Q* to *GG*, inclusive, during the years 1930–1933.

(13) Duration of privilege terminated when given number of shares have been exchanged for bonds: *Example:* Porto Rican-American Tobacco Co. Secured Convertible 6s, due in 1942.

(14) Issues convertible into common stock (most customary arrangement): *Example:* International Telephone & Telegraph Corp. Convertible 4½s, due 1939.

(15) Issues convertible into preferred stock: *Example:* Long Island Lighting Co. Convertible Debenture 5½s, Series *A*, due in 1952.

(16) Issues convertible into Class *A* Stock which is junior to one or more preferred issues: *Example:* International Hydro-electric System Convertible Debenture 6s, due in 1944.

(17) Bonds convertible into other bonds (usually short-term notes into longer term issues, to give the holder the option of repayment or of continuing his investment): *Examples:* Interborough Rapid Transit Co. 7s, due in 1932. Now in default. Hackensack Water Co. Five-year Secured Convertible 5s, due 1938.

(18) Bonds convertible into units of two or more issues: *Examples:* Alleghany Corp. Collateral Trust Convertible 5s, due 1944. Chicago, Milwaukee, St. Paul & Pacific R. R. Co. Convertible Adjustment 5s, Series *A*, due in 2000.

(19) Bonds convertible into either of two classes of preferred stock or into either of two classes of common stock: *Example:* U. S. Dairy Products Corp. Convertible 6½s, Series *B* and *C*, due in 1934 and 1935 respectively.

(20) Other special and more complicated arrangements of the foregoing type: *Examples:* Pressed Steel Car Co. Convertible 6s, due in 1933. Associated Gas & Electric Co. $8 Interest-bearing Allotment Certificates, 5½% Convertible Investment Certificates, 6% Convertible Debenture Certificates and various other issues listed in the 1931 Manuals.

(21) Bonds convertible into securities of a company other than the obligor: *Examples:* Chesapeake Corp. Convertible 5s, due 1947. Porto Rican-American Tobacco Co. Convertible 6s, due 1942. Dawson Railway & Coal Co. First 5s, due 1951.

(22) Bonds convertible into an issue which is itself convertible: *Examples:* This was true of a number of the Associated Gas & Electric Co. issues in 1931 and 1932. See Manuals. East

Coast Utilities Co. 6% Notes, due in 1932, which were convertible into a convertible Class *A* stock.

(23) Bonds convertible into an issue which is both participating and convertible: *Examples:* National Service Companies 6% Notes, due in 1932; Domestic Industries, Inc., Convertible 6½s, due 1940.

(24) Issues convertible into bonds or preferred stock which carry stock-purchase warrants: *Examples:* Wakenva Coal Co., Inc., Collateral Trust 6½s, due 1947; Fox Metropolitan Playhouses, Inc., Convertible 6½s, due 1932; Public Utilities Consolidated Corp. Secured 6s, due 1938.

(25) Bonds convertible into participating issues: *Example:* International Hydro-electric Convertible 6s, due 1944; Community Telephone Co. Convertible 6s, due 1949.

(26) Bonds convertible into stock upon surrender of the bond and payment of a further sum in cash: *Example:* Celotex Co. Convertible 6s, due 1936.

(27) Bonds convertible into stock upon surrender of the bond, with option of obtaining larger amount of stock upon payment of further sum in cash: *Example:* American Telephone & Telegraph Co. Convertible 4½s, due 1939.

(28) Bonds convertible into stock at a flat price throughout the life of the privilege: *Example:* Atchison, Topeka & Santa Fe Ry. Convertible 4½s, due 1948.

(29) Bonds convertible into stock at prices which vary in accordance with a specified time schedule: *Example:* Baltimore & Ohio R. R. Co. Convertible 4½s, due 1960.

(30) Bonds convertible into stock at prices which vary in accordance with the extent to which the holders have exercised the privilege: *Examples:* Anaconda Copper Mining Co. Convertible 7s, due 1938. Issue no longer outstanding. Dodge Brothers, Inc., 6s, due 1940.

3. Bonds with stock-purchase warrants: These also give the holder an interest in assets and earnings beyond the participation therein afforded to the holder of a standard bond. The variations in the terms of warrant-bearing issues are about as numerous as those pertaining to convertible bonds. The following is a partial list. The terms of the privilege are usually subject to alteration through operation of antidilution clauses.

(1) Privilege terminates with maturity of the bond: *Example:* Investors Equity Co. Debenture 5s, Series *B*, due 1948.

(2) Privilege expires prior to the maturity of the bond: *Example:* Remington Rand, Inc., Debenture 5½s, Series *A*, due 1947.

(3) Duration of the subscription privilege subject to termination through exercise of right to redeem the issue: *Example:* Royal Dutch Co. Debenture 4s, Series *A*, due 1945.

(4) Warrant entitles holder to purchase stock at a flat price throughout the life of the privilege: *Example:* Shell Union Oil Corp. 5s, due 1949.

(5) Warrant entitles holder to buy stock at prices which vary in accordance with a specified time schedule: *Example:* Union Oil Co. of California Debenture 5s, due 1945.

(6) Warrant entitles holder to buy stock at prices determined by the order in which the warrants are exercised: *Example:* Central States Electric Corp. Optional Debenture 5½s, Series due 1954.

(7) Warrant entitled holder to purchase stock at the higher of two prices, one being dependent upon the cost of foreign exchange in terms of dollars: *Example:* Isotta-Fraschini First 7s, due 1942.

(8) Bonds carrying warrants for a bonus in common stock: *Example:* Shawmut Bank Investment Trust Senior Debenture 4½s and 5s, due in 1942 and 1952, respectively.

(9) Bonds carrying two different warrants entitling holder to purchase stock on separate scales of prices at the same time: *Example:* Southern National Corp. Debenture 6s, due 1944.

(10) Bonds carrying warrants entitling holder to exchange the bond for stock on one scale of prices and to subscribe for additional stock on another scale of prices: *Example:* Insull Utility Investments, Inc., Debenture 6s, Series *B*, due 1940.

(11) Bonds carrying warrants for purchase of constant number of shares at a flat price: *Example:* Shell Union Oil Corp. Debenture 5s, due 1949.

(12) Bonds carrying warrants for purchase of a varying number of shares at a flat price: *Example:* El Paso Natural Gas Co. First 6½s, due 1943.

(13) Bonds carrying warrants for purchase of a varying number of shares at various prices: *Example:* Adolf Gobel, Inc., 6½s, Series *A*, due 1935.

(14) Bonds carrying warrants entitling the holder to purchase a participating Class *A* stock: *Example:* Central Public Service Corp. 5½s, due 1949.

(15) Bonds carrying warrants entitling the holder to purchase a convertible preferred issue: *Example:* Lexington Telephone Co. First 6s, due 1944.

(16) Bonds carrying warrants entitling the holder to purchase a block or unit of stock consisting of two or more issues: *Example:* Utilities Power & Light Corp. Debenture 5s, due 1959.

(17) Bonds with warrants which are detachable only at the time the warrant is exercised or upon redemption of the bond: *Example:* Southern Pacific Co. 4½s, due 1969.

(18) Warrant detachable upon exercise but becoming void when issue is called: *Example:* Royal Dutch Co. Debenture 4s, Series *A*, due 1945.

(19) Detachability of the warrant subject to formal approval by the corporation: *Example:* El Paso Natural Gas Co. 6½s, due 1943.

(20) Bonds carrying warrants which became detachable upon issuance of bonds: *Example:* Fiat Debenture 7s, due 1946.

(21) Bonds carrying warrants which became detachable only after the lapse of a specified period following issuance of the bonds: *Example:* Capital Administration Co. Debenture 5s, Series *A*, due 1953.

(22) Warrants exercisable only upon payment of the subscription price in cash: *Example:* Abraham & Straus, Inc., Debenture 5½s, due 1943.

(23) Warrants permitting payment of the subscription price either in cash or in face value of the bonds to which they were attached: *Example:* Rand Kardex Bureau, Inc., 5½s, due 1931.

4. Bonds with mixed privileges:

(1) Convertible bonds with stock-purchase warrants: *Examples:* Central Public Service Corp. Convertible 5½s, due 1949; Intercontinents Power Co. Convertible 6s, due 1948.

(2) Convertible bonds which were also participating: *Example:* American & Overseas Investing Corp. Convertible Participating Debenture 5s, due 1932.

(3) Participating bonds with warrants attached: *Example:* Associated General Utilities Co. 5% Participating Income Bonds, due 1956.

5. Bonds with a voice in management:

(1) Classic example of bond with full voting power was the $10,-000,000 issue of 5% bonds of United States Shipbuilding Company issued in 1902 to Mr. Charles M. Schwab in partial payment for his stock in Bethlehem Steel Co. The voting rights attaching to these bonds enabled Mr. Schwab to control the United States Shipbuilding Co. See W. Z. Ripley, *Trusts Pools and Corporations*, pp. 428–429, New York, 1916. Other examples of bonds with full voting power are as follows: Erie Railroad Co. First Consolidated Prior Lien 4s, due 1996; Chicago, Terre Haute & Southeastern Ry. Co. Income 5s, due 1960.

(2) Bonds with restricted voting power: *Examples:* Third Avenue Ry. Adjustment Income 5s, due 1960; Hudson & Manhattan R.R. Co. Adjustment Income 5s, due 1957.

PREFERRED STOCKS WHICH DEVIATE FROM THE STANDARD PATTERN

A. Preferred issues which deviate from the standard pattern with respect to the holder's interest in income:

1. Optional-dividend issues: A number of recent preferred issues grant the holder the option of receiving his dividend either in cash at a stated rate or in fractional shares of common stock. *Examples:* Blue Ridge Corp. $3 Convertible Preferred; Commercial Investment Trust Corp. $6 Convertible Preferred, Optional Series of 1929.

2. Participating preferred issues:

(1) Participation with the common stock without priority as to dividends: *Example:* Bucyrus Erie Co. Convertible Second Preferred. Sole preference is to $7.50 on liquidation.

(2) Participation with the common begins immediately after the preferred has received its prior dividend: *Example:* Celanese Corp. of America 7% Participating Preferred.

(3) Participation begins after the preferred has received its prior dividend and the common has received the same amount per share: *Examples:* Westinghouse Electric & Manufacturing Co. 7% Preferred; Chicago, Milwaukee, St. Paul & Pacific $5 Preferred.

(4) Staggered participation: *Example:* Chicago & Northwestern Ry. Co. 7% Participating Preferred.

(5) Participation with common begins after preferred has received its prior dividend and common has received smaller amount per share: *Example:* United Carbon Company 7% Preferred.

(6) Participation begins after the preferred has received a prior dividend and the common has received a larger amount per share: *Examples:* Southern California Edison Co., Ltd., 5% Participating Original Preferred; Porto Rican-American Tobacco Company Class *A* Participating Preferred.

(7) Participation unlimited: *Example:* A. M. Byers Co. 7% Preferred.

(8) Multiple participation: *Example:* White Rock Mineral Springs Co. 5% Second Preferred.

(9) Participation limited to a stated proportion of profits without specific upper limit in dollars: *Examples:* Celluloid Corp. First Participating $7 Preferred; Celanese Corp. of America 7% Participating Preferred.

(10) Participation limited to a stated proportion of profits, subject to a definite maximum per share; *Examples:* Kendall Co. $6 Participating Preferred, Series *A*; George A. Fuller Co. $6 Prior Participating Preferred and $6 Participating Second Preference Stock.

(11) Participation limited to a maximum sum in dollars: *Examples:* Bayuk Cigars, Inc., 7% Participating Preferred; Hershey Chocolate Corp. Convertible Preference Stock.

3. Convertible preferred issues: Such issues afford the holder the opportunity to increase his income return through shifting from the limited-income status of the straight preferred stock to the unlimited contingent-income position of the common stockholder. Variations in the duration and other terms of the privilege resemble those given above for convertible bonds and are not reported in full detail below.

(1) Duration of conversion privilege unlimited: *Example:* New York, New Haven & Hartford R. R. Co. 7% Preferred.

(2) Duration of conversion privilege limited: *Example:* Tide Water Associated Oil Co. 6% Convertible Preferred.

(3) Duration not subject to termination through redemption: *Example:* Hershey Chocolate Corp. Convertible Preference Stock.

(4) Duration subject to termination through redemption: *Example:* United Biscuit Co. of America 7% Convertible Preferred.

(5) Conversion privilege terminated through redemption of the issue to be extended through issuance of warrants to represent privilege: *Example:* Freeport Texas Co. 6% Convertible Preferred.

(6) Conversion price varies with extent to which holders exercise the privilege: *Examples:* Engineers Public Service Co. $5 Convertible Preferred; Tide Water Oil Co. 5% Convertible Preferred.

(7) Conversion price varies in accordance with a stated time schedule: *Example:* Shell Union Oil Corp. 5½% Convertible Preferred.

(8) Preferred stock convertible into bonds at the company's option in case earnings meet certain minimum requirements: *Example:* Knoxville Gas Co. 6% Preferred.

(9) Preferred which is both convertible and participating: *Examples:* Hershey Chocolate Corp. Convertible Preference Stock; Cincinnati Ball Crank Co. Participating Convertible Preference Stock.

(10) Preferred which is convertible into a participating issue: *Example:* Arnold Print Works $7 Cumulative Preferred.

(11) Preferred which is convertible at the option of the corporation into units of common and another class of preferred stock: *Example:* Reading Co. 4% Second Preferred.

(12) Preferred issue only part of which can be converted: *Example:* International Paper & Power Co. 7% Preferred.

4. **Preferred stock with stock-purchase warrants:** Such issues also afford the holder the right to modify his interest in corporate income beyond the stated dividend rate on the preferred stock alone. These warrants vary in much the same way as do those attached to bonds discussed above, and a detailed consideration of the variations would therefore be repetitious. Some of the leading variations are given below with examples of each.

(1) Preferred stocks issued with perpetual warrants: *Example:* American & Foreign Power Co. $7 Second Preferred.

(2) Preferred stocks issued with warrants which expire after a given date: *Example:* Engineers Public Service Co. $5.50 Preferred.

(3) Warrants attached to preferred stock entitling the holder to purchase stock at a flat price throughout the life of the warrant: *Example:* Alleghany Corp. 5½% Preferred.

(4) Preferred stock issued with warrants entitling the holder to purchase stock at prices varying in accordance with a time schedule: *Example:* General Realty & Utilities Corp. $6 Preferred.

(5) Preferred stock issued with warrants for purchase of stock at prices varying with the extent to which the privilege is exercised: *Example:* Central States Electric Corp. 6% Preferred.

(6) Preferred stock issued with warrants for the purchase of additional shares of the same issue: *Example:* National Public Utilities Corp. Class *A*.

(7) Preferred stock issued with warrants for the purchase of common stock: *Example:* American & Foreign Power Co. $7 Second Preferred.

(8) Preferred stock issued with warrants for the purchase of an issue senior to the common stock: *Example:* Revere Copper & Brass, Inc., 7% Preferred.

(9) Preferred stock issued with warrants for the purchase of stock in another corporation: *Example:* Solvay American Investment Corp. 5½% Preferred.

(10) Preferred stock issued with warrants which are exercisable only by payment of the subscription price in cash: *Example:* The Maytag Co. $3 Preference Stock.

(11) Preferred stock which may be surrendered at par in payment for common stock obtained through exercise of stock-purchase warrants: *Example:* Electric Power & Light Corp. $7 Second Preferred, Series *A*.

(12) Preferred stock issued with detachable warrants: *Example:* Oliver Farm Equipment $6 Class *A* Prior Preferred.

(13) Preferred stock issued with warrants which were detachable only in case of redemption of the senior issue or upon exercise: *Example:* United Aircraft & Transport Corp. 6% Preferred *A*.

(14) Convertible preferred stock with stock-purchase warrants: *Example:* National Public Service Corp. $3.50 Convertible Preferred.

(15) Participating preferred stock with stock-purchase warrants: *Example:* Kendall Co. $6 Participating Preferred, Series *A*.

(16) Convertible issue with warrants and with option to receive dividends in cash or in common stock: *Example:* Electric Shareholdings Corp. $6 Optional Dividend Series Preferred.

5. Miscellaneous deviations with reference to the interest of the holder in income:

(1) Accumulated unpaid dividends draw interest at 5%: *Example:* Pittsburgh Coal Co. 6% Participating Preferred.

(2) Preferred stock without preferences of any kind (really common): *Example:* Great Northern Ry. Preferred.

(3) Preferred stock whose dividend requirements rank ahead of bonds of the company: *Example:* Chicago Rapid Transit Co. 7.8% and 7.2% Prior Preferred, Series *A* and *B*.

(4) Preferred stock with dividend payments mandatory through guarantee: *Example:* Standard Oil Export Corp. $5 Preferred; Pittsburgh, Fort Wayne & Chicago Ry. Co. 7% Preferred.

(5) Preferred dividend rate adjusted on a definite scale in relation to profits: *Example:* Budd Wheel Co. 7% Participating Preferred.

(6) Preferred dividend rate dependent upon rate paid on other issues: *Example:* Southern California Edison Co., Ltd., 5% Original Preferred.

(7) Preferred issues with definitely graduated dividend rates: *Examples:* American Power & Light Co. $5 Preferred, prior to 1931;

Southeastern Power & Light Co. $4 Preferred, prior to 1931; General Baking Corp. (Md.) $6 Convertible Preferred, prior to 1931.

(8) Preferred stock entitled to extra dividends in directors' discretion: *Example:* International Match Corp. $2.60 Participating Preferred.

(9) Dividend rate increased to compensate for loss of conversion privilege: *Example:* Warner Brothers Pictures, Inc., $2.20 Convertible Preferred. Dividend was raised to $3.85 in 1930 when conversion privilege expired.

(10) Preferred issue entitled to all residual earnings after a limited dividend was paid to it and to the common: *Example:* American Brake Shoe & Foundry Co. 7% Preferred, prior to its retirement in 1920.

B. Preferred issues which deviate from the standard pattern with respect to the holder's interest in assets:

1. Preferred issues with no claim to assets whatever: *Example:* Midland Steel Products Co. $2 Noncumulative Preferred.

2. Preferred issues with no preference to assets: *Example:* Chicago & Northwestern Ry. Co. 7% Participating Preferred; New York Dock Co. 5% Preferred.

3. Preferred issue with abnormally low preference to assets in relation to par value and dividend rate: *Example:* Electric Storage Battery Co. Participating Preferred.

4. Preference to assets subject to voluntary modification on the part of the holder by shifting his position:
   (1) Convertible issues discussed above are of this type.
   (2) Warrant-bearing issues also fall into this class where the holder has the right to surrender his preferred in exercising his subscription privilege. Examples are given above.

5. Preferred issues with a right to share in surplus assets beyond a stated sum payable on liquidation:
   (1) Preferred stocks with participating dividends which share in surplus assets: *Examples:* White Rock Mineral Springs Co. 5% Second Preferred; Hershey Chocolate Corp. Convertible Preference Stock.
   (2) Preferred stocks which do not carry participating dividend rights but which have the right to share in surplus assets: *Example:* Wabash Ry. Co. 5% Convertible Preferred B.

6. Preferred stocks with participating dividends but without right to share in surplus assets: *Examples:* Westinghouse Electric & Manufacturing Co. 7% Participating Preferred; United Carbon Co. 7% Preferred.

7. Preferred stock with mortgage lien: *Example:* Georgia Southern & Florida Ry. Co. 5% First Preferred.

8. Preferred stock callable at very high premium: *Examples:* American Radiator & Standard Sanitary Corp. 7% Preferred, callable at 175; Dennison Manufacturing Co. 8% Cumulative Debenture Stock, callable at 160.

9. Preferred stock subject to compulsory redemption: *Examples:* Marlin Arms Corp. 7% Preferred, issued in 1915 and all retired early in 1917 in accordance with compulsory provisions; Houdaille-Hershey Corp. Class *A* Stock, is a preferred stock according to its terms and must be redeemed at $45 a share on July 1, 1953.

10. Preferred stocks guaranteed as to principal in event of liquidation: *Examples:* Standard Oil Export Corp. 5% Preferred; Armour & Co. of Delaware 7% Preferred.

C. Preferred issues which deviate from the standard pattern with respect to the holder's right to vote: Preferred stock usually has either no voting power or voting power shared with the common in such proportions that the latter is able to outvote the former. For detailed study of the voting rights of preferred stock see W. H. S. Stevens, *Stockholders' Voting Rights and the Centralization of Voting Control,* 40 *Quarterly Journal of Economics* 353–392 (May, 1926). Following is a partial list of deviations:

1. Classic example of voting control by preferred stockholders: Rock Island Co. of New Jersey Preferred. See W. Z. Ripley, *Railroads: Finance and Organization,* pp. 524–532, New York, 1915.

2. Preferred stock with exclusive voting power: *Example:* International Silver Co. 7% Preferred. Common stock had no voting power prior to Jan. 1, 1902. Even thereafter the preferred stockholders' votes outnumbered those of the common stockholders.

3. Preferred stock with controlling amount of voting power: *Example:* American Brake Shoe & Foundry Co. 7% Preferred.

4. Preferred stock with preponderant voting power due to multiple voting: *Example:* American Tobacco Co. 6% Preferred.

5. Two preferred issues of same company with combined voting power in excess of that of common: *Example:* Associated Dry Goods Corp. 6% and 7% Preferred issues.

6. Preferred issues attaining exclusive voting power upon occurrence of certain events—usually failure to pay dividends for specified period: *Examples:* Studebaker Corp. 7% Preferred; Bayuk Cigars, Inc., 7% Participating Preferred; Endicott-Johnson Corp. 7% Preferred; Willys-Overland Co. 7% Preferred.

7. Preferred issues which attain the right to elect a majority of the directors in the event of certain defaults: *Examples.* Consolidated Cigar Corp. 6½% Prior Preferred and 7% Preferred; General Tire & Rubber Co. 6% Preferred *A*; Consolidated Oil Corp. 8% Preferred.

COMMON STOCKS WHICH DEVIATE FROM THE STANDARD PATTERN

The task of classifying common stocks from the standpoint of their deviations from the standard pattern as outlined in the text is rendered less difficult by the foregoing grouping of non-standard senior securities. This is due to the fact that many of the issues listed above are non-standard because they infringe in one way or another upon the traditional province of the common stockholder. Many of the irregular characteristics of common stocks may, therefore, be illustrated by reference to companies which have issued the senior securities mentioned above, which by their terms transgress upon the conventional rights and interests of the common stockholders.

A. Common stocks which deviate from the standard pattern with respect to their interest in profits:

1. Common stocks with direct limitations upon the right to receive dividends:

    (1) A. O. Smith Corp. Common. So long as the preferred stock outstanding amounted to $1,500,000 or more, common dividends were limited to $1.20 per share per annum.

    (2) American Brake Shoe & Foundry Co. Common. Prior to 1920 the common dividend was limited to $7 per share per annum.

    (3) Marlin Arms Corp. Common. During 1915 to 1917 the Corporation had outstanding a $3,500,000 issue of 7% Preferred and no common dividends could be paid until it was retired in full.

2. Common stocks with limited interest in earnings due to the right of other classes of securities to participate therein:

    (1) See companies with various classes of participating senior securities listed above, and companies issuing classified participating securities below.

3. Common stocks with an interest in earnings which is subject to voluntary modification by the holders of securities convertible into common stock:

    (1) See companies with various classes of convertible senior securities listed herein.

4. Common stocks the extent of whose interest in earnings is subject to voluntary modification by the holder of stock-purchase warrants:

    (1) See companies listed herein which have issued stock-purchase warrants in connection with senior securities;

    (2) See also those companies which have issued stock-purchase warrants and options for the purchase of common stock without reference to financing through senior securities: *Examples:* Fourth National Investors Corp.; Public Utility Holding Corp. of America; United Corp.; American & Dominion Corp.

5. Common stocks with dividend payments that are mandatory through guarantees or through the terms of leases: *Examples:* Michigan Central R. R. Common; Pittsburgh, Fort Wayne & Chicago Ry. Common.

6. Common (or Capital) stock with a claim to income which is senior to some other issue: *Examples:* Green Bay & Western R. R. Co. Capital Stock, senior to "Income Debentures, Series *B*"; Eastern Utilities Associates Common Stock, senior to "Convertible Stock."

B. Common stocks which are non-standard with respect to their interest in assets:

1. Common stocks junior to bonds having an indefinite claim to assets on redemption or liquidation: *Example:* Siemens & Halske A. G. Common.

2. Common stocks junior to issues which are convertible into common or which have the right to participate with the common in assets:
   (1) See common stocks of companies with convertible and participating issues listed above;
   (2) See common stocks of companies with convertible and participating classified issues below.
3. Common stocks of companies which have issued stock-purchase warrants and options for the purchase of common stock on specified terms: *Examples:* See companies mentioned herein which have issued such warrants or options. Stock-purchase warrants may be a device for siphoning both earnings and assets from the common stockholder.
4. Common stocks junior to preferred issue which has no claim to assets: *Example:* Midland Steel Products Co. Common.
5. Common stock with a claim to assets which is both limited in amount and is senior to a (so-called) debenture issue as well as concurrent with the claim of another (so-called) debenture issue: *Example:* Green Bay & Western R. R. Co. Capital Stock.

C. Common stocks which are non-standard from the point of view of the right of the holders to control the policies of the corporation through exercise of the voting power:
   1. Common issues with limited control due to intrusion of senior issues with controlling or exclusive voting power: *Examples:* See those given above.
   2. Nonvoting common stocks: *Examples:* Great Atlantic & Pacific Tea Co. of America Nonvoting Common; American Tobacco Co. *B.*

D. Classified stocks: This group is considered under the general heading of common stocks because the majority of classified stocks represent a subdivision of what is ordinarily the common stockholders' interest in assets, earnings, and control. There are many instances, however, in which one of the two or more "classified" issues is a genuine preferred stock in everything but title. The ensuing classification (which is far from exhaustive) indicates that the use of one or another of the titles "Class *A*," "Class *B*," or "Common" implies no specific division of the rights and responsibilities of the holders.
   1. The title "Class *A* " may indicate a preference of various kinds over the Class *B* or Common stock, *viz:*
      (1) As to dividends only: *Example:* Coca-Cola Co. *A.*
      (2) As to assets only: *Example:* Northern States Power Co. (Del.) *A.*
      (3) As to assets and dividends: *Example:* Continental Baking Corp. *A.*
      (International Paper & Power Co. Common Class *A* is entitled to receive aggregate dividends of $12 before the Class *B* or Class *C* receive anything. After such payment, Class *A* and Class *B* become one class and are entitled to further preferential dividends totalling $12 per share, after which all three classes become identical.)
   2. Frequently Class *A* shares have conversion or participating rights, in addition to a preference. *Examples:* Curtiss Wright Corp. *A*

(Convertible); Middle States Petroleum Corp. *A* (Participating); McCord Radiator & Mfg. Co. *A* (Convertible and Participating). (Participation may relate to earnings, to assets, or to both.)

3. Class *A* stocks which are essentially preferred issues may be callable (*e.g.* General Outdoor Advertising Co. *A*) or have a sinking-fund provision (*e.g.* Austin Nichols & Co. *A*).

4. Frequently the sole difference between classes refers to voting rights. Various combinations occur, *viz:*

   (1) Class *A* votes; Class *B* does not: *Example:* American Cyanamid Co.

   (2) Class *B* votes; Class *A* does not: *Example:* United Light & Power Co.

   (3) Common votes; Class *B* does not: *Example:* Liggett & Myers Tobacco. Co.

MISCELLANEOUS NON-STANDARD SECURITIES

A. Management stock, founders stock, employees' stock. Management and founders stock has voting control, and generally represents a relatively small capital investment. There are usually some other special features, distinguishing it from ordinary voting common.

   Examples: The Charles E. Hires Co. The management stock of this company is not transferable on the books of the company. It is convertible into Class *B* stock, which is equivalent thereto except that it is nonvoting and transferable.

   Dennison Manufacturing Co. This company has both management stock and employees' stock. They are identical except that management stock has (normally) exclusive voting power. Both issues are exchangeable for senior ranking Class *A* stock (or under certain conditions, for preferred shares senior to the Class *A* stock) upon termination of the owner's employment.

   New York Shipbuilding Corp. This company has founders stock which is entitled as a class to 35% of all net profits after preferred dividends accruing from Jan. 1, 1929. The remaining 65% belongs to an issue of nonvoting participating stock.

B. Restricted shares. These shares are not entitled to dividends initially, but gain the right to dividends on certain conditions, which are generally related to the amount of earnings and to the passage of time. *Example:* Trico Products Corp. (described on page 460).

C. Debentures in name but common stock in fact: *Example:* Green Bay & Western R.R. Co. Income Debentures *B.*

D. Debenture stocks: *Examples:* Bush Terminal Co. 7% Debenture Stock; E. I. du Pont de Nemours & Co. 6% Debenture Stock; Dennison Mfg. Co. 8% Debenture Stock.

   The title "Debenture Stock" is a meaningless expression in America and is somewhat misleading in that it suggests a fixed, unsecured obligation when, as a matter of fact, the title usually designates an ordinary preferred issue. The title in England, Canada, and elsewhere has a different significance.

E. Dividend participation certificates: *Example:* General Gas & Electric Corp. (Del.) Dividend Participation Certificates. These were certificates representing participation rights incorporated in participating preferred stocks of the predecessor company. They were not entitled to any assets in liquidation, but shared with the junior stock issues in dividend distributions after certain prior dividends had been paid. Dividends thereon were paid in 1929. They were exchangeable for Class *A* stock of the corporation during 1929. Unexchanged certificates were retired at $30 each on Nov. 1, 1929. Compare the circumstances under which Midland Steel Products Co. $2 Noncumulative Preferred was created.

F. Bankers shares: *Example:* Cities Service Co. Bankers Shares—a device for subdividing high-priced shares. Possibly a device for unloading high-priced shares at an attractive figure.

G. Allotment certificates: *Example:* Associated Gas & Electric Co. $8 Interest-bearing Allotment Certificates. These were convertible into stock at the option of the company and into other securities at the option of the holder. The company called them for conversion in 1932. See 1931 Manuals.

H. Adjustment preferred: *Example:* Norfolk & Western Ry. Co. 4% Adjustment Preferred. Has no relation to the adjustment or income-bond idea. Probably refers to its issuance in reorganization.

I. Secured debentures: *Example:* American Community Power Co. Secured Debenture 5½s, due 1953. Judged by the standards of American terminology the title is a contradiction in terms.

J. Common dividends being paid while no dividends paid on the preferred: *Example:* American Ship Building Co. during 1932–1933. This apparently anomalous situation was probably the result of the United States Cast Iron Pipe Co. case referred to in the text. Report of the president for the year ended June 30, 1932, stated that "Dividends on the 7% noncumulative preferred stock have been paid during the past year out of funds applicable to preferred stock, which were accumulated in past years in which preferred dividends were not paid. As these funds are practically exhausted, no further dividends can be paid on the preferred stock until earned."

K. Interest on an income bond paid during receivership: *Example:* Wabash Railway Noncumulative Debenture 6s, Series *B*, due 1939. Interest was paid thereon during 1932 and 1933 while fixed-interest obligations were in default. The fact that the income issue was a small one *underlying* substantial amounts of fixed-interest obligations may explain this anomaly.

NOTE 4

On Sept. 14, 1928, a banking group offered "American Certificates" for sale at $28.14, each "Certificate" representing Kroner 20, par value of Kreuger & Toll Co. Participating 5% Debentures, dated July 1, 1928. Expressed in dollars at the then present parity of exchange, each "American Certificate" represented Debentures having a par value of $5.36 (Kr. 1

equal to $0.268). The following features justified classification of the issue as of the common-stock type:

1. The underlying Debentures bore interest at 5%, payable annually, and were entitled to additional interest at the rate of 1% for each 1% by which the dividend paid or declared on the ordinary shares in any fiscal year exceeded 5%.

2. The issue price of the "American Certificates" was 5¼ times the par value of the related Debentures. At the regular (*i.e.*, the non-participating) interest rate of 5% the yield on the offering price would be less than 1%.

3. The owner was dependent for a reasonable income upon the participating feature of the Debentures, and this in turn was governed by the dividend paid on the stock. Only about *one-fifth* of the income and principal value of this security could be ascribed to the bond contract; the remaining four-fifths had all of the contingent and variable features of a common-stock commitment. This division may be set forth as follows:

(Per unit of 20 Kroner)

Item	Bond component	Stock component	Total
Principal............................	$5.36	$22.78	$28.14
Income in 1928......................	0.27	1.07	1.34

NOTE 5

Convincing evidence of the investment character of National Biscuit Co. Preferred is found in the price history and dividend record of the issue. The annual dividend of $7 per share has been paid regularly since organization of the company in 1898. The issue has not sold below par ($100) since 1907. The average of the annual high and low prices since 1907 (through Dec. 31, 1932) was $125.25, on which the annual dividend of $7 has yielded 5.59%. A similar average for the entire history of the issue on the New York Stock Exchange (1899–1932) gives a price of $119.50 and a yield of 5.855%. This average covers a range of 79½ in 1900 and 153¼ in 1931. In only 5 out of the 34 years since the issue was first listed on the New York Stock Exchange has it sold at a price below par. With the exception of the panic year 1907, the issue has not sold below par since 1903.

NOTE 6

Twenty-five million dollars of Seaboard-All Florida Railway First Mortgage 6% Gold Bonds, Series *A*, due Aug. 1, 1935, were originally offered in 1925 at 98½ and interest. The bonds were joint and several obligations of the Seaboard-All Florida Ry., Florida Western & Northern R.R. Co., and the East & West Coast Ry. They were further secured by an unconditional guarantee with respect to both principal and interest, through endorsement by the Seaboard Air Line Ry. Co., which leased the properties of the several roads at a minimum annual net rental equal to the annual interest charges on all bonds outstanding under the mortgage.

The proceeds from the sale of these bonds were used mainly to redeem outstanding first-mortgage obligations of the lessor roads and to construct about 217 miles of new trackage along the east and west coasts of Florida. Thus the bonds had a first lien on approximately 475 miles of newly constructed and established lines.

The Seaboard-All Florida Ry. went into the hands of receivers on Feb. 2, 1931, following an earlier receivership for the Seaboard Air Line Ry. Co. and a default in interest due on these bonds.

The fortunes of the Seaboard Air Line Ry. Co. and of its subsidiary, Seaboard-All Florida Ry., had synchronized in recent years with the rise and fall of the Florida real-estate boom. Although the buyers of these bonds provided $24,625,000 to defray the cost of acquiring and constructing Florida railway properties, by December 1931 their bonds were selling as low as 1 cent on the dollar, the market appraising the value of their investment at only $250,000.

NOTE 7

See, for example, the history of the following underlying issues of the old Brooklyn Rapid Transit System on which interest was defaulted in 1919 and the default cured later by full cash payment and reinstatement of the issues under the mortgages as they existed at the date of default:

1. Brooklyn, Bath & West End R.R. Co. General Mortgage 5s, due in 1933. (References: *Poor's Public Utilities:* 1925, p. 1814; 1923, pp. 2640–2641.)

2. Atlantic Avenue R.R. Co. of Brooklyn Improvement Mortgage 5s, due in 1934. (References same as above.)

3. Nassau Electric R.R. Co. First Mortgage 5s, due in 1944. (References: *Poor's Public Utilities:* 1925, p. 1813; 1923, pp. 2640–2641.)

4. Brooklyn, Queens County & Suburban R.R. Co. First Mortgage 5s due in 1941. (References: *Poor's Public Utilities:* 1925, p. 1810; 1923, pp. 2640–2641.)

In the following cases the defaulted interest was paid up in part by the issue of preferred stock of the Brooklyn-Manhattan Transit Corporation, the successor company which emerged from the 1923 reorganization of the old B.R.T.

1. Nassau Electric R.R. Co. First Consolidated Mortgage 4s, due in 1951. (References: *Poor's Public Utilities:* 1925, p. 1814; 1923, pp. 2640–2641.)

2. Brooklyn Queens County & Suburban R.R. Co. First Consolidated Mortgage 5s, due in 1941. (References: *Poor's Public Utilities:* 1925, p. 1811; 1923, pp. 2640–2641.)

Somewhat unusual treatment was accorded to the holder of Coney Island & Brooklyn R.R. Co. Consolidated 4s, due in 1955, owing to the efforts of the Equitable Life Assurance Society of the United States which was the beneficial holder of all publicly held bonds of this issue. Defaulted interest was cured by cash payment and the holder of the old bonds received not only the same lien as that which was held before, but a higher rate of interest on new bonds exchanged for the old, as well as the benefit of having the mortgage closed in its behalf. [References: *Poor's Public Utilities:*

1923, pp. 2640–2641; 1925, pp. 1811–1812; 1931, p. 138; *Moody's Public Utilities*, 1929, p. 171; 117 *Chronicle* 552 (Aug. 4, 1923); 129 *Chronicle* 1438 (Aug. 31, 1929)].

Another example is afforded by the Murray Body Corp. First Mortgage 6½s, due in 1934. Here there were certain interest and sinking-fund defaults which were later cured by cash payment and the issue assumed by the reorganized corporation. (References: *Poor's Industrials:* 1925, 1926, and 1927, *passim;* 121 *Chronicle* 3013; 122 *Chronicle* 3094; 123 *Chronicle* 334.)

NOTE 8

For examples of this rare occurrence, see the following issues:

1. Indianapolis, Decatur & Springfield Ry., First Mortgage 7s, due in 1906. (References: 55 *Chronicle* at p. 66 of "Investor's Supplement," Nov. 26, 1892; 55 *Chronicle* 938; 57 *Chronicle* 815; 59 *Chronicle* 28.)

2. Colorado Yule Marble Co. First Mortgage 6s, due in 1920. (References: 103 *Chronicle* 760; 109 *Chronicle* 374; Smythe, R. M., *Valuable Extinct Securities*, p. 102, New York, 1929.)

3. Guerin Mills, Inc., First Mortgage 7s, due in 1937. (References: *Poor's Industrials*, 1928, p. 2436; *ibid.*, 1929, p. 645; 127 *Chronicle* 960.)

That the sum payable to the holders of foreclosed mortgage bonds is ordinarily well below par is not only a familiar fact to those who have come into practical contact with the problems of actual foreclosure, but is indicated by the decree-values listed in various compilations of issues which have been defaulted. The most comprehensive of these is the *Marvyn Scudder Manual of Extinct or Obsolete Companies*, Vols. I to IV. New York, 1926, 1928, 1930, 1934. See also: Smythe, R. M., *Valuable Extinct Securities*, New York, 1929; Smythe, R. M., *Obsolete American Securities and Corporations*, Vol. II, pp. 12–17, New York, 1911.

NOTE 9

The Missouri, Kansas & Texas Railway Company went into the hands of receivers in 1915. Prior thereto the First 4s of 1990 had sold as high as 104¼ in 1905 and as late as 1914 had sold at 91⅞. Before the financial difficulties leading to the 1915 receivership, the record of this issue was distinctly that of a high-grade, investment bond. During the 11 years 1903 to 1912, inclusive, the lowest price at which it sold was 98½ (in the panic year 1907).

The receivership was a protracted one, and throughout the period there was a delay of six months in the payment of each interest coupon. Although technical default was avoided, the bonds were dealt in "flat" on the New York Stock Exchange and their status as investments suffered severely. They sold below 60 in 1917 and in 1919, and reached their lowest price of 52⅛ in 1920 while the receivership was still in effect. In 1921, the year in which the plan of reorganization was announced, the issue sold as low as 56, and it was not until 1927 that it regained a semblance of its former prestige as an investment by selling above 90. Thus the first lien did not protect the holder from a substantial market decline during the period of financial difficulty.

The same sort of picture is presented by the record of Brooklyn Union Elevated R.R. First 5s, due in 1950, described in Chap. II of the text. This was an underlying lien on essential parts of the elevated lines of the Brooklyn Rapid Transit Co. which went into the hands of receivers on Dec. 31, 1918, and was reorganized as the Brooklyn-Manhattan Transit Corp. in 1923. The issue ranked as a first-grade investment from 1903 to 1917 and never sold below 90 during this period, except in the panic of 1907 when it dropped to 85, and in 1917 when the receivership appeared imminent. Although the issue was not disturbed by the reorganization, it sold as low as 55 in 1920, while the receivership was still in effect, and did not regain its former standing until 1926, three years after the termination of the receivership. (References: *Reorganization Plan* dated Mar. 15, 1922; *Poor's Public Utilities*, 1923, p. 2637; 116 *Chronicle* 1646.)

NOTE 10

Sources of the figures in this table and the content of the specific series are as follows:

1. Railway Operating Revenues for all Class I Railroads in the United States, including switching and terminal companies. Compiled from *Statistics of Railways in the United States*, Interstate Commerce Commission, Washington, D. C., 1922–1932.

2. Net Railway Operating Income (net after rents) for all Class I Railroads in the United States. Sources same as (1) above.

3. Average yield on fifteen high-grade railroad bonds and fifteen high-grade utility bonds, respectively, as compiled by Standard Statistics Co., Inc. Published in the *Standard Statistical Bulletin* (1931 Base Book) and subsequent monthly *Statistical Bulletins*.

4. Gross and net earnings (after operating expenses and taxes) of gas and electric companies, compiled by the U. S. Department of Commerce from reports of 95 public-utility companies or systems operating gas, electric light, heat, power, traction, and water services and comprising practically all of the important utility organizations in the United States, exclusive of telephone and telegraph companies. Series was discontinued Dec. 31, 1930. Source: *Survey of Current Business (1931 Annual Supplement)*, pp. 196–197, U. S. Department of Commerce, Washington, D. C.

5. New series (overlapping the U. S. Department of Commerce figures on public-utility gross which were discontinued Dec. 31, 1930) to show recent trend of utility gross. Compiled by the authors from several series published in the *Survey of Current Business* covering: (a) *Electrical World* series on gross revenues from the sale of electrical energy which is computed to cover 100% of the industry, exclusive of street railways; (b) American Gas Association series covering total revenues from the sale of manufactured and natural gas to consumers (exclusive of sales of natural gas for manufacture of carbon black and other purely industrial uses); (c) American Electric Railway Association figures on the operating revenues of street railways; and (d) Interstate Commerce Commission figures on total operating revenues of Class A telephone companies (those having annual operating revenues in excess of $250,000).

6. Combined net operating income of 104 telephone companies and net earnings of 65 other public-utility companies, compiled by the Federal Reserve Bank of New York and published in the *Monthly Review of Credit and Business Conditions, Second Federal Reserve District*, April 1, 1933, p. 30. (Federal Reserve Agent, New York.)

NOTE 11

GULF STATES STEEL COMPANY SINKING FUND DEBENTURE 5½s, DUE 1942
($14,000,000 offered in 1927 at 98¾; $2,000,000 offered in 1930 at 98)

Year	Net available for fixed charges	Number of times earned 1929 fixed charges	Price range for bonds
1922	$ 958,208	4.15	(Not issued
1923	1,576,521	6.90	until 1927)
1924	979,315	4.29	
1925	1,036,778	4.54	
1926	799,793	3.50	
1927	881,607	3.86	97½–94
1928	1,154,923	5.06	92⅝–87⅜
1929	1,538,779	6.73	99 –94¾
1930	*489,902(d)*	*(d)*	100⅞–89
1931	*658,538(d)*	*(d)*	90 –26
1932	*318,078(d)*	*(d)*	57½–21

Numerous other examples could be given, of which we present only two.

MARION STEAM SHOVEL COMPANY FIRST SINKING FUND 6s, DUE 1947
($3,600,000 offered in 1927 at 99½)

Year	Net available for fixed charges	Number of times fixed charges earned	Price range for bonds
1922	$1,012,759*	4.68†	(Not issued
1923	1,357,494*	6.35†	until 1927)
1924	974,905*	4.52†	
1925	837,844*	3.88†	
1926	867,130*	4.01†	
1927	654,433*	3.03†	100 –97
1928	583,717	2.78	102 –98¾
1929	737,495	3.62	99½–81
1930	*488,904(d)*	*(d)*	88¾–46
1931	*583,267(d)*	*(d)*	47 –21
1932	*689,766(d)*	*(d)*	55 –21

* For predecessor companies as reported to New York Stock Exchange.
† Based on 1927 charges.

McCRORY STORES CORPORATION DEBENTURE 5½s, DUE 1941
($6,000,000 offered in December 1926, at 98)

Year	Net available for fixed charges	Times 1931 bond interest earned	Times 1931 fixed charges earned	Price range for bonds
1922	$1,185,070	4.35	2.72	(Not issued
1923	1,671,093	6.13	3.84	until 1926)
1924	1,988,987	7.30	4.58	
1925	2,298,684	8.44	5.29	
1926	2 850,236	10.46	6.56	
1927	3,285,267	12.06	7.56	101¼-97
1928	3,004,149	11.02	6.91	102⅝-98½
1929	2,957,550	10.85	6.80	99⅜-92
1930	2,492,092	9.11	5.71	100½-93¾
1931	1,417,566	5.20	3.26	100 -74
1932	312,900	1.15	0.72	91 -52
1933*	....	.....	....	63 -21⅝

* Petition in bankruptcy filed in January 1933.

NOTE 12

The following exhibits illustrate the point made in the text:

BOTANY CONSOLIDATED MILLS, INC., FIRST MORTGAGE 6½s, DUE 1934

Year	Available for fixed charges	Times fixed charges were earned	Price range for bonds
1916–1923	$2,982,145*	5.73†	(Issued in 1924)
1924	2,119,608	5.43	95 bid
1925	991,778	1.67	96½-94
1926	*3,882,687*(d)	(d)	96 -80½
1927	184,741	0.17	92 -80
1928	*510,926*(d)	(d)	83⅛-59
1929	*1,926,149*(d)	(d)	73⅛-40
1930	*2,356,770*(d)	(d)	48¾-33
1931	*2,773,531*(d)	(d)	36½-15
1932‡	....	.....	19 - 5

* Average for seven fiscal years ended Nov. 30, 1923.
† Charges on the above bonds, issued in 1924.
‡ Receivership, Mar. 28, 1932.

R. Hoe and Company, First Mortgage 6½s, Due 1934

Year	Available for fixed charges	Times fixed charges were earned	Price range for bond issue
1919	$1,058,288	2.73*	(Bonds
1920	680,174	1.76*	issued
1921	1,101,402	2.85*	in
1922	1,301,676	3.36*	1924)
1923	1,316,214	3.40*	
1924	835,167	2.16*	100¾–99¼
1925	390,978	1.01	100¼–94
1926	874,975	2.19	99¼–91½
1927	255,192	0.65	102½–96
1928	300,116	0.79	99¼–85
1929	1,047,447	2.39	94½–74⅝
1930	580,664	1.25	90 –65
1931	*310,948(d)*	*(d)*	68 –20
1932†	....	.....	30 – 6⅛

* Number of times 1925 fixed charges were earned. Fixed charges (prior to flotation of $4,500,000 First Mortgage 6½s, due in 1934, late in 1924) were very small in relation to earnings.
† Receivership, Apr. 22, 1932.

Long-Bell Lumber Corporation (No parent company funded debt)
(Consolidated figures)

Year	Available for fixed charges	Times fixed charges were earned	Price range for Long Bell Lumber Co. First 6s, due 1946
1922	$5,047,310	7.08	(Offered at 97
1923	6,742,616	6.22	in 1926)
1924	5,493,046	3.69	
1925	6,099,748	4.74	
1926	4,358,519	2.66	96–93
1927	1,274,270	0.68	95–85
1928	2,215,212	1.12	93–88
1929	1,992,785	1.02	93–75
1930	*1,725,937(d)*	*(d)*	82–65
1931	*3,142,062(d)*	*(d)*	60–17
1932*	*3,456,064(d)*	*(d)*	16– 6

* Defaults on all bond issues in 1932.

NATIONAL RADIATOR CORPORATION DEBENTURE 6½s, DUE 1947

Year	Available for fixed charges	Times fixed charges were earned	Price range for bonds
1922	$1,020,675	1.3*	(Offered at par
1923	2,456,076	3.1*	in August
1924	3,405,764	4.4*	1927)
1925	3,488,980	4.5*	
1926	3,472,185	4.4*	
1927†	896,948†	3.24†	101 −99½
1928	*587,124(d)*	(d)	101 −73⅛
1929	*490,370(d)*	(d)	82¼−20
1930	*761,855(d)*	(d)	40 −14
1931‡	. . . .	. . . . .	25⅞− 5
1932	. . . .	. . . . .	25 − 7

* Full charges on 6½s, due in 1947, which were issued in 1927.
† From Aug. 30 to Dec. 31. Figures for balance of year not available. Total for year estimated at about $1,800,000 or 2.3 times charges for the year.
‡ Receivership, Oct. 9, 1931.

NOTE 13

For example: Mexican Light & Power Co. First 5s, due in 1940, were not in default in June 1933 and were selling at 50, whereas the issues of the Republic of Mexico listed on the New York Stock Exchange were all in default and were selling at from 4 to 6 cents on the dollar at that time; Chile Copper Co. Debenture 5s, due in 1947, were selling at 67 in June 1933, whereas the Republic of Chile 6s were in default since 1931 and were selling at prices ranging from 11 to 12 cents on the dollar; Rio de Janeiro Tramway, Light & Power Co. First 5s, due in 1935, were at 87 in June 1933, whereas the bonds of the City of Rio de Janeiro were in default since 1931 and were selling at 22, having sold below 10 cents on the dollar earlier in the year; Pirelli Co. of Italy Sinking Fund Convertible 7s, due 1952, were selling above par in June 1933, whereas the Kingdom of Italy External Sinking Fund 7s, due in 1951, were selling at 95, neither issue being in default.

NOTE 14

For example, the Sept. 1, 1932 coupon on Alpine-Montan Steel Corp. First 7s, due in 1955, was not paid because of foreign exchange restrictions imposed by the Austrian government, although the corporation was financially able to make the payment. The Aug. 1, 1932 coupon on Rima Steel Corp. First 7s, due in 1955, was not paid owing to a decree of the Hungarian government suspending payments abroad in foreign currencies on Hungarian financial obligations, from and after Dec. 23, 1931. The principal of Deutsche Bank 6% Notes, due Sept. 1, 1932, was not paid at maturity owing to exchange restrictions imposed by the German government. Holders were offered immediate payment in marks to be left in Germany, or payment on Sept. 1, 1935, in dollars with an immediate payment of a cash premium of 2% in dollars.   A similar compromise was worked out with respect to Saxon Public Works, Inc., 5% Notes due July 15, 1932.

NOTE 15

For a detailed treatment of the investment qualities and record of equipment-trust obligations the student is referred to Kenneth Duncan, *Equipment Obligations*, Chap. VII, New York, 1924.   A case history of defaults on equipment obligations and their treatment in railroad reorganizations since 1900 will be found at pp. 229–239 of this excellent treatise.   To quote briefly from Duncan, writing in 1924 (pp. 199–200), "In only three instances has it been necessary for the holders of equipment securities to accept a compromise in the form of receiving other securities instead of cash, in only two instances did they have to retake the equipment and sell it, and in no case did payment finally fail to be made, either in cash or in other securities which could later have been sold for as much as the principal of the equipment obligations on which default has occurred."   See also A. S. Dewing, *Financial Policy of Corporations*, Bk. I, Chap. VII, New York, 1926.

A briefer synopsis of the treatment of equipment obligations in railroad receiverships is reproduced below from a study by Freeman & Co., specialists in equipment obligations, which was published on June 15, 1932.

RAILROAD RECEIVERSHIPS——ULTIMATE OUTCOME OF EQUIPMENT TRUST ISSUES

It is significant that the sale of almost four billion dollars in Railroad Equipment Trust Certificates to American Investors during the past fifty years or so has been characterized by no loss of any consequence.   This is a record that cannot be equalled by any other form of investment medium with the exception of United States Government issues.   The period 1888 to 1930 includes the depressions of 1897, 1902, 1907, etc.   This record has been such as to warrant confidence in the ability of Equipment Trusts to continue their record through the present depression.

The following is a list of railroad receiverships and the ultimate outcome of Equipment Trust issues:

1886—DENVER RIO GRANDE R. R.   Notes exchanged for mortgage bonds and preferred stock which later were worth forty per cent. more than Equipment Trust with bondholders' consent.

1888—CHESAPEAKE & OHIO.   Equipments undisturbed—interest rates on other securities reduced.

1892—CENTRAL RAILROAD & BANKING CO. of GEORGIA.   Undisturbed—paid in full.

1892—SAVANNAH, AMERICUS & MONTGOMERY.   Undisturbed—paid in full.

1892—TOLEDO ST. LOUIS & KANSAS CITY R. R.   Undisturbed—paid in full.

1895—ATCHISON TOPEKA & SANTA FE.   Receiver reserved $1200 mortgage bond to retire each $1,000 Equipment at maturity.

1895—NEW YORK, LAKE ERIE & WESTERN.   Receiver certificates issued to pay Equipments.

1895—UNION PACIFIC.   Undisturbed—mortgage bonds reserved to pay Eq. at maturity.

1896—PHILADELPHIA & READING.   Equipments paid—partly by assessment.

1896—NORTHERN PACIFIC.   Undisturbed—paid regularly.

1899—COLUMBUS HOCKING VALLEY & TOLEDO RY.—Interest paid promptly and ten per cent. of principal retired regularly in accordance with new agreement.

1900—KANSAS CITY, PITTSBURGH & GULF—New first mortgage bonds issued to pay Equip.

1905—CINCINNATI, HAMILTON & DAYTON—Undisturbed.

1905—PERE MARQUETTE—Undisturbed—sold additional Equipment Trusts during receivership to yield 6 %.

1908—SEABOARD AIR LINE—Receivers' certificates sold to pay off maturing Equip.
1908—DETROIT, TOLEDO & IRONTON—Full recovery of principal except for deduction of legal fees and expenses.
1910—BUFFALO & SUSQUEHANNA—Equipment sold.   No loss.
1915—WABASH RAILROAD—   Option of cash or 6 % Equipment Trusts.
1916—MINNEAPOLIS & ST. LOUIS   Paid in full—undisturbed.
1916—MISSOURI PACIFIC   Paid in full—undisturbed.
1916—NEW ORLEANS TEXAS & MEXICO   Paid in full—undisturbed.
1916—ST. LOUIS—SAN FRANCISCO   Paid in full—undisturbed.
1916—WESTERN PACIFIC—   Paid in full—undisturbed.
1916—WHEELING & LAKE ERIE   Paid in full—undisturbed.
1917—WABASH PITTSBURGH TERMINAL   Paid in full—undisturbed.
1918—CHICAGO PEORIA & ST. LOUIS   Temporary default.   Payment resumed in 1919.
1920—WASHINGTON VIRGINIA R. R.   New management paid all arrears.
1921—MISSOURI KANSAS TEXAS   Paid in full—undisturbed.
1921—ATLANTA BIRMINGHAM & ATLANTIC—Cash offering in settlement.
1922—CHICAGO & ALTON—   Paid in full—undisturbed.
1923—MINNEAPOLIS & ST. LOUIS   Still in receivership—full payment being made.
1927—CHICAGO MILWAUKEE & ST. PAUL   Paid in full—undisturbed.
1931—WABASH RAILWAY—   Being paid in full.
1931—FLORIDA EAST COAST RAILWAY—   Being paid in full.
1931—SEABOARD AIR LINE RAILWAY—Principal payments up to 1934 extended to 1935.   All interest now being paid regularly.
1932—MOBILE & OHIO—   Being paid in full.
June 15, 1932.

Subsequent to the publication of the above list the Wabash Railway obtained from its equipment-trust holders a three-year extension of its 1933 and 1934 maturities, the Florida East Coast Ry. negotiated an extension of certain 1932 maturities, and the Mobile & Ohio was about to effect similar extensions in June 1933.

NOTE 16

In view of their investment record, equipment trust obligations sold at unduly high yields in 1932–1933.   Yields obtainable from this class of security in June 1933 are indicated by the following excerpts from the *Fitch Bond Record, Review Section,* dated June 20, 1933.

Road and series	Current basis, %	
	Bid	Ask
Atlantic Coast Line "D"	5.50	4.50
Baltimore & Ohio R.R. "A"	6.75	5.50
Canadian Pacific Ry. "B"	7.25	6.00
Central of Georgia Ry. "Q"	14.00	9.00
Chesapeake & Ohio Ry. "T"	4.80	4.00
Chicago & North Western Ry. "J"	12.00	8.00
Chicago Great Western R.R. "A"	12.00	9.00
Chicago, Milwaukee, St. Paul & Pacific R.R. "E"	14.00	9.00
Chicago, Rock Island & Pacific Ry. "L"	13.50	9.50
Erie R.R. Co. "LL"	8.75	7.25
Great Northern Ry. "B"	6.00	5.00
Illinois Central R.R. "F"	7.00	6.00

Road and series	Current basis, %	
	Bid	Ask
Louisville & Nashville R.R. "D"....................	5.50	4.50
Missouri Pacific R.R. "D".........................	12.50	9.00
New York Central R.R. "7-1920"....................	6.50	5.50
New York, Chicago & St. Louis R.R. "5-1923".........	12.00	9.00
New York, New Haven & Hartford R.R. "5-1925"......	6.50	5.50
Norfolk & Western Ry. "4½-1924"..................	4.25	3.75
Northern Pacific Ry. "4½-1925"....................	6.00	5.00
Pennsylvania R.R. "A"............................	4.50	3.75
Pere Marquette Ry. "6-1920"......................	12.00	9.00
Reading Co. "4½-1924"............................	4.65	4.00
St. Louis-San Francisco Ry. "BB"....................	14.00	10.00
Southern Pacific Co. "E"...........................	6.25	5.25
Southern Ry. "W"................................	11.00	8.50
Union Pacific R.R. "A"............................	4.50	3.75
Wabash Ry. "F"..................................	14.00	10.00

NOTE 17

An *Interim Report* of the Real Estate Securities Committee of the Investment Bankers Association of America (dated May 12, 1931, and printed in full in *Investment Banking*, June 1931, at pp. 7–10) estimated the total volume of real-estate bonds outstanding at $10,000,000,000, divided into classes as follows:

Class 1. Loans less than 75% of present revaluation in good standing, with good record............. $ 2,000,000,000

Class 2. Loans that have had no evidence of trouble but are over 75% of present value of security and appear to be able to work out without foreclosure or loss............................ 2,000,000,000

Class 3. Loans generally in excess of 75% of present value of security where foreclosure or workout with small loss is probable (*losses 10 to 25%*).. 2,500,000,000

Class 4. Items which when originally made were 80 to 100% loans. Such loans are now 125 to 150% items, *with losses from 25 to 60%* when foreclosure and sale are completed.............. 3,000,000,000

Class 5. In this group are the gross errors of judgment. Incompleted, ill-conceived and misplaced buildings, including many leasehold and second-mortgage bond issues. Losses in this class will run *from 60 to 100%* and items should often be entirely abandoned..................... 500,000,000

Total... $10,000,000,000

In its *Annual Report,* rendered in November 1931 before the Twentieth Annual Convention of the Investment Bankers Association of America, the Committee revised the foregoing estimates as follows: "The exact amount of outstanding real-estate bonds is difficult to ascertain due to the large number of small issues of which no record has been kept. The Federal Reserve Board at Washington estimates that there may be a present maximum volume outstanding of $6,000,000,000. This figure is considerably lower than the one estimated in our May report. We believe, however, $6,000,000,000 is approximately correct. It is the liquidation of this volume of real-estate bonds which presents one of the major problems confronting real estate.

"Due to the decline in urban real-estate values, it is estimated that approximately 60% of the outstanding real estate-bond issues are more or less in distress" (*Proceedings of the Twelfth Annual Convention of the Investment Bankers Association of America, 1931, p. 130*).

The character of the distress above referred to was indicated by the chairman of the committee in his introductory remarks when submitting the report. He said: "Now, it is estimated that about 60% of the real-estate bonds which have been issued are more or less in distress. Some only show slight trouble, either in temporary default or non-payment of taxes; others are under the process of reorganization or are in foreclosure" (*ibid.,* p. 128).

The growth in the volume of real-estate bonds actually in default with respect to interest and/or principal payments is shown by the following compilation[1] by Dow, Jones & Co., Inc., as of November 1 in the respective years. Only issues sold to and held by the public are included.

1928	$ 36,229,000
1929	59,755,000
1930	137,463,000
1931	327,968,000
1932	739,326,000
1933	995,017,000

NOTE 18

A harrowing example of this kind is furnished by the "Hudson Towers" at 72d Street and West End Avenue in New York City, projected as a hospital-hotel. Unfortunately for our purposes, the history of this enterprise is not so fully recorded as might be desired, but the main outlines of its career are known. This 27-story building was erected as a hotel, sanitarium, and hospital, catering to patients and their families. It was thus a specialized type of structure (see 124 *Chronicle* 1987). The land actually cost $395,000, and engineers estimated that the building would cost $1,300,-000 to construct. In order to facilitate the sale of $1,650,000 of first-mortgage bonds, the land and building combined were "appraised" at $2,600,000, thus making the bonds "legal for trust funds" under the New York law. This occurred in 1923. Subsequently the building passed through various hands by sale and resale, prior to its completion, and in 1927 second-mortgage bonds amounting to $1,150,000 were sold to the public.

[1] *The Wall Street Journal, Evening Edition,* Dec. 27, 1933.

The project was never completed; and in August 1932 the property was sold for $200,000 on foreclosure of the first mortgage. The outcome from the standpoint of the nonassenting first-mortgage bondholder is indicated by the announcement of the Irving Trust Co. in June 1933 that it was prepared to pay $8.14 on account of each $1,000 principal amount of undeposited first-mortgage bonds. Thus, less than 1 cent on the dollar was realized on liquidation. [References: *Moody's Banks & Finance* 1928, p. 1905; *ibid.*, 1929, p. 2012; *ibid.*, 1930, p. 1557; *ibid.*, 1931, p. 2834; *ibid.*, 1932, p. 2688; *ibid.*, 1933, p. 1214; *Moody's Manual Service* (Banks and Finance), letter dated Aug. 6, 1932, p. 1288; *Standard Corporation Records, F-K volume*, 1932, buff page 3581; *ibid.*, *Daily News Section*, June 14, 1933, at p. 7727; Barbeau, Ernest A., *The Mortgage Bond Racket*, pp. 24–26. Real Estate Bond Research Bureau, Albany, N.Y., 1932.]

NOTE 19

Note the following comment by the Industrial Securities Committee of the Investment Bankers Association of America in its 1928 report (*Proceedings of the Investment Bankers Association of America*, 1928, p. 91).

"Several circulars were examined in which an offering of preferred stock was made based upon a business housed in a building on leasehold property. The reference to the fact of a leasehold rental being a prior charge was made in very small type and in a most inconspicuous way. The investor glancing at the circular could easily derive the impression that the dividend on the preferred stock was a first charge on the earnings. Unfortunately, investors, as a rule, do not read circulars carefully, and the average investor would scarcely have noticed the mention made of the leasehold charge. In our opinion these figures should be set forth in just the same manner in which an interest charge on bonds would be placed."

The argument is equally valid, of course, in the case of a bond issue which is preceded by leasehold rental charges.

A leading example of a leasehold issue which encountered difficulty on account of the ground rental is presented by the Waldorf-Astoria Corp. (New York) First Mortgage Leasehold 7s, due in 1954.

Of the Waldorf issue $11,000,000 were sold to the public in October 1929. The ground rental began at $300,000 a year, but jumped to $600,000 at the end of two years and was graduated upward thereafter to a maximum of $800,000 per year. In addition there were certain building and sinking-fund rentals required to be treated as *operating expenses*, although they were fixed and determinable in amount. The statement in the offering circular that the fixed charges on the First Leasehold 7s were covered over 4.5 times (according to an estimated income account) was therefore misleading, as the rental charges were soon to exceed the interest on the bonds and were lumped in with the operating expenses in such a way as to conceal their true character and effect. If the buyer of the First Leasehold 7s had capitalized the prior charges at 6% he would have discovered that the $11,000,000 issue was junior to about $23,000,000 of prior claims.

Early in 1932 it became necessary to negotiate with the landlord (a subsidiary of the New York Central R.R.) with respect to the ground-rental payments which were in default. A plan of readjustment was proposed whereby the landlord was to make certain concessions with respect to the order and amounts in which ground rentals are to be payable in the future, and in return the bondholders were asked to assent to a modification of the indenture whereby their holdings would be transformed into income bonds carrying contingent charges. Full details of this issue and its record are available in: 129 *Chronicle* 2395; 134 *Chronicle* 2733; 135 *Chronicle* 1667, 2181, 2501; *Moody's Banks and Finance* 1933, pp. 2573-2574.

The bonds in this case declined to a low price of $3\frac{1}{4}$ in 1932.

A very similar situation developed with respect to the Hotel Pierre issue, details of which are obtainable in: 128 *Chronicle* 2819; 133 *Chronicle* 2111; 134 *Chronicle* 1967, 4504; 135 *Chronicle* 4566; 136 *Chronicle* 334, 501, 668, 852; *Moody's Banks and Finance* 1933, p. 2764, under "2 East 61st Street Corporation." Here the reorganization plan was more drastic. The original bonds sold in this case at a low price of 1 cent on the dollar in 1932 and 1933.

Tower Building Company (Chicago) First Leasehold $6\frac{1}{2}$s were offered to the public in 1926 at par. The amount was $1,900,000. The leasehold called for annual payment of a ground rent starting at $190,000 (and increasing thereafter). These heavy leasehold payments were subsequently defaulted; the lease was forfeited in 1931, and the bonds lost all value.

A similar disastrous fate befell the holders of 170 Broadway Corporation (New York) First Leasehold $6\frac{1}{2}$s, due 1949.

NOTE 20

Ratios of railroad maintenance expenditures to gross operating revenues for Class I railroads, based on the five-year period 1926-1930, inclusive, are as follows:[1]

Region	Maintenance of way, %	Maintenance of equipment, %	Total, %
Entire United States	13.69	19.51	33.20
New England	15.28	17.82	33.10
Great Lakes	12.49	20.90	33.39
Central Eastern	12.53	20.66	33.19
Pocahontas	13.40	20.02	33.42
Southern	14.47	20.01	34.48
Northwestern	14.49	18.28	32.77
Central Western	13.95	18.10	32.05
Southwestern	15.85	17.97	33.82

The variations as between the different regions, as indicated above, are distinctly smaller than they were prior to 1920. The maintenance expenditures of numerous roads fell conspicuously below the above standards during 1931 and 1932. For example, the Illinois Central ratios for 1932 were as follows: maintenance of way, 8.36%; maintenance of equipment, 19.48%.

[1] *Statistics of Railways in the United States*, Interstate Commerce Commission, Washington, D. C., 1926 to 1930, inclusive.

Sharp differences as between roads in the same geographical district also developed, as is indicated by the following:

Year and road	Maintenance of way, % of gross	Maintenance of equipment, % of gross	Total, % of gross
1926–1930 average for Southwestern region...........	15.85	17.97	33.82
Atchison:			
1929	15.79	18.13	33.92
1930	15.66	20.05	35.71
1931	13.15	21.98	35.13
1932	11.52	23.69	35.21
St. Louis—Southwestern:			
1929	19.97	16.26	36.24
1930	15.32	15.66	30.98
1931	10.93	14.56	25.49
1932	14.65	16.87	31.52
Southern Pacific:			
1929	12.63	17.46	30.09
1930	12.66	17.16	29.82
1931	12.41	17.21	29.62
1932	11.86	18.57	30.43

NOTE 21

The Chesapeake & Ohio Ry. Co. between the years 1921–1929 furnishes an example of unusually heavy maintenance expenditures. This is reflected in the following figures, which may be compared with the standard maintenance ratios for the Pocahontas region given in the preceding note.

Year	Ratio of maintenance of way to gross, %	Ratio of maintenance of equipment to gross, %	Total, %
1921	14.51	23.87	38.38
1922	12.70	27.01	39.71
1923	12.60	28.10	40.70
1924	14.40.	27.90	42.30
1925	15.20	25.30	40.05
1926	14.23	22.89	37.12
1927	14.37	22.38	36.75
1928	13.47	22.29	35.76
1929	14.39	22.36	36.75
1930	13.55	19.55	33.10
1931	12.88	18.99	31.87

The existence of large current earnings of subsidiaries not paid over to the parent company is illustrated by the following figures with reference to Louisville & Nashville R.R. Co., 51% of whose common shares are owned by Atlantic Coast Line R.R. Co.

Year	Earned per share	Paid per share	Balance after common dividends	Atlantic Coast Line's equity in L. & N.'s undistributed earnings
1922	$14.72	$7.00	$ 5,558,019	$2,834,590
1923	11.54	5.00	7,648,935	3,900,957
1924	12.08	6.00	7,112,794	3,627,525
1925	15.98	6.00	11,680,711	5,957,163
1926	16.60	7.00	11,232,111	5,728,377
1927	14.29	7.00	8,536,241	4,353,483
1928	12.24	7.00	6,133,220	3,127,942
1929	11.73	7.00	5,536,543	2,823,636

A similar though less striking picture is presented by the Chicago, Burlington & Quincy, which during the years 1922 to 1929, inclusive, earned substantially more than it paid out in dividends. This was especially true in the years 1924, 1928, and 1929, although the situation was reversed and dividends in excess of earnings were paid in 1930, 1931, and 1932. The Great Northern Ry. Co. and the Northern Pacific Ry. Co. each owns about 48% of the Burlington common.

NOTE 22

For examples of enterprises wholly or partially industrial in character but masquerading under the "public utility" title see: United Public Service Co., organized in 1927 and engaged in the electric light and power, natural and artificial gas, ice plant and cold storage businesses; Southern Ice & Utilities Co., organized in 1916 and engaged in the ice, ice cream, creamery, and cold storage warehouse businesses; The Utilities Service Co., organized in 1928 to acquire and operate 20 telephone companies in small towns and four ice companies in large towns or cities; Central Atlantic States Service Corp., organized in 1928 and engaged in the ice, coal, and cold storage businesses; Westchester Service Corp., organized in 1928 and engaged in the coal, ice, fuel oil, and building-supply businesses; National Service Cos., organized in 1928 as a holding company for enterprises of the Westchester Service Corp. type, engaged in the ice, fuel, and allied industries. Examination will reveal that these companies have capital structure of the public-utility type despite the fact that their operations are largely or wholly industrial in character.

NOTE 23

At various times the Investment Bankers Association of America has commented through its several committees upon the impropriety of bond circulars which either omit reference to depreciation entirely, or else conceal the actual amount of the depreciation charge through including it in some blanket item in the income account. The following quotations will serve to illustrate:

"There are many honest differences of opinion about depreciation and about the proper policy to provide for it, but whatever policy is adopted, the investor is entitled to know what it is. A circular of a corporation issue which does not mention depreciation leaves out an important factor in the affairs of the company in which the investor is asked to place his funds" ("Report of Special Committee on the Preparation and Use of Bond Circulars," printed in the *Proceedings of the Investment Bankers Association of America*, 1925, p. 274).

"The attention of our membership is particularly directed to the treating of the subject of depreciation. Some few circulars omit the balance sheet entirely, but in most instances this occurs in circulars where it is not particularly vital. However, the practice is quite common to show earnings before depreciation and taxes and then say nothing about the amount of depreciation taken. Inasmuch as it is our endeavor to present to the investor as complete a picture as is possible in an ordinary circular, it would seem that unless the earnings before depreciation are given, the amount of depreciation taken, and amount remaining for bond interest and taxes, leaving the balance to go to surplus, the investor has not all of the facts in the case. If the investor understands a balance sheet and is at all familiar with manufacturing, the manner in which depreciation is taken and its amount will tell him quite a story as to the management of the concern in question. Some circulars show earnings after depreciation and taxes but no earnings before such deductions. It is the opinion of both the Industrial Securities and Business Conduct Committee Chairmen that the ideal picture to the investor would be presented if the circular showed earnings before depreciation, the amount of depreciation and the earnings after depreciation, as separate items" (*Interim Report of the Business Conduct Committee* of the Investment Bankers Association of America *Bulletin*, March 1927, p. 3).

NOTE 24

EXAMPLE OF TREATMENT OF MINORITY INTEREST IN COMPUTING INTEREST COVERAGE FOR PUBLIC-UTILITY HOLDING-COMPANY BONDS

The report of the United Light & Railways Co. (Del.) for 1932 included the results of American Light & Traction Co. of which it owned some 54% of the common stock. The earnings applicable to the 46% minority were about $2,804,000. This minority interest may be treated in three ways, *viz.:*

Method A (which is the customary method). The minority interest is deducted after the parent company's interest charges. Under this method the minority item does not affect the bond-interest coverage in any way.

Method B (which is accurate, but complicated). Subsidiary earnings and charges are included only to the extent of the parent company's ownership. In other words, both the earnings and the fixed charges are reduced by the percentage applicable to the minority holdings of common stock.

Method C (which is recommended). The minority interest is deducted from net earnings (in the same way as an expense item) *before* figuring the interest coverage. This will result in a smaller interest coverage than under Method B, but the understatement will be moderate.

The three methods applied to United Light & Railways Co. report for 1932 will give the following results:

Item	Method A (Customary)	Method B (Accurate)	Method C (Conservative)
Gross earnings.	$67,550,000	$50,970,000†	$67,550,000
Net earnings.	23,318,000	18,100,000†	23,318,000
Minority interest.	. . . .	. . . .	2,804,000
Balance for fixed charges. . . .	23,318,000	18,100,000†	20,514,000
Fixed charges*.	15,453,000	13,039,000†	15,453,000
Minority interest.	2,804,000	. . . .	. . . .
Balance for parent company stocks.	5,061,000	5,061,000	5,061,000
Number of times fixed charges earned.	1.51	1.39	1.33

* Subsidiary interest and preferred dividends and parent-company interest.
† Excluding minority interest (46%) in American Light & Traction figures.

NOTE 25

DISCUSSION OF R. G. RODKEY'S MONOGRAPH ON "PREFERRED STOCKS AS LONG-TERM INVESTMENTS"[1]

The author arrives at his favorable conclusions with regard to industrial preferred stocks not preceded by bonds through the following procedure:

He made up a test list of ten such preferred stocks as of Jan. 11, 1908, representing those companies whose common shares were most actively traded in during 1907; also a similar list, but made up of the preferred issues which were themselves most active in 1907. Two supplementary lists were selected with Jan. 1, 1921 as a starting point. It was assumed that $1,000 was invested in each issue (and the same amount in the common stock for purposes of comparison). It was further assumed that wherever a company placed a bond issue ahead of the preferred the latter was sold and replaced by another not preceded by bonds. The author then traces, year by year until Jan. 1, 1932, the aggregate market value and dividend income on his various funds. The result may be summarized as shown at the top of page 654.

The dividend data show a very satisfactory return on the capital invested in the preferred shares, although the return on the common shares was considerably greater. Mr. Rodkey points to the high income return and particularly to the maintenance of the principal value, as shown in his tables, as evidence of the attractive character of such issues as long-term investments.

[1] *Michigan Business Studies*, Vol. IV, No. 3, 1932.

Test	Value Jan. 1, 1908		Value Jan. 1, 1921		Value Jan. 1, 1932	
	Preferred	Common	Preferred	Common	Preferred	Common
Test 5..............	$10,000	$10,000	$11,617	$ 27,003	$11,035	$36,279
Test 6..............	10,000	10,000	17,462	62,441	11,113	16,663
Total 2 tests........	$20,000	$20,000	$29,079	$89,444	$22,148	$52,942
Test 5a.............	....	....	10,000	10,000	9,705	11,657
Test 6a.............	....	....	10,000	10,000	9,682	16,822
Total 4 tests........	....	....	$49,079	$109,444	$41,535	$81,421
Average per $10,000...	$10,000	$10,000	$12,270	$ 27,361	$10,384	$20,355

Some of the objections to his methods and to his logic may be briefly set forth as follows:

1. The satisfactory showing of each individual preferred issue was completely dependent on an *advance* in the value of the common. In practically every case where the common declined the preferred declined with it, showing a complete lack of investment stability in the face of adverse conditions.

2. This vulnerability of preferred stocks is demonstrated strikingly by the action of the Rodkey lists during 1932. At the low prices of the year, the following values were shown as compared with the original investment of $10,000 in each test.

Test	Value at low prices, 1932	
	Preferred	Common
Test 5..	$ 7,727	$17,991
Test 6..	5,385	6,989
Test 5a.......................................	6,847	5,220
Test 6a.......................................	6,656	7,265
	$26,615	$37,465
Average...............................	$ 6,654	$ 9,366

At these low points the common stocks had an aggregate value close to the original cost but the preferred shares had lost one-third of their original price.

3. Rodkey's lists are not typical of preferred stock investment. The dates of purchase are taken in the midst of severe market depressions (1908 and 1921). The issues, selected according to activity, include a number

obviously speculative at the time, as shown by their low quotation. His final results show a substantial proportion of severe shrinkages offset by a few very large advances. His method of switching is one which would never be followed in practice, and in his Test 6 led to the extraordinary result that on Jan. 2, 1932 two of his $1,000 commitments in preferred stocks turn out to have a combined value of $7,417 while the remaining $8,000 of purchases had dwindled in value to $3,697.

4. If Rodkey's figures are analyzed from the standpoint of the experience of the typical investor during the period covered (instead of basing the results on the low prices arbitrarily used as a starting point), then his railroad-bond lists will display greater stability than the "non-bonded" industrial preferreds. In other words, taking the *average* prices for 1921–1924 as a base, the comparative Jan. 2, 1932 values work out as follows:

Date	Railroad common stocks	Railroad bonds	Industrials without bonded debt	
			Common	Preferred
Average, 1921–1924	100	100	100.0	100.0
Jan. 2, 1932	51	100	114.7	89.5

The fact that the railroad-bond prices were no lower in January 1932 than the 1921–1924 average, although the railroad-common-stock prices had been cut in half during the interim, shows an ability on the part of those bonds to withstand adverse developments (up to that date, at least) which is a real sign of investment quality. On the other hand, the industrial preferred stocks had declined some 10% even though the corresponding common stocks had advanced 15%. This supports our previous contention that only very prosperous conditions, making for a *large* increase in common-stock prices, will maintain the average value of a preferred-stock list. A variety of ups and downs, even though they may balance each other in their effect upon the common stocks, will cause preferred stocks more damage than benefit and will result in a shrinkage in their *average* value.

5. The author espouses the view that better results than shown by his tests would have been realized if the issues had been selected in accordance with careful investment practice instead of by the criterion of mere market activity. This is more than doubtful. The issues making the best statistical exhibit at the date of purchase would all have commanded relatively high prices, and they would not have had as good an opportunity to benefit from the growth of American industry as was enjoyed by the more active and speculative (*i.e.*, lower priced) preferred issues, several of which were included in Mr. Rodkey's lists.

NOTE 26

Calculation of the margin of safety protecting preferred dividends has received surprisingly scant attention at the hands of most writers of textbooks on investment. In some cases this is due to the exclusion of preferred stocks from the category of investment (*e.g.*, the writings of Lawrence Chamberlain), but in most instances no such explanation can be offered. The exceedingly large volume of preferred stock outstanding in recent decades suggests that some discriminating point of view and technique must have been developed for choosing between issues of this type, and it is surprising that more attention has not been given to the matter by those who write books on the "science" of security selection.

In most instances in which the subject receives attention the prior-deductions method of calculation is either explicitly recommended or implicit in the discussion. For example, Carl Kraft and Louis P. Starkweather in their *Analysis of Industrial Securities*, New York, 1930, use this misleading method of calculation in their rather extensive illustrative analysis of Jones Bros. Tea Co. without examining the resultant ratios critically. See p. 127, ratio 20-(b), and pp. 130–132, 162, especially the 1926 and 1927 exhibits.

J. E. Kirshman in his revised *Principles of Investment*, New York, 1933, refers to the coverage on Federal Water Service Corp. Preferred as having been earned "several times over within the past few years," which is a correct statement only in case the prior-deductions method of calculation is used. The combined fixed charges and preferred dividends were never covered more than 1.37 times during the years 1928–1932, inclusive (see pp. 155–156, 437). Likewise, D. F. Jordan repeatedly states the desired margin of safety for preferred stocks in terms of the number of times the preferred dividends alone are earned. See his *Investments*, rev. ed., pp. 157, 160, 162, 167, 185, 192, New York, 1933. Curiously enough, he sees the fallacy of this method in the case of preferred stocks of public-utility holding companies and recommends the total-deductions (over-all) method of calculation (see p. 169).

Badger and Herschel, on the other hand, forcefully call attention to the fallacy of the prior-deductions method of calculating coverage for preferred dividends and recommend the total-deductions calculation as standard procedure. See R. E. Badger, *Investment Principles and Practices*, pp. 308, 309, 413–414, New York, 1928; A. H. Herschel, *The Selection and Care of Sound Investments*, pp. 217–222, New York, 1925.

NOTE 27

The case originated as a move on the part of the State of Ohio to annul the charter of the Hocking Valley Ry. Co. on the ground of an unlawful conspiracy to suppress competition in coal traffic. It was alleged in the original suit that this suppression of competition resulted from the complete ownership of the stock of the Kanawha & Hocking Coal & Coke Co. by the Hocking Valley Ry. Co. and the Toledo & Ohio Central Ry. Co., which roads jointly guaranteed payment of principal and interest on the bonds as part consideration for an exclusive traffic arrangement. The court held that it was *ultra vires* the railroad corporations to own a controlling amount of stock in the coal company and to make the exclusive traffic agreement. See *State* v. *Hocking Valley Railway Company*, 31 *Ohio C. C.* 175 (1909).

Subsequently action was taken by the United States under the federal antitrust laws, resulting in the divorcement of the coal properties from the railroads in 1915. In neither of these decisions was there a *holding* that the guaranty as between the guarantor and the bondholders was void. Both cases were decided on the ground of unlawful conspiracy in restraint of trade, without reference to the rights of the bondholders. The Ohio court stressed the fact that it did not pass on the rights of the bondholders and the federal court expressly held that it did not enjoin the defendant railroad from recognizing its liability as guarantor on the bonds.

Meanwhile, on July 1, 1915, interest due on the bonds was defaulted and a protective committee began action in the New York Supreme Court to compel the railroads to meet their obligations as guarantors. The defendants sought to escape liability upon the defense that they had no legal power to execute guaranties of coal-company bonds, laying stress upon the earlier decisions in the Ohio and federal courts wherein the guaranties were claimed to have been held void. The New York court brushed this argument aside, however, saying that there was no such holding in either case as between the guarantors and the bondholders. In October 1916 the contentions of the protective committee were fully sustained by the New York court against the Toledo & Ohio Central Ry. Co., joint guarantor of the bonds, but the case against the Hocking Valley Ry. was delayed by various defenses set up by that defendant. The defaulted interest on the bonds was paid up on Jan. 1, 1918, and interest thereafter was paid regularly when due. All litigation involving the guaranty on the bonds was settled in October and November 1919 by decrees of the United States District Court for the Southern District of Ohio and the Court of Appeals of Ohio.

Meanwhile, in December 1916, the New York Central R.R. Co. had been authorized by the court to purchase the bonds deposited with the protective committee at par and interest, and practically all of them were so purchased. Following this a reorganization plan was devised and promulgated in November 1919, whereby one-third of the outstanding bonds were retired at par for cash and the remaining two-thirds were exchanged for new 6% first-mortgage bonds of the reorganized company.

The ultimate favorable outcome from the standpoint of the bondholders was due to a series of victories won in the courts whereby the validity of the guaranty was sustained. The history of this case can be traced fairly readily through the following sources: *Poor's Manual (Railroads)*, 1902, pp. 363, 402; 1913, p. 734; 1914, p. 1057; 1916, p. 400; *Moody's Manual*, 1912, p. 3627; 1913, p. 272; *Poor's Manual (Industrials)*, 1915, p. 2729; 1916, p. 2571; 1917, pp. 2417–2418, 2606; 1918, p. 622; *Moody's Manual (Industrials)*, 1918, pp. 1595–1596; 1920, pp. 2711–2712: 101 *Chronicle* 50, 134; 102 *ibid.* 1166, 2167; 103 *ibid.* 848, 1302, 1795, 1892, 1985, 2347; 104 *ibid.* 667, 768; 105 *ibid.* 611; 106 *ibid.* 2125; 109 *ibid.* 1529, 2175, 2361.

NOTE 28

Davison Chemical Co. refused to meet its guaranty of the principal of Silica Gel Corp. $6\frac{1}{2}\%$ Notes at maturity in 1932, and offered the holders its own obligations in lieu thereof. But this maneuver was essentially similar to other endeavors to effect "voluntary" extensions or modifications of fixed obligations, and did not involve any repudiation of the guaranty as such. Noteholders not accepting the plan of exchange, and suing for their money, were paid in full, although immediately thereafter Davison Chemical went

into receivership.   The history of the Western Pacific Ry. bond guaranty
by the Denver & Rio Grande R.R. is the classic example of enforcement of
a guaranty to the ultimate limit of transferring ownership of the guarantor
railroad from its stockholders to the owners of the guaranteed bonds.   For
a brief history of this case see Arthur S. Dewing, *Financial Policy of Corporations*, pp. 154–156, New York, 1926.

Another striking example is afforded by the Savoy-Plaza Corp. Debenture
5½s, due 1938, which in effect were guaranteed by U. S. Realty & Improvement Co. and by Childs Co.   This guaranty resulted in a series of purchase
offers made to the debenture holders, and finally in the calling of the entire
issue at 102¼ in October 1932.   At the same time the *First Mortgage* 6s
of the same enterprise, not guaranteed, were about to default and had sold
as low as *5 cents on the dollar*.

NOTE 29

The statements in the text may be verified by a detailed examination of
the price records from which the following have been drawn as illustrations.
On Oct. 31, 1929, the Kansas City Terminal 4s, due 1960, sold at 86¼ to
yield 4.9%, whereas, on the same day the General 4s of the Chicago, Rock
Island & Pacific Ry., due in 1988, sold at 90 to yield 4.5%.   Four years
later, on Nov. 22, 1933, the Kansas City Terminal bonds sold at 86¼,
although the Rock Island General 4s had declined to 42, a price yielding
about 10%.   On Dec. 8, 1927, the Terminal bonds sold at 93⅞ and the
Chicago, Milwaukee & St. Paul Ry. General 4s, due 1989, sold at 93 to yield
somewhat less than the former.   On Feb. 24, 1933, the Terminal bonds
were selling at 90, to yield about 4.65%, whereas the St. Paul General 4s
had declined to a price of 38 and a yield of around 11%.   Between Nov. 7,
1927 and June 15, 1932 the Terminal 4s declined from 93 to 82¾ (yields
of 4.4% and 5.18%, respectively) while Missouri-Kansas-Texas R.R.
Prior Lien 4s, due 1962, declined from 93 to 31⅛ (yields of 4.39% and over
15%, respectively).

NOTE 30

The New York and Harlem R.R. situation presents some interesting
aspects of leases and guarantees.

1. The major part of the property is leased to the New York Central for
401 years at a rental equivalent to bond interest and $5 dividends on the
preferred and common stock.   The bond interest and principal are both
specifically guaranteed by the N.Y. Central, but there is no specific guaranty
of dividends.   However, dividends have been paid regularly under the
lease since 1873.

2. The street railway properties were leased separately to N.Y. Rys. Co.
for a rental equivalent to an additional $2 per share on both classes of stock.
When N.Y. Rys. Co. became bankrupt, the lease was terminated and the
traction lines taken back and operated by the N.Y. & Harlem.   In 1932 a
new lease of these properties for 999 years was negotiated with N.Y. Rys.
Corp. (successor to the former lessee).   The only consideration was a lump
payment of $450,000, so that this transaction appears virtually identical
with a sale of the street railway lines for the sum mentioned.

3. Some N.Y. & Harlem stockholders endeavored to obtain large additional payments from the N.Y. Central on the ground that the valuable "air rights" (or rights to build over the Harlem's right-of-way) were not covered by the lease and had to be paid for separately. The speculative glamor of this suit raised the price of the shares to as high as 505 in 1928, representing less than a 1% dividend return. The suit was dismissed in 1932, by which time the price had fallen to 82¼.

The Mobile & Ohio situation has some similar features of interest, *viz.*:

1. In 1901, Southern Ry. Co. issued "Mobile & Ohio Stock Trust Certificates" in exchange for nearly all the Mobile & Ohio capital stock. It agreed to pay 4% on these certificates in perpetuity.

2. Mobile & Ohio became prosperous and from 1908 to 1930 paid the Southern Ry. 140% in dividends. The Interstate Commerce Commission and the State of Alabama endeavored to compel the Southern to give up control of the Mobile on the ground that it violated antitrust laws. At the same time holders of Stock Trust Certificates started action looking either to the return of the deposited stock or to obtaining larger dividends on their certificates. The price of these advanced to 159½ in 1928, in anticipation of the legal moves.

3. The collapse of earnings after 1929 forced Mobile & Ohio into receivership in 1932. Interest due Sept. 1 on its bonds was defaulted, but on Oct. 1 holders of the *stock* trust certificates received the 2% due from Southern Ry. and these payments were continued throughout 1933. Previously however, the price of the Certificates had fallen as low as 3½, but this reflected mistrust of Southern's financial capacity rather than any question regarding the legality of the obligation to pay the 4% dividend.

NOTE 31

INDUSTRIAL OFFICE BUILDING COMPANY REORGANIZATION

The history of this enterprise illustrates in striking fashion the difference between the theoretical rights and the actual experience of a first-mortgage bondholder. In 1926 the company erected an office building in Newark, N.J. The cost of land and building was apparently about $3,800,000, but the land value was marked up from $300,000 to $2,000,000 through the familiar process of "appraisal." The cost of the building was defrayed through sale of the following securities:

6% first-mortgage bonds	$3,150,000
7% unsecured notes	450,000
Preferred stock	450,000
Common stock	100,000

(The mark-up of the real estate gave the common stock a "book value" of about $1,800,000.)

Following a period of poor earnings, interest was defaulted on June 1, 1932, and a receiver was appointed. Shortly thereafter a reorganization plan was drawn up, providing as follows:

1. The first-mortgage 6% bonds due 1947 were to be exchanged for first-mortgage 5% *income* bonds, also due 1947.

2. The 7% unsecured notes due 1937 were to be exchanged for 7% unsecured *income* notes, due 1948.

3. The 8% preferred stock was to be exchanged for new 8% preferred.

4. The common stock was to be exchanged for new common.

5. All these exchanges were to be made par for par or share for share.

The plan was carried out by the purchase of the property at foreclosure sale for $100,000 by the Reorganization Committee. First-mortgage bondholders who did not accept the new securities received in cash only $56.43 per $1,000 bond.

In this readjustment the bondholders gave up their fixed claim to interest, receiving no compensation of any kind therefor, while the stockholders gave up nothing at all. (Dividends are to be postponed until after two-thirds of the bonds have been retired, but such retirements inure to the benefit of the stockholders and this provision does not really represent a sacrifice on their part.) This was an extraordinarily one-sided composition or "compromise" —the more so since the bondholders were clearly entitled to take direct possession of the property. The Reorganization Committee defended their generosity to the stockholders on the ground that it was desirable to retain the services (at a salary) of the largest stockholder as manager of the property. In effect the real owners of the building took a preferred-stock issue (*i.e.*, income bonds) for their capital and gave up all the junior equity to the management. This seems a staggering price to pay for the supervision of an office building.

It may be objected that our criticism is somewhat far-fetched, since the building was unlikely to return more than the interest on the income bonds in any case, so that the equity retained in full by the stockholders was scarcely worth arguing about. But it is highly fallacious to measure the *potential* earnings by the results shown in an unparalleled depression. Viewing the proposition over the long-term future, there were several different kinds of possibilities which might make the stock equity valuable. Among them were the following:

1. The return of prosperity and even of a new real-estate boom.

2. Substantial inflation of the currency, which would reduce the burden of the bonded debt.

3. Some special favorable development affecting the neighborhood or the building. It happened that immediately after the Reorganization Plan was consummated, the New York Stock Exchange made every arrangement to transfer its business to Newark, and this very office building was spoken of as the home of the Curb Exchange. Had this actually come about, a large profit would have been realized entirely by the old stockholders of this formerly bankrupt enterprise. This profit should properly have belonged to the bondholders, because they took all the risk of future loss (as shown by the decline of the market price of the issue to 4 in February 1933).

Attention should be called to the fact that this property, valued at $5,500,000, was sold at foreclosure for $100,000, netting the undepositing bondholders about 5 cents on the dollar. (The issue had been floated at 100 in 1927.) That this was a grossly inadequate price is clear from the fact that *net earnings after taxes* for the first half of 1932 had been $67,000.

In the writers' view, the transfer of property at a negligible price in pursuance of a reorganization scheme of this sort is more inequitable than the "freezing out" of stockholders or other owners in the ordinary bankruptcy proceeding. The right of the creditors to levy on the assets often works great hardship, but it can scarcely be called unfair in the light of the specific terms of the loan agreement and the original possibilities of profit to the stockholder from the use of the borrowed funds. But in the Industrial Office Building example, the judicial process was availed of to deprive the individual bondholder of the remedy which he had been assured he would have in the event of default—*viz.*, either the taking over of the property on his behalf, or the distribution to him of his share of the cash value of the property realized in a bona fide sale.

NOTE 32

We believe that the two examples following should be preserved as a warning to the analyst against excessive reliance upon (*a*) the protective covenants in the indenture, and (*b*) the statistical exhibit when selecting industrial bonds.

I. Willys Overland Co. Ten-Year First 6½s, due September 1933. Amount of original issue, $10,000,000.

A. Protective provisions:

1. A direct first mortgage upon all the fixed assets now owned or hereafter acquired (except for new purchase-money liens), and secured also by pledge of all stocks owned in the principal subsidiary companies. The subsidiaries were prohibited from creating mortgages or funded debt unless same were pledged to secure this issue.
2. A sinking fund of 10% of the issue each year ($1,000,000 per annum) was to retire 90% of the issue prior to maturity.
3. Net current assets must at all times equal at least 150% of the outstanding bonds.
4. Cash dividends were to be paid only out of earnings subsequent to Sept. 1, 1923, and only if the current assets after deducting such dividend are no less than 200% of current liabilities, and net current assets are not less than 200% of the outstanding bonds at par.

B. Statistical exhibit, Dec. 31, 1928:

1. Interest had been earned 12 times in 1928; an average of over 11 times in 1923–1928; and at least 3½ times in each of the past six years.
2. The market value of the preferred and common stock on Dec. 31, 1928 was $110,000,000 or 22 times the bond issue of $5,000,000.
3. The consolidated net current assets on Dec. 31, 1928 were $28,700,000, or more than five times the outstanding bonds.
4. The consolidated net tangible assets applicable to the bonds were over 14 times the amount of the issue.

C. History Subsequent to 1928: In the four years 1929–1932 the consolidated surplus decreased from $39,600,000 to $400,000. Of this shrinkage, $6,000,000 represented dividends paid and the balance was due to operating and other losses. Coincidentally, the net current assets of $28,700,000 were converted into a net excess of current liabilities amounting to $2,400,-000, a total shrinkage of over $30,000,000.

The operations of the sinking fund reduced the bond issue to only $2,000,-000 at the end of 1931, but the sinking-fund installment due July 1932 was not met. In February 1933 receivers were appointed. Interest on the bonds due March 1933 was defaulted and the principal was also defaulted in September 1933.

The bonds, which had sold as high as 101½ in 1931 and at 92 in 1932, declined to 24 at the end of 1933.

It is to be noted that no action was taken by the Trustee or by the bond-holders at the time of default in the sinking fund in July 1932, nor at the time the working capital first declined below the stipulated minimum. Prompt defensive measures then might have compelled payment of the relatively small bond issue. A bondholders' protective committee was formed after the receivership. Finding reorganization plans impracticable, it favored liquidation; but it then found legal difficulties in the way of fore-closing on its lien. (A payment of $250 per $1,000 bond was to be made in partial liquidation during 1934.)

II. Berkey and Gay Furniture Co. First 6s, due serially 1927–1941. Amount of original issue $1,500,000.

A. Protective provisions:

1. Secured by a first lien on fixed property valued at some $4,400,000, or over 290% of the original issue. Additional bonds could be issued up to $1,000,000 against pledge of additional property, but at a rate not exceeding 50% of the cost thereof.

2. The net current assets were to be maintained at $2,000,000, and current assets were required to equal twice current liabilities.

3. The serial maturity was equivalent to a sinking fund averaging $70,000 annually, which would retire two-thirds of the issue prior to maturity.

B. Statistical exhibit, Dec. 31, 1927.

1. Interest had been earned over three times in 1927; an average of about 4½ times in 1922–1927; and not less than three times in any year of the six-year period.

2. Net current assets were $3,698,000, or 2½ times the $1,460,000 of bonds outstanding.

3. Total tangible assets applicable to the issue were $8,500,000 or about $6,000 per bond.

C. History Subsequent to 1927: Between Jan. 1, 1929 and July 31, 1931 the company reported losses aggregating nearly $3,000,000. In 1930 alone the working capital shrank from $2,900,000 to $650,000. By July 1931 an excess of current liabilities was shown. Interest on the bonds was defaulted in November 1931. Receivers were appointed in February 1932. The installment of the bonds due May 1932 was defaulted. A decree directing foreclosure under the mortgage was issued in April 1933. The bonds, which had sold at par in 1928 and as high as 65 in March 1931, were worth only one cent on the dollar at the end of 1933.

A protective committee was formed for the bond issue following the default in bond interest. It is difficult to say whether prompter action on behalf of the bondholders would have availed anything in this disastrous situation. But certainly they should have bestirred themselves at the end of 1930, when the working capital covenant had been violated, and not stood idly by until the default in interest payments nearly a year later.

NOTE 33

Evidence of the growth in financing through privileged issues is provided in the following figures for the *total number of privileged issues outstanding* as listed in *Moody's Manuals* for the years indicated. Both bonds and stocks are included.

Year	Total number of privileged issues outstanding	Convertible	Participating	With warrants
1925	434	434	(Not given)	(Not given)
1926	613	503	(Not given)	110
1927	1,129	537	410	182
1928	1,551	716	539	296
1929	2,091*	932	638	521*
1930	2,661†	1,173†	882†	606
1931	2,668	1,214	862	592
1932	2,559	1,213	829	517
1933	2,338	1,120	791	427

* Includes 119 issues (with warrants) of investment trusts and other financial companies. This class not included in totals for previous years.
† Increase largely accounted for by inclusion this year for the first time of convertible and participating issues of investment trusts and other financial institutions not previously included in the Moody lists.

NOTE 34

The application of the antidilution formula to the somewhat complicated case of Chesapeake Corp. Convertible Collateral 5s, due 1947, is based on the following state of facts. The bonds, issued in May 1927, are secured by the pledge of Chesapeake & Ohio Ry. Co. common stock, into which they were made convertible after May 15, 1932. The indenture contained the customary antidilution provisions and stated that for the purpose of computing new conversion prices 1,190,049 shares of Chesapeake & Ohio common were to be deemed to be outstanding as of the date of issuance of the bonds. Subsequently Chesapeake & Ohio issued new shares as follows:

(a) 296,222 shares at $100 per share to holders of record on Apr. 30, 1929.

(b) 46,066.5 shares issued in 1930 in exchange for Hocking Valley Ry. Co. common stock. Working back from the company's reports it appears that the Hocking Valley stock was appraised at $7,076,710.18, or at the rate of $153.62 for the C & O stock issued in exchange.

(c) 382,211 shares at $100 per share to holders of record on June 12, 1930.

Finally, on July 31, 1930, the par value of Chesapeake & Ohio common was reduced from $100 per share to $25 per share, and four new shares were issued in exchange for each old share theretofore outstanding.

On the basis of these facts the computation of the conversion price in the early part of 1933 was as follows:

$$C' = \frac{(1,190,049 \times \$220) + (296,222 \times \$100) + (46,066.5 \times \$153.62) + (382,211 \times \$100)}{\dfrac{1,190,049 + 296,222 + 46,066.5 + 382,211}{4 \text{ (due to 4 for 1 split on 7/31/30)}}} = \$43.97$$

Base figure   Offer of 4/30/29   Hocking Valley   Offer of 6/12/30

NOTE 35

Consolidated Textile Corp. Three-Year 7% Convertible Debentures, due 1923, had a conversion privilege of this type. The indenture provided that "The rate at which common stock of the company shall be delivered on any such conversion shall be upon the basis of 22 shares of such common stock for each $1,000 Note, and eleven shares of such common stock for each $500 Note, or, if any additional common stock of the company is at any time issued by it for less than $46 per share, the rate of conversion shall be reduced to the price in money or in fair value of property at or for which such common stock is issued . . . and if any further stock is subsequently issued at a lower price the conversion rate shall be still further reduced, and so on from time to time, with a cash adjustment of interest and dividend accrued."

These Debenture Notes were issued in April 1920. In November of that year additional stock was offered to stockholders at $21 per share and the conversion price was accordingly reduced to $21 per share from about $46 per share. The privilege never attained a substantial value, the stock not having sold above 46½ prior to November 1920 and failing to exceed 21⅞ subsequent to the lowering of the conversion price in November. The issue was called at 102½ in October 1921.

NOTE 36

The $67,000,000 of American Telephone & Telegraph Co. Convertible 4½s, due 1933, which were offered to shareholders in 1913 are an example of this comparatively rare condition. The bonds were convertible into common stock at $120 per share from Mar. 1, 1915 to Mar. 1, 1925. The indenture provided that the stock obtainable on conversion was to be "part of the authorized capital stock of the Telephone Company *as such authorized capital stock shall be constituted at the time of such conversion*" and did not contain the usual antidilution clauses. It is interesting to note that both the preceding and subsequent convertible issues of American Telephone & Telegraph Co. did contain an antidilution clause. See, for example, the indentures securing the convertible 4s issued in 1906 and the convertible 4½s issued in 1929.

Over half of the 4½s, due 1933, were converted in 1915, the first year in which the privilege was exercisable, and the balance was rapidly reduced thereafter through conversion. In 1925, when the privilege expired, $1,899,400 remained unconverted, and these were called at par in 1931. Meanwhile, prior to 1925, several privileged subscriptions were offered to shareholders and this may account for the rapid conversion of this issue unprotected against dilution through shareholders' "rights," although the higher yield on the stock under an $8 and $9 dividend rate doubtless was a factor.

Another example which is not quite so clearly in point is that of the Brooklyn Union Gas Co. Convertible 5½s, due 1936. These were offered in December 1925 with the right to convert into 20 shares of common stock on or after Jan. 1, 1929. The indenture was somewhat ambiguously worded to the effect that "in the event of a change in the character of the stock of the Company prior to the maturity of the bonds, so as to increase or decrease the number of shares which the stockholders would be entitled to receive for their stock, then the number of shares which the holders of these bonds shall receive upon conversion shall be correspondingly increased or decreased." This left the matter in doubt as to whether protection against all forms of dilution was afforded or whether protection was given against stock dividends, stock-splits, and reverse split-ups only. It was perhaps for this reason that very large arbitrage spreads existed between the bonds and the stock prior to Jan. 1, 1929, when actual conversion could occur, although here again the higher yield from dividends on the equivalent amount of stock may have accounted in part for the discrepancies. Relevant data are appended below.

Date	Price of common	Equivalent price for bonds	Price of bonds	Spread in dollars per $1,000 bond
3/19/26	71½	143	129	$140
6/18/26	81	162	145	170
9/17/26	91	182	155	270
12/24/26	93½	187	162½	245
3/18/27	91	182	159	230
6/17/27	115	230	197	330
9/23/27	142	284	224	600
12/30/27	152	304	265	390
3/30/28	153	306	272	340
6/29/28	143	286	262	240
9/28/28	166	332	309	210
12/28/28	187½	375	375	0

NOTE 37

Dodge Brothers, Inc., Convertible Debenture 6s, due 1940, illustrate the increase in conversion price which occurs when shares in the issuing corporation are exchanged for a smaller number of shares in a merger with another corporation. The bonds, issued in 1925, were convertible into Class A stock of Dodge Brothers, Inc., up to a maximum of $30,000,000 out of a total issue of $75,000,000. Conversion was set at the rates fixed in the following schedule:

First $5,000,000 converted, 1 share of *A* stock for $30 of bonds at par.
Second $5,000,000 converted, 1 share of *A* stock for $35 of bonds at par.
Third $5,000,000 converted, 1 share of *A* stock for $40 of bonds at par.
Fourth $5,000,000 converted, 1 share of *A* stock for $50 of bonds at par.
Fifth $5,000,000 converted, 1 share of *A* stock for $60 of bonds at par.
Sixth $5,000,000 converted, 1 share of *A* stock for $70 of bonds at par.

The indenture provided that in case of merger or consolidation the purchaser must assume the bonds and provide for their conversion into the same kind and amount of shares as were issuable in the merger or consolidation with respect to the number of shares of Class *A* stock to which the holder of the bond was entitled from time to time upon conversion.

The first $15,000,000 of the bonds were converted into Dodge Brothers Class *A* stock prior to the merger of that company with Chrysler Corp. in July 1928, and the assumption of the remaining bonds by the latter. In this acquisition five shares of the Class *A* stock into which the bonds were convertible were exchanged for one share of Chrysler Corp. common. Hence, in accordance with the indenture provisions, the fourth $5,000,000 of bonds are now convertible at the rate of four shares of Chrysler common for each $1,000 bond (a conversion price of $250 per share for Chrysler). Likewise, the fifth and sixth units are convertible into Chrysler common at $300 and $350 per share, respectively.

NOTE 38

Spanish River Pulp & Paper Mills, Ltd., First Mortgage 6s, due in 1931, were issued in 1911 as a straight bond without profit-sharing privileges. A default in interest payments occurred in 1915–1916, resulting in a compromise between the bondholders and the company. Under this agreement the overdue interest payments of 1915–1916 were postponed until October 1922; sinking-fund payments were temporarily suspended; and the holders of these and certain bonds of affiliated companies were given the right to receive during the life of their bonds a pro rata share of 10% of the amount allocated in any year for dividends on the preferred and common stocks of the Spanish River Co.

Year	Number of times interest earned	Market range for the bonds
1919	2.62	105½– 97
1920	3.03	97½– 93
1921	4.39	87 – 86¼
1922	2.39	115 – 93¼
1923	3.46	105 – 95
1924	4.37	104 – 97
1925	3.85	106¼–106¼
1926	3.96	108 –105
1927	3.36	108¾–108½
1928	Bonds called at 110	

As a result of this arrangement the bondholders not only received 10% of all cash dividends paid on the Spanish River Co. Preferred and Common until the bonds were retired in 1928, but they also received 10% of the Preference Stock issued in July 1920 as a 42% stock dividend to liquidate accruals on the preferred stock.

The investment quality of these bonds subsequent to 1918 is indicated by the figures shown on page 666.

NOTE 39

The technique of an intermediate hedging operation is illustrated by the following transactions made in 1918–1919, involving the purchase of a $1,000 Pierce Oil Corp. 6% Note, due 1920 and the sale of common stock against it. The Pierce Oil note was convertible at any time into 50 shares of common stock. (Accrued interest on the note is excluded.)

Date	Purchase	Range for month	Sale	Range for month
Oct. 1918......	1M 6% note at 100½ = $1,008	99½–101½	25 common at 19 = $ 470	16¼–19⅛
Dec. 1918......	25 common at 16 = $ 403	15¾– 17		
Jan. 1919......	....	....	25 common at 19 = $ 470	16 –19⅜
May 1919......	....	....	25 common at 28 = $ 696	24¾–28⅝
Dec. 1919......	50 common at 17½ = $ 881	17 – 20⅝	1M note at 100 = $1,000	Called at 100
	$2,292		$2,636	
Profit.........	$ 344			

Low price for note, October 1918 to December 1919, was 99½.

These five transactions may be analyzed as follows:

1. Purchase of note and sale of half of related stock against it, at price not far from parity. This permitted a covering profit if the stock declined and a profit through sale of the other half if the stock advanced.

2. A decline in the stock permitted the covering profit.

3. Recovery of the stock permitted the original position to be restored.

4. Advance of the stock permitted sale of second half at price to assure profit on the operation.

5. Renewed decline in the stock permitted repurchase of shares sold at profit while note was disposed of at par.

Because the near maturity of the note issue (coupled with the reasonably strong financial condition of the company) could be counted upon fairly well to keep its price up, it was not necessary to sell out the note at Step 2. It could be held in the hope that the sale of the stock could be repeated.

NOTE 40

We have already indicated in Chap. XIV that 95% of all preferred stocks listed on the New York Stock Exchange failed to maintain an investment price level in 1932. A study of 4,605 domestic and foreign bond issues (including foreign governmental issues, but excluding domestic governmental bonds) listed in the *Fitch Bond Record* of Jan. 1, 1933, indicates the following distribution at the low price for each issue for 1932:

Class (by price range)	Number	% of total	Cumulative %
0– 10	725	15.74	15.74
11– 20	566	12.29	28.03
21– 30	424	9.21	37.24
31– 40	338	7.34	44.58
41– 50	366	7.95	52.53
51– 60	412	8.95	61.48
61– 70	384	8.34	69.82
71– 80	415	9.02	78.84
81– 90	383	8.32	87.16
91–100	417	9.06	96.22
Over 100	175	3.78	100.00
Total......	4,605	100.00	

NOTE 41

The corporation statutes of most continental countries prescribe certain compulsory reserves, one of the functions of which is to facilitate maintenance of regular dividends. These reserves are accumulated from annual profits but ordinarily do not reach large proportions. The power to declare dividends usually resides in the stockholders assembled at the "general meeting" which is an annual affair, although provision for interim dividends is also made.

In England the Companies Act does not limit the dividend-declaring function to the annual "general meeting" of the shareholders; but the recommended form of by-laws (Table *A* of the statute) provides for this mode of declaration and it is the general custom in framing articles of association to stipulate that "the company in general meeting" or "the directors with the sanction of a general meeting," may declare annual dividends. See *First Schedule, Table A of the Companivs Act*, 1929, 19 & 20 Geo. V., Chap. 23. A discussion of British dividend law and policies is available in *Palmer's Company Law*, 13th ed., pp. 222–233, 628, London, 1929.

The following data on dividend practices of foreign corporations are offered in support of the statement in the text:

ROYAL DUTCH COMPANY FOR THE WORKING OF PETROLEUM WELLS IN THE
NETHERLANDS INDIES
(Figures in florins, 000 omitted)

Year	Available for ordinary stock	Paid on ordinary stock
1920	Fl. 129,062	Fl. 128,291
1921	100,820	99,652
1922	85,853	85,186
1923	82,059	80,364
1924	86,251	84,464
1925	92,833	92,564
1926	98,024	96,844
1927	99,328	98,905
1928	99,920	98,905
1929	123,089	120,870
1930	90,229	85,616

BATAAFSCHE PETROLEUM COMPANY
(Principal producing subsidiary of Royal Dutch)
(Figures in florins, 000 omitted)

Year	Net for reserves	Reserves	Balance for ordinary	Paid on ordinary
1914	Fl. 45,140	Fl. 10,772	Fl. 34,418	Fl. 35,000
1915	44,277	9,100	35,178	35,000
1916	77,655	39,149	34,016	34,000
1917	102,786	54,445	42,841	43,000
1918	126,170	28,763	90,392	90,000

SIEMENS & HALSKE A.G. (Germany)
(Figures in reichsmarks, 000 omitted)

Year	Net profit	Dividends	Directors' statutory bonus	Special reserve	Balance
1925	Rm. 6,245	Rm. 5,460	Rm. 66	Rm. 750	Rm. *31(d)*
1926	12,730	9,100	299	3,500	*168(d)*
1927	16,401	10,920	415	5,000	67
1928	15,936	12,740	531	2,500	166
1929	16,036	12,901	538	2,500	97
1930	13,622	13,382	560	....	*321(d)*
1931	8,615	8,603	255	....	*243(d)*
1932	2,474	6,193	111	....	*3,830(d)*

BRITISH-AMERICAN TOBACCO COMPANY, LTD.
(Figures in pounds sterling, 000 omitted. Fiscal years ending 9/30)

Year	Net income	Preferred dividends	Ordinary dividends	Surplus	% earned on ordinary stock	% paid on ordinary stock
1921	£4,323	£225	£3,842	£ 256	25.95	24
1922	4,401	225	4,012	165	26.02	25
1923	4,495	225	4,015	255	26.57	25
1924	4,866	225	4,259	382	28.88	26½
1925	5,145	225	4,488	433	30.61	$27^{1}\!\!\frac{1}{12}$
1926	6,196	225	4,957	1,014	25.43	25
1927	6,354	225	5,875	254	26.08	25
1928	6,564	225	5,883	456	26.92	25
1929	6,358	225	5,892	241	26.01	25
1930	6,502	555	5,895	51	25.22	25
1931	5,334	585	4,717	33	20.14	20
1932	5,438	585	4,717	137	20.58	20

In the case of General Electric, Ltd., the American policy of retaining a large proportion of the earnings has apparently been followed. The greater part of these surplus earnings, however, were carried to "Reserve Account," as the following figures indicate:

GENERAL ELECTRIC COMPANY, LTD.
(Figures in pounds sterling, 000 omitted)

Year	Net income	Preferred dividends	Ordinary dividends	Reserves	Balance to surplus
1919	£382	£ 85	£114	£...	£183
1920	490	108	123	150	110
1921	613	229	211	...	172
1922	207	233	106	...	131(d)
1923	311	249	106	...	43(d)
1924	396	252	106	...	38
1925	585	252	158	172	2
1926	629	252	161	120	96
1927	584	252	169	140	23
1928	612	252	225	130	4
1929	631	252	225	130	24
1930	715	252	316	130	17
1931	632	252	225	133	22
1932	582	252	180	130	19

NOTE 42

A series of discrepancies in the relative prices of securities of the Interborough Rapid Transit Co. (New York) securities, described herewith, will exemplify the type of opportunity for analytical work of definite character which is recurrently presented in the securities markets.

1. In November 1919 the 4½% bonds and the preferred stock of Interborough Consolidated Corp. both sold at 13. The bonds (called Interborough Metropolitan 4½s) were in default and the company was in receivership. The bondholders were entitled to claim all the assets, which had substantial value; the stockholders were without equity of any sort. In the subsequent reorganization the preferred and common shares were extinguished completely, while the 4½% bondholders received new securities eventually worth considerably more than 13% of the face amount of the bonds.

2. In January 1920, Interborough Rapid Transit Co. 7% notes, due September 1921, sold at 64½, while the same company's First and Refunding 5s, due 1966, sold at 53¼. Each 7% note was secured by deposit of about $1,562 of 5% bonds, and was convertible into about $1,144 of 5% bonds. At the relative prices the notes were far more desirable than the bonds because: (a) the notes enjoyed better security; (b) they yielded a larger return; and (c) their conversion privilege permitted the owner to benefit from any advance in the price of the 5% bonds.

The notes were extended for one year at 8%; and in 1922 the holders were offered $100 in cash and $900 in 7% secured, convertible notes, due 1932. Those not accepting either offer were able to compel payment in full. An exchange from 5s into 7s at the prices above indicated would have shown a substantial profit at various times in 1921 and 1922.

3. In the early part of 1929, Interborough Rapid Transit Company capital stock repeatedly sold at a higher price than Manhattan Ry. Co. "Modified Guaranty" stock (e.g., 55½ for I.R.T. vs. 54 for Manhattan Mod. Gty. in March 1929). This price relationship was illogical because:

a. "Manhattan Modified" was entitled to cumulative annual dividends of 5%, and to payment of 6¼% accumulated, before Interborough stock received anything.

b. "Manhattan Modified" was further entitled to receive a total of 7% in the event that Interborough received 6%.

c. Interborough could not receive more than 7% prior to 1950.

d. Dividends of 5% were actually being paid on Manhattan, while Interborough was not receiving anything.

It should have been manifest that the Manhattan shareholders were certain to receive at least as high a dividend as the Interborough shareholders for the next 21 years. By August 1929, the price disparity was corrected, for the "Manhattan Modified" stock sold 16 points higher than Interborough (39¼ against 23).

4. In October 1933, I.R.T. 5% bonds and 7% notes both sold at 65. This disparity was discussed in detail in Chap. I and referred to again in Chap. LI (p. 603).

5. In December 1932, Manhattan Ry. "Unmodified" shares sold at 18 while the "Modified" shares sold at 6⅝. The stock was originally entitled to dividends of 7%, guaranteed unconditionally by Interborough. The modified shares were subject to an agreement under which payment of dividends was contingent on earnings. However, the Plan of Modification (adopted in 1922) provided that in the event of defaults by the Interborough in the payment of taxes and bond interest under the Manhattan lease the original terms of the guaranty would be restored with respect to the modified shares. The Interborough was in receivership, and default under the Manhattan lease was highly probable. Hence the price relationship between the two classes of Manhattan stock appeared unjustified in the light of the facts.

NOTE 43

The following is a representative list of preferred and common stocks which sold for less than their *net current assets* per share at their low prices during 1931 and the first four months of 1932. Most of these issues sold at still lower prices later in 1932.

Company	1931– April 1932		Net quick per share Pre- ferred	Net quick per share Com- mon	1932–1933 low price	
	Low price Pre- ferred	Low price Com- mon			Pre- ferred	Com- mon
Allis-Chalmers.........	....	6½	......	$11	....	4
Amer. Agric. Chem.....	....	4⅜	......	43	....	3½
California Packing.....	....	5⅜	......	8	....	4¼
Diamond Match.......	19½	10⅝	$ 48	14	20½	12
Endicott-Johnson......	98⅜	23½	276	37	98	16
Liquid Carbonic.......	....	11¾	......	23	....	9
Mack Truck..........	....	12	......	36½	....	10
Mid-Continent Petrol..	....	3¾	......	8	....	3¾
Montgomery Ward.....	59	6½	462	16	41	3½
Nat'l Cash Register....	....	7⅛	......	15	....	5⅛
U.S. Indus. Alcohol....	....	19¼	......	23½	....	13¼
U.S. Pipe & Foundry...	12¼	8¾	26	10½	11½	6⅛
Wesson Oil...........	44¼	9½	74	.....	40	....
Westinghouse Air Brake	....	9½	......	11	....	9¼
Westinghouse Electric..	60¼	19⅞	1,164	34½	52½	15⅝

A similar list of stocks which at their low price during the first five months of 1932 sold at or below their *cash assets* per share is given on page 673.

Company	Low price Jan.– May 1932	Cash assets per share	Net quick assets per share	1932– 1933 low price
Amer. Car & Foundry*........	20	$ 50	$108	15
Amer. Locomotive*...........	30¼	41	63	17⅛
Amer. Steel Foundries*.......	58	128	186	34
Amer. Woolen*..............	15½	30½	85	15½
Congoleum-Nairn............	7	7	12	6½
Howe Sound.................	5¾	10	11	4⅞
Hudson Motor...............	2⅞	5½	7	2⅞
Hupp Motor	1½	5½	7½	1½
Lima Locomotive............	8½	19	36	8½
Magma Copper..............	4½	9	12	4¼
Marlin Rockwell.............	5¾	11½	13	5¾
Motor Products.............	11	15½	19	7⅜
Munsingwear................	10	17	34	5
Nash Motors................	8	13½	14	8
New York Air Brake.........	5	5	9	4¼
Oppenheim Collins...........	5	9½	15	2½
Reo Motor..................	1½	3	5½	1⅜
Standard Oil of Kansas.......	7	8½	14	7
Stewart Warner..............	1⅞	3½	7	1⅞
White Motor................	7	11	34	6⅞

* Preferred stock.

These examples have been taken from several articles by one of the authors dealing with this phenomenon. See Graham, Benjamin: "Inflated Treasuries and Deflated Stockholders," *Forbes*, June 1, 1932, p. 11; "Should Rich Corporations Return Stockholders' Cash?" *Forbes*, June 15, 1932, p. 21; "Should Rich but Losing Corporations Be Liquidated?" *Forbes*, July 1, 1932, p. 13. The 1932–1933 low prices are added to complete the picture.

NOTE 44

The analyst must frequently calculate the relative values of subscription rights and the common stock covered thereby. To facilitate this calculation we append two simple formulas.

Let R = value of right.

M = market price of stock.

S = subscription price of stock.

N = number of rights needed to subscribe to one share.

Formula A, applicable before stock sells "ex-rights" (*i.e.*, the purchaser of the stock will be entitled to receive the rights).

$$R = \frac{M - S}{N + 1}$$

Formula B, applicable after the stock sells "ex-rights" (*i.e.*, the purchaser of the stock does not get the rights, which are retained by the holder of record).

$$R = \frac{M - S}{N}$$

Example: Rights are given to buy one share of stock at 50 for each five shares held. Stock is selling at 64 "with rights" ("rights on" or "cum rights").

$$\text{Value of right} = \frac{\$64 - \$50}{5 + 1} = \$2.33$$

Example: Same offer; stock is selling "ex-rights" at 90.

$$\text{Value of right} = \frac{\$90 - \$50}{5} = \$8$$

These calculations are subject, however, to necessary refinements to reflect: (1) any dividend to be received by the old stock but not on the new shares; and, contrariwise, (2) any saving in interest by reason of not having to pay for the new stock until the rights expire.

NOTE 45

Two Examples of Corporate Pyramiding

First Example: The essential character of the Insull pyramid may be brought out by the following partial summary:

		Liabilities senior to common stock (Dec. 31, 1931)
Company 1 (Top Company)	Corporation Securities Co. An investment company of specialized character. Its chief holdings were in Co. 2—$59,000,000 and Co. 3—$42,000,000, out of total portfolio of $145,000,000.	Bank loans, etc....... $33,000,000 Funded debt......... 24,000,000 Preferred stock....... 37,000,000
Company 2	Insull Utility Investments, Inc. Also a specialized investment co. Its chief holdings were in Co. 3—$64,000,000 out of total portfolio of $252,000,000. (It also held $32,600,000 of stocks of Co. 1.)	Bank loans, etc....... $53,000,000 Funded debt......... 58,000,000 Preferred stock....... 46,000,000
Company 3	Middle West Utilities Co. A public utility holding company controlling a number of sub-systems. Gross business of system in 1931 was $173,000,000. Chief subsidiary was Co. 4.	Parent company: Bank loans, etc....... $35,000,000 Funded debt......... 40,000,000 Preferred stock....... 61,000,000 /Note: Public's holdings of: Subsid. bonds.... $283,000,000 Subsid. preferred. 152,000,000 Subsid common . 10,000,000/

		Liabilities senior to common stock (Dec. 31, 1931)
Company 4	National Electric Power Co. A public utility holding company controlling several subsystems. Gross business in 1931 was $68,000,000. Chief subsidiary was Co. 5.	Parent company: Bank loans, etc....... Not reported separately Funded debt......... $10,000,000 Pfd. & Class A stock.. 36,000,000
Company 5	National Public Service Corp. A public utility holding company controlling four subsystems. Gross business in 1931 was $36,000,000. Chief subsidiary was Co. 6.	Parent company: Bank loans, etc....... Not reported separately Funded debt......... $20,000,000 Pfd. & Class A stock.. 30,000,000
Company 6	Seaboard Public Service Co. A public utility holding company controlling six subsystems. Gross business in 1931 was $16,000,000. Chief subsidiary was Co. 7.	Parent Company: Funded debt......... None Preferred stock....... $9,000,000
Company 7	Virginia Public Service Co. A public utility operating and holding company. Gross business in 1931 was $7,600,000.	Funded debt......... $37,000,000 Preferred stock....... 10,000,000

Note that a pyramided structure of six successive holding companies was built above the various operating companies in this system. The complete collapse of this structure is shown by the fact that every one of these six superposed holding companies was thrown into bankruptcy. For description, charts and discussion of the Insull Group see James C. Bonbright, and Gardiner C. Means, *The Holding Company*, pp. 108–113, New York, 1932.

Second Example: The United States and Foreign Securities Corp. set-up provides a fairly simple demonstration of the workings of a pyramided structure in the general investment trust field.

This company was organized in 1924. The public bought $25,000,000 of $6 First Preferred at par (the company receiving $24,000,000), and the organizing bankers bought $5,000,000 of $6 Second Preferred at par. The 1,000,000 shares of Common Stock, representing a purely nominal investment (10 cents per share), were divided: 25% to the public, and 75% to the organizers. Thus the latter supplied one-sixth of the capital, subordinated to the other five-sixths, and received a three-quarters interest in the surplus profits. Toward the end of 1928, the holding company form of pyramiding was utilized by the formation of a second company, U.S. & International Securities Corp., a $60,000,000 enterprise. The public contributed $50,-000,000 of the capital, receiving $5 First Preferred Stock at 100, plus one-fifth of the Common. United States & Foreign Securities Corporation contributed $10,000,000, receiving $5 Second Preferred at 100, plus four-fifths of the Common. This arrangement gave the organizers of the original company control over the additional funds subscribed without further investment on their part. Because of a $30,000,000 appreciation in the resources of U.S. & Foreign Securities Corp., the end of 1928 found the contributors of the original $5,000,000 now controlling $110,000,000 of capital (including subscriptions callable) and entitled to about 78% of the surplus profits or enhancement thereof.

The actual operation of this arrangement from the standpoint of both book value ("break-up value") and market quotations is shown by the following tabulation:

A. PERIOD 1924–1928

Item	Total	Public's	Organizers'
Original investment..............	$ 30,000,000*	$25,000,000	$ 5,000,000
Book value, December 1928.....	60,000,000	32,000,000	27,000,000
% increase in book value........	100%	30%	450%
Maximum market value of U.S. & Foreign capitalization†........	100,000,000	42,000,000	57,000,000
% Increase in market value.....	233%	70%	1,040%

* Company received $29,000,000.
† First Preferred @ 100; Second Preferred estimated @ 80; Common @ 70.

B. PERIOD 1928–1933

Results are shown per $100 of original investment, because of decrease in First Preferred Stock outstanding due to repurchases by the company.

Date	Public's investment	Organizers' investment
Book value:		
Dec. 31, 1928.....	$130*	$550
Dec. 31, 1932.....	100*	35
Sept. 30, 1933.....	107*	111

* First Preferred at par, plus liquidating value of attached common.

Date	Public's investment	Organizers' investment	1st Pfd.	2d Pfd. (est.)	Common
Market price:					
High, 1929..	$170	$1,150	100	80	70
Low, 1932...	27½	11½	26	10	1½
Nov., 1933..	73	185	64	50	9

These figures show typical results for a highly speculative capital structure under both favorable and unfavorable developments. It will be noted that the variations in book or break-up value were greatly intensified in both directions by the excessive optimism and pessimism of the public's attitude toward investment trust securities. It is significant to observe also that when a book value about equal to the original investment per share was reestablished in 1933, the market registered a substantial depreciation for the public's part of the capital and a corresponding premium for the organizers' interest.

Note 46

A COMPARISON OF MISSOURI, KANSAS & TEXAS AND ST. LOUIS-SAN FRANCISCO

(Circular issued in January 1922)

Introduction.

The new securities of the Missouri, Kansas & Texas Railway present a number of attractive opportunities for both the investor and the speculator. The pending Reorganization Plan, which has recently been declared operative, reduces the fixed charges of the system to a very conservative figure, so that the bond interest should be regularly covered with a substantial margin. Furthermore, the road's excellent exhibit under current adverse conditions gives promise of a substantial earning power available for the junior securities.

The protracted receivership of the M. K. & T. will ultimately be found to have strengthened the position of the new issues. For during this period large expenditures were made for the physical rehabilitation of every part of the system. The resulting improvement in roadway and equipment has in turn led to greater operating efficiency, so that its transportation costs during the past year have been considerably lower than the average of other roads.

In analyzing the value of the new M. K. & T. securities, it is inevitable that comparison be made with the St. Louis-San Francisco. The two systems are highly similar in location, character of traffic, and financial structure. In fact the reorganization of Missouri, Kansas & Texas has been closely patterned after that of the 'Frisco, which was consummated in 1916.

The similarity of capitalization of the two roads is illustrated by the following table, comparing the current price and yields of various issues:

TABLE I

	St. Louis-San Francisco				Missouri, Kansas & Texas			
	Rate, %	Due	Price about	Yield, %	Rate, %	Due	Price about	Yield, %
Prior lien bonds........	4	1950	69½	6.35	4	1970	65	6.35
	5	1950	83½	6.25	5	1970	78	6.50
	6	1928	96½	6.55	6	1932	92	7.15
Adjustment bonds†.....	6	1955	73½	8.16 }	5	1967	45	11.11*
Income bonds†........	6	1960	55½	10.81 }				
Preferred stock........	(6)	....	38	....	(7)	....	25½	
Common stock........	..	....	21½	....	..	....	8¼	

* Assuming full interest paid.
† Straight yields given.

In the following pages we discuss the general situation of the two companies, with respect to capitalization and operating results, and then present a detailed comparison of the corresponding security issues. Our analysis indicates that M. K. & T. will possess two underlying advantages over the St. Louis-San Francisco:

I. Its fixed charges are lower in proportion to gross earnings.

II. Its operating efficiency is greater.

Through these important points of superiority, M. K. & T. should be enabled to provide a larger degree of protection for its bonds, and a greater relative earning power for its stocks. Basing our conclusions on a study of the two systems, we recommend the following exchanges to holders of St. Louis-San Francisco securities:

A—From 'Frisco Prior Lien 4s, 5s and 6s into the corresponding M. K. & T. Prior Lien issue, at their lower prices.

B—From 'Frisco Income 6s at 55½ into M. K. & T. Adjustment 5s at 45.

C—From 'Frisco Common Stock at 21½ into M. K. & T. Preferred Stock at 25½.

Moreover, judging the M. K. & T. issues on their individual merits, we regard the prior Lien Bonds as well-secured high yielding investments; and the Adjustment Bonds, Preferred Stock and Common Stock as affording attractive speculative opportunities.

The Missouri, Kansas & Texas and the St. Louis-San Francisco operate chiefly in the same states and at many points are in close competition.

TABLE II.—MILEAGE OPERATED DECEMBER 31, 1920

State	M. K. & T.	St. Louis-San Francisco
Missouri.............	544	1,720
Kansas..............	487	626
Texas...............	1,721	495
Oklahoma............	1,036	1,517
Other States.........	19	898
Total..............	3,807	5,256

Hence the character of traffic of the two systems is fairly similar, except that the 'Frisco carries considerably more coal and lumber and proportionately less oil. The rates per mile for both freight and passenger business are almost identical. M. K. & T. however averages a substantially heavier train load and longer haul.

TABLE III.—CALENDAR YEAR 1920

Item	M. K. & T.	St. Louis-San Francisco
Average revenue train load......	442 tons	398 tons
Average haul per revenue ton. ...	248 miles	187 miles

These two advantages no doubt account in good part for the much lower transportation costs of the M. K. & T. in 1921.

Capitalization.

The security issues of the two companies will compare as follows:

TABLE IV.—COMPARATIVE CAPITALIZATION

Item	M. K. & T.	'Frisco
Equipments and underlying issues	$ 7,248,000	$ 86,782,000
Prior lien bonds................	93,073,000	121,748,000
Adjustment bonds..............	57,500,000	39,220,000
Income bonds...................	....	35,192,000
Preferred stock.................	24,500,000	7,500,000
Common stock................	783,155 shares (no par)	504,470 shares (par $100)
Fixed interest charges..........	4,917,717	9,248,374
Contingent interest charges......	2,875,000	4,750,912
Total interest charges.........	$ 7,792,717	$ 13,999,286

The above figures for St. Louis-San Francisco are taken from the last available report, as of December 31st, 1920. Those for M. K. & T. are based on the assumption that all the old securities are exchanged under the provisions of the Reorganization Plan. It is probable, however, that some of the present senior liens, especially the First 4s, due 1990, will still remain outstanding. In such event, the amount of the underlying bonds, as stated above, would be increased and that of Prior Lien issues decreased— the aggregate remaining practically unchanged. The prospects are that the fixed interest charges will actually amount to somewhat less than the total given in the Plan, since the company will save ½ of 1% annually on such of the $40,000,000 of 1st 4s as are not exchanged.

TABLE V.—COMPARATIVE GROSS EARNINGS AND INTEREST CHARGES PER MILE OPERATED

	M. K. & T.		'Frisco	
	Per mile	% of gross	Per mile	% of gross
Gross earnings*...............	$16,870	100.0	$16,730	100.0
Fixed interest.................	1,300	7.7	1,790	10.7
Contingent interest...........	760	4.5	920	5.5
Total interest...............	$ 2,060	12.2	$ 2,710	16.2

* 1921 figure, December estimated.

The "Contingent Interest Charges" represent the requirements of the Income and Adjustment Bonds, which need be paid only if earned. This elastic provision is a source of strength for both roads, as it will enable them to reduce their interest payments in critical years without financial disturbance.

Table V indicates the advantage that will be gained by M. K. & T. through the drastic scaling down in its fixed interest charges. The latter will require only 7.7c. out of each dollar of receipts, a ratio so low as to guarantee a large margin of safety for the Prior Lien Bonds under ordinary conditions. In this respect M. K. & T. is seen to enjoy an important advantage over St. Louis-San Francisco, its interest charges—both fixed and contingent—being proportionately lower.

Earning Power.

In comparing the earning power of two enterprises, it is customary to take the average of reports covering a number of years. In the present case, however, the disturbing influence of Federal control makes such a procedure impracticable. For the figures of earlier years are too remote, and those from 1917 to 1920 are too abnormal, to afford a sound basis for analysis. It is necessary, therefore, to lay chief emphasis upon the most recent operating results. Statements for the eleven months ended November 30th, 1921 have just been published. By adding one-eleventh to these figures the following approximation to the full year's income account may be obtained:

TABLE VI.—INCOME ACCOUNT CALENDAR YEAR 1921 (ONE MONTH ESTIMATED)

	M. K. & T.		'Frisco	
	Income	% of gross	Income	% of gross
Mileage operated.............	3,784	.....	5,165	
Gross revenues...............	$63,842,000	100.0	$86,521,000	100.0
Maintenance.................	24,635,000	38.6	26,874,000	31.1
Other operating expenses......	25,072,000	39.3	37,275,000	43.1
Taxes.......................	2,731,000	4.3	3,790,000	4.4
Rentals, etc., less other income	1,654,000	2.6	1,065,000*	1.2
Balance for interest..........	9,750,000	15.2	17,517,000	20.2
Fixed interest................	4,918,000	7.7	9,248,000	10.7
Contingent interest...........	2,875,000	4.5	4,750,000	5.5
Balance for stocks............	1,957,000	3.0	3,519,000	4.0
Pfd. div. requirements........	1,715,000	2.7	450,000	.5
Balance for common..........	242,000	0.3	3,069,000	3.5

* 1920 figures partly used.

In analyzing the above figures, it is necessary to pay particular attention to the much heavier expenditures for maintenance made by M. K. & T.

Out of each dollar of receipts, the latter road devoted 38.6c. to upkeep, against only 31.1c. in the case of 'Frisco. It is well understood that the amounts spent on maintenance are largely a matter of arbitrary determination by the management and hence afford a method for more or less artificially controlling the net earnings. As compared with other roads in the same territory, it would seem that 'Frisco has been undermaintained and M. K. & T. overmaintained during the past year. The result of this diverse policy has been to make St. Louis-San Francisco's net earnings appear considerably larger and those of "Katy" considerably smaller, than on a normal basis of upkeep expenditure.

If in the case of both roads the latter had been taken at 35% of gross—apparently a reasonable figure—the net earnings of M. K. & T. would have been $2,300,000 *greater* and those of 'Frisco $3,280,000 *smaller* than the results actually reported.

How radically such a revision would affect the position of the various securities is shown by the following analysis.

TABLE VII.—EARNING POWER 1921

Item	Actual results		Adjusted results (maintenance ratio equalized at 35%)	
	M. K. & T.	'Frisco	M. K. & T.	'Frisco
Fixed interest earned	1.94 times	1.89 times	2.51 times	1.54 times
Total interest earned	1.25 times	1.25 times	1.55 times	1.02 times
Earned on preferred per share.........	$8.00	$46.92	$17.39	$3.19
Earned on common per share.........	0.30	6.08	3.25	Nil

The Prior Lien Bonds.

Although the M. K. & T. Prior Lien issue are selling several points lower than the corresponding 'Frisco bonds, the above table shows that they are better secured. For, despite the much heavier maintenance expenditure of "Katy," its fixed interest requirements were earned in 1921 with fully as large a margin. If proper allowance is made for the difference in upkeep, then the superior showing of M. K. & T. becomes very marked.

The Income and Adjustment Bonds.

The interest on the M. K. & T. Adjustment 5s will be cumulative after 1925, while the St. Louis-San Francisco Income 6s are permanently non-cumulative. During the next three years at least one-half of the income available for the M. K. & T. Adjustments must be paid in interest. On the base of the earnings of 1921, it is probable that the income bondholders will receive the full 5% for this year.

These M. K. & T. and 'Frisco issues yield the same return, if full interest is paid. The "Katy" bonds are closer to the rails, being directly junior to the Prior Lien issues, while the 'Frisco Income 6s are subject also to the Adjustment Mortgage. As indicated by Table VII, the M. K. & T. Adjustments should have the benefit of a considerably larger earning power under normal operating conditions.

M. K. & T. 7% Preferred.   (*Cumulative after January 1, 1928*)

Because of the similarity in market price, this issue is comparable with 'Frisco common rather than 'Frisco preferred.   M. K. & T. Preferred makes an excellent exhibit in respect to current earnings, and appears not only distinctly preferable to St. Louis-San Francisco common, but also an independently attractive speculative purchase.

M. K. & T. Common.

While dividends on the issue are doubtless very remote, it should quickly reflect marketwise any improvement in the general railroad situation or in the position of Missouri, Kansas & Texas.   At its present price of $8¼ per share, it possesses unusual speculative opportunities as a low priced railroad issue.

A COMPARISON OF ATCHISON, SOUTHERN PACIFIC, AND NEW YORK CENTRAL

(Circular issued in April 1922)

Introduction.

Recent weeks have witnessed a revival of interest in high-grade railroad shares.   This activity is of particular significance because it is based on both investment and speculative considerations.   The continued advance in the bond list has first been followed by corresponding strength in the preferred issues, and is now directing attention to the investment type of common stocks—namely, those with long-established dividend records.

From the speculative standpoint also, railroad shares of the better class are becoming increasingly attractive.   Indications point clearly to a great improvement in net earnings during 1922, as compared with 1921.   Already substantial increases in car loadings are being reported, and the improvement should be intensified by the industrial revival expected later in the year.   Of even greater importance is the continued reduction of operating expenses, which is gradually leading to a return of a normal ratio of net earnings to gross receipts.

The high-grade railroad common stocks therefore deserve consideration by both investor and speculator.   We present herewith the results of an examination of the present status and recent record of three of the prominent issues of this type—Atchison, Southern Pacific, and New York Central. Some of the most important data are summarized in the following brief table:

COMMON STOCK

Road	Price about	Dividend rate, %	Yield, %	Earnings per share		Fixed charges earned 1921
				1921	Average 1914–1921	
Atchison.........	100	6	6.00	$14.69	$12.89	4.00 times
Southern Pacific..	90	6	6.67	7.25*	8.35*	2.13* times
New York Central	91	5	5.50	8.92	6.64	1.44 times

* Partly estimated. See text.

These figures indicate clearly the pre-eminence of Atchison from the standpoint of earning power and financial strength. As compared with New York Central, it shows a higher dividend return, larger earnings, and a much smaller proportion of bonded debt. While Southern Pacific and Atchison both pay 6% in dividends, Atchison has shown such pronounced superiority in earning power as to justify fully its ten-point higher quotation.

In addition to its remarkable record of earnings the following features in Atchison's exhibit deserve special note:

1. Its wealth of cash assets.
2. Its valuable oil properties.
3. Its low and steadily decreasing funded debt.

The record of the three companies is analyzed in greater detail in the following pages. Based upon a careful study of the available data, we submit the following conclusions:

A—That Atchison should be purchased at the present time, either as an attractive investment or for conservative speculative profit.

B—That Atchison is intrinsically more desirable than Southern Pacific, because of its substantially greater earning power.

C—That investment holdings of New York Central might well be exchanged into Atchison, in order to obtain a higher dividend yield, larger average earning power, and greater financial stability.

From the speculative standpoint, it is proper to point out that the small amount of New York Central stock, in relation to its bonded debt and gross revenues, may result in a more rapid increase in profits per share under favorable conditions. Conversely, however, a relatively small decline in net earnings can seriously reduce the balance available for the stock.

Corporate Structure.

In analyzing the position of a railroad company, it is often necessary to consider not only its own operations, but also those of subsidiary or affiliated lines in which it has a substantial investment. Atchison and Southern Pacific publish reports covering the results of the entire system, but New York Central has large stock holdings in a number of important lines which report their operations separately. The aggregate mileage of these controlled companies actually exceeds that of the New York Central proper. Each year the subsidiaries carry a substantial amount to surplus, a good part of which really accrues to New York Central stock, but is not reflected in the parent company's return. To afford a proper basis for judging the value of New York Central shares, we shall analyze its earning power as indicated both by its own statement and by a consolidated report embracing all its subsidiaries. An added reason for using the latter method is found in a recent statement that the New York Central intends to acquire the outstanding minority shares of the controlled companies, in order to merge their operations with its own.

The following table lists the separately operated subsidiaries of the New York Central, together with their mileage and the percentage of stock held within the system.

NEW YORK CENTRAL SYSTEM

Company	Mileage	% of stock owned
N. Y. Central R. R..............	6,069	
Cincinnati Northern..............	245	56.9
C. C. C. & St. Louis..............	2,421	50.1
Indiana Harbor Belt.............	120	60.0
Kanawha & Michigan............	176	100.0
Lake Erie & Western.............	738	50.1
Michigan Central.................	1,865	89.8
Pittsburgh & Lake Erie...........	224	50.1
Toledo & Ohio Central...........	492	100.0
Total system..................	12,350	

As regards Southern Pacific also, the exhibit of previous years must be revised, in order to reflect the adjustments that have followed from the recent segregation of the oil properties. Allowance is to be made for the elimination of the former oil income, the exchange of convertible bonds into stock and the receipt of $43,000,000 in cash through the sale of the Pacific Oil shares.

Earning Power.

Particular interest attaches to the results during 1921 because they are the most recent available and also because they represent the first full year of independent operation. A summarized income account for 1921 appears at top of page 685.

The fixed charges and non-operating income of Southern Pacific are estimated on the basis of the 1920 report, as adjusted to reflect the segregation of the oil lands.

It will be seen at once that Atchison makes the best exhibit, not alone in earnings per share, but especially in the small ratio of fixed charges to available income. The combined income account of New York Central and its subsidiaries indicates very substantial profits per share, but due consideration must be given here to the large proportion of its total capitalization represented by bonds and rental agreements.

INCOME ACCOUNT 1921
(In Thousands of Dollars)

Item	Atchison	Southern Pacific	N. Y. Central R. R.	N. Y. Central System
Mileage..................	11,678	11,187	6,077	12,350
Gross revenue............	$228,925	$269,494	$322,538	$535,821
Net after rents............	41,268	39,823	56,679	90,615
Other income.............	11,082	8,000*	15,665	17,251
Total income.............	$ 52,350	$ 47,823	$ 72,344	$107,866
Fixed charges, etc.........	13,018	22,800*	50,048	71,519
Preferred dividends........	6,209	....	....	500
Applicable to minority stock..................	....	....	....	4,302
Balance for common.......	33,123	25,023	22,296	31,545
Per share...............	14.69	7.25	8.91	12.62†

* Estimated. See text.
† Per Share N. Y. Central Stock.

The conclusions indicated by the 1921 figures are confirmed by a consideration of the record of each company since 1914. We give below the

ANNUAL EARNINGS PER SHARE OF COMMON STOCK 1914–1921

Calendar year	Atchison		Southern Pacific*		N. Y. Central R. R.		N. Y. Central System	
	Operating basis	Guaranteed basis	Operating basis	Guaranteed basis	Operating basis	Guaranteed basis	Operating basis	Guaranteed basis
1921	$14.69	......	$ 7.25	.....	$ 8.92	.....	$12.62	
1920	12.54	$13.98	1.89	$8.61	12.34(d)	$5.49	14.65(d)	$9.68
1919	15.41	16.55	7.03	8.40	6.23	7.97	10.73	8.62
1918	10.59	9.98	10.63	8.38	6.59	7.16	13.39	8.34
1917	14.50	......	13.96	.....	10.24	.....	13.25	.....
1916	15.36	......	11.00	.....	18.26	.....	23.50	.....
1915	10.99	......	8.90	.....	11.08	.....	13.80	.....
1914	9.03	......	6.01†	.....	4.10	.....	3.69	.....
Average: Operating basis	$12.89		$8.33		$6.64		$ 9.54	
Guaranteed basis		13.14		9.06		9.16		11.69

* See text.
† Year ended June 30.

annual earnings per share during this period. For 1918, 1919 and 1920, two results are presented, based both on the actual operations and on the government rental and guarantee. The Southern Pacific figures are adjusted as indicated above.

Not the least remarkable feature of the above exhibit is the regularity with which Atchison's net has been maintained at a high rate since 1915, despite the unusual conditions affecting the carriers as a whole during a good part of this period. The contrast with New York Central and Southern Pacific is especially sharp in the transition year 1920.

Another significant feature is the substantial increase in Atchison's non-operating income, which rose from $4,311,000 in 1918 to $15,100,000 in 1919 and $9,842,000 in 1920. A good part of these profits was derived from its oil properties, the importance of which seems to have been insufficiently recognized.

Operating Statistics.

The superior earning power of Atchison as compared with both Southern Pacific and New York Central, rests to some extent on a smaller capitalization in relation to gross receipts, but more particularly upon lower operating expenses. The appended table shows clearly the advantage enjoyed by Atchison in the field of transportation costs:

ANALYSIS OF OPERATING EXPENSE

Per cent of gross receipts expended for:	Atchison			Southern Pacific			N. Y. Central R. R.		
	1921	1918–20	1913–17	1921	1918–20	1913–17	1921	1918–20	1913–17
Maintenance..........	36.9	36.0	30.1	33.9	34.3	25.4	31.9	36.3	29.9
Transportation, etc....	38.7	39.6	34.1	45.0	45.1	38.8	45.1	47.6	40.0
Total Operating Expenses........	75.6	75.6	64.2	78.9	79.4	64.2	77.0	83.9	69.9

It will be observed that Atchison has been consistently liberal in its maintenance expenditures. As compared with the similarly located Southern Pacific, Atchison has regularly devoted a larger percentage of its revenues to upkeep, and a much smaller percentage to transportation charges.

Capitalization Structure.

The proportion of stocks to bonds is largest for Atchison and least for New York Central. The capitalization of the latter system appears rather ill-balanced, so that relatively small changes in net income result in wide fluctuations in the balance available for each share of stock. In prosperous years this preponderance of bonded debt results in a large apparent earning power for the stock, but in periods of depression it may constitute a serious burden.

SECURITIES HELD BY PUBLIC
(Thousands omitted)

Class of issue	Atchison (Dec. 31, '21)	% of total	Southern Pacific (Jan. 14, '21)	% of total	N. Y. Central Railroad (Dec. 31, '20)	% of total	N. Y. Central System (Dec. 31, '20)	% of total
Bonds and guaranteed stocks..	$289,888	45.3	$473,644	57.9	$ 840,110*	77.1	$1,156,261*	77.5
Preferred stocks.	124,173	19.4	....	....	....	....	9,998	0.9
Minority stocks.	....	....	....	....	....	....	74,302	4.9
Common stocks.	225,398	35.3	344,780	42.1	249,597	22.9	249,597	16.7
Total........	$639,459	100.0	$818,424	100.0	$1,089,707	100.0	$1,490,158	100.0

* Includes Securities of Leased Companies, and $66,700,000 for cash rentals capitalized at 5 %.

Conclusion.

The unique status of Atchison in the railroad field is perhaps best illustrated by its treasury position. Despite the fact that the Company has sold virtually no bonds during the past eight years, it held on December 31st last over $52,700,000 in cash and Government Bonds, while its current liabilities totalled $28,279,000.

The combination of large earning power and strong financial condition justifies the expectation of an eventual increase in the dividend rate.

CHICAGO, MILWAUKEE & ST. PAUL
ST. LOUIS SOUTHWESTERN

(Circular issued in September 1922)

Ten years ago the grouping of these railroads for analysis would have appeared far-fetched. St. Paul was then one of the most substantial systems in the country. St. Louis Southwestern was decidedly inferior in prestige, earning power, dividend record, and presumably in future prospects. The relative standing of the two companies was measured by the average quotation for their common shares—above par for St. Paul, below 30 for St. Louis Southwestern. Today the market brackets them together by ascribing the same value to their stocks, both preferred and common—about 52 and 33 respectively. This radical change in market position reflects a corresponding reversal in earning power, as is evident from the condensed income account for 1913 and 1921 shown on page 688.

Admitting at the outset that St. Paul's operating expenses in 1921 were so abnormally large as to detract greatly from the significance of the *net earnings*, the relative increase in *gross receipts* and *fixed charges* is undoubtedly of prime importance. St. Louis Southwestern has made good progress, because its gross has expanded twice as fast as its charges. But St. Paul has retrogressed, because its fixed charges have grown three times as rapidly as its traffic. Striking as are these figures, it is possible they convey a misleading impression as to the future prospects of the two lines. It may be true, as many contend, that the dismal showing of St. Paul in 1921 represents about the last phase of a painful transition period from its old prosperity to a new and even greater success. Or, conversely, some may claim that the great improvement recorded by St. Louis Southwestern is but a temporary reflection of accidental conditions. To reach dependable conclusions, therefore, a thoroughgoing analysis of the leading physical and financial elements underlying each road is requisite; and this we attempt in the following pages.

Item	Chicago, Milwaukee & St. Paul			St. Louis Southwestern		
	Years ended		In-crease, %	Years Ended		In-crease, %
	June 30, 1913	Dec. 31, 1921		June 30, 1913	Dec. 31, 1921	
Mileage operated.........	9,613 miles	10,808 miles	12.4	1,609 miles	1,808 miles	12.4
Gross receipts..	$94,084,000	$146,766,000	56.0	$13,297,000	$25,113,000	88.8
Net after taxes..	27,376,000	9,763,000	.....	3,600,000	4,881,000	
Fixed charges, etc. (net).....	9,235,000	23,112,000*	150.0	1,714,000	2,404,000	40.3
Balance for dividends........	18,141,000	13,349,000(d)	.....	1,886,000	2,477,000	....
Fixed charges earned.......	2.96 times	.42 times	.....	2.10 times	2.03 times	....
Earned per share preferred.....	$15.60	Nil	.....	$9.48	$12.53	....
Balance per share common	8.60	Nil	.....	5.45	9.16	....

* Eliminating from Non-Operating Income $2,278,000 due from U. S. Government, applicable to prior years.

On page 689 we submit as of especial interest, an estimate of the average future earning power of the two systems under presumably normal operating conditions. In basing the gross revenues upon the 1921 figures, we have conservatively assumed that increases in traffic will be offset by reductions in rates.

If these figures fairly reflect the situation, they would justify a definite decision in favor of St. Louis Southwestern over St. Paul. The former would enjoy a greater margin of safety over interest charges; it would earn three times as much on its preferred stock; and it would show a balance of nearly $10 per share on its common, against no earnings whatever for St. Paul's junior shares.

The superior exhibit of St. Louis Southwestern is not due to any temporary fluctuations in traffic or operating expenses. It arises primarily from a basic advantage in *capitalization structure*. The combined fixed charges (net) and preferred dividends of St. Paul consumed 21.1% of its 1921 gross; the corresponding requirements of St. Louis Southwestern are only 13.6%

of gross. This saving of 7½% of gross marks all the difference between inability to cover preferred dividends and a substantial balance available for the common stock.

ESTIMATED AVERAGE FUTURE INCOME ACCOUNT

Item	St. Paul		St. Louis Southwestern	
	Income	% of gross	Income	% of gross
Gross receipts (1921 basis)	$150,000,000	100.0	$25,000,000	100.0
Net after taxes (20% of gross)................	30,000,000	20.0	5,000,000	20.0
Fixed charges, etc., net (1921 basis)...........	*23,500,000	15.7	2,400,000	9.6
Balance for dividends.....	6,500,000	4.3	2,600,000	10.4
Fixed charges earned.....	1.28 times	.....	2.08 times	.....
Earned per share preferred	$4.32	.....	$13.07	.....
Balance per share common	Nil	.....	9.81	.....

* See footnote to table on previous page. Figures for Chi. T. H. & S. E. (leased July 1, 1921) are included for a full year.

Basing our views on the salient points referred to above, and upon the supporting analysis presented herewith, the preferred and common shares of St. Louis Southwestern appear to us decidedly more attractive than those of St. Paul, now selling at similar prices.

We invite inquiries regarding the relative status of the various bond issues of the two systems.

Growth of Traffic.

The fairness of our estimate of future earnings, given above, may be challenged on the ground that it fails to take into account the much better prospects of St. Paul for increasing its revenues. It has been generally believed that St. Paul enjoys an unusually favorable location, and that with the development of the territory tapped by its Puget Sound extension, its traffic would show a phenomenal increase.

The results over a ten year period should concededly afford some index of the correctness of this theory. If we compare the figures for 1921 and 1913 (the Puget Sound having been absorbed in the latter year), it is startling to note that the traffic density has experienced a substantial shrinkage. In fact, despite the addition of 1,200 miles to its lines, St. Paul actually moved fewer passengers and less freight in 1921 than in 1913, and carried them a shorter distance. Had it not been for the sharp advance in rates, its gross receipts would have been less than those of 9½ years before.

In the meantime, St. Louis Southwestern has made a decided gain in traffic. If we go back to 1920, the year of maximum tonnage, we find that the Southern line's volume of business per mile has increased four times as fast as St. Paul's.

CHANGES IN TRAFFIC DENSITY, 1913–1921 (REVENUE TON MILES PER MILE OF ROAD)

Year	St. Paul		St. Louis Southwestern		All roads	
	Total	% change	Total	% change	Total	% change
1913	892,000	....	542,000	....	1,335,000	
1920	1,071,000	+20	1,022,000	+89	1,748,000	+31
1921	770,000	−14	730,000	+35		

Fixed Charges.

As contrasted with the sluggish trend of St. Paul's traffic, its fixed charges have expanded enormously. While but 12.4% has been added to mileage, interest on bonded debt has increased 65.7%. The situation is emphasized if account is taken of rentals and other deductions, as well as of non-operating income. The combined balance of these items has changed from a credit of $2,202,000 in 1913 to a debit of $4,164,000 in 1921. The final result is that the *net* figure of all deductions, less offsets, has increased exactly 150% in the past ten years. This is undeniably an adverse development of prime importance.

Turning for contrast to St. Louis Southwestern, we note a very moderate advance (11.4%) in interest charges—less, in fact, than in mileage. An increase of about $218,000 in rentals, etc., and decrease of the same amount in "other income," brought the rise in the net deductions to 40%. This is below the increase in physical traffic, and less than one-half the growth in revenues.

Operating Expenses.

While prior to the period of Government control the trend of operating expenses merited careful study, the extreme variations of these figures in recent years have greatly reduced their value in forecasting future results. If, nevertheless, we group the statements of the past ten years in three periods—Pre-Control, Control, and Post-Control—an interesting sidelight is thrown upon the operating problems of the two roads (see p. 691).

The striking feature of this exhibit is first the steady success of St. Louis Southwestern in keeping its actual transportation costs fairly close to the pre-war percentage of gross; and secondly, the difficulty experienced by St. Paul in reducing its corresponding expenses from the abnormal figure to which they rose under Government administration. The 10% advantage in this respect reported by "Cotton Belt" in 1918–1920 might be overlooked as accidental, but the persistence of the same ratio since the beginning of 1921 cannot be without significance.

The maintenance figures given on page 691 are inconclusive. From 1913 to 1917, St. Paul appears to have been undermaintained. During Government control the upkeep expenditure of both roads attained the same high

DEVELOPMENT OF OPERATING EXPENSES
(In Percentage of Gross Receipts)

Item	Average 1913–1917		Average 1918–1920		Average 1921–June 1922	
	St. Paul	St. L. S. W.	St. Paul	St. L. S. W.	St. Paul	St. L. S. W.
Maintenance..........	27.02	31.36	42.31	42.57	37.74	35.03
All other expenses and taxes.............	45.95	44.05	56.92	46.30	57.07	47.04
Total...............	72.97	75.41	99.23	88.87	94.81	82.07

figure, from which St. Louis Southwestern has been receding to normal somewhat more rapidly than the Western system.

In our estimate of future earning power, we have placed St. Paul's expenses at relatively no higher than "Cotton Belt's." This assumption, it would appear from the above data, implies considerable optimism as to the possibilities of its undoubtedly capable management.

Financial Position.

With the present bond market and a U. S. Treasury willing to aid the roads with weak credit, the question of current assets and liabilities does not seem as important as once it was. By borrowing no less than $55,000,-000 from the United States, St. Paul has been able to meet its deficits, increase its property account, and retain a fair cash position. St. Louis-Southwestern has no notes payable, and, in fact, has somewhat reduced its bonded debt since 1913. At the same time, it has transformed an excess of $540,000 in current liabilities on June 30, 1915, into a credit balance of current assets amounting to $5,726,000.

Within the next two and one-half years, St. Paul will have to meet maturing obligations totalling $53,000,000 in addition to $10,000,000 of notes due the U. S. on January 1 next. But outside of small equipment obligations, St. Louis Southwestern has no indebtedness maturing before June 1, 1932.

Physical Factors.

A. Character of Traffic.

Although St. Louis Southwestern styles itself "The Cotton Belt Route," its traffic does not lean heavily upon this staple. In fact, only 6.2% of its total tonnage in 1921 consisted of cotton and its products. St. Paul depends somewhat more upon wheat and corn, the two crops making up 9.3% of last year's traffic. A noteworthy element is the gradual decrease in the percentage of higher grade tonnage (manufactures and miscellaneous) on St. Paul and the equally pronounced change in the opposite direction on the St. Louis Southwestern.

B. *Train Load and Average Haul.*

Item	St. Paul			St. Louis S.W.		
	1913	1921	Change, %	1913	1921	Change, %
Av. revenue train load (tons)...............	357	483	+35.3	300	465	+55.0
Average haul per ton (miles)...............	246	243	− 1.2	238	252	+ 5.9

"Cotton Belt" is seen to have increased its train load more rapidly than St. Paul, and to have gradually lengthened its average haul so as to outstrip the larger road. Heavy train loading and long hauls are both important factors in cutting transportation costs.

C. *Equipment.*

The relative status of the two roads in this respect is best illustrated by the fact that in every year St. Louis Southwestern has received a substantial net income from hiring out surplus equipment, while St. Paul has with equal consistency been paying out considerable sums net for the use of foreign rolling stock.

Summary.

The facts presented above are especially impressive because of the unanimity with which they favor St. Louis Southwestern over St. Paul. On almost every point considered, it may be said that whereas ten years ago St. Paul shaped up better than the Southern line, in 1921 the conditions were directly reversed. This statement would apply (on a per mile basis) to gross receipts, operating ratio, interest and other charges, character of traffic, treasury position and other items.

It is difficult to escape the conclusion that both the preferred and common shares of St. Louis Southwestern are entitled to a higher market valuation than those of St. Paul.

We wish to stress particularly the speculative possibilities of St. Louis Southwestern common. The withholding of dividends on the 5% non-cumulative preferred since 1914 has enabled the company to plough back large sums into the property, for the ultimate benefit of the common stock. Since June 30, 1912, surplus earnings so reinvested have amounted to $12,875,000—or nearly $80 per share of common (present price 34).

NOTE 47

MEMORANDUM FOR HOLDERS OF VICTORY BONDS

(Circular issued in May 1921)

We desire to point out to owners of Victory 4¾s, due June 1, 1923, the advantage to be gained through their exchange at current prices into an equivalent amount of Liberty Fourth 4¼s, due 1938.

At this writing the Victory 4¾s are selling at about $97.70, and the Liberty 4¼s at about $87.20. The straight income return on both issues is the same—4.86%. Differently stated, each $400 of Victory notes can be exchanged for $450 of Liberty Fourth 4¼s, on an even basis of both cost and income return.

But the Liberty bonds have a great advantage over the Victory Notes from the standpoint of prospective market appreciation. The possible advance of the Victory Notes is strictly limited to two points, since their near maturity (1923) precludes their selling at any considerable premium. The Liberty bonds, however, are selling at so substantial a discount from par (over 12½%), that it is not only possible but quite probable that there will be an important advance during the next few years.

To use perhaps an extreme example, if we suppose that by 1923 all Victory and Liberty bonds have returned to par, the rise in the Fourth Liberty bonds would amount to over twelve points against only two points for the Victories. By making the proposed exchange, the investor would then realize $450 for each $400 of Victory Notes now owned. In any event, the Liberty 4¼s need to advance only two points in the next two years to make the suggested exchange profitable.

In this connection we would point out that all indications favor an impending advance in high grade bond prices. The tendency toward lower interest rates is already apparent, as is evidenced by the reduction in the Federal rediscount rate. For this reason, long term investments are now quite generally preferred over short term notes, and consequently the income return to be obtained on the former is considerably less than that on near maturities. But in the case of the Victory issue, these short term notes can be exchanged for long term Liberty bonds without any reduction in straight income return.

Liquidation in the Liberty issues has been drastic and until recently continuous, but this period now appears about ended. Bonds bought with borrowed money have for the most part been paid for or sold; weak holdings have been nearly eliminated, and the Liberty issues may now be regarded as largely in the hands of real investors. This greatly improved technical position should result in a substantial advance in price, in response to any buying activity.

A further advantage to be gained from the proposed exchange lies in the exemption of Liberty bonds (up to certain limits) from *surtax* as well as normal tax; whereas, the Victory notes are exempt only from normal tax.

For these two important reasons—prospects of much greater price appreciation and superior tax exemption—we recommend that holdings of Victory notes be now transferred into an equivalent amount of Liberty Fourth 4¼s.

We shall be glad to supply further information regarding this suggestion and in particular to discuss with individual investors the current saving in taxes to be gained from the exchange.

NOTE 48

"Investors Guide Stock Reports," a department of Standard Statistics Co., Inc., issued the following two bulletins in October and December 1933.

B (N.Y.S.E.)   BALDWIN LOCOMOTIVE WORKS   69

Stock	Rating	Dividend	Price	Date	Yield
Common	Hold II	None	11⅛	12/21/33	None
$7 Preferred	Hold, P.S.*	None	34⅞		None
Warrants	Hold II	....	7		....

COUNSEL: Constructive developments in sight serve to neutralize the adverse effect in the COMMON of the eventual exercise of stock purchase warrants. The PREFERRED has long term speculative attraction.

POSITION & PROSPECT: Although Baldwin's operating expenses have been held to a minimum, the lack of locomotive orders in 1933 is likely to be reflected in another net loss for the year. Consolidated bookings have recently exhibited moderate expansion and the 1934 outlook for the company has been considerably improved by loans, which have been granted to a number of roads by the PWA for the purchase of new equipment, including 30 locomotives. Applications are now pending from other carriers for loans for equipment which will include 133 locomotives. Thus, there are definite indications that a start has been made by the carriers to modernize their tractive power, a program which is likely to be in full swing later in 1934. Baldwin, with its strong trade position, may be expected to obtain a goodly share of the business. While effective earnings on the common are still sometime off, especially since the stock is subject to considerable dilution by the indicated eventual exercise of warrants attached to the consolidated mortgage bonds permitting the purchase, at $5 of 480,000 additional common shares, it appears that common per share losses should show progressive abatement from now on. FINANCIAL POSITION is strong.

BACKGROUND: Baldwin Loco. Works is one of the two largest builders of steam locomotives. It also manufactures forgings and castings, hydraulic and special machinery, engines, air conditioning units, refrigeration equipment, etc. The company has a stock interest in General Steel Castings and owns valuable Philadelphia real estate.

CAPITALIZATION: Funded debt, $15,500,000. 7% cum. pfd. ($100 par) 200,000 shares, red. at $125. Common (no par) 843,000 shares. Preferred dividend accumulations total $17.50 per share at present.

	Earnings		Dividends		Price range	
	Com.	Pfd.	Com.	Pfd.	Com.	Pfd.
1933	Est. $5.24(d)	Est. $15.50(d)	None	None	17⅝– 3½	60 – 9½
1932	6.50(d)	20.39(d)	None	None	12 – 2	35 – 8
1931	6.55(d)	20.61(d)	$0.87½	$3.50	27⅞– 4⅝	104½–15
1930	1.94	15.18	1.75	7.00	38 –19⅜	116 –84

Caution—This information has been obtained from sources believed to be reliable but is not guaranteed.

INVESTOR'S GUIDE STOCK REPORTS

(Copyrighted and Published by Standard Statistics Co., Inc.,
345 Hudson St., N.Y.)

* P. S. = Preferred-Speculative.

BRY (N.Y.S.E.)   BEATRICE CREAMERY CO.   76

Stock	Rating	Dividend	Price	Date	Yield
Common	Switch	None	12½	10/17/33	None
$7 Preferred	Switch	$7	72		9.9%

COUNSEL: In view of near term uncertainties, holdings of the COMMON and PREFERRED shares should be switched to issues with more promising prospects.

POSITION & PROSPECTS: Dairy operations remain under the handicap of the industry's unfavorable statistical position. Milk production is well in excess of consumption requirements, and this situation not only has resulted in the building up of record sized stocks of butter and cheese but also has prevented sustained price strength in these commodities. Price advances on fluid milk, instigated mainly by state milk control boards or AAA marketing agreements, have been passed on almost entirely to farmers. In addition, earnings of the company for the six months ended August 31, last, were adversely affected by increased costs under the NRA and by unsatisfactory ice cream sales during the peak months of July and August. Share returns for the period amounted to $4.47 on the preferred and $0.28 on the common, against $6.34 and $0.82, respectively, for the like interval a year earlier. Because of seasonal factors, an even smaller profit is indicated for the final half. Recovery promises to be slow until the excessive milk supplies are eliminated. FINANCIAL POSITION is strong.

BACKGROUND: Beatrice is the third largest unit in the dairy products industry. Formerly deriving the major portion of its earnings from butter, the company in recent years has considerably expanded its activities in ice cream and milk; in addition, it distributes cheese, eggs, and poultry. Properties are located mainly in the Middle West, but extension into eastern and Pacific Coast markets also has been effected.

CAPITALIZATION: Funded debt, none. 7% cum. preferred ($100 par) 107,851 shares. Common ($25 par) 377,719 shares.

	Earnings*		Dividends†		Price range†	
	Com.	Pfd.	Com.	Pfd.	Com.	Pfd.
1933	$0.84(d)	$ 4.03	None	7.00‡	27 – 7	85 – 45
1932	3.54	19.30	$2.50	7.00	43½–10½	95 – 62
1931	7.12	32.49	4.00	7.00	81 –37	111 – 90
1930	7.31	34.02	4.00	7.00	92 –62	109¼–101¼

* Years ended February 28.
† Calendar years.
‡ Continuance possible.

Caution—This information has been obtained from sources believed to be reliable but is not guaranteed.

INVESTOR'S GUIDE STOCK REPORTS

(Copyrighted and Published by Standard Statistics Co., Inc.,
345 Hudson St., N.Y.)

It is evident that the advice to hold Baldwin Locomotive and to sell Beatrice Creamery shares was based predominantly upon the view that the prospects of the locomotive business were good and those of the dairy industry were poor. With respect to the former it is implied that the improvement will continue for a number of years; in the case of Beatrice Creamery it is not clear whether the statement that "recovery promises to be slow" presages a delay of months or of years.

The approach of the securities analyst towards these two common issues, if based upon the principles and technique developed in this book, would be quite different from—in fact, almost the direct opposite of—that indicated in the "Stock Reports" given above. The analyst's initial reasoning as to Beatrice Creamery would run somewhat as follows: "Current conditions are known to be unfavorable and the near-term prospects are generally considered unfavorable also. The price of the stock has declined substantially. Is it possible that the shares may have intrinsic or permanent value considerably in excess of the current low price, which is governed by the current situation?"

In the case of Baldwin Locomotive, his reasoning might well run in the contrary direction:

"The company's prospects are decidedly better for 1934 than they were for 1933 and 1932. However, the stock is selling at five times the low price of 1932. Are these prospects favorable enough and dependable enough to make the common stock attractive at its current price, in view of the very unsatisfactory record for the past ten years?"

In developing the answer to these questions a statistical analysis somewhat along the following lines would be in order. (The data are not presented as a "comparison" of Baldwin and Beatrice in the ordinary sense, but rather as an aid in arriving at *separate* analytical conclusions in respect to each issue.)

Item	Baldwin Locomotive	Beatrice Creamery
A. *Capitalization:*		
Bonds at par.......	$15,500,000	
Preferred stock at market.........	7,000,000	$ 7,750,000
Total senior issues..	$22,500,000	7,750,000
Common stock at market.........	9,400,000	4,700,000
Warrants at market	3,400,000	
Total common-stock issues..........	12,800,000	
Total capitalization	35,300,000	12,450,000
B. *Recent Income Account:*	12 mo. ended Sept. 1933	12 mo. ended Aug. 1933
Sales.............	7,730,000	44,045,000
Net before depreciation and interest	1,000,000(d)	1,831,000
Depreciation.......	1,850,000	1,605,000
Interest...........	1,160,000	
Preferred dividend requirement.....	1,400,000	750,000
Balance for common	5,410,000(d)	524,000(d)

C. *Earnings Record* (000 omitted):*

Year	Baldwin Locomotive			Beatrice Creamery			
	Sales	Earned on total capital	Earned for common	Sales	Earned on total capital	Earned for common	Earned per share of common
1933	$ 7,730	$2,850(d)	$5,410(d)	$44,045	$ 226	$ 524(d)	$ (d)
1932	10,579	2,941(d)	5,478(d)	46,264	434	323(d)	(d)
1931	20,436	2,982(d)	5,523(d)	54,059	2,101	1,363	$3.54
1930	49,872	4,202	1,637	82,811	3,354	2,626	7.12
1929	42,797	3,093	900	83,682	2,489	1,971	7.31
1928	37,214	600	1,104(d)	53,307	1,523	1,103	6.31
1927	49,011	3,400	1,685	52,744	1,223	890	6.66
1926	65,569	5,800	4,049	33,974	1,006	735	5.97
1925	27,876	500(d)	2,225(d)	35,050	1,003	760	6.18

* Baldwin: Year ended Sept. 30, 1933, and calendar years preceding. Figures are on a comparable basis, except those for 1925. Figures for 1925–1928 are corrected to reflect the average depreciation of $1,022,000 per annum, as discussed in Chap. XXXIV. Earnings on total capital for 1928 are approximate.

Beatrice: 1933 means year ended Aug. 31, 1933. 1932 means year ended Feb. 28, 1933, and similarly for 1925–1931. Profit of $389,000 on sale of securities made by Beatrice in 1928 is excluded.

D. *Results for "Normal Period" 1925–1930:*

Average earnings for total capitalization of Baldwin
Locomotive works.........................about $2,900,000

Average earnings for common stock and warrants of
Baldwin...................................... 824,000

Average earnings per share of Baldwin common (assuming warrants exercised and 6% earned on the amount received by the company)........................ $ 0.73

Maximum earnings per share of Baldwin common (as adjusted)...................................... $ 3.17

Average earnings per share of Beatrice common....... $ 6.59

Maximum earnings per share of Beatrice common...... $ 7.31

NOTE: owing to the continuous expansion of Beatrice Creamery between 1925 and 1932, involving the issuance of additional shares, the earnings *per share* of common must be considered as more significant than the amounts earned for the common stock as a whole.

E. *Balance Sheet Figures* (Dec. 31, 1932):

Item	Baldwin	Beatrice
Current assets...................	$13,900,000	$9,410,000
Current liabilities................	1,200,000	748,000
Net current assets................	$12,700,000	$8,662,000
Tangible asset value per share of common.....................	$26.50	$48.75

NOTE: Baldwin's working capital figures are adjusted to exclude the interest of the Midvale Company minority stockholders. The asset value of Baldwin common is adjusted on the assumption that the warrants are exercised. The asset value of Beatrice common has not been adjusted for a write-down of fixed assets in 1933, the amount of which had not been reported.

A study of these quantitative exhibits yields no reason to believe that Baldwin Locomotive common stock is intrinsically attractive at about $11 per share. The only markedly favorable items are the earnings of the single year 1926, and the book value; but neither of these may be considered particularly significant. Superficially, the issue appears to possess a factor of "leverage," or speculative capitalization structure, based upon the presence of a large amount of senior securities. In fact, however, this leverage could become of real value only if the profits exceeded any figure realized since 1926.

In the case of Beatrice Creamery the statistical showing is impressive on two important counts. The first is the consistently large earnings per share in the six years 1925–1930, amounting regularly to almost 50% on the current price of 12½. The second is the very large sales of the enterprise per dollar of common stock at market. Even at the low prices of dairy products in 1933 there were nine dollars of sales for each dollar of common stock. In 1929 the ratio was about eighteen to one. Manifestly there is need of only a very small profit per dollar of business done to yield a large percentage of earnings on the present price of the stock.

Certain other analytical features of the Beatrice exhibit are of interest, *viz.:*

1. The capitalization structure gives the common stock especially favorable speculative possibilities from the technical point of view. All of the relatively large senior capital is represented by preferred stock, which carries no danger of financial embarrassment.

2. The large tangible asset value in relation to the market price is not without significance. While this point must not be taken too seriously, it has a bearing on the question whether the company is likely to earn a reasonable amount on the common shares over the long future. Although a write-down of the fixed assets was in contemplation, this conclusion would hold also on the revised basis.

3. Assuming the write-down to be justified, it would imply that the depreciation charges in recent years had been larger than necessary. In the year ended February 1934, the depreciation charge was reduced to about $1,400,000, compared with $1,900,000 in the previous year. Had this rate applied for the 12 months ended February 1933, the company would have shown some earnings for its common stock in that year.

4. The working capital position is strong for this type of enterprise, and in relation to the market price of its shares.

Qualitative Considerations.

A. *Baldwin Locomotive:* It would appear difficult to form any dependable conclusion as to the long-term prospects, or the normal earning power, of this enterprise. The industry is a basic one, and the exceedingly low rate of locomotive buying for some years past would undoubtedly point to a large accumulated demand. Nevertheless, the business has shown itself to be erratic in the extreme, and views as to its future performance must be more in the nature of conjecture than intelligent prediction.

B. Beatrice Creamery: The business of this company would seem to possess an underlying stability as well as permanence. The demand for dairy products is certainly not subject to the variations existing in the demand for locomotives. While periods of oversupply may affect selling prices drastically, the resultant difficulties are not more serious than are found in countless other lines of business. There is reason to believe that the dairy industry will grow over the long future as it has in the long past. The recession of demand during 1929–1933 was a natural phenomenon of deep depression, and it would hardly appear to hold ominous significance for the years to come. Beatrice Creamery is not so favorably situated as the two larger companies (Borden's and National Dairy Products), which enjoy greater diversification and a profitable business in trade-marked brands. Yet the probabilities would point strongly to a recovery of the earning power of Beatrice Creamery to somewhere near its former well-established level, when general conditions are once again propitious.

An individual prediction of this kind may go astray, for to some extent it must be at the mercy of the future. But it is our view that conclusions based upon this type of reasoning will yield more profitable results—on the average and over the long pull—than the type of "market counsel" represented by the bulletins quoted at the beginning of this final note.[1]

---

[1] Our criticism of certain individual methods followed by Standard Statistics Company, Inc., should not be construed as reflecting upon the work of this outstanding organization in general. On the contrary, it deserves high praise for the accuracy and completeness of its reporting and for the enterprise and open-mindedness it has always shown in developing its scope and technique.

INDEX